SECOND EDITION

FLUID MACHINERY

Application, Selection, and Design

SECOND EDITION

FLUID MACHINERY

Application, Selection, and Design

Terry Wright
Philip M. Gerhart

CRC Press is an imprint of the
Taylor & Francis Group, an **informa** business

CRC Press
Taylor & Francis Group
6000 Broken Sound Parkway NW, Suite 300
Boca Raton, FL 33487-2742

Printed in the United States of America on acid-free paper
10 9 8 7 6 5 4 3 2 1

International Standard Book Number: 978-1-4200-8294-4 (Hardback)

Library of Congress Cataloging-in-Publication Data

Wright, Terry, 1938-
Fluid machinery : application, selection, and design / authors, Terry Wright, Philip M. Gerhart. -- 2nd ed.
p. cm.
"A CRC title."
Includes bibliographical references and index.
ISBN 978-1-4200-8294-4 (hardcover : alk. paper)
1. Turbomachines. 2. Fluid mechanics. I. Gerhart, Philip M. II. Title.

TJ267.W75 2010
621.406--dc22 2009042279

Visit the Taylor & Francis Web site at
http://www.taylorandfrancis.com

and the CRC Press Web site at
http://www.crcpress.com

Contents

Preface

This book is in fact the second edition of *Fluid Machinery: Performance, Analysis and Design* by Terry Wright. The subtitle change is thought to more adequately reflect the emphases of both the current work and the first edition. Unlike most books on the subject, which seem to emphasize only design of turbomachinery, at least half of the current work is dedicated to the more widespread engineering tasks of application of turbomachines and selection of the proper machine for a particular application.

The preface to the first edition lays out the philosophy for the work in some detail and there is no need to repeat it here as there have been few changes. What has changed is the micro-organization of the material. There has been substantial reorganization within the chapters, hopefully allowing a more logical flow for the learner. It would be fair to say that the reader, be it a university student or a practicing engineer, has been at the front of our minds during this revision.

A particularly vigorous effort has been made with the mathematical symbols, with the aim of keeping notation consistent within and across chapters. At times, this has required us to abandon customary usage in the field. Probably, the best example is the use of the vector-diagram angle β. Traditional use has β as the angle between the relative vector and the blade speed for radial flow machines and as the angle between the relative vector and the axial (throughflow) direction in axial machines. We decided to keep β consistently as the relative velocity–speed angle in order to avoid confusion.

As might be expected, we have updated the material where appropriate. This presents a particular challenge in the area of computational fluid dynamics (CFD) in turbomachinery. There has certainly been astounding growth in this field over the years since the first edition was published in 1999; however, even the leaders in this field state that a good design must begin with an essentially one-dimensional layout of the type that we emphasize. As a result, we only "point the way" toward using CFD for turbomachinery design and analysis.

A particularly strong feature of the book is the inclusion of a significant number of exercise problems at the end of the chapters. There are nearly 350 problems, with about a third of them new to this edition. A solutions manual is also available to instructors.

Preface to the First Edition

The purpose of this book is to provide a fairly broad treatment of the fluid mechanics of turbomachinery. Emphasis is placed on the more utilitarian equipment, such as compressors, blowers, fans, and pumps, that will be encountered by most mechanical engineers as they pursue careers in industry. This emphasis is intended to allow the text to serve as a useful reference or review book for the practicing engineer. Both gas and hydraulic turbines are considered for completeness, and the text inevitably includes material from the large literature on gas turbine engines. These machines traditionally have been treated as aerospace equipment and are considered at length in the literature (Oates, 1984; Wilson, 1984; Oates, 1985; Bathie, 1996; Lakshiminarayana, 1996; Mattingly, 1996). Although recent developments in power generation for either load-peaking, distributed generation or process cogeneration have significantly increased the chances that an engineering graduate will encounter gas turbine engines, this text will focus primarily on the more commonly encountered industrial equipment.

The performance parameters of fluid machinery are carefully developed and illustrated through extensive examples. The relationship of the inherent performance of a machine, in terms of the flow rate, head change, and sound power or noise generation through the rotating impeller, is discussed and treated as it relates to the fluid system with which the machine interacts. The dependence of machine performance on the resistance characteristics of the fluid system is emphasized throughout by examining the machine and the system simultaneously through the text. The characteristic sound pressure and sound power levels associated with a fluid machine are treated in this text as a basic performance variable—along with flow and pressure change.

The fundamental relationship between the shape and internal geometry of a turbomachine impeller and its inherent performance is treated from the beginning of the text. In the early chapters, the shape and size of a machine are related through the concepts of similarity parameters to show how the head and the flow combine with shape and size to yield unique relationships between the geometry and performance. The development of these "specific" speed, noise, and size relations is set out in an empirical, traditional manner as correlations of experimental data. The concepts are used to achieve a basic unification of the very broad range and variety of machine types, shapes, and sizes encountered in engineering practice.

In the later chapters, the theme of geometry and performance is continued through the approximate treatment of the flow patterns in the flow passages of the machine. The fundamental consideration of the equations for mass and angular momentum leads to the governing relations for turbomachinery flow and performance. Again, the process is related to the machine geometry, size, speed, and flow path shape. This higher level of detail is related as closely as possible to the overall consideration of size and speed developed earlier.

Following extensive examples and design exercises for a broad range of equipment, applications, and constraints, the later chapters begin tightening the rigor of the calculational process. The simplifying assumptions used to develop the earlier illustrations of fundamental performance concepts are replaced with a more complex and more rigorous analysis of the flow fields. The more thorough treatment of the flow analyses provides a more realistic view of the complexity and difficulties inherent in understanding, analyzing, and designing turbomachinery components.

Following the development of greater rigor and greater calculational complexity, near the end of the text, some of the more vexing problems associated with turbomachinery analysis and design are introduced as advanced design topics. The influence of very low Reynolds numbers and high levels of turbulence intensity is considered as they influence the design and geometric requirements to generate specified levels of performance. Limitations on performance range and acceptable operation are introduced in these later chapters through consideration of compressibility, instability, and stalling phenomena and the inherent degradation of machine and system interaction.

Some attention is given throughout the text to the need to apply advanced analytical techniques to turbomachinery flow fields in a final design phase. However, the more approximate techniques are emphasized throughout most of the book. Here, the sense of a preliminary design approach is employed to promote a basic understanding of the behavior of the machinery and the relation between performance and geometry. The final chapter provides an overview of the calculational techniques that are being used to provide a rigorous, detailed analysis of turbomachinery flows. Beginning with a reasonable geometry, these techniques are used to examine the influence of detailed geometric refinements and allow the designer to achieve something of an optimized layout and performance for a machine. The chapter is used to emphasize the importance of current and future computational capabilities and to point the reader toward a more rigorous treatment of fluid mechanics in machine design.

Throughout most of the text, the examples and problem exercises are either partially or totally concerned with the design or selection process. They deal with system performance requirements or specifications, along with size, speed, cost, noise, and efficiency constraints on the problem solution. The purpose of this pragmatic design approach to turbomachinery applications is to expose the reader—either a student or a practicing engineer—to the most realistic array of difficulties and conflicting requirements possible within the confines of a textbook presentation. By using examples from a fairly large range of industrial applications, it is hoped that the reader will see the generality of the basic design approach and the common ground of the seemingly diverse areas of application.

Authors

Philip M. Gerhart holds a BSME degree from Rose-Hulman Institute of Technology and MS and PhD degrees from the University of Illinois at Urbana-Champaign. He is a registered professional engineer in Indiana and Ohio. He was a professor of mechanical engineering at the University of Akron from 1971 to 1984, chair of the Department of Mechanical and Civil Engineering at the University of Evansville from 1985 to 1995, and has been dean of the College of Engineering and Computer Science since 1995.

Dr. Gerhart has written two books and more than 35 scholarly papers and reports. He has been principal investigator on grants from the United States Army, NASA, the National Science Foundation, and the Electric Power Research Institute. He has served as a consultant to several firms in the power and process industries. He serves as an associate director of the Indiana Space Grant Consortium.

Dr. Gerhart is a member of the American Society for Engineering Education and a fellow member of the American Society of Mechanical Engineers. He served as ASME's vice-president for Performance Test Codes from 1998 to 2001. He has served many years on the Performance Test Codes Standards Committee and the technical committees on fans and fired steam generators. He was awarded the ASME's Performance Test Codes Gold Medal in 1993 and the Silver Beaver award from the Boy Scouts of America in 2001.

Terry Wright holds BS, MS, and PhD degrees in aerospace engineering from the Georgia Institute of Technology and is a registered professional engineer (retired) from Alabama. He initially joined the Westinghouse Research Laboratories and served there for many years as a research scientist and fellow engineer. Much of his effort in this period was in working with the Sturtevant Division of the Westinghouse Corporation, involved with their design and manufacture of turbomachinery.

Dr. Wright became a professor of mechanical engineering at the University of Alabama at Birmingham in the mid-1980s and was active in teaching and mentoring in fluid mechanics and applications in turbomachinery and minimization of turbomachinery-generated noise. While at the university, he consulted with industrial manufacturers and end users of turbomachinery equipment. In addition to his academic and research activities, he also served as chairman of the department of mechanical engineering through most of the 1990s.

He has acted as a technical advisor to government and industry, has published over 90 research and industrial reports (of limited distribution), and has also published over 40 papers in engineering journals, pamphlets, and proceedings of the open literature. He has been active on technical committees on turbomachinery and turbomachinery noise in the American Society of Mechanical Engineers.

Dr. Wright has served as an emeritus professor of the University of Alabama at Birmingham and is active in writing and society activities. He continues to interact with the manufacturers and industrial users of turbomachinery equipment and is a current advisor/consultant to the ASME PTC 11 committee on fan inlet flow distortion.

1

Introduction

1.1 Preliminary Remarks

For convenience of review and quick reference, this introduction includes the basic fundamentals of thermodynamics and fluid mechanics needed to develop and manipulate the analytical and empirical relationships and concepts used in turbomachinery. The standard nomenclature for turbomachinery will be used where possible, and the equations of thermodynamics and fluid mechanics will be particularized to reflect practice in the industry.

1.2 Thermodynamics and Fluid Mechanics

The physics and properties of the common, simple fluids considered in this book include those of many gases, such as air, combustion products, dry steam, and others, and liquids, such as water, oils, petroleum products, and other Newtonian fluids transported in manufacturing and energy conversion processes. These fluids and the rules governing their behavior are presented here in terse fashion for only the simple working fluids, as needed for examples and problem solving in an introductory context.

More complete coverage of complex fluids and special applications is readily available in textbooks on thermodynamics and fluid mechanics, as well as in more specialized engineering texts and journals. See, for example, White (2008); Fox et al. (2009); Munson et al. (2009); Gerhart et al. (1992); Van Wylen and Sonntag (1986); Moran and Shapiro (2008); Baumeister et al. (1978); the journals of the American Society of Mechanical Engineers and the American Institute of Aeronautics and Astronautics, and the *Handbook of Fluid Dynamics and Turbomachinery*, Schetz and Fuhs, 1996. Here, there will be no extensive treatment of multiphase flows, flows of mixtures such as liquid slurries or gas-entrained solids, fluids subject to electromagnetic effects, or ionized or chemically reacting gases or liquids.

1.3 Units and Nomenclature

Units will generally be confined to the International System of units (SI) and British Gravitational system (BG) fundamental units as shown in Table 1.1. Unfortunately, turbomachinery performance variables are very frequently expressed in industry-specific units that will require conversion to the fundamental unit systems. Units for some of these performance parameters, based on pressure change, throughflow, and input or extracted power, are given in Table 1.2. Conversion factors between the more common units are available in Appendix B.

As seen in Table 1.2, often the units are based on instrument readings, such as manometer deflections, or electrical readings rather than the fundamental parameter of interest. As an engineer, one must deal with the nomenclature common to the particular product or industry at least some of the time. Hence, this book will include the use of gpm (gallons per minute) for liquid pumps, hp [horsepower = (ft × lb/s)/550] for shaft power, and some others as well. However, this book will typically revert to fundamental units for analysis and design and convert to the traditional units if necessary or desirable.

TABLE 1.1

Fundamental Units in SI and BG

Length	Meter (m)	Foot (ft)
Mass	Kilogram (kg)	Slug (slug)
Time	Second (s)	Second (s or sec)
Temperature	Kelvin (K)	°Rankine (°R) or °F
Force	Newton (N); $N = kg \times m/s^2$	Pound (lb); $lb = slug \times ft/s^2$
Pressure	Pascal (Pa); $Pa = N/m^2$	lb/ft^2
Work	Joule (J); $J = N \times m$	$ft \times lb$
Power	Watt (W); $W = J/s$	$ft \times lb/s$

TABLE 1.2

Traditional (Industrial) Performance Units in BG and SI

Pressure	lb/ft^2 (psf); in. wg; in. Hg; lb/in^2 (psi)	N/m^2 (Pa); mm H_2O; mm Hg
Head	foot (ft)	m; mm
Volume flow rate	ft^3/s (cfs); ft^3/min (cfm); gal/min (gpm)	m^3/s; l/s (liter $l = 10^{-3}$ m^3); cc/s
Mass flow rate	slug/s; lbm/s	kg/s
Weight flow rate	lb/s; lb/hr	N/s; N/min
Power	Watts; kW; hp; ft × lb/s	N × m/s; J/s; kJ/s; Watts; kW

1.4 Thermodynamic Variables and Properties

The variables and properties frequently used include the state variables: pressure, temperature, and density (p, T, and ρ). They are defined, respectively, as: p is the average normal stress in the fluid; T is a measure of the internal energy in the fluid (actually, a measure of the kinetic energy of molecular motion); and ρ is the mass per unit volume of the fluid (in thermodynamics, the specific volume, $v = 1/\rho$, is more often used). These are the fundamental variables that define the state of the fluid. When considering (as one must) work and energy inputs to the fluid, then the specific energy (e), internal energy (u), enthalpy ($h \equiv u + p/\rho$), entropy (s), and specific heats of the fluid (c_p and c_v) must be included. Consideration of fluid friction and heat transfer will involve two transport properties: the viscosity (μ) and the thermal conductivity (κ). In general, all of these properties are interrelated in the state variable functional form, for example, $\rho = \rho(p, T)$, $h = h(T, p)$, and $\mu = \mu(T, p)$; that is, they are functions of the state properties of the fluid.

The internal energy u is a measure of the thermal energy of the fluid; the specific energy, e, includes thermal energy as well as potential and kinetic energies, such that $e = u + V^2/2 + gz$. Here, g is the gravitational force per unit mass, commonly referred to as the "acceleration due to gravity," V is the local velocity, and z is a coordinate above a specified datum (positive in the upward direction).

For gases, this book restricts attention to those gases whose behavior can be described as thermally and calorically perfect. That is, the properties are related according to $p = \rho RT$, $u = c_v T$, and $h = c_p T$ with R, c_v, and c_p constant properties of the particular fluid. R is defined in terms of the molecular weight, M, of the gas; $R = R_u/M$. R_u is the universal gas constant, 8310 $m^2/(s^2\,K)$ in SI units and 49,700 $ft^2/(s^2\,{}^\circ R)$ in BG units. Appendix A provides limited information on these and other fluid properties for handy reference in examples and problem solving. R, c_p, and c_v are related by $R = c_p - c_v$ and by the specific heat ratio, $\gamma = c_p/c_v$. Other relations between the gas constants are $c_v = R/(\gamma - 1)$ and $c_p = \gamma R/(\gamma - 1)$. For most turbomachinery air-moving applications, one can accurately assume that c_v, c_p, and γ are constants, although for large temperature excursions, such as that might occur in a high-pressure compressor or gas turbine, c_v, c_p, and γ increase with temperature.

In addition to these thermodynamic state variables, it is necessary to define the "real-fluid" transport properties: dynamic and kinematic viscosity. Dynamic viscosity, μ, is defined in fluid motion as the constant of proportionality between shear stress and strain rate in the fluid. For simple shear flows, the relationship is $\tau = \mu(\partial V/\partial n)$, where n is the direction normal to the velocity V. Dynamic viscosity has the units of stress over velocity gradient (i.e., Pa × s or lb × s/ft^2). Dynamic viscosity is virtually independent of pressure for most fluids, yet it can be a fairly strong function of the fluid temperature. Typical variation of dynamic viscosity in gases can be approximated in a power law form such as $\mu/\mu_0 \approx (T/T_0)^n$. For air, the values $n = 0.7$ and $T_0 = 273\,\mathrm{K}$ can be used with $\mu_0 = 1.71 \times 10^{-5}\,\mathrm{kg/m\,s}$. Other approximations are available in the literature, and data are presented in Appendix A. The kinematic viscosity, $\nu \equiv \mu/\rho$, is useful in incompressible flow analysis and is convenient in forming the Reynolds number, $Re = \rho V d/\mu = V d/\nu$.

Liquid viscosities decrease with increasing temperature, and White (2008) suggests the following as a reasonable estimate for pure water:

$$\ln\left(\frac{\mu}{\mu_0}\right) = -1.94 - 4.80\left(\frac{273.16}{T}\right) + 6.74\left(\frac{273.16}{T}\right)^2, \qquad (1.1)$$

with $\mu_0 = 0.001792\,\mathrm{kg/ms}$ and T in Kelvin (within perhaps 1% error).

When dealing with the design and selection of liquid-handling machines, the prospect of "vaporous cavitation" within the flow passages of a pump or turbine is an important consideration. The critical fluid property governing cavitation is the vapor pressure of the fluid. The familiar "boiling point" of water (100°C ~ 212°F) is the temperature required to vaporize water at *standard atmospheric pressure*, 101.3 kPa. The vapor pressure of water (or any other liquid) is lower at lower temperatures, so boiling or cavitation can occur with a reduction in the pressure of a liquid, even at temperatures near atmospheric. As pressure is prone to be significantly reduced in the entry (or suction) regions of a pump, if the fluid pressure becomes more or less equal to the vapor pressure of the fluid, boiling or cavitation may commence there. The vapor pressure is a strongly varying function of temperature. For water, p_v ranges from nearly zero (0.611 kPa) at 0°C to 101.3 kPa at 100°C. Figures A.3 and A.4 showing variation of p_v with T are given in Appendix A for water and some fuels. A rough estimate of this functional dependence (for water) is given as $p_v \approx 0.61 + 10^{-4}T^3$ (p_v in kPa and T in °C). This approximation (accurate to only about 6%) illustrates the strong nonlinearity that is typical of liquids. Clearly, fluids at high temperature are easy to boil with pressure reduction and cavitate readily.

The absolute fluid pressure associated with the onset of cavitation depends, in most cases, on the local barometric pressure. Recall that this pressure varies strongly with altitude in the atmosphere and can be modeled, using elementary hydrostatics, as $p_b = p_{SL}\exp\left[(-g/R)\int \mathrm{d}T/T(z)\right]$, integrated from sea level to the altitude z. The function $T(z)$ is accurately approximated by the linear lapse rate model, $T = T_{SL} - Bz$, which yields $p_b = p_{SL}(1 - Bz/T_{SL})(g/RB)$. Here, $B = 0.0065\,\mathrm{K/m}$, $p_{SL} = 101.3\,\mathrm{kPa}$, $g/RB = 5.26$, and $T_{SL} = 288\,\mathrm{K}$. This relation allows approximation of the absolute inlet-side pressure for pump cavitation problems with vented tanks or open supply reservoirs at a known altitude.

1.5 Reversible Processes, Irreversible Processes, and Efficiency with Perfect Gases

In turbomachinery flows, not only must the state of the fluid be known, but also the process or path between the end states is of interest in typical expansion and compression processes. Because turbomachinery flows are essentially adiabatic, the ideal process relating the end-state variables is the isentropic process. Recalling the combined first and second law of thermodynamics,

$$T\,ds = dh - \frac{dp}{\rho}, \tag{1.2}$$

and using the perfect gas relations (c_p and c_v constant, $dh = c_p\,dT$, $du = c_v\,dT$, $R = c_p - c_v$, and $p = \rho RT$), this becomes

$$ds = \frac{c_p\,dT}{T} - \frac{R\,dp}{p}. \tag{1.3}$$

On integration between end states 1 and 2, one can write the change in entropy as

$$s_2 - s_1 = c_p \ln\left(\frac{T_2}{T_1}\right) - R \ln\left(\frac{p_2}{p_1}\right). \tag{1.4}$$

If the fluid flow process is adiabatic and reversible (without heat addition or friction), the process is isentropic, $s_2 - s_1 = 0$ and

$$\left(\frac{p_2}{p_1}\right) = \left(\frac{T_2}{T_1}\right)^{\gamma/(\gamma-1)} = \left(\frac{\rho_2}{\rho_1}\right)^{\gamma}. \tag{1.5}$$

This relationship between the fluid properties has the form of the well-known *polytropic process*, for which

$$\left(\frac{p_2}{p_1}\right) = \left(\frac{T_2}{T_1}\right)^{n/(n-1)} = \left(\frac{\rho_2}{\rho_1}\right)^{n}. \tag{1.6}$$

Many different processes can be described by varying the exponent n; for example, $n = 1$ represents an isothermal process, $n = 0$ represents an isobaric (constant pressure) process, $n = \gamma$ represents an isentropic process, and $n = \infty$ represents an isochoric (constant volume) process. Any *reversible* process with $n \neq \gamma$ would involve heat transfer.

All real processes are irreversible because of fluid friction, turbulence, mixing, and so on. A real process is characterized by an *efficiency*, η:

> For a compression (pumping) process, $\eta \equiv$ (work input in ideal [reversible] process/work input in real process).
>
> For an expansion (turbine) process, $\eta \equiv$ (work output in real process/work output in ideal [reversible] process).

Limiting consideration for the time being to compression processes, the *isentropic efficiency* is defined by

$$\eta_{s,\text{compression}} \equiv \frac{w_s}{w} = \frac{c_p T_1[(p_2/p_1)^{(\gamma-1)/\gamma} - 1]}{c_p(T_2 - T_1)} = \frac{(p_2/p_1)^{(\gamma-1)/\gamma} - 1}{(T_2/T_1) - 1}. \tag{1.7}$$

The isentropic efficiency compares the work input along the actual (irreversible) path of compression with the work that would be input in an ideal compression that follows a different thermodynamic path between the initial and final pressures. Alternatively, the so-called *polytropic efficiency* compares the real and ideal work inputs along the actual (adiabatic but irreversible) path of compression. This efficiency is defined by considering a specific point on the compression path as

$$\eta_{p,\text{compression}} \equiv \frac{1}{\rho}\frac{dp}{dh} = \frac{\gamma - 1}{\gamma}\frac{T}{p}\frac{dp}{dT}.$$

To express η_p in terms of the endpoints of the compression process, η_p is assumed constant and the equation is integrated to give

$$\frac{p_2}{p_1} = \left(\frac{T_2}{T_1}\right)^{\eta_p\gamma/(\gamma-1)}.$$

This equation is equivalent to Equation 1.5, if we put

$$\frac{n}{n-1} = \frac{\eta_p\gamma}{\gamma - 1}, \tag{1.8}$$

in other words

$$\eta_p = \frac{n}{n-1}\frac{\gamma-1}{\gamma}. \tag{1.9}$$

Isentropic and polytropic efficiencies can be related by the following pair of equations:

$$\eta_s = \frac{(p_2/p_1)^{(\gamma-1)/\gamma} - 1}{(p_2/p_1)^{(\gamma-1)/\eta_p\gamma} - 1}, \tag{1.10a}$$

$$\eta_p = \frac{\gamma-1}{\gamma}\frac{\ln(p_2/p_1)}{\ln\left(1 + [(p_2/p_1)^{(\gamma-1)/\gamma} - 1]/\eta_s\right)}. \tag{1.10b}$$

For expansion processes (turbines), the equations are

$$\eta_s = \frac{(T_1/T_2) - 1}{1 - (p_2/p_1)^{(\gamma-1)/\gamma}}, \tag{1.11a}$$

$$\eta_p = \frac{n-1}{n}\frac{\gamma}{\gamma-1}, \tag{1.11b}$$

$$\eta_s = \frac{1 - (p_2/p_1)^{\eta_p(\gamma-1)/\gamma}}{1 - (p_2/p_1)^{(\gamma-1)/\gamma}}, \tag{1.11c}$$

$$\eta_p = \frac{\gamma}{\gamma-1}\frac{\ln[1 - \eta_s(1 - [p_2/p_1]^{(\gamma-1)/\gamma})]}{\ln(p_2/p)}. \tag{1.11d}$$

In many cases (e.g., liquid pumps and low pressure fans), fluid density changes are either absent or negligibly small. In such cases, the distinction between isentropic and polytropic efficiencies vanishes and one uses the hydraulic or aerodynamic efficiency, defined for pumping machinery as the ratio of pressure change over density to the actual work done on the fluid (η_a (or η_H) $= (\Delta p/\rho)/w$) and for work-producing machinery as the ratio of actual work to the change in pressure divided by density. In other cases, in which density change is small but significant, the polytropic process model is used but the polytropic efficiency is approximated by the aerodynamic efficiency. This will be discussed in a later chapter.

1.6 Equations of Fluid Mechanics and Thermodynamics

Fluid mechanics analyses use the natural laws that govern Newtonian physics. That is, the flow must satisfy: conservation of mass, $\mathrm{d}m/\mathrm{d}t = 0$; Newton's second law of motion, $\boldsymbol{F} = \mathrm{d}(m\boldsymbol{V})/\mathrm{d}t$ (here, the bold letters indicate the vector character of the terms); which in terms of angular momentum is $\boldsymbol{M} = \mathrm{d}\boldsymbol{H}/\mathrm{d}t = \mathrm{d}(\Sigma(\delta m)\boldsymbol{r} \times \boldsymbol{V})/\mathrm{d}t$, where δm is the mass of each term being included in the sum; and conservation of energy, $\mathrm{d}Q'/\mathrm{d}t - \mathrm{d}W/\mathrm{d}t - \mathrm{d}E/\mathrm{d}t = 0$. In the energy equation, the first law of thermodynamics, Q' is the heat transferred to the fluid, W is the work done by the fluid, and E is the energy of the fluid. These equations, along with the second law of thermodynamics and state equations mentioned above, complete the analytical framework for a fluid flow.

In the study of fluid mechanics, these basic forms are converted to a control volume formulation using the Reynolds transport theorem. Conservation of mass for *steady flow* becomes

$$\iint_{cs} \rho(\boldsymbol{V} \cdot \boldsymbol{n})\,\mathrm{d}A = 0, \tag{1.12}$$

where "cs" indicates integration over the complete surface of the control volume and $(\boldsymbol{V} \cdot \boldsymbol{n})$ is the scalar product of the velocity with the surface unit normal vector, $\boldsymbol{n}$ (i.e., it is the "flux term"). For any single surface, the mass flow rate is

$$\dot{m} = \int \rho(\boldsymbol{V} \cdot \boldsymbol{n})\mathrm{d}A. \tag{1.13}$$

Equation 1.12, the so-called continuity equation, says simply that in order to conserve mass, what comes into the control volume must leave it. For simple inlets and outlets, with uniform properties across each inlet or outlet, the continuity equation becomes

$$\sum(\rho VA)_{\text{out}} - \sum(\rho VA)_{\text{in}} = 0. \tag{1.14}$$

For incompressible flow (ρ = constant), the continuity equation reduces to

$$\sum(VA)_{\text{out}} = \sum Q_{\text{out}} = \sum(VA)_{\text{in}} = \sum Q_{\text{in}}, \tag{1.15}$$

where Q is the volume flow rate VA and $\dot{m} = \rho VA = \rho Q$.

For steady flow, Newton's second law becomes

$$\sum \boldsymbol{F} = \iint_{\text{cs}} \boldsymbol{V}\rho(\boldsymbol{V} \cdot \boldsymbol{n})\,\mathrm{d}A, \tag{1.16}$$

retaining the vector form shown earlier. Again, for simple inlets and outlets

$$\sum \boldsymbol{F} = \sum \dot{m}\boldsymbol{V}_{\text{out}} - \sum \dot{m}\boldsymbol{V}_{\text{in}}. \tag{1.17}$$

The incompressible form can be written with $\dot{m} = \rho Q$ as

$$\sum \boldsymbol{F} = \rho \sum\left(Q\boldsymbol{V}_{\text{out}} - \sum Q\boldsymbol{V}_{\text{in}}\right). \tag{1.18}$$

For a steady, compressible flow, through a control volume, the conservation of energy equation is

$$\dot{Q} - \dot{W}_{\text{sh}} = \iint_{\text{cs}} \rho\left(h + \frac{V^2}{2} + gz\right)(\boldsymbol{V} \cdot \boldsymbol{n})\,\mathrm{d}A, \tag{1.19}$$

where $\dot{Q}$ is the rate of heat transfer and $\dot{W}_{\text{sh}}$ is the rate of shaft work.

Using Equation 1.2 and the second law of thermodynamics, a *mechanical energy equation* can be developed; that is

$$-\dot{W}_{\text{sh}} - \dot{\Phi} = \dot{m}\left(\int_1^2 \frac{\mathrm{d}p}{\rho} + \frac{V_2^2 - V_1^2}{2} + gz_2 - gz_1\right), \tag{1.20}$$

where $\dot{\Phi}$ is the dissipation of useful energy by viscosity and the integral is taken along the thermodynamic process path between the inlet and the outlet.

The well-known Bernoulli equation for steady, incompressible, frictionless flow without shaft work is frequently a very useful approximation to more realistic flows and can be developed from the mechanical energy equation by dropping work and loss terms and assuming constant density; that is

$$\frac{p_1}{\rho} + \frac{V_1^2}{2} + gz_1 = \frac{p_2}{\rho} + \frac{V_2^2}{2} + gz_2 = \text{constant} = \frac{p_T}{\rho}, \tag{1.21}$$

where p_T (sometimes written as p_0) is the total pressure of the flowing fluid. In turbomachinery flows, where work always takes place in the flow process, Bernoulli's equation is not valid when the end states are located across the region of work addition or extraction and the total pressure rises or falls.

If shaft work or frictional losses are to be included, then the Bernoulli equation must be replaced by the incompressible mechanical energy equation,

$$\frac{p_1}{\rho} + \frac{V_1^2}{2} + gz_1 = \frac{p_2}{\rho} + \frac{V_2^2}{2} + gz_2 + \frac{w_{\text{sh}}}{\rho} + \frac{\phi_{\text{v}}}{\rho}. \tag{1.22}$$

Here, w_{sh} is the shaft work per unit mass (positive for work output) and ϕ_{v} is the viscous dissipation per unit mass. This equation is frequently rewritten in "head" form as

$$\frac{p_1}{\rho g} + \frac{V_1^2}{2g} + z_1 = \frac{p_2}{\rho g} + \frac{V_2^2}{2g} + z_2 + h_{\text{sh}} + h_{\text{f}}, \tag{1.23}$$

where each term in the equation has units of length. h_{sh} and h_{f} are the shaft head addition or extraction and the frictional loss, respectively.

Equations 1.12 through 1.23 are the basic physical relationships for analyzing the flow in turbomachines and their attached fluid systems are considered in this book. If further review or practice with these fundamentals is needed, see White (2008); Fox et al. (2004); Gerhart et al. (1992); Schetz and Fuhs (1996); and Baumeister et al. (1978).

1.7 Turbomachines

This book will be restricted to the study of fluid mechanics and thermodynamics of turbomachines. This requires that a clear definition of turbomachinery be established at the outset. Paraphrasing from other authors (Balje, 1981; White, 2008), turbomachines can be defined as follows:

> A *turbomachine* is a device in which energy is transferred to or from a continuously moving fluid by the action of a moving blade row. The blade

row rotates and changes the stagnation pressure of the fluid by either doing work on the fluid (as a pump) or having work done on the blade row by the fluid (as a turbine).

This definition excludes a large class of devices called positive displacement machines. They have moving boundaries that either force the fluid to move or are forced to move by the fluid. Examples include piston pumps and compressors, piston steam engines, gear and screw devices, sliding vane machines, rotary lobe pumps, and flexible tube devices. These are not turbomachines, according to the definition, since flow does not move continuously through them and they will not be further considered here. See Balje (1981) for more detailed material on these types of machines. Also no treatment will be provided here for the very broad areas of mechanical design: dynamics of rotors, stress analysis, vibration, bearings and lubrication, or other vital mechanical topics concerning turbomachinery. Others works may be consulted for further study of these important topics (Rao, 1990; Beranek and Ver, 1992; Shigley and Mischke, 1989). Because of its overriding importance in selection and siting of turbomachines, the subject of noise control and acoustics of turbomachinery will be included in our treatment of the performance and fluid mechanics of turbomachines.

A variety of names are used for the component parts of a turbomachine. The rotating element is variously called the *rotor*, the *impeller*, the *wheel*, and the *runner*. Whatever it is called, the rotating element carries a number of *blades*. Sometimes, turbine blades are called *buckets*. Nonrotating blade rows that direct or redirect the flow are called *stators* or *vanes*. If they accelerate the flow, such as in a turbine, they may be called *nozzles*, and fixed blades that decelerate the flow might be called *diffusers*. Finally, the rotating elements are mounted on a *shaft* and the working parts of the machine are typically enclosed in a *casing*.

1.8 Classifications

Much has been written on classifying turbomachinery, and a major subdivision is implied in the definition stated above. This is the power classification, identifying whether power is added to or extracted from the fluid. Pumps, which are surely the most common turbomachines in the world, are power addition machines and include liquid pumps, fans, blowers, and compressors. They operate on fluids such as water, fuels, waste slurry, air, steam, refrigerant gases, and a very long list of others. Turbines, which are probably the oldest type of turbomachines, are power extraction devices and include windmills, waterwheels, modern hydroelectric turbines, the exhaust side of automotive engine turbochargers, and the power extraction end of an aviation gas turbine engine. Again, they operate on a seemingly endless list of

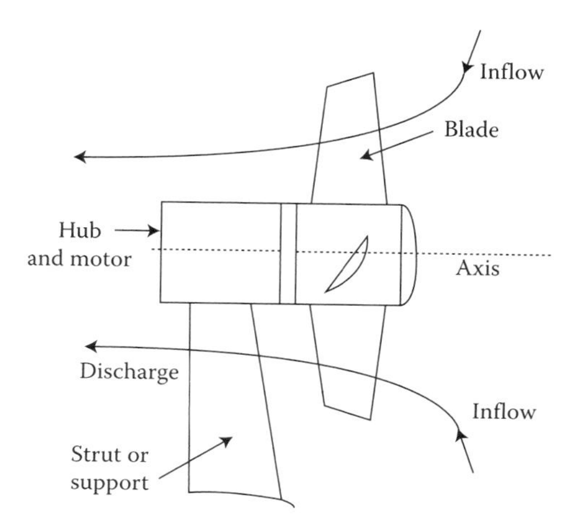

FIGURE 1.1 Example of an open flow turbomachine. No shroud or casing defines the limits of the flow field.

fluids, including gases, liquids, and mixtures of the two, as well as slurries and other particulate-laden fluids.

The manner in which the fluid moves through and around a machine provides another broad means of classification. For example, some simple machines are classed as open or open flow, as illustrated in Figure 1.1. Here, there is no casing or enclosure for the rotating impellers and they interact rather freely with the flowing stream—most often, the atmosphere. Consistent with the power classification, the propeller is an open flow pumping device, and the windmill is an open flow turbine. Figure 1.2 shows examples of enclosed or encased turbomachines where the interaction between

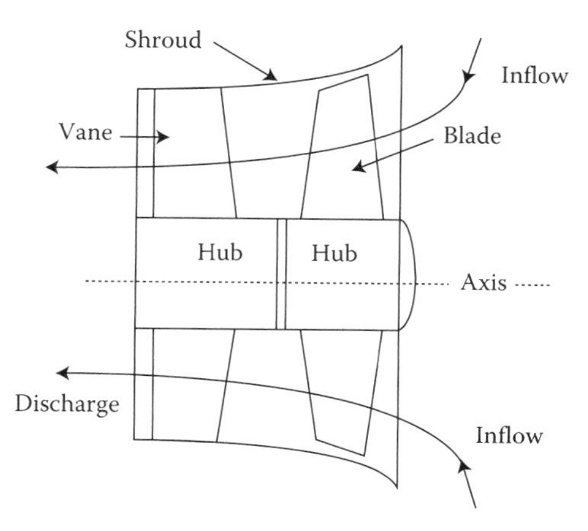

FIGURE 1.2 Example of enclosed or encased turbomachine with the shroud controlling the outer streamlines.

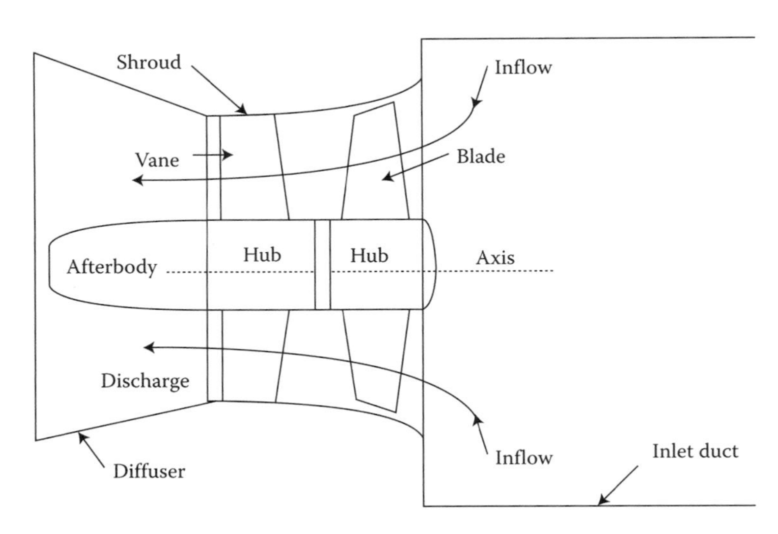

FIGURE 1.3 Layout of an axial fan with the major components.

the fluid and the device is carefully controlled and constrained by the casing walls. Again, these examples are power classified—in this case, as pumps, fans, or compressors.

Since all turbomachines have an axis of rotation, the predominant organization of the mass flow relative to the rotating axis can be used to further refine the classification of machines. This subdivision is referred to as flow path or throughflow classification and deals directly with the orientation of the streamlines that carry the mass flow. In *axial flow machines*—pumps, fans, blowers, or turbines—the fluid moves through the machine on streamlines or surfaces approximately parallel to the axis of rotation of the impeller. Figure 1.3 shows an axial (or axial flow) fan characterized by flow parallel to the fan axis of rotation.

Radial flow machines, with flow that is predominantly radial in the working region of the moving blade row, are illustrated in Figure 1.4, which shows a cutaway or sectioned view. The device is a radial flow (or "centrifugal") fan or pump and illustrates typical geometrical features of such machines. If the flow direction was reversed, the geometry would be typical of radial inflow turbines.

Since nothing is ever as simple or straightforward as one would like, there must be a remaining category for machines that fail to fit the categories of predominantly axial or predominantly radial flow. *Mixed flow machines* such as pumps, fans, turbines, and compressors may all fall into this class and are illustrated by the compressor shown in Figure 1.5. Flow direction for a pump is generally from the axial path to a conical path moving upward at an angle roughly between 20° and 65°. Again, reversing the direction of flow yields a path that is typical of a mixed flow turbine.

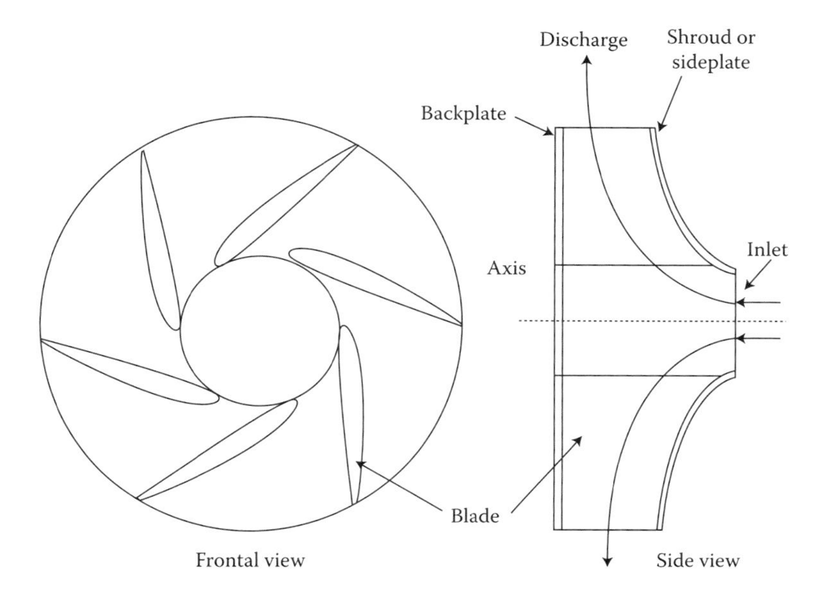

FIGURE 1.4 A radial throughflow turbomachine. The flow moves through the blade row in a primarily radial direction.

The simple flow paths shown here may be modified to include more than one impeller. For radial flow machines, two impellers can be joined back-to-back so that flow enters axially from both sides and discharges radially as sketched in Figure 1.6. These machines are called *double suction* (for liquid pumps) or *double inlet* (for gas movers such as fans or blowers). The flow

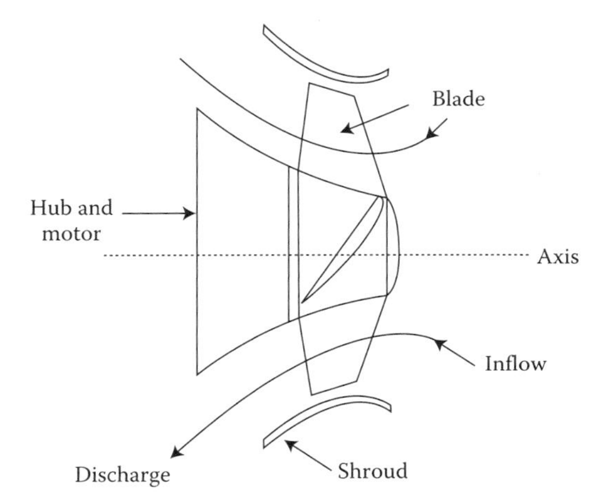

FIGURE 1.5 A mixed throughflow machine. The flow can enter axially and exit radially or at an angle from the axis, as shown.

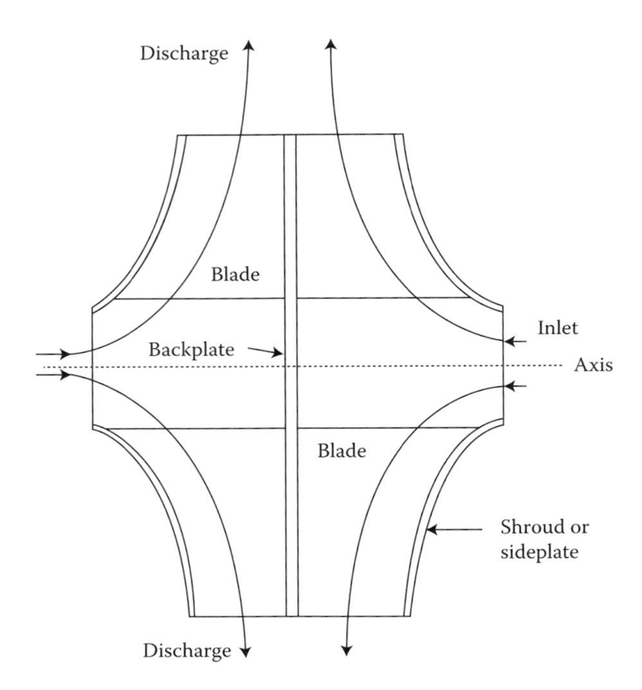

FIGURE 1.6 A double-inlet, double-width centrifugal impeller.

paths are parallel to each other, and flow is usually equal in the two sides, with equal energy addition occurring as well.

A *double-flow* design is often used in large (axial flow) steam turbines; in these devices, the fluid enters at the center and splits to flow toward both ends in the axial direction. Other machines might consist of two or more impellers in axial flow, radial flow, or even in mixed flow configurations. In these machines, the flow proceeds serially from one impeller to the next, with energy addition occurring at each stage. These multistage machines are illustrated in Figure 1.7.

Further breakdown of these classifications can include the compressibility of the fluid in the flow process. If the density is virtually constant in the entire flow process, as in liquid pumps and turbines, the *incompressible flow* label can be added. For gas flows, if there are large absolute pressure changes or high speeds or large Mach numbers involved that lead to significant changes in density, the machines can be labeled as *compressible flow* or simply as compressors. This book will try to keep the range of names for turbomachines as nearly unified as possible and make distinctions concerning gas–liquid and compressible–incompressible when it is convenient or useful to do so.

A set of photographs of turbomachines is included at the end of this chapter (Figures 1.19 through 1.26). They should help to relate the various flow paths to actual machines.

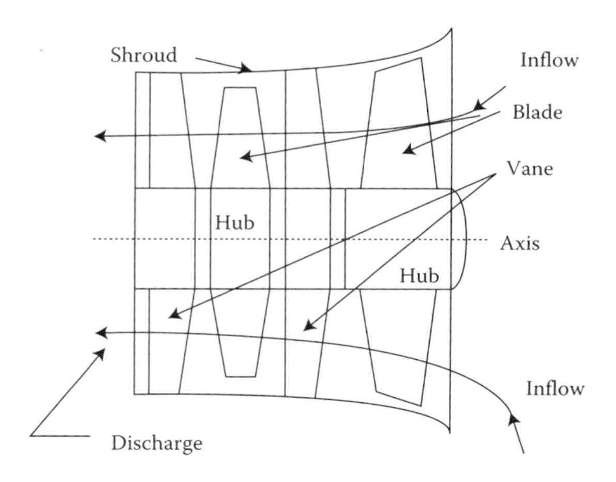

FIGURE 1.7 A two-stage axial fan configuration.

1.9 Turbomachine Performance and Rating

The *performance parameters* for a turbomachine are typically taken to be (1) the fluid flow rate through the machine, (2) a measure of the specific energy change of the fluid (pressure rise or drop, head, or pressure ratio), (3) shaft power, and/or (4) efficiency. In this book, the sound (or noise) generated by the machine is also treated as a performance variable. These performance parameters are related to each other and to the machine *operating parameters*, which are typically (1) rotational speed, (2) fluid density, and (sometimes) (3) fluid viscosity. The relationships between these parameters are the most important information about any machine; these relationships are loosely termed the "performance" of the machine.

Engineering information on the performance of a turbomachine is almost always determined experimentally, in a process called *performance test*. Performance testing is sometimes called "performance rating" or simply "rating," especially if done by the machine manufacturer in their shop or laboratory. Consideration of the process of rating a machine gives valuable insight into machine performance. A schematic layout of a rating facility is shown in Figure 1.8. This figure illustrates one of several possible layouts that may be used to rate a fan. Performance measurement is typically done with equipment and procedures specified by recognized standards to ensure accuracy, acceptance, and reproducibility of test results. For fans, the appropriate standard is "Laboratory Methods of Testing Fans for Rating" (AMCA, 1999). If a field test were desired, the standard might be ASME-PTC 11 Fans (ASME, 2008).

Referring again to Figure 1.8, shown on the left is a centrifugal fan; this is the machine being rated. This fan draws air from the room into the unrestricted

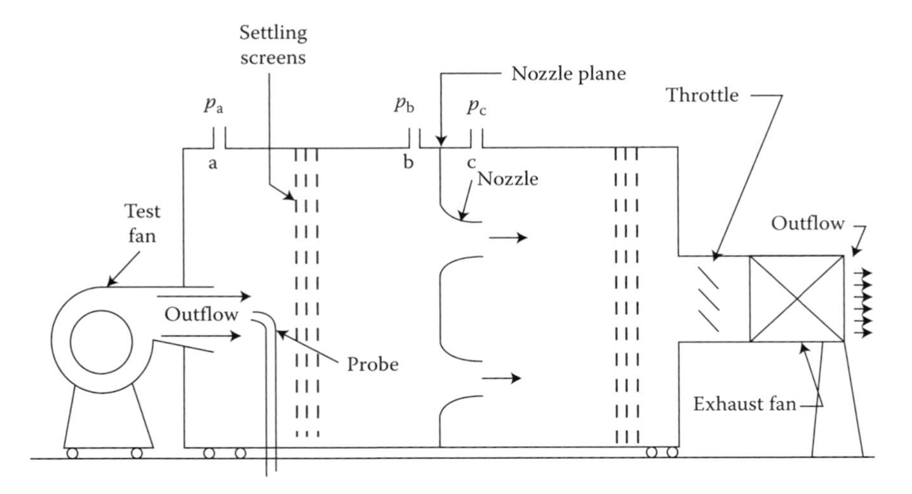

FIGURE 1.8 Schematic of a fan performance test facility, based on AMCA 210. (From AMCA, 1999. *Laboratory Methods of Testing Fans for Aerodynamic Performance Rating*, ANSI/AMCA 210-99, ANSI/ASHRAE 51-99, Air Moving and Conditioning Association. With permission.)

intake and discharges the air into the sealed flow box, or (more formally) plenum chamber. There are numerous pressure taps located along the flow path through the plenum (shown as "a," "b," and "c"). If compressibility is significant, the alternate instrumentation for determining total pressure and temperature at the fan discharge would be the Pitot-static probe shown, with a thermocouple for local temperature measurement. At "a," (or on the Pitot-static probe), the discharge pressure of the fan is monitored. Because the fan is doing work on the air passing through it, the pressure, p_a (total or static), will be greater than the room ambient pressure (as measured outside the plenum or near the intake). The fan discharge pressure is used to identify the pressure increase imparted by the fan to the air by calculation of the total pressure change ($p_{02} - p_{01}$). These pressures include the kinetic energy term associated with the fan discharge velocity. If the pressure change across the machine is sufficiently small (less than about 1% of the barometric pressure), the static pressure change ($\Delta p_s = p_a - p_{ambient}$) is used. This value is called the *static pressure rise* of the fan. The fan *total pressure rise* would include the velocity pressure of the discharge jet (only) to yield $\Delta p_T = \Delta p_s + \rho V_j^2/2$, where ρ is the ambient density. (This use of a total-to-static pressure rise as a performance variable is unique to fans.)

The jet of air being discharged by the fan is spread out (or "settled") as it moves through the first chamber of the box and the row of resistive screens through which the air must pass. The amount of resistance caused by these screens is specified by the test standard to ensure that the flow is smoothly distributed across the cross-section of the plenum chamber on the down-stream side of the screens. In other test arrangements such as constant area pipes or ducts, the flow settling may rely on flow straighteners such as nested tubes, honeycomb, successive perforated plates, or fine-mesh screens. Any particular arrangement must yield a nearly uniform approach velocity as the flow nears the inlet side of the flow metering apparatus. The meter might be a set of precision-built or calibrated flow nozzles (e.g., ASME long radius nozzles), as sketched, mounted in the center plane of the plenum chamber (see ASME, 2004; Holman and Gadja, 1989; Granger, 1988; or Beckwith et al., 1993 for details of the construction of these nozzles). The meter may also be a nozzle or a sharp-edged orifice plate in a pipe or duct or sometimes a precision Venturi meter.

The pressure tap at point "b" supplies the value upstream of the flow nozzles and the tap at "c" gives the downstream pressure. If compressibility is significant, the total pressure and temperature at the flow meter must be established to provide accurate information for the calculation of the mass flow rate. The difference in pressure across the meter, perhaps read across the two legs of a simple U-tube manometer or pressure transducer, supplies the differential pressure or pressure drop through the nozzle, $\Delta p_{\mathrm{b-c}} = p_{\mathrm{b}} - p_{\mathrm{c}}$. This differential pressure is proportional to the square of the velocity of the air being discharged by the nozzles. The product of the velocity and the nozzle area provides the volume flow rate handled by the fan.

Another item of performance data required is a measure of the power being supplied to the test fan. This can be determined by a direct measure of the electrical power, in watts, being supplied to the fan motor, or some means may be provided to measure torque to the fan impeller along with the rotating speed of the shaft. The product of the torque and speed is, of course, the actual power supplied to the fan shaft by the driving motor. To complete the acquisition of test data in this experiment, it remains to make an accurate determination of the air density. In general, the density at the inlet of the blower and at the flow meter must be known.

Fluid velocity at the discharge of the nozzle is calculated from

$$V_{\mathrm{n}} = c_{\mathrm{d}}\left(\frac{2\Delta p_{\mathrm{b-c}}}{\rho(1-\beta^4)}\right)^{1/2}, \tag{1.24}$$

where c_{d} is the nozzle discharge coefficient used to account for viscous effects in the nozzle flow and β is the ratio of nozzle diameter to the diameter of the duct upstream of the nozzle. (In the system illustrated in Figure 1.8, $\beta \approx 0$.) c_{d} is a function of the diameter-based Reynolds number for the nozzles. One correlation (Beckwith et al., 1993) for c_{d} is

$$c_{\mathrm{d}} = 0.9965 - 0.00653\left(\frac{10^6}{Re_{\mathrm{d}}}\right)^{1/2}. \tag{1.25}$$

The Reynolds number is $Re_{\mathrm{d}} = V_{\mathrm{n}}d/\nu$ (ν is the fluid kinematic viscosity).

The inlet density must be accurately determined from the barometric pressure, p_{amb}, ambient dry-bulb temperature, T_{amb}, and the ambient wet-bulb temperature, T_{wb}. The correction to the air density, ρ, can be carried out using a psychometric chart (see, e.g., Moran and Shapiro, 2008; AMCA, 1999). The psychometric chart from the AMCA standard is reproduced in Appendix A. The density at the flow meter can be determined from the inlet value and the total pressure ratio according to

$$\rho_{meter} = \rho_{inlet}\left(\frac{p_{02}}{p_{01}}\right)\left(\frac{T_{01}}{T_{02}}\right), \tag{1.26}$$

where the pressure and temperature are in absolute units. For "incompressible" test conditions, the two densities will be essentially the same.

To complete the data gathering from the fan, the noise generated by the test fan should be measured using a microphone and suitable instrumentation to provide values of L_w, the sound power level in decibels. Turbomachinery noise will be considered in detail in Chapter 4; it is mentioned here because of the frequently overriding importance of this performance parameter.

The efficiency of the fan is calculated from the main performance variables as the output fluid power divided by the shaft input power, P_{sh}. This is the traditional definition for pumps, fans, and blowers, even into the compressible regime. It should be noted that this yields the so-called *overall efficiency* wherein the difference between fluid power and shaft power (the *losses*) includes mechanical losses in both bearings and seals, and aerodynamic/hydraulic losses in the flow through the impeller. Mechanical losses can be isolated by defining a mechanical efficiency, $\eta_M = P_{fluid}/P_{sh}$. The (overall) efficiency is

$$\eta_{To} = \dot{m}\frac{([p_{02} - p_{01}]/\rho)}{P_{sh}}. \tag{1.27}$$

This is equivalent, for incompressible flows, to

$$\eta_T = \frac{Q(\Delta p_T)}{P_{sh}}. \tag{1.28}$$

The "T" subscript on η corresponds to the use of total pressure rise in the fluid output power calculation. Use of Δp_s yields the commonly used static efficiency η_s.

The flow rate, pressure rise, sound power level, efficiency, and input power taken together define a specific performance point for the machine being tested. This particular point of operation is obtained through the effects of a downstream throttle or an auxiliary exhaust blower, or both, as shown in Figure 1.8. Both devices allow the pressure rise and flow rate of the device to be varied by increasing or decreasing the overall resistance imposed on the fan. Reduced resistance (a more open throttle or lower back pressure) will allow more flow (Q or $\dot{m}$) at a reduced pressure rise (Δp_T or Δp_s). Successive adjustment of the throttle or back pressure provides a series of performance points that cover the full performance range of the fan for the particular speed of operation and fluid being handled. Traditionally, the results are plotted on a series of curves, as shown in Figures 1.9a–c. For pumping machinery, the flow rate, Q, is normally taken as the independent (x-axis) variable. These curves are called the machine's *performance curves.* If the fan were to be operated at a different speed, a different set of curves, similar in shape but different in magnitude, would result.

There are several important things to notice from the performance curves for this fan. First, there is a point of maximum efficiency, called the best efficiency point (BEP) defined by the η versus Q curve. The corresponding values of Q, Δp, P_{sh}, and L_W, along with the maximum value of η, define the BEP. To the right of the BEP (at higher flow rate), Δp decreases with increasing flow, yielding a negative slope that represents a usable, stable range of performance for the fan. For this fan, somewhere to the left of the BEP (at lower flow rate), the curve of Δp versus Q goes through a zero slope condition, followed by a region of positive slope. (The slope of the curve is exaggerated for effect.) In this positive slope region, the fan would operate unstably while Δp drops, η declines sharply, and L_W increases sharply. Simply put, everything goes wrong

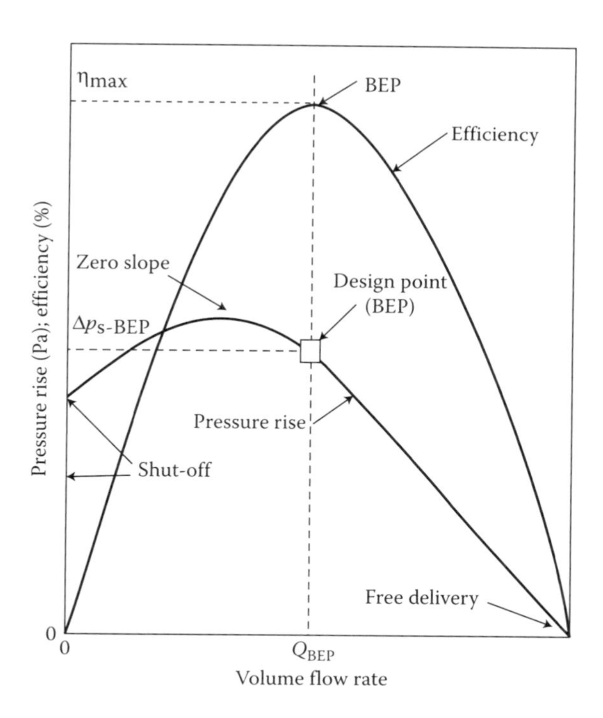

FIGURE 1.9a Efficiency and pressure rise curves with nomenclature.

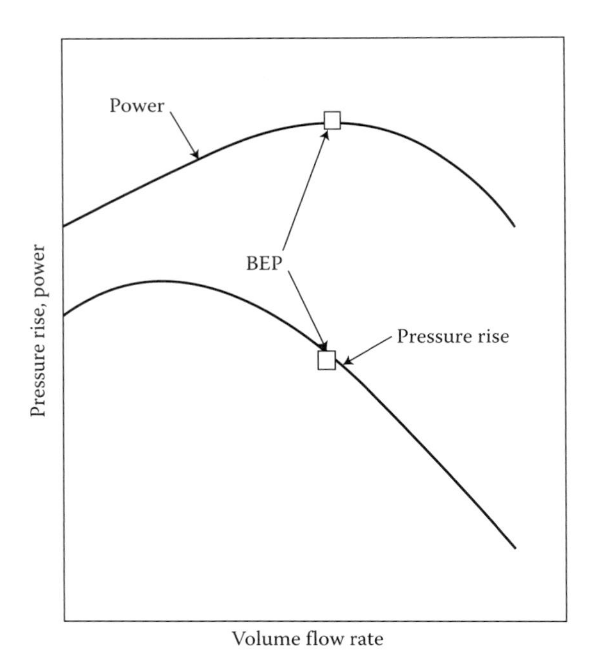

FIGURE 1.9b Input power curve (pressure rise curve repeated for reference).

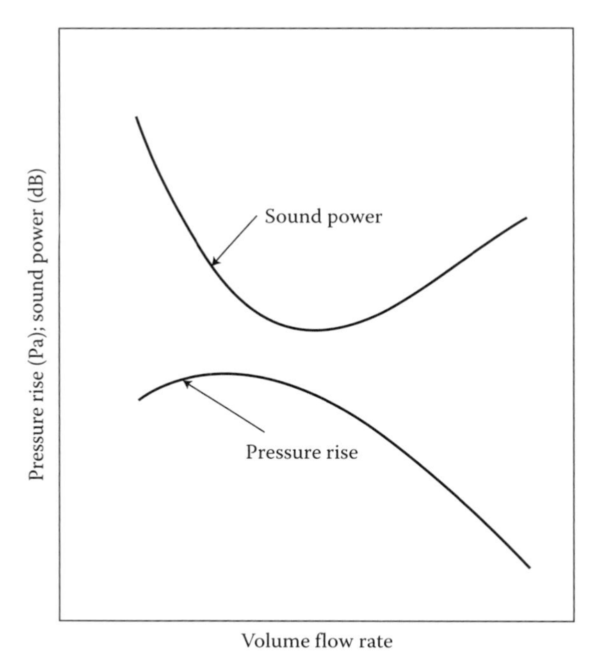

FIGURE 1.9c Sound power curve (pressure rise curve repeated for reference).

at once, and this region represents a virtually unusable regime of operation for this fan. This zone is called the *stalled region* and must be strictly avoided in selection and operation. (Stall will be discussed further in Chapter 10.) At the extreme left of the curve, when $Q = 0$, the fan is still producing some pressure rise but no flow. This limit condition is called the "blocked" or "shutoff" point. With $Q = 0$, the efficiency is zero as well. At the other extreme, $\Delta p = 0$ at the right end of the curve. This limit is called the "free-delivery" or "wide open" point and represents a maximum flow rate for the fan at the given operating speed. The efficiency is also zero at the free-delivery point because Δp is zero. It can be seen that the efficiency generally drops away from the BEP sharply on the left and more gradually on the right. The same can be said for the sound power level, L_W, with a gradual increase in noise to the right of BEP and a sharp increase to the left.

Example: Performance Test on a Fan

Consider a fan test carried out according to these procedures. The plenum chamber data gathered for a single performance point are

$$p_a = 10 \text{ in. wg}; \quad p_b = 9.8 \text{ in. wg}$$

$$p_c = 4.0 \text{ in. wg}; \quad L_W = 85 \text{ db}$$

$$P_m = 3488 \text{ W(motor power)}$$

$$\eta_e = 0.83 \text{ (motor efficiency)}.$$

Ambient air data are

$$T_{amb} = 72°\text{F}; \quad T_{wb} = 60°\text{F}; \quad p_{amb} = 29.50 \text{ in. Hg}.$$

Flow is measured by five 6-in. diameter nozzles. The total flow area is thus

$$A_{total} = 5 \times \frac{\pi}{4} \times 0.5^2 = 0.1963 \text{ ft}^2.$$

The ambient air density is calculated using the psychometric chart in Appendix A. The wet-bulb depression is

$$\Delta T_{d-w} = T_{amb} - T_{wb} = (72 - 60)°\text{F} = 12°\text{F}.$$

At the top of the psychometric chart, one enters with 12° suppression and drops vertically to the downward sloping line for $T_{dry\ bulb} = 72°$F, then horizontally to the left to intersect the upward sloping line for $p_{amb} = 29.5$ in. Hg. Finally, drops down vertically to the abscissa and reads the value for the weight density, $\rho g = 0.0731$ lb/ft^3, so that the mass density is

$$\rho = \frac{\rho g}{g} = \frac{0.0731}{32.17} = 0.00227 \text{ slug/ft}^3.$$

The dynamic viscosity at $T_{amb} = 72°F$ is $\mu = 3.68 \times 10^{-7}$ slug/ft · s, so

$$\nu = \frac{\mu}{\rho} = 1.621 \times 10^{-4}\ \text{ft}^2/\text{s}.$$

Next, calculate the nozzle pressure drop as

$$\Delta p_{b-c} = p_b - p_c = (9.8 - 4.0)\ \text{in. wg} = 5.8\ \text{in. wg}.$$

Converting this value back to basic units,

$$\Delta p_{b-c} = 5.8\ \text{in. wg} \times \left(\frac{5.204(\text{lb/ft}^2)}{\text{in. wg}}\right) = 30.18\ \text{lb/ft}^2.$$

Now the data are substituted into the equation for nozzle velocity, noting that there is no Reynolds number as yet to calculate c_d (since velocity is unknown). Start with $c_d \approx 1.0$ and make an iterative correction. A first guess is then

$$V_n = c_d \left(\frac{2\Delta p_{b-c}}{\rho(1-\beta^4)}\right)^{1/2}$$
$$= 1.0 \times \left(2 \times 30.18\ \text{lb/ft}^2/0.00227\ \text{slug/ft}^3(1-0^4)\right)^{1/2}$$
$$= 163.1\ \text{ft/s}.$$

Next, calculate the Reynolds number

$$Re_d = \frac{V_n d}{\nu} = \frac{163.1 \times 0.5}{1.621 \times 10^{-4}} = 5.09 \times 10^5.$$

Using this value estimate c_d:

$$c_d = 0.9965 - 0.00653\left(\frac{10^6}{Re_d}\right)^{1/2}$$
$$= 0.9965 - 0.00653\left(\frac{10^6}{5.09 \times 10^5}\right)^{1/2} = 0.9874.$$

Adjust the velocity for $c_d < 1.0$ by multiplying the first guess by 0.987 to get $V_n = 161.0$ ft/s. A new Reynolds number is calculated as

$$Re_d = 0.987 \times 5.09 \times 10^5 = 5.02 \times 10^5,$$

then, a new c_d is calculated as 0.9873. This change in c_d is negligible, so accept the value of $V_n = 161.0$ ft/s. The volume flow rate becomes

$$Q = V_n \times A_{tot}$$
$$= 161.0\ \text{ft/s} \times 0.1963\ \text{ft}^2$$
$$= 31.57\ \text{ft}^3/\text{s}.$$

The traditional unit for Q is cubic feet per minute (cfm),

$$Q = 31.57\,\text{ft}^3/\text{s} \times 60\,\text{s/min} = 1896\,\text{cfm}.$$

The remaining item of data is the electrical power measured at the fan drive motor. The fan shaft power is obtained by multiplying by the given motor efficiency:

$$P_{sh} = \eta_e \times P_m,$$

so

$$P_{sh} = 0.83 \times 3448\,\text{W} = 2862\,\text{W} = 2110\,\text{ft} \cdot \text{lb/s} = 3.84\,\text{hp}.$$

Finally, calculate the fan static pressure rise and fan static efficiency:

$$\Delta p_s = p_a - p_{amb} = 10\,\text{in. wg} - 0 = 10\,\text{in. wg},$$

$$\eta_s = \frac{Q\Delta p_s}{P_{sh}} = 31.57\,\text{ft}^3/\text{s} \times \frac{10\,\text{in. wg} \times 5.20\,\text{in. wg/psf}}{2110\,\text{ft} \cdot \text{lb/s}} = 0.778.$$

Summarizing the performance point

$$Q = 31.57\,\text{ft}^3/\text{s} = 1896\,\text{cfm},$$

$$\Delta p_s = 52.0\,\text{lb/ft}^2 = 10.0\,\text{in. wg},$$

$$P_{sh} = 2110\,\text{ft} \cdot \text{lb/s} = 3.84\,\text{hp},$$

$$L_W = 85\,\text{dB},$$

$$\eta_s = 0.778.$$

If the total efficiency or a correction for compressibility (discussed in Chapter 3) is desired, additional measurements of total pressure and total temperature in the discharge jet are required. Suppose these data are

$$p_{02} = 11.5\,\text{in. wg and } T_{02} = 76.7°\text{F}.$$

Then

$$\Delta p_T = p_{02} - p_{01} = 11.5\,\text{in. wg} - 0 = 11.5\,\text{in. wg}.$$

The (incompressible) total efficiency would simply use the Δp_T value to yield

$$\eta_T = \frac{\Delta p_T Q}{P_{sh}} = 0.895.$$

1.10 Rating and Performance for Liquid Pumps

Liquid pumps are perhaps the most common type of turbomachine in use. Performance curves and rating methods for pumps are similar to those for fans, but there are some significant differences. A typical performance rating setup for liquid pump testing is illustrated in Figure 1.10. Standards for testing and rating pumps include ASME-PTC-8.2 *Centrifugal Pumps* (ASME, 1990) or HI-1.6 *Test Standard for Centrifugal Pumps* (HI, 1994). The test pump is installed in a piping network forming a closed loop with a pressure- and temperature-controlled reservoir. The primary performance information is based on pressure readings at the pump suction and delivery flanges, and another set of pressure readings taken upstream and downstream of a flow measurement device, typically a flow nozzle or a sharp-edged orifice plate, positioned downstream of the pump. In addition, power-measuring instrumentation is required for the pump torque and speed or, alternatively, to monitor the electrical power input to a calibrated drive motor. Usually, the reservoir temperature and pressure and the temperature and pressure at the suction flange of the pump will also be monitored. These data are useful in determining the cavitation characteristics of the pump (as will be discussed in Chapter 3).

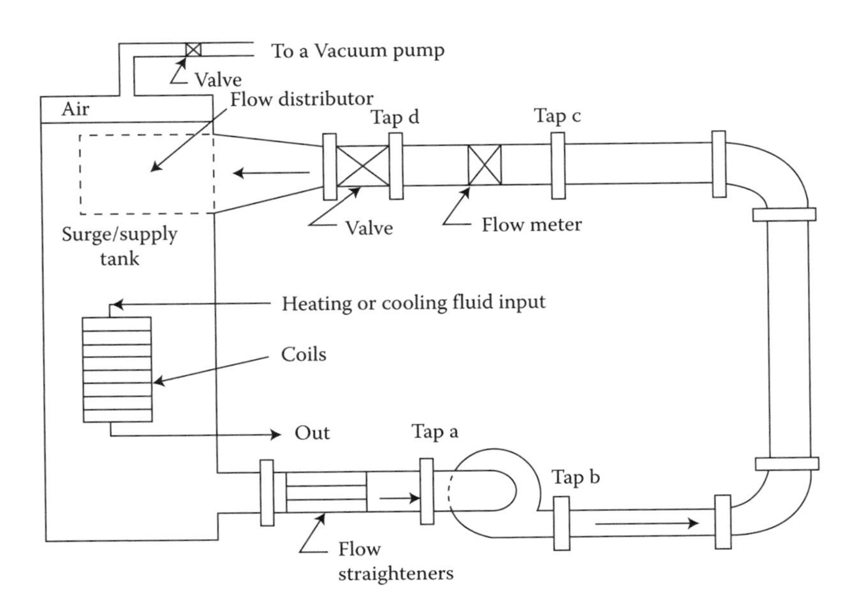

FIGURE 1.10 A liquid pump rating facility.

An important part of the flow loop is the downstream valve used to control the system resistance imposed on the pump. This valve will usually be a "noncavitating" type to avoid instabilities and noise in the pump flow.

As shown in Figure 1.10, the pressures are read from taps "a," "b," "c," and "d" on gages, transducers, or mercury manometers. For increased accuracy, the flow meter pressure drop should be read with a differential-sensing indicator. For illustration, assume that the power input to the motor, P_m, is determined with a wattmeter and the motor efficiency, η_m, is known. The pressure rise across the pump, $(p_b - p_a)$, is used together with the flow rate and areas to calculate the pump head

$$H = \frac{p_b - p_a}{\rho g} + \frac{V_b^2 - V_a^2}{2g} + z_b - z_a. \tag{1.29}$$

The liquid density is evaluated from the temperature T_a.

The pressure differential across the flow meter, here assumed to be a sharp-edged orifice, is used to calculate the volume flow rate in the same manner as flow nozzles are used to determine the air flow in the fan performance test setup. The necessary characteristics for the orifice flow are the Reynolds number and the value of β for the plate. β is defined as $\beta = d/D$, the ratio of the open diameter of the orifice to the pipe's inner diameter. The discharge coefficient for the orifice is then given as

$$c_d = f(Re_D, \beta), \tag{1.30}$$

where the Reynolds number is defined as

$$Re_D = \frac{VD}{\nu}, \tag{1.31}$$

where V is the pipe velocity, D is the pipe diameter, and ν is the kinematic viscosity of the liquid. The flow rate is calculated as

$$Q = c_d A_t \left(\frac{2(p_d - p_c)}{\rho(1 - \beta^4)} \right)^{1/2}. \tag{1.32}$$

One of the several available correlations for c_d is given (ASME, 2004), for standard taps, as

$$c_d = f(\beta) + 91.71\beta^{2.5}\, Re_D^{-0.75}, \tag{1.33}$$

$$f(\beta) = 0.5959 + 0.0312\beta^{2.1} - 0.184\beta^8, \tag{1.34}$$

$$A_t = \left(\frac{\pi}{4}\right) d^2. \tag{1.35}$$

Example: Pump Rating

An example will help to illustrate the procedures for data acquisition and analysis. Sample data required to define a single point on the characteristic performance curve for a pump are

$$T_a = 25°\text{C}; \quad p_a = 2.5\,\text{kPa}; \quad P_m = 894\,\text{watts}$$
$$p_d = 155.0\,\text{kPa}; \quad p_c = 149.0\,\text{kPa}; \quad \eta_e = 0.890$$
$$D = 10\,\text{cm}; \quad d = 5\,\text{cm}.$$

The inlet temperature yields a water density of 997 kg/m^3 and a kinematic viscosity of $0.904 \times 10^{-6}\,\text{m}^2/\text{s}$. As usual, one cannot calculate an *a priori* Reynolds number to establish c_d, so an initial value of c_d is estimated and then iterated to a final result. If one assumes that Re_d is very large, then

$$c_d \cong 0.5959 + 0.0312(0.5)^{2.1} - 0.184(0.5)^8 = 0.6025.$$

The first estimate for volume flow rate is

$$Q = c_d A_t \left(\frac{2(p_d - p_c)}{\rho(1 - \beta^4)} \right)^{1/2}$$
$$= (0.6025)(0.001963\,\text{m}^2) \left[\frac{(6000\,\text{N/m}^2)}{(997\,\text{kg/m}^3)(1 - 0.5^4)} \right]^{1/2}$$

or

$$Q = 0.004238\,\text{m}^3/\text{s},$$

so that $V_t = 2.16$ m/s. As the first refinement, the Reynolds number is

$$Re_D = \beta\, Re_d = \beta \left(\frac{V_t d}{\nu} \right) = 0.5 \left(\frac{2.16 \times 0.05}{0.904 \times 10^{-6}} \right) = 5.973 \times 10^4.$$

Recalculating c_d yields

$$c_d = 0.6025 + \frac{91.7(0.5)^{2.5}}{(5.974 \times 10^4)^{0.75}}$$
$$= 0.6025 + 0.0042 = 0.6067.$$

This 0.6% change is small, so evaluate Q as

$$Q = 1.006 \times 0.00424\,\text{m}^3/\text{s} = 4.26\,\text{l/s}.$$

Volume flow rate in liters per second, while not in fundamental SI units, is common practice in metric usage.

Figure 1.10 suggests that the pipe size (and hence fluid velocity) and centerline elevation change across the pump, but no data are available, so the pump head is estimated from the pressure differential measured across the pump

$$H = \frac{p_b - p_a}{\rho g} + \frac{V_b^2 - V_a^2}{2g} + z_b - z_a \approx \frac{p_b - p_a}{\rho g}$$

$$= \frac{(155{,}000 - 2500)(\text{N/m}^2)}{997 \times 9.81\ \text{N/m}^3} = 15.59\ \text{m}.$$

The fluid power is

$$P_{fl} = \rho g Q H = (9.807\ \text{m/s}^2)(997\ \text{kg/m}^3)(0.00426\ \text{m}^3/\text{s})(15.59\ \text{m}),$$
$$P_{fl} = 649\ \text{W}.$$

Calculating the shaft power from the motor power and efficiency

$$P_{sh} = P_m \eta_e = 796\ \text{W},$$

allows calculation of the pump efficiency according to

$$\eta = \frac{P_{fl}}{P_{sh}} = \frac{649\ \text{W}}{796\ \text{W}} = 0.815.$$

This is once again an overall efficiency, accounting for both mechanical and hydraulic (flow path) losses. The overall efficiency is related to the mechanical efficiency, η_M, and the hydraulic efficiency, η_H (defined in Section 1.5), by

$$\eta = \eta_M \times \eta_H. \tag{1.36}$$

In summary, the test point for the pump is defined by

$$Q = 0.00426\ \text{m}^3/\text{s},$$
$$H = 15.59\ \text{m},$$
$$P_{sh} = 796\ \text{W},$$
$$\eta = 0.815.$$

As in the fan test, the valve downstream of the pump can be opened or closed to decrease or increase the flow resistance and provides a range of points to create the performance curves (also called the *characteristic curves*) for the pump. The characteristic curves for a pump are similar to those for fans (Figure 1.9a–c) with a few differences. The fluid energy parameter is usually the pump head, H, as opposed to the pressure rise. It is not typical to present sound power as a pump performance parameter. Finally, the units for flow Q are more typically gpm or l/min for pumps.

1.11 Compressible Flow Machines

Liquid pumps and hydraulic turbines are essentially incompressible flow machines; a pressure change of ~21 MPa (3000 psi) would be required to produce a 1% density change in water. Fans that handle a highly compressible fluid often create a pressure change large enough to cause a small density change (a pressure change of 1400 Pa [5.6 in. wg] will produce a density change of 1% in air); however, the effects of compressibility in fans can be handled with a "correction factor" (discussed in Chapter 3). For machines such as compressors and steam and gas turbines, significant fluid density changes occur between the inlet and the outlet. There are several differences between performance and rating of incompressible flow machines and these compressible flow machines.

Probably the most obvious change in performance variables is to use the mass flow rate, $\dot{m}$ in place of the volume flow rate. If a volume flow rate is desired, it is customary to base the volume flow rate on the machine inlet stagnation density ($Q_{in} = \dot{m}/\rho_{01}$). Another obvious performance parameter is the shaft power, P_{sh}, which clearly has the same significance for both compressible and incompressible flow machines.

The most significant changes are in the fluid specific energy and in the fluid input or output power. First, consider the pressure and temperature. When dealing with compressible flow machines, it is customary to combine the fluid properties with the fluid velocity/kinetic energy by using the stagnation properties (also called total properties) (see White, 2008; Gerhart et al., 1992):

$$T_0 = T + \frac{V^2}{2c_p} = T\left(1 + \frac{V^2}{2c_pT}\right) = T\left(1 + \frac{\gamma - 1}{2}Ma^2\right), \tag{1.37}$$

$$p_0 = p\left(\frac{T_0}{T}\right)^{\gamma/(\gamma-1)} = p\left(1 + \frac{V^2}{2c_pT}\right)^{\gamma/(\gamma-1)} = p\left(1 + \frac{\gamma - 1}{2}Ma^2\right)^{\gamma/(\gamma-1)}, \tag{1.38}$$

$$\rho_0 = \frac{p_0}{RT_0}, \tag{1.39}$$

where V is the fluid velocity and Ma is the Mach number. It is assumed that the fluid can be modeled as a perfect gas.

Using stagnation properties, the conservation of energy equation (Equation 1.19) becomes

$$\dot{Q} - P_{sh} = \dot{m}c_p(T_{02} - T_{01}) \tag{1.40}$$

and the mechanical energy equation (Equation 1.20) becomes

$$-P_{\text{sh}} - \dot{\Phi} = \dot{m}\left(\int_1^2 \frac{\mathrm{d}p_0}{\rho_0}\right). \tag{1.41}$$

(*Note*: In these equations, the "thermodynamic" symbol for work rate, $\dot{W}$, has been replaced by P for "power," the symbol used in this book.) For all practical purposes, turbomachines are adiabatic ($\dot{Q} = 0$), so the specific work on/by the fluid, from Equation 1.40 is

$$-w = c_{\text{p}}(T_{02} - T_{01}). \tag{1.42}$$

The ideal process would be frictionless, as well as adiabatic, and the fluid power is obtained by putting $\dot{\Phi} = 0$ and using the isentropic process equation ($\rho_0 = \text{constant} \times p_0^{1/\gamma}$) to perform the integral in Equation 1.41, giving

$$-P_{\text{fl}} = \dot{m}\left(\frac{\gamma}{\gamma-1}\frac{p_{01}}{\rho_{01}}\left[\left(\frac{p_{02}}{p_{01}}\right)^{(\gamma-1)/\gamma} - 1\right]\right) = \dot{m}c_{\text{p}}T_{01}\left[\left(\frac{p_{02}}{p_{01}}\right)^{(\gamma-1)/\gamma} - 1\right], \tag{1.43}$$

where perfect gas relations from Section 1.4 have been used. The fluid specific energy follows by dividing by the mass flow rate

$$-\frac{P_{\text{fl}}}{\dot{m}} = c_{\text{p}}T_{01}\left[\left(\frac{p_{02}}{p_{01}}\right)^{(\gamma-1)/\gamma} - 1\right]. \tag{1.44}$$

This parameter corresponds to gH for a pump and $\Delta p/\rho$ for a fan. The specific energy for a compressible fluid depends on the initial temperature of the fluid and on the pressure ratio across the machine. It is convenient to define an *isentropic head*, H_{s}, by

$$gH_{\text{s}} \equiv c_{\text{p}}T_{01}\left[\left(\frac{p_{02}}{p_{01}}\right)^{(\gamma-1)/\gamma} - 1\right]. \tag{1.45}$$

The (isentropic) efficiency of a compressor is the ratio of the fluid-specific energy to the work done on the fluid. Using Equations 1.42 and 1.44

$$\eta_{\text{c}} = \frac{(p_{02}/p_{01})^{(\gamma-1)/\gamma} - 1}{(T_{02}/T_{01}) - 1}, \tag{1.46}$$

which is identical to Equation 1.7, except that stagnation pressures and temperatures are used so that a "total" efficiency results.

For turbines, the isentropic efficiency is the ratio of actual work to ideal work, so

$$\eta_t = \frac{1 - (T_{02}/T_{01})}{1 - (p_{02}/p_{01})^{(\gamma-1)/\gamma}}. \tag{1.47}$$

The definitions of polytropic efficiency (Equations 1.10 and 1.11) can also be used by replacing the static pressure with the stagnation pressure.

The performance curves for compressors and turbines are qualitatively similar to those for incompressible flow machines (pumps, fans, hydraulic turbines); however, the performance parameters are mass flow rate ($\dot{m}$), stagnation pressure *ratio* (p_{02}/p_{01}), shaft power (P_{sh}), and isentropic efficiency. Figure 1.11 shows typical performance characteristics for a compressor, including curves for several different shaft speeds, and Figure 1.12 shows similar curves for a gas turbine. Technically, each plot is restricted to a specific value of inlet temperature (T_{01}). Also seen in these figures is a phenomenon that does not occur in incompressible flow machines, *choking*. Choking occurs in high-speed flow when the Mach number reaches 1 at some point within the machine and the mass flow reaches a limiting, maximum value (White, 2008; Gerhart et al., 1992). The figure also illustrates an alternate method for presenting efficiency data: as contours of constant efficiency.

Example: Compressible Flow Efficiency

Apply the compressible flow definitions for compressor (isentropic) efficiency and polytropic efficiency to the fan test data from Section 1.9.

The applicable data are

$$T_{01} = T_{amb} = 72°\text{F}; \quad p_{01} = p_{amb} = 29.5\ \text{in. Hg};$$
$$p_{02} = 11.5\ \text{in. wg}; \quad T_{02} = 76.7°\text{F}.$$

Converting to absolute SI units

$$T_{01} = 315\ \text{K}; \quad p_{01} = 99.91\ \text{kPa}; \quad p_{02} = 102.77\ \text{kPa}; \quad T_{02} = 318.8\ \text{K}.$$

Substituting into Equation 1.46 gives

$$\eta_c = \frac{(p_{02}/p_{01})^{(\gamma-1)/\gamma} - 1}{(T_{02}/T_{01}) - 1} = \frac{(102.77/99.91)^{(1.4-1)/1.4} - 1}{(318.8/315.9) - 1}$$
$$= 0.8828.$$

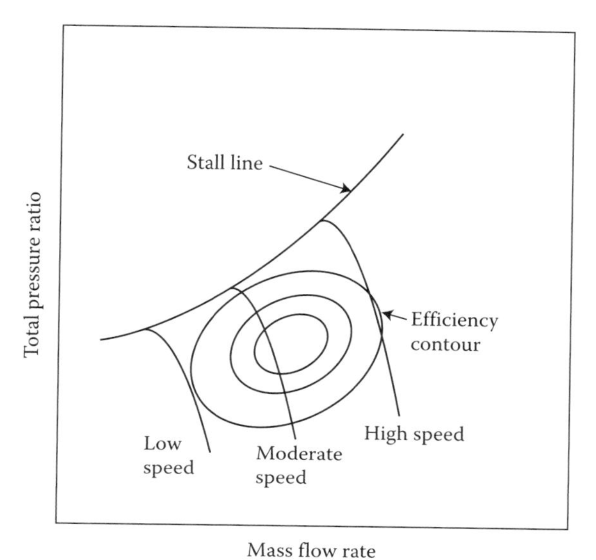

FIGURE 1.11 Performance curves for a compressor.

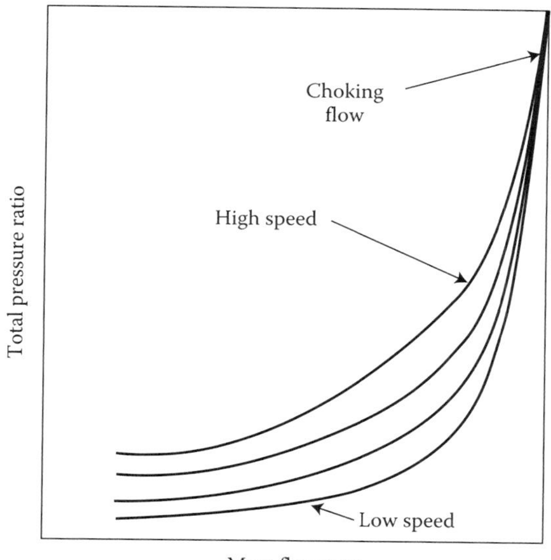

FIGURE 1.12 Performance curves for a compressible fluid turbine.

The polytropic efficiency is

$$\eta_p = \frac{\gamma - 1}{\gamma} \frac{\ln(p_{02}/p_{01})}{\ln\left[1 + \left(([p_{02}/p_{01}]^{(\gamma-1)/\gamma} - 1)/\eta_s\right)\right]}$$
$$= \frac{1.4 - 1}{1.4} \frac{\ln(102.77/99.91)}{\ln\left[1 + \left(([102.77/99.91]^{(1.4-1)/1.4} - 1)/0.8828\right)\right]}$$
$$= 0.8824.$$

For this low pressure ratio machine, isentropic and polytropic efficiencies are essentially identical. The difference between them will increase as the pressure ratio increases. The total efficiencies calculated here are slightly different from the value calculated in Section 1.9, mainly due to compressibility effects. There is also a difference because in Section 1.9, mechanical losses were included in the efficiency because the shaft power was used, but here mechanical losses are not considered.

1.12 Typical Performance Curves

Any specific turbomachine has its own unique performance curve, as determined by its size, rotating speed, design details, and fluid handled. Experience has shown, however, that the shape of the performance curve can be roughly correlated with the machine type. The curves are rendered independent of speed and size (i.e., independent of the exact magnitudes of the performance variables) by dividing each variable by its value at the machine's BEP (also called the "design point" or "point of rating"). Figures 1.13a and 1.13b show typical curve shapes for pumps, roughly divided into "axial flow," "radial flow," and "mixed flow" types. Figures 1.14a and 1.14b show typical curves for fans, here divided into "axial flow," "backward curve," "radial tip," and "forward curve," the latter three types being subcases of radial flow fans. Figures 1.15a and 1.15b show performance curves for radial flow ("centrifugal") and axial flow compressors. The effect of speed must be included explicitly for compressors and compressible fluid turbines, even in this dimensionless form. Finally, typical curves for a hydraulic turbine (radial flow type) are shown in Figure 1.16.

One can deduce several important things from these curves. The pressure/head versus flow curves for axial flow pumps and axial, backward curve, and radial tip fans all have a region to the left of the point of rating (lower flow) where the curve has a positive slope. Operation in this region should be avoided because of the possibility of instability. This problem is especially severe in compressors; hence the curves are not even drawn into this region, beyond the "limit of stable operation." Some radial flow pumps

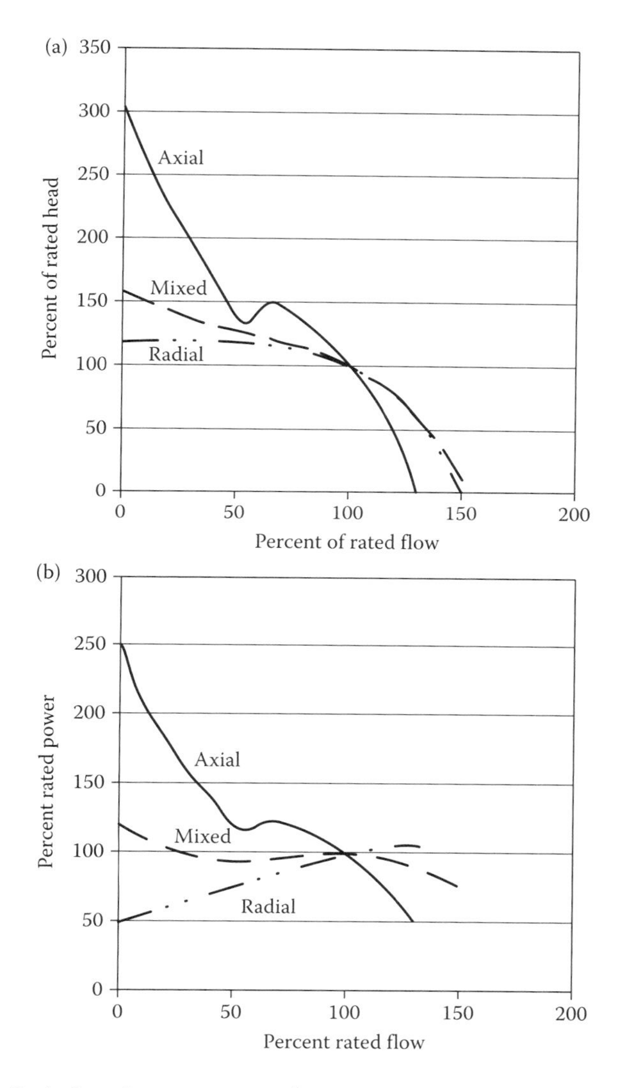

FIGURE 1.13 Typical performance curves for pumps, (a) head–flow curves and (b) shaft power–flow curves (note values in percent of design).

also have a tendency toward this characteristic. For similar reasons, operation in the region where the curve "wiggles" (axial pump and axial fan) should be avoided.

Other important conclusions can be drawn from the power curves. The most desirable power characteristic is for maximum power to occur near the point of rating; so that a drive motor sized for operation at the point of rating will not be overloaded at higher or lower flows. From the curves, an axial flow

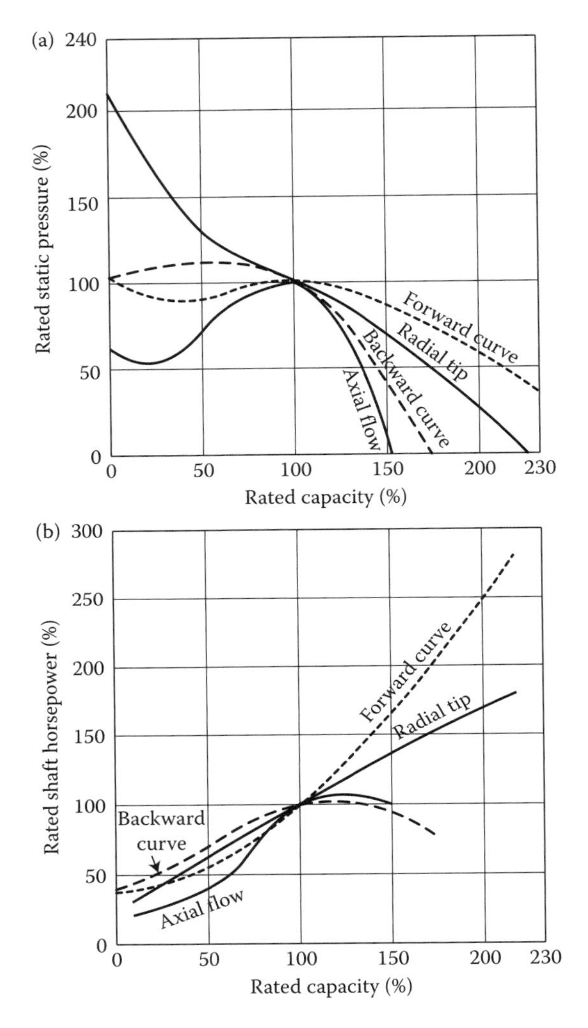

FIGURE 1.14 Typical performance curves for fans, (a) pressure–flow curves and (b) shaft power–flow curves (note values in percent of design).

pump at reduced flow, and both radial tip and forward curve fans at high flow would have this problem, as would a centrifugal compressor at high flow.

The power curves also have implications for start-up. Machines with power characteristics that rise steeply at low flow rates should be started with the connected system "wide open," whereas machines with minimum power requirements near zero flow should be started against a shutoff system.

There are also issues with flow delivery. Axial flow devices have a relatively small range of operation between stall and free delivery, whereas the various radial flow machines are capable of flows nearly double the rated value.

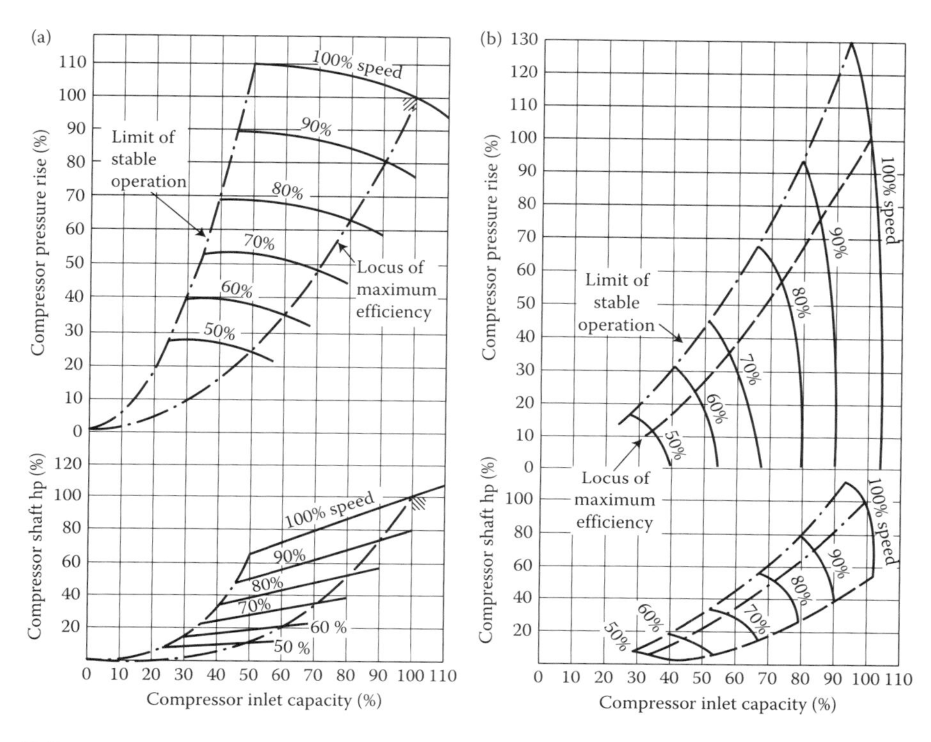

FIGURE 1.15 Typical performance curves for compressors, (a) centrifugal (radial flow) and (b) axial flow (note values in percent of design). (Compressed Air and Gas Institute. With permission.)

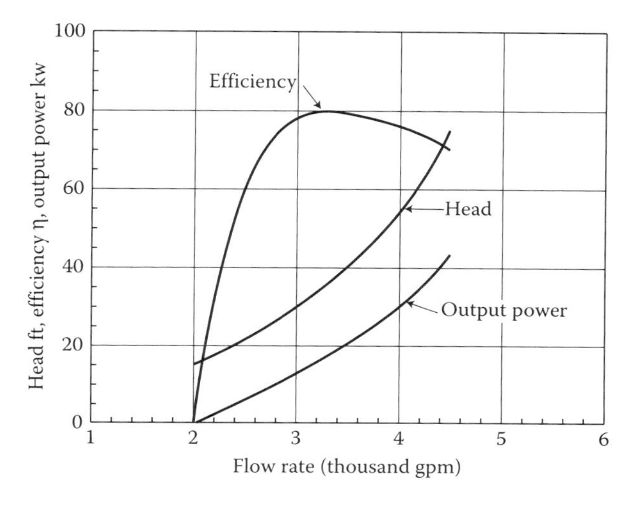

FIGURE 1.16 Typical performance curve for a radial flow hydraulic turbine.

1.13 Machine and System

In application, a fan or pump would be connected to a "flow system" rather than the simple throttle or valve used for rating. This system would typically consist of, say, a length of ductwork or pipe containing elbows, grilles, screens, valves, heat exchangers, and other resistance elements. The pumping machine supplies energy to the fluid and the flow system consumes the energy through fluid friction losses as well as any "permanent" fluid energy increases, such as increasing the fluid elevation. A "resistance curve" or "system curve" is a relation between energy used, flow rate, and system characteristics and can be described in terms of a pressure loss Δp_f (subscript f for "friction") or a head loss h_f:

$$\frac{\Delta p_\mathrm{f}}{\rho} = gh_\mathrm{f} = \sum\left[\left(f\frac{L}{D} + K_\mathrm{m}\right)\frac{V^2}{2}\right]. \tag{1.48}$$

In this equation, f is the Darcy friction factor; L/D is the ratio of the characteristic length to the diameter of the ductwork; and K_m is the loss coefficient for a given system element such as an elbow in the duct. V^2 depends on Q because $V = Q/A_\mathrm{duct}$ so that

$$\frac{\Delta p_\mathrm{f}}{\rho} = gh_\mathrm{f} = \sum\left[\left(f\frac{L}{D} + K_\mathrm{m}\right)\frac{1}{A_\mathrm{duct}^2}\right]\left(\frac{Q^2}{2}\right) \approx \text{constant} \times Q^2, \tag{1.49}$$

where the "constant" is a (nearly) fixed function of the system (it may vary slightly with Reynolds number). A similar equation may be written for either Δp_f or h_f by absorbing g and/or ρ into the "constant."

When the pumping machine is installed in the system, the flow must be the same through both and the energy supplied by the machine must equal the energy used by the system. The (unique) *operating point*, then, is that point where the machine performance curve and the system curve intersect, as shown in Figure 1.17. Several system curves are shown in the figure; if one adds length or roughness (which increases friction factor), or closes a valve, one increases the value of C and gets a steeper parabola. As the steeper parabola crosses the characteristic curve of the machine, the intersection occurs at the necessarily higher value of Δp (or H) and the value of Q is commensurately reduced. Similarly, reducing the pipe length or roughness, opening a valve, or otherwise decreasing resistance causes the system parabola to be less steep (smaller resistance constant), and the intersection with the machine curve occurs at a lower pressure/head and higher flow rate.

The system curve described in Equation 1.49 and Figure 1.18 assumes that there is no net energy change imparted to the fluid. In a liquid pumping system, it may be necessary to account for significant changes in elevation (hydrostatic head) imposed on the pump system. An air-handling system

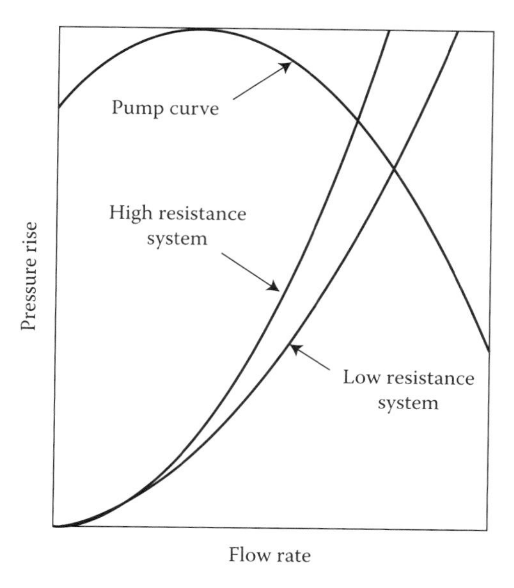

FIGURE 1.17 Pump or fan performance curves with system curves of increasing resistance.

with a fan may deliver air to a region maintained at a positive pressure, such as a "clean room" or a pressurized furnace. In these cases, the system curve can be characterized by $h_{sys} = CQ^2 + \Delta z$ or $\Delta p_{sys} = CQ^2 + \Delta p_{net}$. The additional term is independent of Q and results in a vertical offset of the system curve

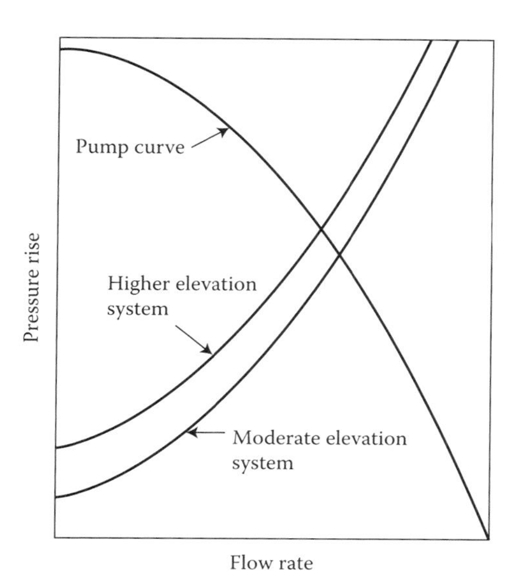

FIGURE 1.18 The influence of elevation or net pressure change on operating point.

as shown in Figure 1.18. The same comments apply to variation of resistance, as well as possible changes in the value of Δz or Δp_{net} with changes in the geometry of the piping system. Recall that the hydrostatic (Δz) terms are nearly always negligible in air/gas systems.

One can determine the operating point either graphically, as indicated in Figures 1.17 and 1.18, or analytically. In most cases, the machine characteristic curve is available as a graph and the system curve is available as an equation. Alternatively, either or both may be available as a table of values. Graphical solution is accomplished by plotting the machine curve and the system curve on the same set of axes. This method provides a strong visual indication of the operating point but precision can be limited.

To determine the operating point analytically, it is usually necessary to fit an equation to the machine characteristic curves. These curves can be complicated, as shown in Figures 1.13 and 1.14; however, consideration is usually limited to the stable-operation portions of the curves. These portions can usually be fit adequately with second-order polynomials such that, for a pump, $H \approx aQ^2 + bQ + c$, with a, b, and c determined by "least-squares" fitting. The operating point is then determined by solving a quadratic equation such as $aQ^2 + bQ + c = CQ^2 + \Delta z$.

1.14 Summary

This chapter introduces the fundamental concepts of fluid machinery. Following a review of basic fluid mechanics and thermodynamics, a discussion of the conservation equations for fluid flow is presented.

The definition of turbomachinery is given, and the distinctions drawn between turbomachines and positive displacement devices are delineated. Within the broad definition of turbomachines, the concepts of classification by work done on/by the fluid moving through the machine are used to differentiate the turbine family (work-extracting machines) from the pumping family (work-adding machines). Further subclassifications of the various kinds of fluid machinery are based on the path taken by the fluid through the interior of a machine. These are described in terms of axial flow, radial flow, and the in-between mixed flow paths in pumps and turbines.

The concepts of performance capability and rating of fluid machinery are introduced and developed. The techniques used to test the performance of turbomachinery are introduced to illustrate the performance variables. Special attention is given to machines handling compressible fluids. Typical performance curves for various classes of turbomachines are presented. Finally, the interaction between a fan or pump and the system it operates in is illustrated.

We close this chapter with a group of figures that illustrate the kind of equipment the book deals with primarily. Figure 1.19 shows a packaged fan

FIGURE 1.19 An SWSI centrifugal fan showing drive-motor arrangement at left, volute and discharge, and the inlet box at the right end. (From Process Barron. With permission.)

FIGURE 1.20 Centrifugal compressor impeller showing blade layout details. (From Process Barron. With permission.)

FIGURE 1.21 Axial fans showing impellers of different types. (From New York Blower Company, Willowbrook, IL. With permission.)

FIGURE 1.22 Fan installation in a power plant. (Note the dual-inlet ducts to supply flow to both sides of the DWDI Fan.) (From Process Barron. With permission.)

FIGURE 1.23 Centrifugal pumps showing casing layouts and flow paths for single-suction and double-suction. (From ITT Fluid Technology. With permission.)

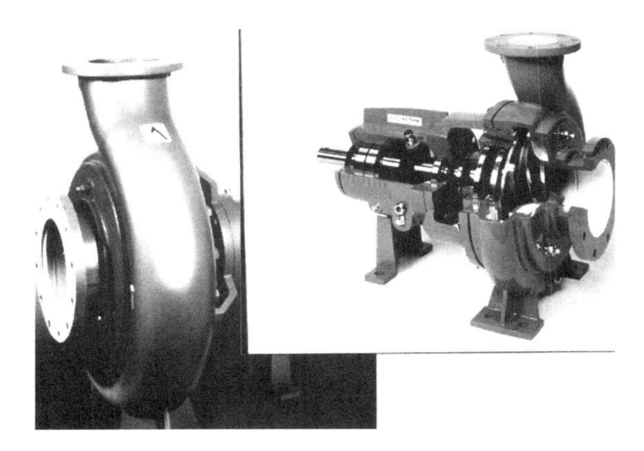

FIGURE 1.24 An end-suction pump with a cutaway view of the flow path, impeller, and mechanical drive components. (From ITT Fluid Technology. With permission.)

FIGURE 1.25 View of a pump installation with two pumps, showing the drive motor arrangements. (From ITT Fluid Technology. With permission.)

FIGURE 1.26 An installation with six double-suction pumps operating in parallel. (From ITT Fluid Technology. With permission.)

assembly with accessories. Figure 1.20 shows a typical compressor impeller, and Figure 1.21 shows several axial fan impellers. Figure 1.22 shows a large fan installation in a power plant, while Figures 1.23 through 1.26 show a variety of pump configurations, installations, and arrangements.

EXERCISE PROBLEMS

1.1. The pump whose performance curves are shown in Figure P1.1 is operated at 1750 rpm and is connected to a 2-in. i.d. galvanized iron pipe 175 ft long. What will be the flow rate and power input?

1.2. The pump of Figure P1.1 is installed in the system of Figure P1.2. What flow rate, power, and efficiency result? All pipes are 2-in. i.d. commercial steel.

1.3. An Allis-Chalmers pump has the performance data given by Figure P1.1. At the pump's BEP, what values of Q, H, and Power (bhp) occur? Estimate the maximum (free-delivery) flow rate.

1.4. The pump of Figure P1.1 is used to pump water 800 ft uphill through 1000 ft of 2-in. i.d. steel pipe. This can be accomplished by connecting a number of these pumps in series. How many pumps will it take? (Hint: Each pump should operate near BEP.)

1.5. Two pumps of the type of Figure P1.1 can be used to pump water with a vertical head rise of 130 ft.

   a. Should they be in series or parallel?

   b. What diameter steel pipe is required to keep both pumps operating at BEP?

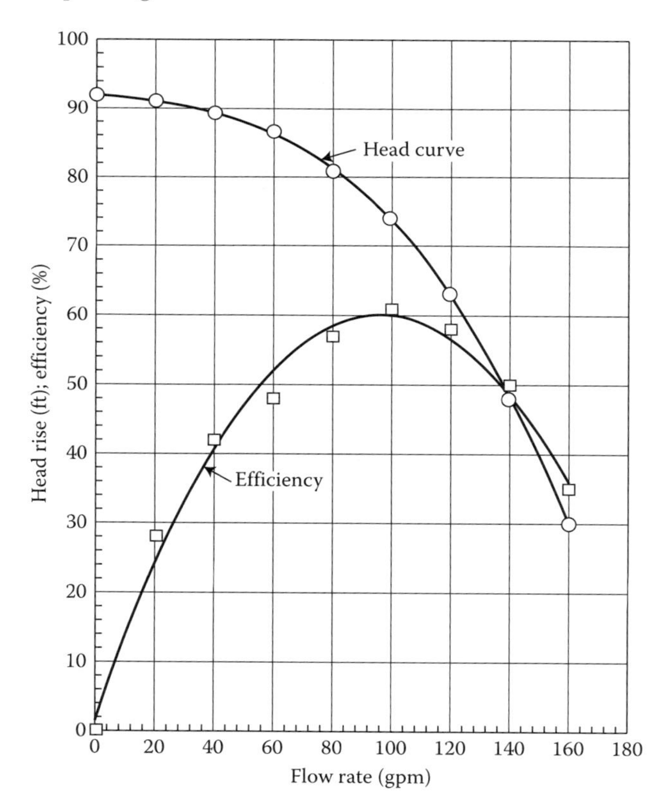

FIGURE P1.1 Performance curves for a pump with a 9-in. impeller and speed 1750 rpm.

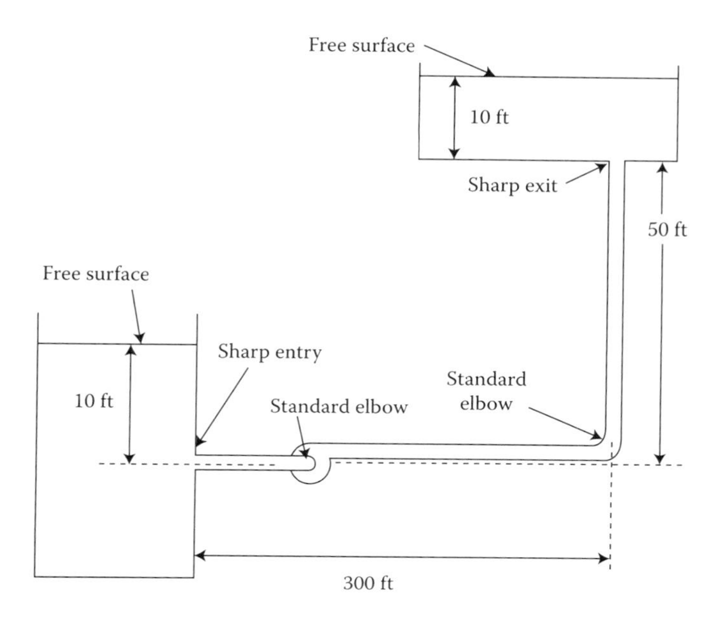

FIGURE P1.2 Pump and system arrangement.

1.6. A fuel transfer pump delivers 400 gpm of 20°C gasoline. The pump has an efficiency of 80% and is driven by a 20-bhp motor. Calculate the pressure rise and head rise.

1.7. If the pump and pipe arrangement of Problem 1.1 now includes an elevation increase of 30 ft, what flow will result?

1.8. A small centrifugal fan is connected to a system of ductwork. The fan characteristic curve can be approximated by the equation $\Delta p_S = a + bQ$, where $a = 30\,\text{lbf/ft}^2$ and $b = -0.40\,(\text{lbf/ft}^2)/(\text{ft}^3/\text{s})$. Duct resistance is given by $h_f = (fL/D + \sum K_m)(V^2/2g)$. The resistance factor is $(fL/D + \sum K_m) = 10.0$, and the duct cross-sectional area is $1\,\text{ft}^2$.

   a. Calculate the volume flow rate of the fan-duct system.

   b. What is the static pressure rise of the fan?

   c. Estimate the hp required to run the fan (the static efficiency of the fan is $\eta_S = 0.725$).

1.9. A slurry pump operates with a mixture of sand and water, which has a density of $1200\,\text{kg/m}^3$ and an equivalent kinematic viscosity of $5 \times 10^{-5}\,\text{m}^2/\text{s}$. The pump, when tested in pure water, generated a head rise of 15 m at a flow rate of $3\,\text{m}^3/\text{s}$. Estimate the head rise in the sand slurry with a flow rate of $3\,\text{m}^3/\text{s}$.

1.10. A three-stage axial fan has a diameter of 0.5 m and runs at 1485 rpm. When operating in a system consisting of a 300-m long, 0.5-m diameter duct, the fan generates a flow rate of $5.75\,\text{m}^3/\text{s}$. Estimate the static pressure rise required for each of the three stages. Use $f = 0.04$ for the duct.

1.11. In order to supply 375 gpm, a group of pumps identical to the pump of Figure P1.1 are operated in parallel. How many pumps are needed if each is to operate at the BEP? Develop a graph for the combined performance of this group of pumps.

1.12. A double-width centrifugal blower supplies flow to a system whose resistance is dominated by the pressure drop through a heat exchanger. The heat exchanger has a K-factor of 20 and a cross-sectional area of 1 m^2. If the blower generates a pressure rise of 1 kPa, what is the volume flow rate (in m^3/s) for each side of the double-sided impeller?

1.13. A medium-sized pump delivers $1.50\,m^3/s$ of SAE 50 engine oil (SG=0.9351). The pump has a total efficiency of 0.679 and is direct-connected to a motor that can supply a shaft power of 25 kW. What head rise can the pump generate at this flow and power?

1.14. The centrifugal fan described in Problem 1.8 must be used to meet a flow-pressure rise specification in metric units. Convert the equation for its pumping performance into SI units and

a. Estimate the flow rate achievable through a 15-cm diameter duct for which the resistance is characterized by $K_{res} = (fL/D + \sum K_m) = 1.0$.

b. Develop a curve of Q versus diameter, D, for $15\,cm < D < 45\,cm$ with $K_{res} = 1.0$.

c. Develop a curve of Q versus K_{res} with a diameter of 10 cm and $1 < K_{res} < 5$.

1.15. Develop a set of performance curves in SI units for the pump data given in Figure P1.1 (Q in m^3/h, H in m, and P in kW). Use these curves to predict flow rate, head rise, and required power if two of these pumps are connected in series to a filter system with a resistance coefficient of $K_m = 50.0$. Assume that the diameter of the flow system is 5 cm and neglect other resistances.

1.16. Estimate the pressure rise of the slurry pump of Problem 1.9 at the stated conditions. Compare this result to the pressure rise with water as the working fluid.

1.17. Use the pump performance curves developed in Problem 1.15 to estimate the BEP power and efficiency in SI units. Estimate the shutoff and free-delivery performance.

1.18. Estimate the flow delivery of the fan of Problem 1.12 if the fan is a single-width version of the given fan.

1.19. What flow rate will the pump of Problem 1.15 generate through 100 m of 5 cm diameter pipe with a friction factor of $f = 0.030$?

1.20. One can arrange the three stages of the axial fan of Problem 1.10 to operate in a parallel arrangement, delivering their individual flows to a plenum that is connected to a duct. If this duct is to be 110 m long, with $f = 0.04$, what size must the duct be to ensure a total flow delivery of $17.25\,m^3/s$? (Neglect any plenum losses.)

1.21. An air compressor operates in ambient conditions such that $p_{01} = 100\,kPa$, $\rho_{01} = 1.2\,kg/m^3$, and $T_{01} = 280\,K$. The compressor performance is $Q = 1\,\text{m}^3/\text{s}$, $\Delta p_T = 25\,\text{kPa}$, $P_{sh} = 28\,\text{kW}$, and $\rho_{02} = 1.4\,\text{kg/m}^3$. Calculate the isentropic efficiency and the polytropic efficiency.

1.22. A compressor pumps natural gas ($\gamma = 1.27$) with an inlet flow condition of 13.9 psia at a temperature of 90°F requiring 38 hp for an inlet flow rate of 1800 cfm. The pressure ratio is 1.25 and the outlet temperature is 135°F. Calculate the isentropic efficiency and the polytropic efficiency.

1.23. The performance of a water pump is to be measured in a test stand similar to the configuration shown in Figure 1.10. The pressure rise from "a" to "b" is inferred from a mercury manometer deflection of 25 cm. The flow meter is a sharp-edged orifice with $\beta = 0.6$ and $d = 10\,\text{cm}$. The pressure change across the meter from "d" to "c" is indicated by a column height change of 15 cm on a second mercury manometer. Find the pressure rise and volume flow rate of the pump (in kPa and m^3/s).

1.24. The pump of Problem 1.23 is driven by an AC induction motor with $\eta_m = 0.85$. The electrical input to the motor, as tested, was 1.20 kW. Calculate the efficiency of the pump at the test point. (The results for Problem 1.23 are $0.0283\,\text{m}^3/\text{s}$ and 30.77 kPa.)

1.25. A 3-ft-diameter axial flow blower was tested in a flow facility like the one shown in Figure 1.8. Four 1-ft diameter nozzles were used to measure the flow. One performance point had $\Delta p_s = 2\,\text{in. wg}$, $\Delta p_n = 2\,\text{in. wg}$, and 7.6 hp supplied to the shaft of the blower. Calculate the values of Q (cfm), Δp_T (in. wg), and η_s and η_T for the blower test point.

1.26. A panel fan, like the one shown in Figure 1.21 (upper left), is used to provide room ventilation in a four-bay garage/shop where minor maintenance is done on vintage sports cars. The large volumes of exhaust gases generated when a "427 Cobra" engine is run are collected in an evacuation system connected to the exhaust pipe. For added safety of the occupants of the shop, the panel fan exchanges the air in the shop at the rate of 10,000 cfm. Resistance to the flow due to inlet and outlet louvers and the discharge velocity pressure of the fan is approximately five times the outlet velocity pressure of the 3-ft diameter fan. The total efficiency of the panel fan is $\eta_T = 0.76$. Estimate the shaft power required in hp.

1.27. A packaged fan assembly, similar to the one shown in Figure 1.19, can provide a pressure rise of 1.5 kPa at a flow rate of $18\,\text{m}^3/\text{s}$ at BEP. The fan has a diameter of 1.07 m and operates at 1175 rpm, with $\rho = 1.21\,\text{kg/m}^3$. Determine the length of 1-m diameter duct ($f = 0.045$) through which this fan can supply $18\,\text{m}^3/\text{s}$ of air.

1.28. The performance curve for the fan of Problem 1.27 can be approximated by a straight line through its BEP (using techniques developed in Chapter 7). The curve is $\Delta p_T = 4488 - 166Q$, with Δp_T in Pa and Q in m^3/s. If this fan is connected to a 1-m diameter duct ($f = 0.045$) whose length is 66 m, what flow rate will result?

1.29. The hydraulic turbine for which the performance is given in Figure 1.16 is a radial inflow design based on an existing pump impeller and housing. The diameter of the impeller is 16.8 in. Based on this diameter, the inflow and outflow losses of the equipment connecting the "pump" to the supply and discharge plumbing may be calculated as $h_f = 10V_d^2/2g$, where the discharge velocity is $V_d = Q/A_{turb} = Q/((\pi/4)D^2)$. If the reservoir can supply water with a gross head of 60 ft:

a. Calculate the net head (after losses) for the turbine. (Match the gross head to the turbine's head flow curve to get an initial estimate of flow rate.)

b. Read the matched volume flow rate from Figure 1.16 and calculate the ideal power.

c. Read the efficiency from the curve of Figure 1.16 and estimate the actual output power, P_{sh}. Compare with the value read from the output versus flow rate curve. Is the difference is within your ability to read accurately from a small graph?

1.30. If the losses for the inlet and outlet equipment used for the turbine in Problem 1.29 can be reduced to $h_f = 1.0\,V_d^2/2g$, determine the net output power for the hydraulic turbine with matched net head and flow rate according to the head-flow curve of Figure 1.16. Compare this value to the result of Problem 1.29 (30.7 kW). Can you supply your home electrical demands with the difference in power?

1.31. Tests are performed on a forced draft fan in a power plant. The fan is a double-width-double–inlet-type. The fan draws air from the atmosphere (barometric pressure = 29.80 in. Hg, temperature = 50°F, and relative humidity = 65%). Measurements are made using a Pitot-static probe at the fan discharge (which measures 10 ft × 10 ft). Measurements of static pressure and velocity pressure are made at the center 100 elemental areas. The data are shown in the tables below.

a. Calculate the volume flow rate (cfm), and the average velocity.

b. Neglecting losses (i.e., $\eta_T = 100\%$), calculate the input power required. Do not neglect the effects of nonuniform velocity.

c. Repeat part (b), but this time, neglect the effects of nonuniform velocity (i.e., assume the kinetic energy correction factor, $\alpha = 1.0$).

1.32. Figure P1.32a shows an instructional hydraulic turbine test apparatus (a Pelton wheel setup on a hydraulics bench). The pump, driven by a constant speed motor, supplies water to the turbine. The pump performance curves are shown on Figure P1.32b. The turbine performance curves are shown in Figure P1.32c. During a test on the turbine, the turbine load is adjusted to maintain constant turbine speed.

Measured Velocity Pressure (in. wg)

Y Measured from duct center	4.5	3.5	2.5	1.5	0.5	0.5	1.5	2.5	3.5	4.5	
4.5	0.642	0.821	0.811	1.538	2.986	1.860	0.597	1.395	0.192	0.621	4.5
3.5	0.376	1.289	2.469	2.329	3.285	1.749	3.306	1.277	2.397	0.739	3.5
2.5	0.564	3.277	2.573	3.890	3.724	2.937	2.632	3.432	3.190	1.457	2.5
1.5	0.796	2.192	4.035	3.875	4.342	3.967	2.728	4.727	4.117	2.651	1.5
0.5	1.352	1.636	4.665	3.921	4.184	4.605	4.464	2.694	3.555	2.832	0.5
0.5	1.453	2.883	3.088	3.961	5.113	4.675	4.249	3.705	2.378	2.538	0.5
1.5	1.544	3.475	4.168	3.753	4.151	4.051	3.829	2.363	1.669	4.080	1.5
2.5	0.848	2.080	1.901	2.223	2.565	2.932	2.262	1.820	1.641	1.481	2.5
3.5	1.079	1.708	1.771	1.387	2.719	3.755	3.910	2.620	0.712	0.630	3.5
4.5	0.092	1.898	1.692	4.935	0.487	1.836	0.513	0.989	2.583	0.714	4.5
	4.5	3.5	2.5	1.5	0.5	0.5	1.5	2.5	3.5	4.5	

X Measured from duct center

Measured Static Pressure (in. wg)

Y Measured from duct center	4.5	3.5	2.5	1.5	0.5	0.5	1.5	2.5	3.5	4.5	
4.5	12.594	12.936	12.637	12.764	12.787	12.603	12.969	12.648	12.502	12.762	4.5
3.5	12.830	12.723	12.962	12.724	12.554	12.856	12.621	12.567	12.522	12.966	3.5
2.5	12.804	12.966	12.575	12.807	12.973	12.811	12.933	12.609	12.972	12.692	2.5
1.5	12.648	12.502	12.875	12.836	12.876	12.894	12.593	12.710	12.876	12.768	1.5
0.5	12.960	12.569	12.886	12.809	12.591	12.562	12.978	12.715	12.577	12.802	0.5
0.5	12.744	12.807	12.736	12.708	12.790	12.552	12.650	12.659	12.663	12.780	0.5
1.5	12.868	12.589	12.575	12.977	12.753	12.624	12.724	12.840	12.828	12.928	1.5
2.5	12.663	12.508	12.672	12.754	12.928	12.772	12.572	12.811	12.949	12.725	2.5
3.5	12.606	12.794	12.941	12.843	12.636	12.609	12.679	12.659	12.733	12.648	3.5
4.5	12.495	12.920	12.656	12.922	12.511	12.804	12.885	12.825	12.772	12.550	4.5
	4.5	3.5	2.5	1.5	0.5	0.5	1.5	2.5	3.5	4.5	

X Measured from duct center

a. Plot head versus flow for the turbine when the turbine operates at a (constant) speed of 1000 rpm. Repeat for 1500 rpm.

b. For a turbine speed of 1000 rpm, determine the flow rate, the pump power input, and the turbine power output. Also, calculate the installation efficiency.

1.33. Write an essay (two to three pages of machine-generated text) on ASME Performances Test Codes. Cover at least the following topics:

a. What are Performance Test Codes?

b. What are they used for?

c. How are they applied in the field of turbomachinery engineering?

d. What are your personal experience(s) with Performance Test Codes?

1.34. An air compressor has the following discharge and inlet conditions: $T_2 = 380°\text{F}, p_2 = 50\,\text{psia}, V_2 = 300\,\text{ft/s}, T_1 = 30°\text{F}, p_1 = 10\,\text{psia}$, and $V_1 = 500\,\text{ft/s}$. Compute (per pound of air) the:

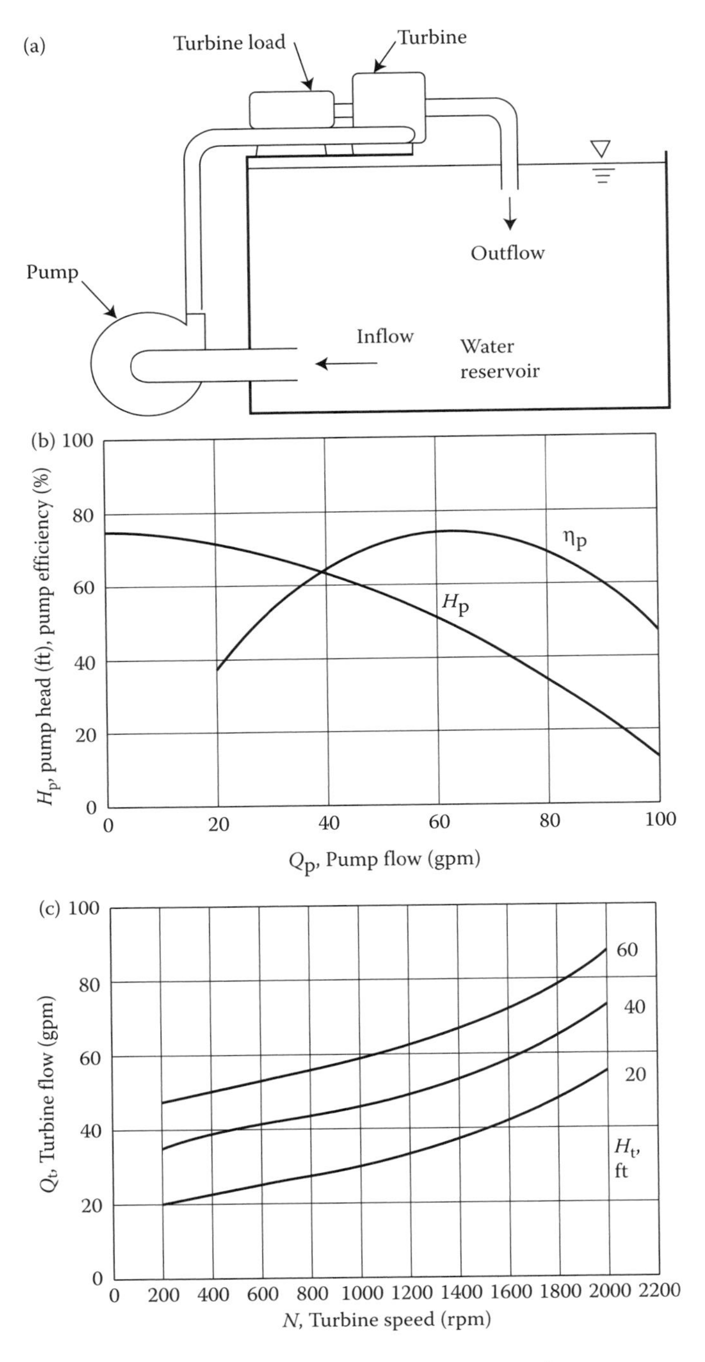

FIGURE P1.32 (a) Schematic of a turbine test setup, (b) pump performance curves, (c) turbine performance curves, and (d) turbine power output curves.

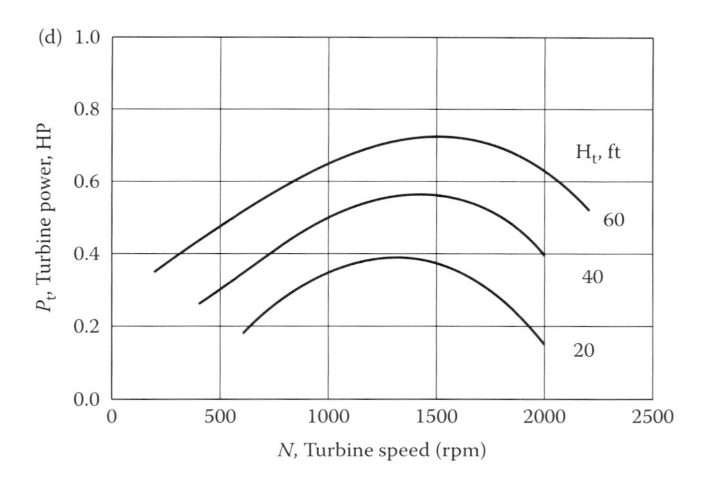

FIGURE P1.32 (continued).

a. Actual work delivered to the air.
b. Ideal work, assuming *isentropic* compression and *neglecting* changes in kinetic energy.
c. Ideal work assuming *polytropic* compression between inlet and outlet total states.
d. Ideal work assuming *polytropic* compression between inlet and outlet *static* states with actual changes of kinetic energy included.

1.35. A low-pressure steam turbine expands steam from inlet conditions 7 psia, 300°F to outlet conditions 3 psia, 175°F. Model the steam as an ideal gas with $\gamma = 1.33$; $R = 85.83$ ft lbf/lbm°R, $c_p = 0.44$ Btu/lbm°R, and calculate
a. Work per pound of steam flow.
b. Isentropic efficiency.
c. Polytropic efficiency.
d. Stage efficiency, assuming that the expansion is done in three stages of equal pressure ratio and stage efficiency.

1.36. A pressure blower increases the pressure of air from 101 to 125 kPa. The temperature is 15°C and the air velocity is the same at the inlet and the outlet of blower. Compare the compression work (useful energy added to the air) for a polytropic process with $n = 1.6$ to that calculated by using a mean density ($\rho_m \equiv \frac{1}{2}(\rho_1 + \rho_2)$) and to that calculated by assuming that the density is constant and equal to the inlet value. (Hint: For the "mean density" calculation, assume that inlet and outlet states are related by a polytropic process with $n = 1.6$.)

1.37. A small fan is tested in a laboratory: the fan inlet area = 1.0 ft^2 and the fan discharge area = 0.6 ft^2. A Pitot-static tube measures inlet average velocity and a manometer measures static pressure differential, Δp, across the fan. The following data are collected during a fan test, with the fan speed of 1750 rpm and the fan wheel diameter of 1.5 ft.

Inlet Average Velocity (ft/min)	Δp (in. wg)	Drive Power (hp)
1000	19.2	4.8
2500	15.5	7.6
4000	6.4	6.8

Plot fan performance curves for total pressure rise, Δp_T (in. wg) and efficiency, η versus flow, Q (cfm) from the test data.

1.38. A centrifugal pump performance test is conducted in a facility illustrated schematically in Figure 1.10. The test fluid is 60°F water and the test pump is driven by an electric motor. The "flow meter" is a sharp-edged orifice plate installed in a 6-in. (nominal) schedule 40 pipe. The orifice diameter is 3.0 in. and the pressure drop across the orifice is 4.6 psi. The pump suction pipe (at "a") is also 6-in. nominal schedule 40. A pressure gage at "a" reads 10 in. of mercury, vacuum. The discharge pipe (at "b") is 4-in. nominal schedule 40. A pressure gage at "b" reads 24.6 psi. The pipe centerline at "a" is at the same elevation as the pump drive shaft and the pipe centerline at "b" is 6 in. below the drive shaft.

 a. Calculate the flow rate in gpm and the pump head in ft.
 b. What is the water hp delivered by the pump?
 c. Make a reasonable estimate of the power required to drive the pump and the electric power requirement for the drive motor.
 d. If the pump is driven by a 4-pole AC electric motor with 2% slip, estimate the torque in the pump drive shaft.
 e. If the pump test fluid (water) is heated by the coils in the tank, at what temperature would the water at "a" vaporize?

1.39. Air flowing in a 4 m × 6 m rectangular duct has the following idealized velocity distribution:

$$V(x,y) = 35\left(1 - \frac{x}{2}\right)^{1/4}\left(1 - \frac{y}{3}\right)^{1/5},$$

where x and y are distances in meters measured from the center of the duct and V is in m/s. Calculate the average velocity, the volume flow rate, and the kinetic energy correction factor. You may use either numerical calculation or analytic calculation. If you use numerical, the discretization error should be <1%.

1.40. Performance testing on a fan yields the data given in the table below. Plot performance curves (total pressure rise, power, and efficiency) for this fan and identify the BEP. Both graphical and least-squares curve fits are required.

Flow Q (cfm)	Total Pressure Rise Δp_T (in. wg)	Shaft Power P (hp)
0	14.85	48.15
10,190	14.80	56.49
20,310	12.24	60.30
28,580	10.33	64.64
40,450	6.61	61.18
507,304	2.39	47.74

1.41. The Taco Corporation markets a series of centrifugal pumps whose performance curves are shown in Figure P1.41. A number of these pumps are to be used in series to pump 500 gpm of water through a 4-in. schedule 40 steel pipe to the top of a 300-ft hill. The equivalent length of the pipe is 375 ft. Specify the number and size of the pumps required and estimate the power requirement. Can you suggest a better scheme for this pumping job? Assume that similar pumps will be used.

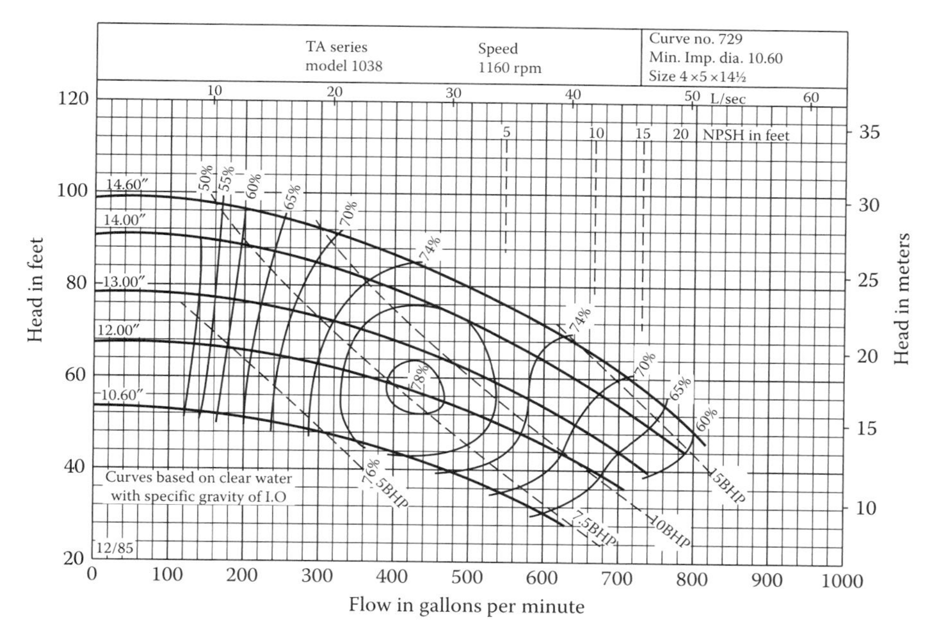

FIGURE P1.41 Performance curves for a centrifugal pump. (Taco, Inc., Cranston RI.)

2

Dimensional Analysis and Similarity for Turbomachinery

2.1 Dimensionality

Recall the study of dimensional analysis from fluid mechanics (see any undergraduate text—White, 2008; Fox et al., 2009; Munson et al., 2009; and Gerhart et al., 1992—and review as necessary). The concept is linked to the principle of dimensional homogeneity and is critical to the correct design of experiments and the extraction of the maximum possible information from a mass of test data. When the concept is applied to turbomachines, it provides a greater depth of understanding of performance and classification of machines, it allows the extrapolation of performance data from one operating condition to another, and it helps to clarify the fundamental reasons behind the wide configuration differences in machines.

The *principle of dimensional homogeneity* is as follows:

> If an equation correctly represents a relationship between variables in a physical process, it will be homogeneous in dimensions. That is, each term will add to the other terms only if they are all of like dimensions.

This is a basic "Truth Test" to determine the fundamental validity of any mathematical relationship that purports to represent a physical process. For example, consider the familiar Bernoulli equation for steady, incompressible, frictionless flow with no shaft work:

$$p + \frac{1}{2}\rho V^2 + \rho g z = p_0. \qquad (2.1)$$

According to the principle of dimensional homogeneity, each term in this equation must have the same dimensions. Since two of the terms (p and p_0) are pressures, the dimensions of each term must be those of pressure: Force/Length^2 $[F/L^2]$ or Mass/(Length $\times$ Time^2) $[M/LT^2]$. In practice, the dimensions are expressed in units; each term in Bernoulli's equation must have units such as N/m^2 [Pa] in SI units or lb/ft^2 in BG units.

Because each term has the dimensions and units of pressure, the terms are often given names that imply that they actually are pressures:

$$\text{Static (or thermodynamic) pressure} + \text{Velocity (or dynamic) pressure} + \text{Hydrostatic pressure} = \text{Stagnation (or total) pressure}; \quad (2.2)$$

however, the second term is actually kinetic energy per unit volume and the third term is potential energy per unit volume. The total pressure is the sum of the three terms on the right-hand side.

2.2 Similitude

The principle of similarity, or similitude, follows from the principle of dimensional homogeneity. The root of the concept lies in the ability to form the variables of a problem into dimensionless groups, also called dimensionless numbers. This can be illustrated using the Bernoulli equation. If we consider a gas flow, then the hydrostatic term is almost always negligible. Putting $\rho gz = 0$, the equation can be rearranged as follows:

$$\frac{p_0 - p}{1/2\rho V^2} = 1.$$

The quantity on the left is a dimensionless group (it has neither dimensions nor units) and is called the *pressure coefficient*. Bernoulli's equation can be reduced to the simple statement that the pressure coefficient is constant.

The principle of similarity is used to relate information obtained by testing a model to the corresponding information that would pertain to the full-scale device, called the prototype. In general, there must be *geometric similarity*, *kinematic similarity*, and *dynamic similarity*. Specifically,

> A model and a prototype are geometrically similar if all physical or body dimensions in all three (Cartesian) axes have the same linear ratio.

For our purposes, this means that all length dimensions, all angles, all flow directions, all orientations, and even surface roughness heights must "scale." In simple terms, the model must "look" exactly like the prototype.

For kinematic similarity:

> The motions of two systems are kinematically similar if similar elements lie at similar locations at similar times.

By including time similarity, one introduces such elements as velocity and acceleration—as in a fluid flow problem. Note that for similitude in a kinematic sense, geometric similarity is presupposed. In fluid mechanics, kinematic similarity means that ratios of velocity (e.g., fluid velocity to blade velocity) and ratios of time scales (frequency or rotational speeds) must be equal.

Finally, for dynamic similarity:

> Dynamic similarity requires that ratios of forces, masses, and energies be equal between a model and a prototype, presupposing geometric and kinematic similarities.

In general, dynamic similarity requires equality of the Reynolds number, Mach number, and Froude number. At times, equality of Weber number and/or cavitation number will be required.

The real objective of a similarity study is to accurately predict the performance or behavior of one object (or machine, or test) from information about the performance of a similar object under a different set of conditions (e.g., size, speed, and fluid involved). To do this, strictly speaking, complete similitude must be precisely maintained. In practice, it is frequently impractical, if not impossible, to achieve complete similarity. For example, it is usually not possible to match both Reynolds and Mach numbers in testing an object for high-speed operation. In such cases, extrapolation of data is fairly accurate if one can achieve or maintain the same "regime" for both model and prototype. Maintenance of the "regime" usually means having a Reynolds number that is larger than some lower limit or that surface roughness is "acceptably" small (perhaps implying a "hydraulically smooth" surface). Figure 2.1 (from Emery et al., 1958) illustrates the occurrence of a high Reynolds number regime. The figure shows the compressor blade row fluid turning angle, θ_{fl}, as a function of Reynolds number. Apparently, if the Reynolds number is greater than about 35×10^4, the exact value of the Reynolds number is irrelevant and one could reasonably assume that operation of this blade row at a Reynolds number of, say, 10^6 would produce the same turning angle as operation at a lower Reynolds number, say 5×10^5. A warning is in order; although these data are well behaved, it is limited, and judgment, experience, and risk are key issues in trying to push such information beyond the range of the original data.

2.3 Dimensionless Numbers and Π-Products

The basis for any similarity analysis is a set of dimensionless numbers, formed from the problem variables, that adequately describe the phenomenon under study. These numbers are determined by dimensional analysis. There are at least four methods for forming the dimensionless numbers; in order of preference these methods are as follows: (1) adapting "standard" dimensionless numbers (e.g., the Reynolds number and the Mach number); (2) by

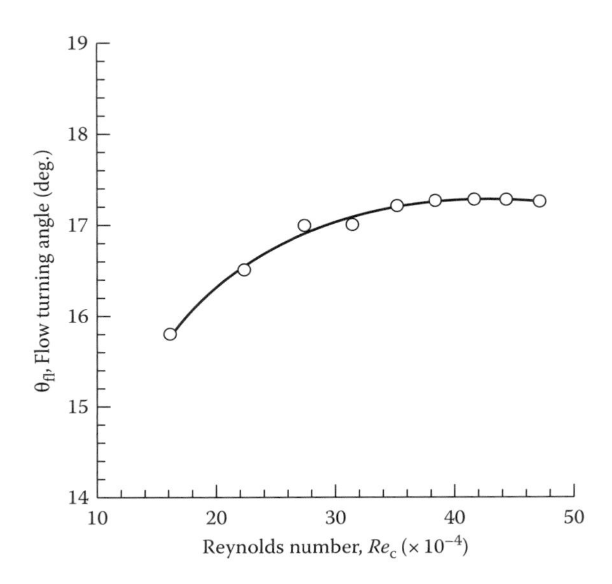

FIGURE 2.1 Variation of the blade row fluid turning angle with the Reynolds number. (From Emery, J. C. et al., 1958, *Systematic Two-Dimensional Cascade Tests of NACA 65-Series Compressor Blades at Low speed*, NACA-TR-1368. With permission.)

inspection; (3) by nondimensionalizing the governing equations, and (4) using the Π-product method. Here we will use a mixture of methods (1), (2), and (4).

First, review the Π method. Recall the basic rules. If one has a set of performance variables based on a number of fundamental dimensions, the object is to group the variables into products that are dimensionless, thus forming new variables that will be fewer in number than the original variable list. The steps in the procedure are as follows:

- List and count the variables (N_V).
- List and count the fundamental units or dimensions (N_u).
- Select a number of variables as "primary" (usually, this group consists of N_u of the problem variables).
- Form Π-products from the primaries plus a remaining variable. Once one obtains ($N_V - N_u$) new dimensionless variables, the variable list is reduced by N_u.

As an example, consider the lift force on an airfoil, which may represent a simple model of a compressor or a turbine blade (Figure 2.2). Following are the dimensional variables:

F_L: lift force (ML/T^2)
ρ: air density (M/L^3)

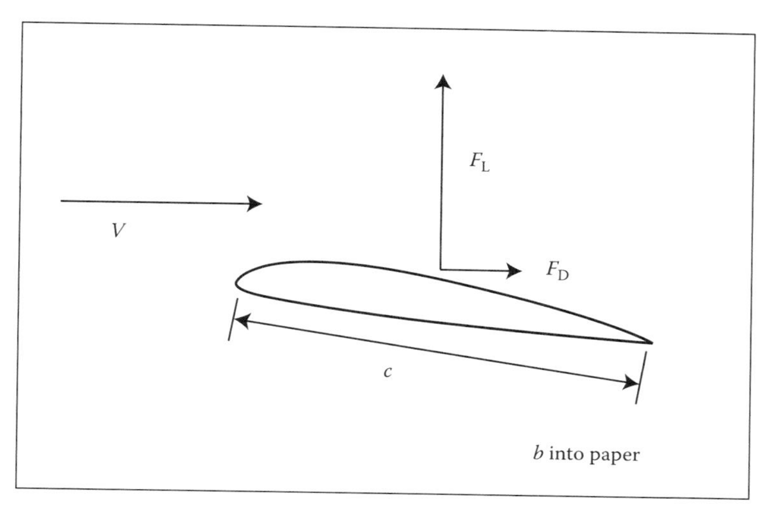

FIGURE 2.2 Lift on an airfoil.

c: chord of the airfoil (L)
μ: viscosity of the air (M/LT)
V: air velocity (L/T).

Following the process outlined above, the steps are as follows:

- List the variables: F_L, ρ, c, μ, $V (N_v = 5)$.
- List dimensions: ML/T^2, M/L^3, L, M/LT, L/T $(N_u = 3)$.
- Choose N_u (3) primary variables: These three variables must not be able to form a dimensionless product by themselves. Choose V and c (since V contains time and c does not) and choose ρ to go with them because it contains mass while V and c do not.
- Form Π-products: $N_v - N_u = 2$; so develop two products:

$$\Pi_1 = F_L \rho^a V^b c^c; \qquad \Pi_2 = \mu \rho^a V^b c^c.$$

Solve for the exponents according to the rule that Πs are dimensionless; therefore, the right-hand side must also be dimensionless. Considering Π_1 yields the equation

$$M^0 L^0 T^0 = \left[\frac{ML}{T^2}\right]\left[\frac{M}{L^3}\right]^a \left[\frac{L}{T}\right]^b [L]^c.$$

Consider each dimension separately

$$M: 0 = 1 + a, \quad a = -1,$$

$$T: 0 = -b - 2, \quad b = -2.$$

$$L: 0 = 1 - 3a + b + c, \quad c = -2.$$

Therefore,

$$\Pi_1 = F_L \rho^{-1} V^{-2} c^{-2} = \frac{F_L}{\rho V^2 c^2}. \tag{2.3}$$

Turning to the second Π-product, the dimension equation becomes

$$M^0 L^0 T^0 = \left[\frac{M}{LT}\right]\left[\frac{M}{L^3}\right]^a \left[\frac{L}{T}\right]^b [L]^c$$

and

$$M: 0 = 1 + a, \quad a = -1,$$

$$T: 0 = -b - 1, \quad b = -1,$$

$$L: 0 = -1 - 3a + b + c, \quad c = -1.$$

Therefore,

$$\Pi_2 = \frac{\mu}{\rho V c}. \tag{2.4}$$

Anyone familiar with fluid mechanics knows that the parameters derived above as Π_1 and Π_2 are never used in these forms. Traditionally, Π_1 is written as the lift coefficient C_L:

$$C_L \equiv \frac{F_L}{1/2 \rho V^2 c^2} \quad \text{or} \quad C_L \equiv \frac{F_L}{1/2 \rho V^2 cb}$$

where b is the span of the airfoil/blade so that cb is the planform area and "1/2" is a nod to Bernoulli's equation.

Inverting Π_2 gives the well-known Reynolds number

$$Re = \frac{\rho V c}{\mu} = \frac{Vc}{\nu}. \tag{2.5}$$

This example shows that using method (1) for establishing dimensionless parameters often saves a lot of work. Anyone familiar with fluid flow would know that the Reynolds number would appear if viscosity were included in the dimensional variables. As an additional note, both the lift coefficient and the Reynolds number are dynamic similarity parameters. The Reynolds number is a ratio of inertia "force" to viscous force and lift coefficient is a ratio of lift force to inertia "force."

2.4 Dimensionless Performance Variables and Similarity for Turbomachinery

Now consider the performance of turbomachines and limit consideration for the time being to incompressible flow machines. There are many ways to choose the (dimensional) fluid specific energy variable (e.g., head H, its energy equivalent gH, and pressure rise Δp_T, or Δp_s). To start, use the following list:

- Q: volume flow rate (L^3/T)
- gH: head (specific energy variable) $[L^2/T^2](gH = \Delta p_T/\rho)$
- P: power (ML^2/T^3)
- N: rotational speed $(1/T)$
- D: diameter (L)
- ρ: fluid density (M/L^3)
- μ: fluid viscosity (M/LT)
- d/D: diameter ratio (dimensionless).

The flow rate is listed first because it is considered the primary independent variable (see, e.g., Figure 1.9a through c). d/D is included as a reminder that geometric similitude is imposed, thus eliminating changes of shape or proportion. A subtle dimensionless ratio that has been omitted for the time being is ε/D, the relative roughness.

There are three basic dimensions involved: mass, length, and time (M, L, and T as before), so one can expect to reduce the seven-dimensional variables to four. The variables ρ, N, and D are a good choice for the primaries for two reasons: (1) they directly represent mass, time, and length and (2) they are operating variables as opposed to performance variables.

Which of the four methods for determining dimensionless parameters should be used? The presence of viscosity suggests a Reynolds number:

$$Re = \frac{\text{Density} \times \text{Size} \times \text{Speed}}{\text{Viscosity}}.$$

What to use for density (ρ), size (D), and viscosity (μ) is clear enough, but what about speed. An essential characteristic of a turbomachine is that there are *two* speeds involved; the speed of the fluid and the speed of the rotor (i.e., of the blade tip), $ND/2$. Because N and D were chosen as primary variables, the speed used in the Reynolds number will be ND (the "2" is dropped for convenience). Therefore, one parameter is the so-called *Machine Reynolds Number*:

$$Re_D = \frac{\rho \times D \times ND}{\mu} = \frac{\rho ND^2}{\mu} = \frac{ND^2}{\nu}. \quad (2.6)$$

Can another parameter be determined by inspection? The presence of two different types of velocity implies that their ratio would be an important kinematic similarity parameter:

$$\Pi_2 = \frac{\text{Characteristic fluid speed}}{\text{Characteristic rotor speed}} = \frac{Q/(\pi/4)D^2}{ND/2}.$$

Dropping all of the constants, the second parameter is the flow coefficient, ϕ:

$$\phi = \frac{Q}{ND^3}. \tag{2.7}$$

The third parameter, involving the fluid specific energy, gH, can be determined either by inspection or by the Π-product method. Since gH is (fluid) energy per unit mass, a likely form for the third parameter is

$$\Pi_3 = \frac{\text{Fluid specific energy}}{\text{Kinetic energy per unit mass}} = \frac{gH}{1/2\,(\text{Velocity})^2}.$$

Using the rotor velocity as consistent with the choice of N and D as primary variables and dropping the constants gives the *head coefficient*, ψ, as the third dimensionless parameter:

$$\psi = \frac{gH}{N^2D^2} = \frac{\Delta p_T}{\rho N^2 D^2}. \tag{2.8}$$

Determining ψ by the Π-product method is left to the reader.

The final parameter, involving the power input, can be developed by the Π-product method:

$$\Pi_4 = P\rho^a N^b D^c.$$

The dimension equation is

$$M^0L^0T^0 = \left[\frac{ML^2}{T^3}\right]\left[\frac{M}{L^3}\right]^a\left[\frac{1}{T}\right]^b [L]^c,$$

giving

$$\begin{aligned}
&\text{M: } 0 = 1 + a; \quad a = -1,\\
&\text{T: } 0 = -3 - b; \quad b = -3,\\
&\text{L: } 0 = 2 - 3a + b + c; \quad c = -5.
\end{aligned}$$

The final parameter is the *power coefficient*, ξ:

$$\xi = \frac{P}{\rho N^3 D^5}. \tag{2.9}$$

The reader is invited to deduce the form for ξ by inspection.

Referring back to the performance curves in Chapter 1, it seems that one important parameter is missing, namely the efficiency η. Recall that $\eta = \rho QgH/P$ and consider

$$\Pi_5 = \frac{\psi\phi}{\xi} = \frac{(gH/N^2D^2)(Q/ND^3)}{P/\rho N^3D^5} = \frac{\rho QgH}{P}. \tag{2.10}$$

The efficiency is an alternate dimensionless parameter that can be used instead of the power coefficient.

Collecting all dimensionless groups together, the performance equations for a family of geometrically similar turbomachines can be written in functional notation as

$$\begin{aligned} \psi &= f_1\left(\phi, Re_D, \frac{d}{D}\right), \\ \eta &= f_2\left(\phi, Re_D, \frac{d}{D}\right), \\ \xi &= f_3\left(\phi, Re_D, \frac{d}{D}\right), \end{aligned} \tag{2.11}$$

where d/D is included as a reminder of geometric similitude.

If the Reynolds number is large, one may hypothesize a "weak" dependence on Re [e.g., look at the behavior of the friction factor, f, versus the Reynolds number, Re, for large Reynolds number for moderate-to-large relative roughness, ε/D, on a Moody diagram (White, 2008)]. Also assuming complete geometric similarity so that d/D can be dropped, the performance equations become, approximately,

$$\begin{aligned} \psi &= g_1(\phi), \\ \eta &= g_2(\phi), \\ \xi &= g_3(\phi). \end{aligned} \tag{2.12}$$

This is really compact!

These relationships are very interesting. What can one do with them? Consider the characteristic curves for a fan (Figure 1.9a and b) generated in the discussion of performance testing in Section 1.9, where Δp_T, η, and P were plotted as functions of Q. For an actual test, values of ρ, N, and D would be known and values of Δp_T, P, η, and Q can be read from the graphs. Then a series of values of ϕ, ψ, ξ, and η can be calculated. If one plots these (e.g., ψ versus ϕ), the dimensionless representation of the fan's performance is obtained. The curves shown in Figure 2.3 are an example. These $\psi - \phi$, $\xi - \phi$, and $\eta - \phi$ curves can be used to generate an infinite number of (dimensional)

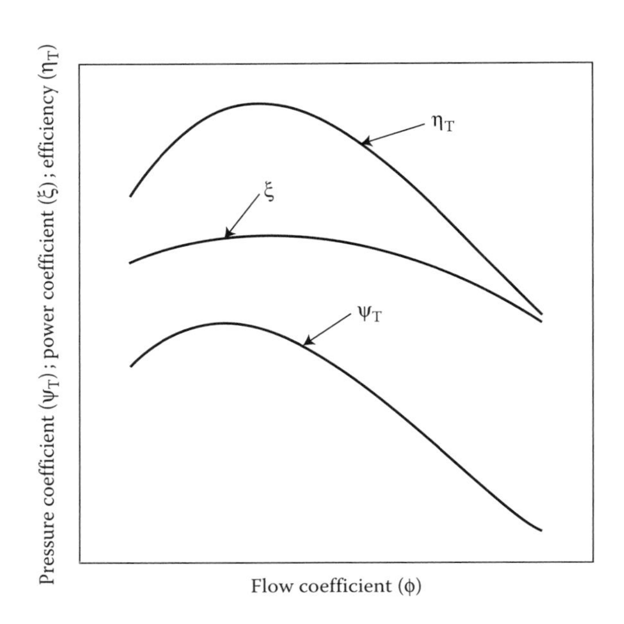

FIGURE 2.3 Dimensionless performance curves.

$\Delta p_T - Q$, $P - Q$, and $\eta - Q$ curves by choosing any values of ρ, N, and D we want. We only need to maintain geometric similitude. The primary restriction lies in the assumption of weak Reynolds number dependence (as well as Mach number independence, which will be discussed in Section 2.5).

Example: Dimensionless Performance Variables

A fan is tested over the following range:

$$0 \le Q \le 100\,\mathrm{m^3/s},$$

$$0 \le \Delta p_T \le 1500\,\mathrm{Pa},$$

$$0 \le P \le 100\,\mathrm{kW},$$

with the design point performance (BEP) of $Q = 80\,\mathrm{m^3/s}$; $\Delta p_T = 1000\,\mathrm{Pa}$; and $P = 90\,\mathrm{kW}$. The fan diameter is $D = 1.2\,\mathrm{m}$, the speed is $N = 103\,\mathrm{s^{-1}}$ (980 rpm), and the air density is $\rho = 1.2\,\mathrm{kg/m^3}$.

The normalizing factors to be used to form the dimensionless variables are

$$ND^3 = 177\,\mathrm{m^3/s},$$

$$\rho N^2 D^2 = 18{,}181\,\mathrm{kPa},$$

$$\rho N^3 D^5 = 3270\,\mathrm{kW}.$$

The design point (BEP) values of ϕ, ψ, ξ, and η_T are calculated as

$$\phi = \frac{80}{177} = 0.452,$$

$$\psi = \frac{1000}{18{,}181} = 0.055,$$

$$\xi = \frac{90}{3270} = 0.0279,$$

$$\eta_T = \frac{1000 \times 80}{90{,}000} = 0.890.$$

As a check,

$$\eta_T = \frac{\psi\phi}{\xi} = \frac{0.055 \times 0.452}{0.0279} = 0.890.$$

What would be the BEP performance of this fan design at a different size and speed? For example, suppose

$$D = 30\,\text{inches} = 2.5\,\text{ft},$$

$$N = 1800\,\text{rpm} = 188.5\,\text{s}^{-1},$$

$$\rho g = 0.074\,\text{lb/ft}^3; \qquad \rho = 0.0023\,\text{slug/ft}^3.$$

Note that the units have been changed to the BG system! Will this matter? Calculate

$$ND^3 = 2945\,\text{ft}^3/\text{s},$$

$$\rho N^2 D^2 = 510.8\,\text{lb/ft}^2 = 98.23\,\text{in. wg},$$

$$\rho N^3 D^5 = 1.504 \times 10^6\,\text{ft} \cdot \text{lb/s}.$$

Now use the previously calculated ϕ, ψ, and ξ to calculate Δp_T, Q, and P:

$$Q = \phi\left(ND^3\right) = 0.452 \times 2945 = 1325\,\text{ft}^3/\text{s} = 79{,}250\,\text{cfm},$$

$$\Delta p_T = \psi\left(\rho N^2 D^2\right) = 0.055 \times 510.8 = 28.1\,\text{lb/ft}^2 = 5.4\,\text{in. wg},$$

$$P = \xi\left(\rho N^3 D^5\right) = 0.0279 \times 1.504 \times 10^6 = 41{,}950\,\text{ft} \cdot \text{lb/s}$$

$$= 76.3\,\text{hp} = 56\,\text{kW}.$$

The efficiency is

$$\eta_T = \frac{\Delta p_T Q}{P} = \frac{1325 \times 28.1}{41{,}950} = 0.890.$$

According to the scaling rules, η_T should be conserved in the resizing process, so this last calculation serves as a check on the other results.

As an additional check, one should examine the relation between the "model" and "prototype" Reynolds numbers. Using a typical value for kinematic viscosity of air ($\nu = 1.6 \times 10^{-4}\ \text{ft}^2/\text{s}$) gives $Re_D = ND^2/\nu = 0.73 \times 10^7$ for the "model" fan and $Re_D = 1.0 \times 10^7$ for the original fan. Are these sufficiently close together or, alternately, sufficiently large to justify neglecting differences in viscous effects? One way to estimate this effect is to compare with the Darcy friction factor. Using the Colebrook formula (White, 2008)

$$f = \frac{0.25}{\left\{\log_{10}\left[(\varepsilon/D)/3.7 + 2.51/Re_D\sqrt{f}\right]\right\}^2}, \quad (2.13)$$

and assuming a very small roughness as a worst case, one obtains $f \approx 0.0085$ for the original fan and $f \approx 0.00813$ for the model fan. For the hydraulically smooth case, the viscous friction losses should be slightly less for the prototype (~4%) and might increase the efficiency by perhaps 0.005, *if* all of the inefficiency were attributable to friction. A later chapter will deal more rigorously with the Reynolds number and roughness influence, but as a general rule, *changes* in values of Re_D when Re_D is above 10^7 have little influence on efficiency. This "threshold" Reynolds number is even lower for rough surfaces.

A little reflection will lead to the conclusion that it is not necessary to explicitly calculate ϕ, ψ, and ξ to transform performance variables from one size and set of operating conditions (D_1, N_1, and ρ_1) to a second size and/or set of operating conditions (D_2, N_2, and ρ_2). One can make the transformation directly using ratios. The following equations hold between any pair of *corresponding points*:

$$\begin{aligned}
&\text{From } \phi\text{: } \frac{Q_2}{Q_1} = \left(\frac{N_2}{N_1}\right)\left(\frac{D_2}{D_1}\right)^3, \\
&\text{From } \psi\text{: } \frac{\Delta p_2}{\Delta p_1} = \left(\frac{\rho_2}{\rho_1}\right)\left(\frac{N_2}{N_1}\right)^2\left(\frac{D_2}{D_1}\right)^2, \\
&\text{From } \psi\text{: } \frac{H_2}{H_1} = \left(\frac{N_2}{N_1}\right)^2\left(\frac{D_2}{D_1}\right)^2, \\
&\text{From } \xi\text{: } \frac{P_2}{P_1} = \left(\frac{\rho_2}{\rho_1}\right)\left(\frac{N_2}{N_1}\right)^3\left(\frac{D_2}{D_1}\right)^5, \\
&\text{From } \eta\text{: } \frac{\eta_2}{\eta_1} = 1;\ \text{that is, } \eta_2 = \eta_1.
\end{aligned} \quad (2.14)$$

Re-examining the previous example, one can transform, say, Δp_T directly by

$$\Delta p_T = 1000\ \text{Pa} \times \frac{0.0023\,(14.62\ \text{kg/slug})}{1.2} \times \left(\frac{188.5}{103}\right)^2 \times \left(\frac{2.5\,(0.3048\ \text{m/ft})}{1.2}\right)^2,$$

$$= 1345\ \text{Pa} = 5.4\ \text{in. wg.}$$

2.5 Compressible Flow Similarity

When significant changes in density are likely to occur through the turbomachine, the use of volume flow rate, Q, and the head H or pressure rise, Δp, become inappropriate. Instead, flow rate is the mass flow rate, $\dot{m}$, in kg/s or slug/s. The fluid specific energy change ("head") is represented by the isentropic change in stagnation enthalpy, Δh_{0s}, or a related variable, such as stagnation temperature or stagnation pressure, as discussed in Section 1.11. If attention is restricted to flow of a perfect gas, one can relate Δh_{0s} to the change in stagnation temperature through the machine as work is done on the fluid; that is,

$$\Delta h_{0s} = c_p \Delta T_{0s}. \tag{2.15}$$

One can still define the *isentropic head*, H_s, according to

$$gH_s = \Delta h_{0s} = c_p \Delta T_{0s}. \tag{2.16}$$

A modified variable list for dimensional analysis must now include the machine inlet stagnation temperature and at least two of the following: the gas constant (R), the specific heats, c_p, c_v, or the specific heat ratio, γ $(= c_p/c_v)$. The density and viscosity must be evaluated at an appropriate reference condition; typically the inlet stagnation state is chosen. The dimensional performance equations can be written in "functional form" as

$$\begin{aligned}
\Delta h_{0s} &= f_1\left(\dot{m}, N, D, \rho_{01}, T_{01}, \mu_{01}, R, \gamma, \frac{d}{D}\right),\\
P &= f_2\left(\dot{m}, N, D, \rho_{01}, T_{01}, \mu_{01}, R, \gamma, \frac{d}{D}\right),\\
\eta &= f_3\left(\dot{m}, N, D, \rho_{01}, T_{01}, \mu_{01}, R, \gamma, \frac{d}{D}\right).
\end{aligned} \tag{2.17}$$

As before, d/D is a reminder of the requirement for geometric similarity. The efficiency could be either the isentropic efficiency or the polytropic efficiency. η, d/D, and γ need not be included in a dimensional analysis because they are already nondimensional. They must of course be considered in the underlying constraints on maintaining similarity and can be used in combination with any dimensionless parameter resulting from a dimensional analysis, perhaps to form new parameters.

An important dimensional variable in the analysis of compressible flow is the speed of sound (Anderson, 1984; White, 2008). It is customary to replace the stagnation temperature, T_{01}, and the gas constant, R, with the speed of sound according to $a_{01} = (\gamma R T_{01})^{1/2}$.

The new variable list is as follows:

- Δh_{0s}: with dimensions (L^2/T^2)
- $\dot{m}$: with dimensions (M/T)
- N: with dimension $(1/T)$
- D: with dimension (L)
- ρ_{01}: with dimensions (M/L^3)
- a_{01}: with dimensions (L/T)
- μ_{01}: with dimensions (M/LT).

There are seven variables requiring the three dimensions: length (L), time (T), and mass (M). One must choose three primary variables to form four Π-products. The same three variables used for incompressible flow are appropriate: ρ_{01}, N, and D. Three of the dimensionless parameters can be written by direct analogy, namely, the (mass) flow coefficient, the "head" coefficient, and the Reynolds number. The fourth (new) parameter would, of course, be a Mach number (Anderson, 1984; White 2008):

$$\Pi_1 = \frac{\Delta h_{0s}}{N^2 D^2} : \text{ Head coefficient,}$$

$$\Pi_2 = \frac{\dot{m}}{\rho_{01} N D^3} : \text{ Mass flow coefficient,}$$

$$\Pi_3 = \frac{ND}{a_{01}} : \text{ Machine Mach number,}$$

$$\Pi_4 = \frac{\rho_{01} N D^2}{\mu_{01}} : \text{ Machine Reynolds number.}$$

To these can be added

$$\Pi_5 = \frac{P}{\rho_{01} N^3 D^5} : \text{ Power coefficient,}$$

$$\Pi_6 = \eta_s : \text{ Isentropic efficiency,}$$

$$\Pi_7 = \eta_p : \text{ Polytropic efficiency.}$$

These parameters can be used as they stand or some or all of them can be rearranged using isentropic flow and perfect gas relations to form the more conventional variables used for high-speed compressors and turbines. The relations to be used are $p/\rho^\gamma =$ Constant for isentropic flow, $p = \rho RT$, $h = c_p T$, and $a_0^2 = \gamma R T_0$. Thus an alternate variable is

$$\psi' = \Pi_1 \times \Pi_3^2 = \frac{\Delta h_{0s}}{a_{01}^2} = \frac{c_p (T_{02s} - T_{01})}{\gamma R T_{01}} = \frac{\gamma}{\gamma - 1}\left(\frac{T_{02s}}{T_{01}} - 1\right). \quad (2.18)$$

Dropping the already dimensionless $\gamma/(\gamma-1)$ and the "1" from the expression yields

$$\psi' = \frac{T_{02s}}{T_{01}}. \tag{2.19}$$

Further, since $T_{02s}/T_{01} = (p_{02}/p_{01})^{\gamma-1/\gamma}$ for isentropic flow, an alternate "head coefficient" is

$$\psi = \frac{p_{02}}{p_{01}}. \tag{2.20}$$

The *stagnation pressure ratio* is the conventional "head coefficient" for high-speed compressors and turbines.

For the flow coefficient, again combine parameters so that

$$\phi' = \Pi_2 \times \Pi_3 = \frac{\dot{m}}{\rho_{01} N D^3} \times \frac{ND}{a_{01}} = \frac{\dot{m}}{\rho_{01} a_{01} D^2}.$$

Introducing $a_{01} = (\gamma R T_{01})^{1/2}$ and $\rho_{01} = p_{01}/RT_{01}$

$$\phi' = \frac{\dot{m}}{\rho_{01}(\gamma R T_{01})^{1/2} D^2} = \frac{\dot{m} R T_{01}}{p_{01} D^2 (\gamma R T_{01})^{1/2}},$$

or in the final form, dropping the γ gives the conventional *mass flow coefficient* for high-speed compressors and turbines:

$$\phi = \frac{\dot{m}(RT_{01})^{1/2}}{p_{01} D^2}. \tag{2.21}$$

The traditional forms of the dimensionless performance functions for high-speed gas compressors and turbines are

$$\frac{p_{02}}{p_{01}} = f\left(\frac{\dot{m}(RT_{01})^{1/2}}{p_{01} D^2}, \frac{ND}{(RT_{01})^{1/2}}, \frac{\rho_{01} N D^2}{\mu_{01}}, \gamma\right) \tag{2.22}$$

and

$$\eta_s = g\left(\frac{\dot{m}(RT_{01})^{1/2}}{p_{01} D^2}, \frac{ND}{(RT_{01})^{1/2}}, \frac{\rho_{01} N D^2}{\mu_{01}}, \gamma\right). \tag{2.23}$$

Traditional use of these compressible parameters frequently involves a simplified form that rests on the assumptions that (1) there is negligible influence of Reynolds number, (2) the impeller diameter is held constant, and (3) there is no change in the gas being used (e.g., air as a perfect gas).

Under these assumptions, R, γ, Re, and D are dropped from the expressions, leaving a set of *dimensional* scaling parameters (that nevertheless have dimensionless-parameter significance):

$$\frac{p_{02}}{p_{01}} = f\left(\frac{\dot{m}(T_{01})^{1/2}}{p_{01}}, \frac{N}{(T_{01})^{1/2}}\right), \tag{2.24}$$

$$\eta_s = g\left(\frac{\dot{m}(T_{01})^{1/2}}{p_{01}}, \frac{N}{(T_{01})^{1/2}}\right). \tag{2.25}$$

At times, this is carried one step further by defining *reduced pressure* δ and *reduced temperature* θ by $\delta \equiv p_{01}/p_{\text{standard sea level}}$; $\theta \equiv T_{01}/T_{\text{standard sea level}}$ to write the performance equations as

$$\frac{p_{02}}{p_{01}} = f\left(\frac{\dot{m}(\theta)^{1/2}}{\delta}, \frac{N}{(\theta)^{1/2}}\right) \tag{2.26}$$

and

$$\eta_s = g\left(\frac{\dot{m}(\theta)^{1/2}}{\delta}, \frac{N}{(\theta)^{1/2}}\right). \tag{2.27}$$

The parameters $\dot{m}(\theta)^{1/2}/\delta$ and $N/(\theta)^{1/2}$ have the units of mass flow and rotational speed, respectively, and are called *reduced flow* and *reduced speed*.

The form for the head-flow characteristic curves for these compressible flows is very similar to that introduced for the low-speed, low-pressure machines. Figures 2.4 and 2.5 show the (dimensionless) ψ versus ϕ curves of a compressor and a high-speed turbine, using the Mach number to differentiate the curves relating to different values of speed. Qualitatively, the most notable difference between these curves and their incompressible counterparts is the very sharp steepening of the characteristics at the highest flow rates. Here, the compressible flow has reached local sonic velocity somewhere in the machine so that the mass flow becomes choked (Anderson, 1984).

2.6 Specific Speed and Specific Diameter

The scaling rules above, whether the conventional low-speed (incompressible) rules or the compressible similarity rules, can be used rather formally to examine the influence of size change and speed change on machine performance. What is missing from these studies is a systematic way to vary both speed and size in order to obtain particular values for head and flow. An example will illustrate this.

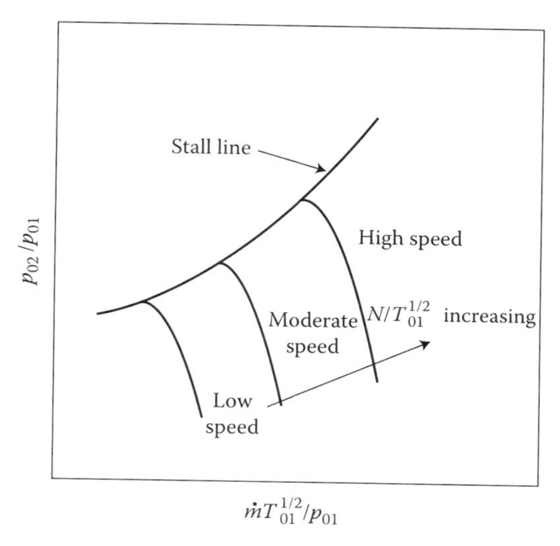

FIGURE 2.4 Compressor characteristic curves.

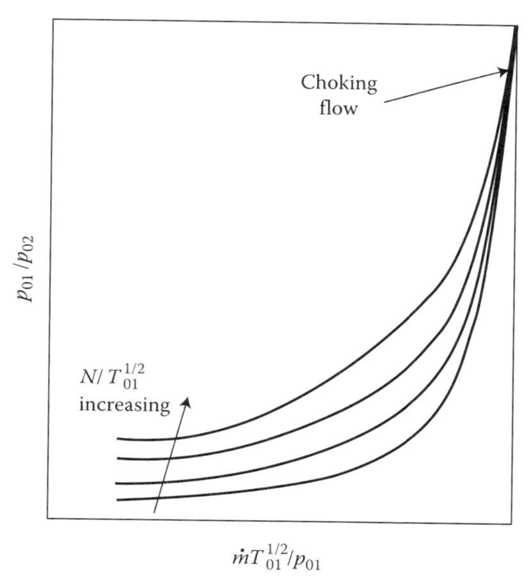

FIGURE 2.5 Turbine characteristic curves.

Example: Determining Turbomachine Size and Speed

A particular type of fan has, at its BEP, $\phi = 0.455$ and $\psi = 0.055$. What size (D) and speed (N) would be required for a fan of this type to deliver 20,000 cfm at 1 in. wg total pressure?

In this case, the flow and pressure are known and the speed and size are sought. From the definitions of ϕ and ψ

$$Q = \phi \times ND^3 \quad \text{and} \quad \Delta p_T = \psi \times \rho N^2 D^2.$$

Rearranging

$$ND^3 = \frac{Q}{\phi} \quad \text{and} \quad N^2 D^2 = \frac{\Delta p_T}{\rho\psi}$$

gives two equations in two unknowns. Solving each equation for N and equating the two expressions, we obtain

$$N = \frac{Q}{D^3\phi} = N = \left(\frac{\Delta p_T}{\rho\psi}\right)^{1/2} \frac{1}{D}.$$

Now we can solve for D:

$$D = \frac{Q^{1/2}}{(\Delta p_T/\rho)^{1/4}} \frac{\psi^{1/4}}{\phi^{1/2}}. \tag{2.28}$$

Substituting back, we obtain

$$N = \frac{Q}{\phi D^3} = \frac{Q}{\phi}\left(\frac{\Delta p_T/\rho}{Q^{1/2}} \frac{\phi^{1/2}}{\psi^{1/4}}\right)^3 = \frac{(\Delta p_T/\rho)^{3/4}}{Q^{1/2}} \frac{\phi^{1/2}}{\psi^{3/4}}. \tag{2.29}$$

Substituting numbers, we obtain

$$D = \left[\frac{(20{,}000/60)^{1/2}}{(1 \times 5.204/0.00223)^{1/4}}\right] \frac{0.055^{1/4}}{0.455^{1/2}} = 1.91 \text{ ft},$$

$$N = \frac{(1.0 \times 5.204/0.0223)^{3/4}}{(20{,}000/60)^{1/2}} \frac{0.455^{1/2}}{0.055^{3/4}} \times \frac{60}{2\pi} = 1045 \text{ rpm}.$$

What has really been done? Two new dimensionless numbers have been derived; they are

$$\text{Specific speed, } N_s = \frac{NQ^{1/2}}{(gH)^{3/4}} = \frac{\phi^{1/2}}{\psi^{3/4}}. \tag{2.30}$$

$$\text{Specific diameter, } D_s = \frac{D(gH)^{1/4}}{Q^{1/2}} = \frac{\psi^{1/4}}{\phi^{1/2}}. \tag{2.31}$$

As the example shows, N_s and D_s can be used to determine size and speed for a particular type of machine from the specification of flow and pressure; however, their significance far exceeds this application. Specific speed and specific diameter are two extremely important dimensionless numbers because they are used to characterize many parameters of turbomachines. Significant facts about specific speed and specific diameter are as follows:

- N_s and D_s are not independent dimensionless parameters; they can be formed from ϕ and ψ. Had Q and gH been chosen as primary variables in the dimensional analysis, N_s and D_s would have appeared as Π_1 and Π_2.

- As they are defined in Equations 2.30 and 2.31, N_s and D_s can vary between zero (at shutoff for N_s and at free delivery for D_s) and infinity (at shutoff for D_s and at free delivery for N_s).
- In normal usage, these two variables are specified *at best efficiency*; $N_s = NQ_{BEP}^{1/2}/(gH_{BEP})^{3/4}$ and $D_s = D(gH)_{BEP}^{1/4}/Q_{BEP}^{1/2}$. Almost always, the phrases "specific speed" and "specific diameter" mean "specific speed at best efficiency" and "specific diameter at best efficiency."
- Values of (true dimensionless) specific speed and specific diameter for turbomachines vary from about 0.2 to about 10. (Positive displacement machines are used for $N_s < 0.2$ and $D_s > 10$.)
- Traditionally, N_s and D_s are calculated for compressible flow machines by using the isentropic head, gH_s, and the inlet stagnation volume flow, $Q_1 = \dot{m}/\rho_{01}$.
- As a result of the restriction to the BEP, a unique value of N_s and a unique value of D_s can be assigned to each (geometrically similar) family of turbomachines.

In practice, engineers sometimes get careless with units and dimensions in calculating specific speed and specific diameter. In the United States, the specific speed for pumps is often evaluated by

$$n_s\,(\text{pump}) = \frac{N\,(\text{rpm})\sqrt{Q\,(\text{gpm})}}{[H\,(\text{ft})]^{3/4}}, \tag{2.32}$$

and the specific speed for fans by

$$n_s\,(\text{fan}) = \frac{N\,(\text{rpm})\sqrt{Q\,(\text{cfm})}}{\left[\Delta p_T\,(\text{in. wg})\right]^{3/4}}. \tag{2.33}$$

Although these specific speeds are not dimensionless, they do offer a couple of advantages: they are convenient to calculate and they produce numerical values that are typical of the actual speed in rpm (i.e., a few hundred to a few thousand).

Example: Specific Speed and Specific Diameter

What are the specific speed and specific diameter of the fan "sized" earlier in this section. Recall that $Q = 20{,}000$ cfm and $\Delta p_T = 1$ in. wg. This fan was sized at $D = 1.91$ ft and $N = 1045$ rpm. This was said to be the BEP.

$$N_s = \frac{NQ^{1/2}}{(\Delta p_T/\rho)^{3/4}} = \frac{1045 \times 2\pi/60 \times (20{,}000/60)^{1/2}}{(1.0 \times 5.204/0.00223)^{3/4}} = 5.95,$$

$$D_S = \frac{D(\Delta p_T/\rho)^{1/4}}{Q^{1/2}} = \frac{1.91(1.0 \times 5.204/0.00223)^{1/4}}{(20{,}000/60)^{1/2}} = 0.73.$$

In this case, these could be calculated more quickly by

$$N_S = \frac{\phi^{1/2}}{\psi^{3/4}} = \frac{0.455^{1/2}}{0.055^{3/4}} = 5.94$$

and

$$D_S = \frac{\psi^{1/4}}{\phi^{1/2}} = \frac{0.055^{1/4}}{0.455^{1/2}} = 0.72,$$

which are the same values, allowing for roundoff error. In practice, it is more typical to have values readily available for flow, pressure (head), speed, and size than for ϕ and ψ, so the direct calculation would be more likely.

2.7 Correlations of Machine Type and the Cordier Diagrams

As noted above, a unique value of specific speed N_S and a unique value of specific diameter D_S can be assigned to each (geometrically similar) family of turbomachine by defining these parameters at the machine's BEP. It follows that, to a good approximation, the value of N_S or D_S can be used to choose the best (i.e., most efficient) type of machine for a given task, independently of the actual speed or size of the machine or the actual magnitude of the flow rate or fluid specific energy change. Logic, confirmed by experience, has shown, for example, that radial flow machines with small inlet "eyes" and narrow blade passages, relative to impeller diameter, perform best at relatively lower flow rates and higher pressures. On the other hand, axial flow machines, which are much more open and lack "centrifugal forces" to do work on the fluid, perform better at higher flow rates and lower pressure changes. The radial flow machine has a low value of N_S (say 0.5) and a high value of D_S (say 5), while the opposite is true for the axial flow machine (perhaps $N_S \approx 6$, $D_S \approx 1.5$).

In practice, machines are available in a more or less continuous spectrum from very narrow radial impellers through mixed flow impellers to very open axial flow types, so the range of specific speeds and specific diameters available is continuous. Figure 2.6 shows how various types of turbomachines correlate with specific speed and specific diameter, and Figure 2.7 shows how impeller shapes for pumps are related to specific speed.

Specific speed and specific diameter are interrelated. In the 1950s, Cordier (1955) carried out an intensive empirical analysis of "good" turbomachines

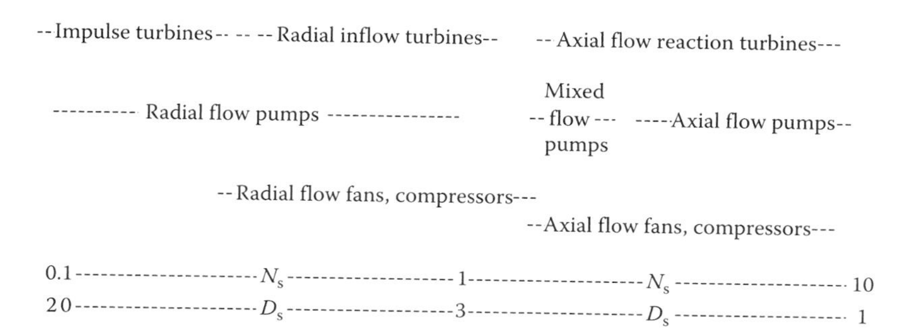

FIGURE 2.6 Specific speed and specific diameter ranges for various types of turbomachines.

using extensive experimental data. He attempted to correlate historical data on turbomachine design and performance in terms of N_S, D_S, and η_T using total pressure/total head to form the dimensionless parameters. He found (subject, of course, to uncertainty and scatter in the data) that turbomachines, which for their type had good to excellent efficiencies, tended to group along a definable curve when plotted with their values of N_S versus D_S. Machines whose efficiencies could be classified as poor for a particular type of device were found to scatter away from the locus of excellent machines in $D_S - N_S$ coordinates. Cordier reasoned that several decades of intense, competitive development of turbomachinery design had led to a set of "rules" of acceptable design practice, and that his curves were, in effect, a practical guideline for both initial design layout and effective selection of good machinery for a specified purpose. The "specified purpose" is, of course, stated in terms of Δp_T (or gH or gH_S), Q (or $\dot{m}$), and ρ or (T_{01} and p_{01}), as appropriate to the situation.

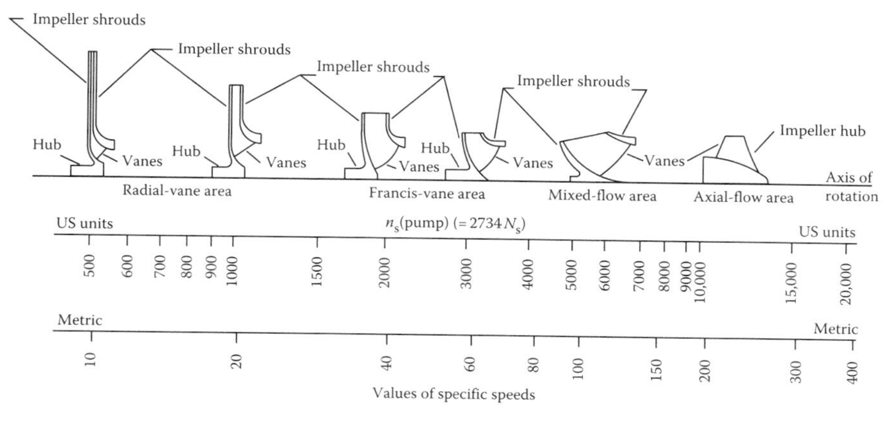

FIGURE 2.7 Correlation of pump impeller shapes with specific speed and specific diameter. (Courtesy of the Hydraulic Institute, Parsippany, NJ. http://www.pumps.org)

The "Cordier concept" was extended by Balje (1981). Among other things, Balje, as well as Csanady (1964), combined Cordier's multiple machine-specific curves into a single curve. The modified Cordier diagram is illustrated in Figure 2.8. The points shown are typical of Cordier's data and are not exhaustive. Recall that "less-than excellent" machines tend to scatter further away from the mean curve and note the log–log scale of the plot.

To a good approximation, the N_S–D_S curve can be approximated by two straight lines in log–log coordinates, as shown. The algebraic form for the fitted curves is

$$\ln(N_S) = a\ \ln(D_S) + b,$$

which gives the working equation

$$N_S = C(D_S)^a.$$

Using linear regression on Cordier's data [as given by Balje (1981)] gives the equations

$$N_S \approx 8.26 D_S^{-1.936}, \quad \text{for } D_S < 3.23 \tag{2.34}$$

and

$$N_S \approx 2.5 D_S^{-0.916}, \quad \text{for } D_S > 3.23. \tag{2.35}$$

For convenience, the inverse of these equations is

$$D_S \approx \left(\frac{8.26}{N_S}\right)^{0.517}, \quad \text{for } N_S > 0.85 \tag{2.36}$$

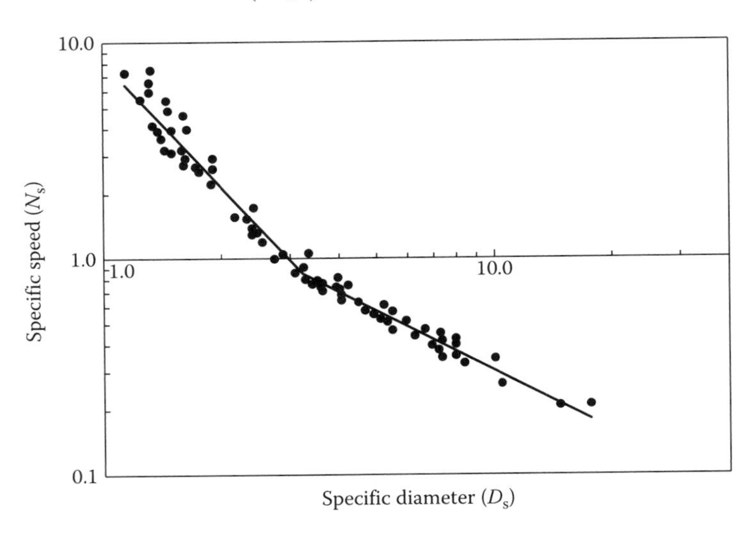

FIGURE 2.8 Cordier diagram for all types of turbomachines. (After Cordier, O., 1955, *Similarity Considerations in Turbomachines*, Vol. 3, VDI Reports; Csanady, G. T., 1964, *Theory of Turbomachines*, McGraw-Hill, New York; and Balje, O. E., 1981, *Turbomachines*, Wiley & Sons, New York.)

and

$$D_s \approx \left(\frac{2.5}{N_s}\right)^{1.092}, \quad \text{for } N_s < 0.85. \tag{2.37}$$

These equations are approximate, with an uncertainty of the order of a few percent.

Cordier and others (e.g., Balje) also found that the (BEP) efficiencies of "good" machines group into definable, if approximate, curves defining total efficiency, η_T, as a function of N_s. Figure 2.9 shows "Cordier Efficiency

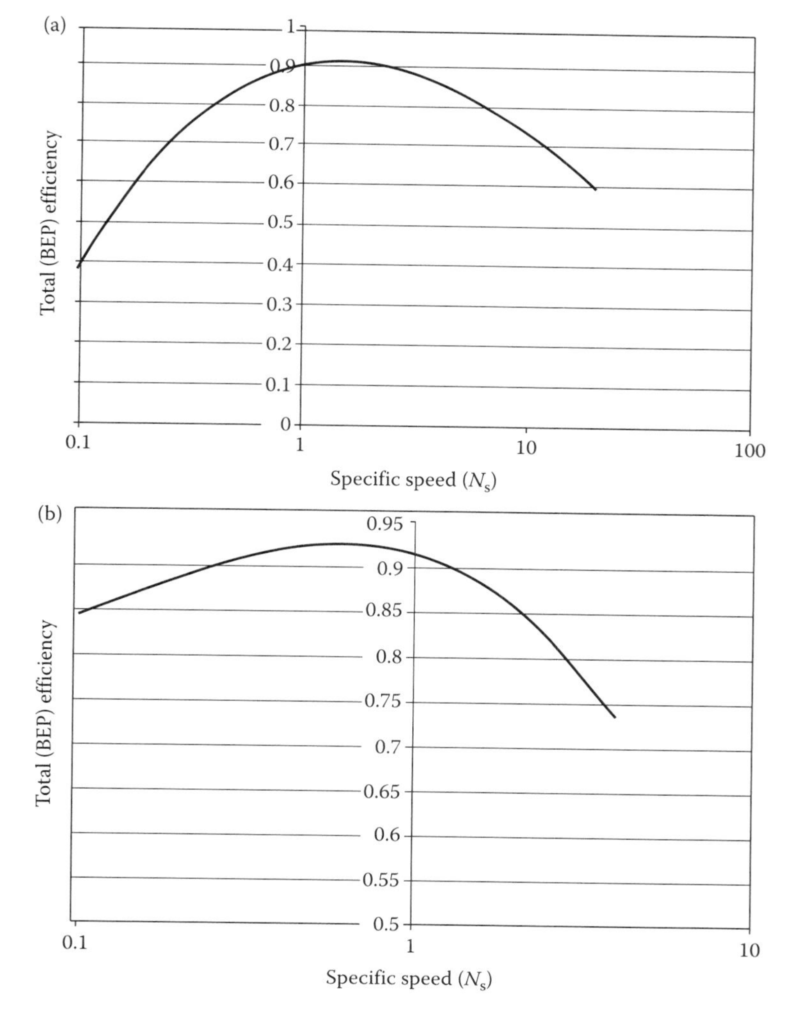

FIGURE 2.9 "Cordier Curves" for total efficiency: (a) pumping machinery and (b) turbines. (After Balje, O. E., 1981, *Turbomachines*, Wiley & Sons, New York.)

Curves" as smooth curves for pumping machinery and for turbines. For high-speed machines with compressible flow, values from the curves should be interpreted as polytropic efficiency.

Several points concerning these efficiency curves must be emphasized:

- Efficiency values taken from these curves represent the "best possible" values that might reasonably be expected.
- The curves are not "performance curves" in the usual sense; they represent only BEP values. Thus, a pumping machine with $N_s = 2$ is inherently more efficient than a machine with $N_s = 6$.
- Each point on the curve represents a different machine family; a machine with $N_s = 2$ is (slightly) different in shape from a machine with $N_s = 2.2$.
- The curves are quite approximate; accuracy is likely no better than a few percent.

The Cordier correlations, $N_s \sim D_s$ as well as efficiency, apply only to single-stage, single-suction/width machines. Information on multistage or double-suction/width machines is obtained by apportioning the head or flow as appropriate.

As the correlations implied by these efficiency curves have considerable uncertainty, values read from the curves are sufficiently accurate for hand calculations. For computer calculation, the following curve fits reproduce the curves quite accurately:

- For pumping machinery

$$\eta_T \approx 0.898 + 0.0511(\ln[N_s]) - 0.0645(\ln[N_s])^2 + 0.0046(\ln[N_s])^3. \tag{2.38}$$

- For turbines

$$\eta_T \approx 0.913 + 0.0442(\ln[N_s]) - 0.0501(\ln[N_s])^2 - 0.0078(\ln[N_s])^3. \tag{2.39}$$

One must take care not to interpret the high level of numerical precision available from the curve fits as high accuracy in the estimated efficiency!

The "Cordier efficiencies" represent, more or less, the upper limit for a given type of turbomachine. If one wishes a more realistic estimate of the efficiency, a variety of "derating" factors might be applied to account for such effects as size, manufacturing tolerances, surface roughness, impeller tip clearance, and the like. This subject will be discussed in a later chapter.

2.8 Summary

This chapter introduced the concepts of dimensionality and reviewed the rule of dimensional homogeneity. The basic tenets of similitude were re-examined to define the rules of similarity. The need for geometric, kinematic, and dynamic similarity was discussed. Methods for determining dimensionless groups were reviewed and discussed.

Using a variety of methods, the fundamental dimensionless geometric and performance variables for turbomachines were developed and a set of rules for scaling the performance of a turbomachine was obtained, subject to the restrictions of geometric similitude and negligible effects from changes in the Reynolds number Re. These scaling rules are as follows:

Incompressible flow:

$$\psi = f_2(\phi),$$
$$\xi = f_3(\phi),$$
$$\eta = f_4(\phi),$$

where

$$\phi = \frac{Q}{ND^3}; \quad \psi = \frac{\Delta p/\rho}{N^2D^2} = \frac{gH}{N^2D^2}; \quad \xi = \frac{P}{\rho N^3 D^5},$$
$$\eta = \frac{Q\Delta p}{P} = \frac{\rho g Q H}{P} = \frac{\phi\psi}{\xi}; \quad Re = \frac{\rho N D^2}{\mu} = \frac{ND^2}{\nu}.$$

Compressible flow:

$$\frac{p_{02}}{p_{01}} = f_1\left(\frac{\dot{m}(RT_{01})^{1/2}}{p_{01}D^2}, \frac{ND}{(RT_{01})^{1/2}}\right),$$
$$\xi = f_2\left(\frac{\dot{m}(RT_{01})^{1/2}}{p_{01}D^2}, \frac{ND}{(RT_{01})^{1/2}}\right),$$
$$\eta = f_3\left(\frac{\dot{m}(RT_{01})^{1/2}}{p_{01}D^2}, \frac{ND}{(RT_{01})^{1/2}}\right),$$

The concepts and definitions of specific speed and specific diameter, N_s [$N_s = NQ^{1/2}/(gH)^{3/4}$] and D_s [$D_s = D(gH)^{1/4}/Q^{1/2}$], were developed as a natural outgrowth of constrained scaling of machine performance. N_s and D_s were shown to be indicators of the most efficient type of machine for a specified performance. Finally, the Cordier correlations between N_s, D_s, and total efficiency, η_T, were introduced.

EXERCISE PROBLEMS

2.1. Repeat Problem 1.1 with the pump running at 1200 rpm. Calculate the specific speed and diameter at BEP.

2.2. Using the pump data from Figure P1.1, what speed would provide a shutoff head of 360 ft? For what speed would the power at best efficiency be 0.55 hp? What speed would yield a flow rate at peak efficiency of 150 gpm?

2.3. A pump that is geometrically similar to the pump of Problems 1.1 and 2.1 has an impeller diameter of 24 in. What will the flow rate, head, and power be at the point of best efficiency?

2.4. A variable-speed wind tunnel fan provides a flow rate of 150,000 cfm against a pressure of 200 lbf/ft^2 when it is running at 1500 rpm. What flow and pressure rise will the fan develop at 2000 rpm? At 3000 rpm?

2.5. A centrifugal air compressor delivers 32,850 cfm of air with a pressure change of 4 psi. The compressor is driven by an 800 hp motor at 3550 rpm.

a. What is the overall efficiency?

b. What will the flow and pressure rise be at 3000 rpm?

c. Estimate the impeller diameter.

2.6. A commercially available fan has BEP performance of $Q = 350\ \text{ft}^3/\text{s}$, $\Delta p_s = 100\ \text{lbf/ft}^2$, and $\eta_s = 0.86$. Density is $0.00233\ \text{slug/ft}^3$, and speed and size are $N = 485\ \text{s}^{-1}$ and $D = 2.25$ ft.

a. At what speed will the fan generate 20 lbf/ft^2 at BEP?

b. What will the flow rate be at that speed?

c. Estimate the fan efficiency at that speed.

d. Calculate the required motor horsepower.

2.7. An airplane propeller of diameter D produces thrust force F_T when rotating at speed N in air of density ρ. The airplane is flying at speed V. Find a dimensionless parameter involving these variables.

2.8. If the total pressure rise of an axial fan is a function of density (ρ), speed (N), and diameter (D), use the Π-product method of dimensional analysis to show that a pressure rise coefficient, ψ_T, can be written as

$$\psi_T = \frac{\Delta p_T}{\rho N^2 D^2}.$$

2.9. An axial fan with a diameter of 6 ft runs at 1400 rpm with an inlet air velocity of 40 ft/s. If one runs a one-quarter scale model at 4200 rpm, what axial inlet air velocity should be maintained for similarity? Neglect Reynolds number effects.

2.10. Use the method of Π-products to analyze centripetal acceleration. With a particle moving on a circular path of radius R at speed V, show that the acceleration (a) is a function of the group V^2/R.

2.11. Repeat Problem 2.10 with angular rotation N (radians/s) substituted for V.

2.12. Use the simple scaling laws to relate the performance of the 12.1- and 7.0-inch diameter impellers whose performance curves are shown in Figure P2.12. What is wrong? Can you find the appropriate exponent for D (not 3) in the flow scaling formula that correctly relates the two impellers? (Hint: Try some cross plots at points related by simple head scaling, and look for an integer exponent other than 3 for the diameter).

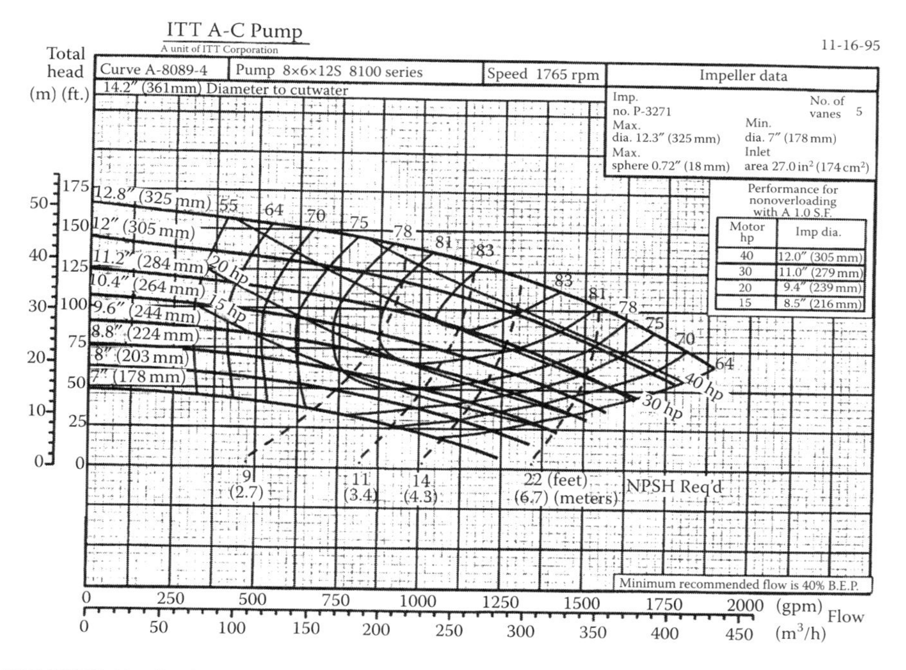

FIGURE P2.12 Performance curves for a family of Allis-Chalmers double-suction pumps. (From ITT Fluid Technology. With permission.)

2.13. Explain and discuss the results of Problem 2.12. Use dimensional analysis to support your discussion. It will help to know that the impellers of different diameters all use a common casing size.

2.14. A pump is driven by a two-speed motor having speeds of 1750 and 1185 rpm. At 1750 rpm, the flow is 45 gpm, the head is 90 ft, and the total efficiency is 0.60. The pump impeller has a diameter of 10 in.

a. What values of Q, H, and η_T are obtained if the pump runs at 1185 rpm?

b. Find the specific speed and the specific diameter of the pump (at 1750 rpm).

2.15. A pump curve and a system curve are shown in Figure P2.15. The pump is operating at $N = 3550$ rpm. If one reduces the pump speed gradually, at what speed will the volume flow rate go to zero?

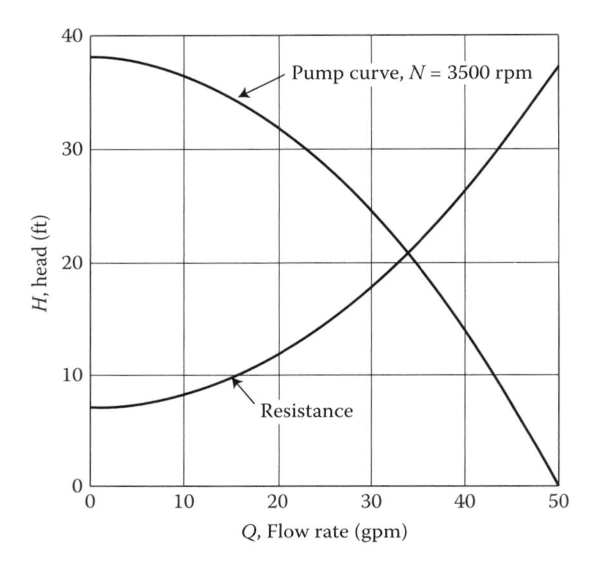

FIGURE P2.15 Variable-speed pump performance curve.

2.16. A pump driven by a two-speed motor delivers 250 gpm at 114 ft of head at 1750 rpm and 10 hp when pumping water. The low-speed mode operates at 1150 rpm. Calculate the flow rate, head rise, and power at the low speed.

2.17. Using the information in Figure P2.17, construct the performance curves of a vane axial fan with $D = 2.5$ m, $N = 870$ rpm, and $\rho = 1.20\ \text{kg/m}^3$.

a. If this fan is supplying air to a steam condenser heat exchanger with a face area of $10\ \text{m}^2$ and an aerodynamic loss factor of $K = 12$, find the operating point of the fan (flow, total pressure rise, and power).

b. Can this fan be used with two such heat exchangers in series? With three?

2.18. We want to use the type of fan represented in Figure P2.17 as an electronic cooling blower with a personal computer. Space available dictates a choice of fan size less than 15 cm. Scale a 15-cm fan to the proper speed to deliver $0.15\ \text{m}^3/\text{s}$ of air at $\eta_T \geq 0.55$.

2.19. A 0.325-m-diameter version of the fan of Figure P2.17 runs on a variable-frequency speed-control motor that operates continuously between 600 and 3600 rpm. If total efficiencies between 0.60 and 0.40

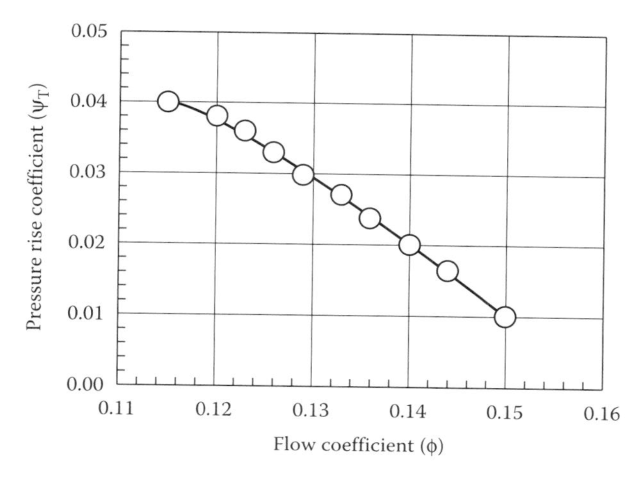

FIGURE P2.17a Dimensionless performance of a vane axial fan, fan characteristic.

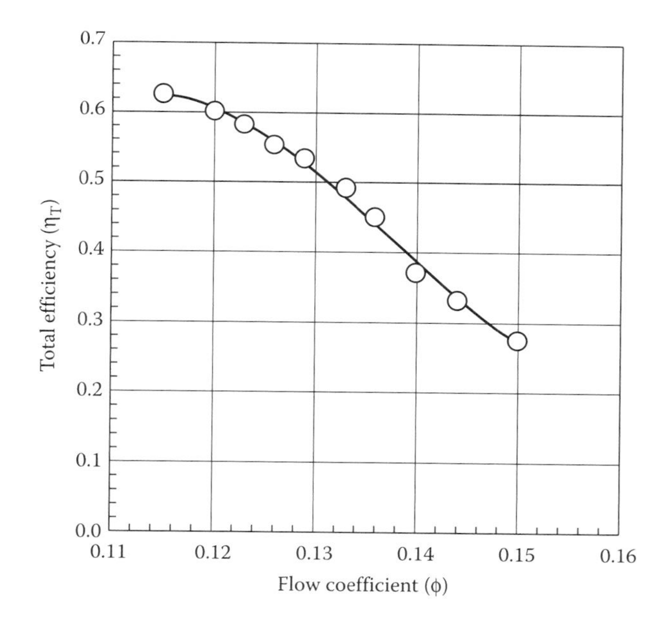

FIGURE P2.17b Dimensionless performance of a vane axial fan, efficiency curve.

are acceptable, define the region of flow-pressure performance capability of this fan as a zone on a Δp_T versus Q plot. What maximum power capability must the motor provide?

2.20. A ship-board ventilation fan must supply 150 kg/s mass flow rate of air ($\rho = 1.22\,\text{kg/m}^3$) to the engine room. The required static pressure rise is 1 kPa. Specify the smallest size for this fan for which the required power is less than 225 kW. Describe the fan in terms of its design parameters.

2.21. A three-speed vane axial fan, when operating at its highest speed of 3550 rpm, supplies a flow rate of 5000 cfm and a total pressure of 30 lbf/ft^2 with $\rho = 0.00233\ slug/ft^3$. Lower speeds of 1775 and 1175 rpm are also available.

a. Estimate the size and efficiency of the fan at 3550 rpm.

b. Estimate the flow rate and pressure rise at the lower speed setting of 1175 rpm.

2.22. Show that $\psi_S = \psi_T - (8/\pi^2)\phi^2$ for an axial flow fan. State your assumptions.

2.23. Use the nondimensional performance curves given in Figure P2.17 for a vane axial fan to generate new nondimensional curves for the *static* pressure and *static* efficiency of this fan. Determine the operating point for maximum static efficiency and estimate the free-delivery flow coefficient for the fan.

2.24. Use the curves of Figure P2.17 to specify a fan that can supply $10\ m^3/s$ of air with a static pressure rise of 500 Pa ($\rho = 1.01\ kg/m^3$). Estimate the required power and total pressure rise at this performance.

2.25. The pump of Problem 1.1 (Figure P1.1) can be used in a side-by-side arrangement with other identical pumps (with suitable plumbing) to provide an equivalent pump of much higher specific speed.

a. Generate a table of H, P, and η for a set of five pumps in parallel and calculate the specific speed and diameter of this equivalent "pump."

b. Sketch the "plumbing" arrangement for the layout.

2.26. Consider the fan developed in Problem 2.17. The condenser has a rectangular face 3 m by 5 m and a K-factor of 20. Size two fans to operate in parallel to cool the condenser. Specify size, speed, pressure rise, flow rate per fan, and total power.

2.27. A small centrifugal pump is tested at $N = 2875$ rpm in water. It delivers $0.15\ m^3/s$ at 42 m of head at its BEP ($\eta = 0.86$).

a. Determine the specific speed of the pump.

b. Compute the required input power.

c. Estimate the impeller diameter.

d. Calculate the Reynolds number and estimate the efficiency according to Cordier.

2.28. A pump handles a liquid slurry of sand and water with a density of $1250\ kg/m^3$ and a kinematic viscosity of 4.85×10^{-5}. The pump was tested in pure water with a BEP pressure rise of 150 kPa at $2.8\ m^3/s$ with $\eta = 0.75$. Estimate the pressure rise and efficiency with a slurry flow rate of $2.8\ m^3/s$.

2.29. A pump running on a two-speed motor delivers $60\ m^3/hr$ at a head of 35 m of water at 1485 rpm and 8.0 kW. The pump efficiency is

$\eta = 0.72$ at the high speed. At the low-speed setting, the pump operates at 960 rpm. Estimate the flow rate, head rise, and power for low-speed operation.

2.30. The performance of a centrifugal blower at its point of best efficiency is $Q = 10\,\text{m}^3/\text{s}$, $\Delta p_S = 5\,\text{kPa}$, and $\eta_S = 0.875$. The inlet air density is $\rho = 1.21\,\text{kg/m}^3$ and the size and speed are $D = 0.7\,\text{m}$ and $N = 475\,\text{s}^{-1}$.

a. At what speed would this blower generate 1 kPa at BEP?

b. What will the BEP flow rate be at this speed?

c. Calculate the required shaft power at the speed found in (a).

2.31. The hydraulic turbine of Problem 1.29 (with performance curves given in Figure 1.16) may be modified to operate with a 4-pole generator at 1800 rpm to yield 60 Hz alternating current. (Assume that the increased rotational stresses in the impeller remain within acceptable limits at the higher speed.) Calculate the required head, flow rate, and resultant power output for design point performance at 1800 rpm.

2.32. The turbine with performance specified in Figure 1.16 has the speed changed to 1800 rpm and is operated at $Q = 6000$ gpm. Determine the head necessary to match the turbine requirements and calculate the output power and efficiency.

2.33. A type of hybrid fan is shown in Figure P5.31 and discussed in Problem 5.31. The fan is available in an array of sizes from $D = 13.5$ in. to $D = 80.7$ in. From considerations of structural strength and critical vibration levels, these fans must be limited in speed. A class I fan must not exceed an impeller tip speed of $U = 200\,\text{ft/s}$, and a class II fan must have $U \leq 250\,\text{ft/s}$ (approximately). Clearly a class II fan must be more robust than a class I design.

a. Calculate the rotational speed limit for a class II fan with 60 in. diameter.

b. Scale the performance curve of Figure P5.32 to this size and speed.

2.34. An *induced draft (ID) fan* is used to "suck" exhaust gas from a steam generator furnace. At its "normal" operating point, the fan turns at speed N_1, generates flow Q_1, and total pressure rise Δp_{T1} with efficiency η_{T1}. If the power to the fan is cut off, it will coast to a stop. Engineers need to know the time required for it to stop turning. Show that the fan speed as a function of time (t) is given by

$$N = \frac{N_1}{1 + Q_1 \Delta p_{T1} t / \eta_{T1} I N_1^2},$$

where I is the mass moment of inertia of the fan rotor/shaft assembly. It may be assumed that the flow-system resistance is parabolic ($\Delta p_T = KQ^2$).

2.35. Measurements on an operating water pump give the following: $N = 1750$ rpm; $\Delta p_T = 43$ psi; and $P = 14$ hp. Estimate the pump efficiency and flow produced. What type of pump is this? Assume that the pump is operating at its BEP.

2.36. Large centrifugal fans are an integral part of the combustion system of fossil fuel-fired power stations. Such fans are used to support the combustion and gas cleanup processes associated with a coal-fired steam generator. *Forced draft (FD) fans* are used to feed fresh air into the steam generator. *ID fans* are used to exhaust gaseous products of combustion from the steam generator and to move the gas through the cleanup equipment such as precipitators, scrubbers, Selective Catalytic Reducers (SCRs), and the like. Since no steam generator operates under exactly the same conditions at all times, it is necessary to vary the flow and pressure supplied by the fans. (This is in addition to using variations in the system to vary flow, such as opening or closing dampers.) One way to obtain changes in fan performance is by using *variable inlet vanes* (Figure 5.16). Variable inlet vanes, considered part of the fan itself, can be set at various angles to "preswirl" the flow, causing changes of fan performance characteristics. Figures P2.36a and b are for pressure and efficiency curves at various vane positions (90° is "wide open") for a fan operating at constant speed. For the following questions, assume that the fan is connected to a system whose resistance curve is parabolic ($\Delta p_T = KQ^2$), with $K = 3.67 \times 10^{-11}$ in. wg/cfm^2. The fan is an FD fan handling air with $\rho = 0.075$ lbm/ft^3.

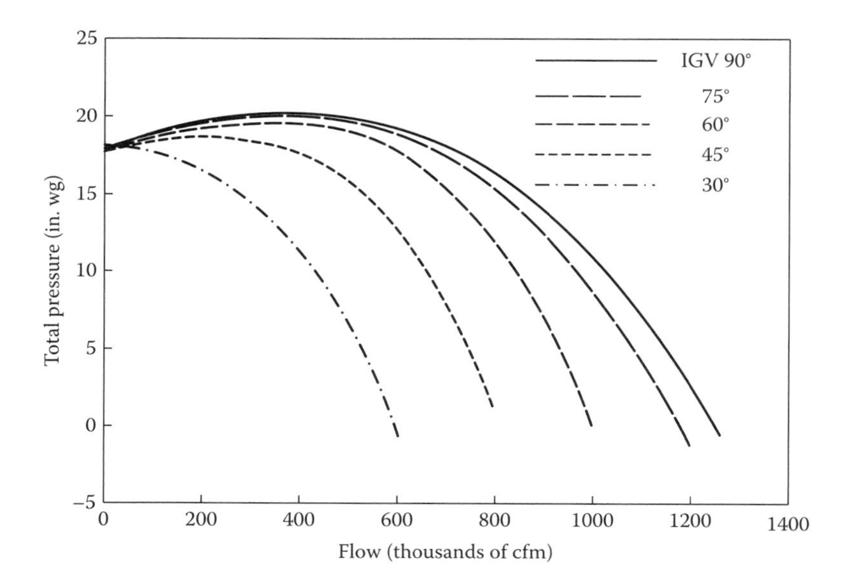

FIGURE P2.36a Performance of a centrifugal fan with variable inlet vanes (710 rpm).

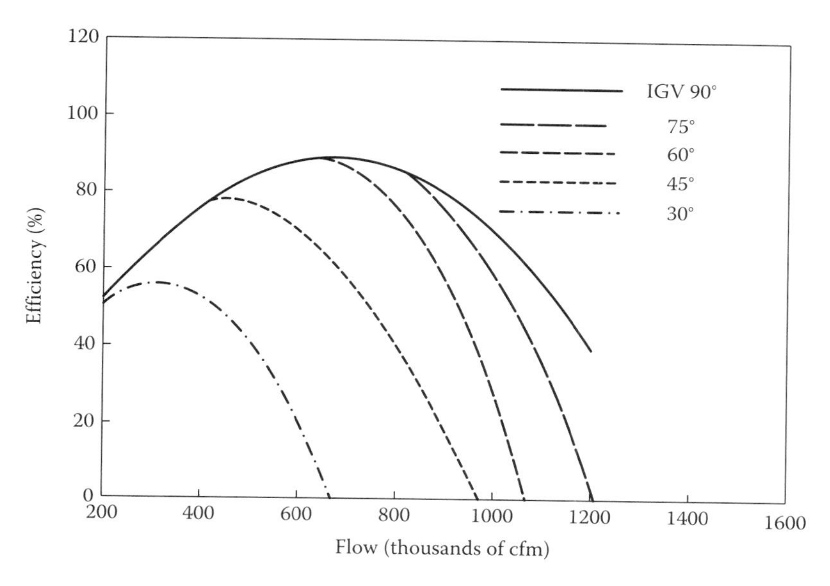

FIGURE P2.36b Total efficiency of a centrifugal fan with variable inlet vanes (710 rpm).

a. What is the flow, pressure, and power when the fan is operated with vanes wide open (90°)?
b. Estimate the vane setting required to produce a flow of 500,000 cfm. What would the power be?
c. Can one use similarity laws to predict fan performance at, say, vane position of 60° from performance at 90°? Explain.
d. A second way to vary fan performance is to leave the vanes (if any) wide open and change the fan speed. Assume that the fan is fitted with a variable-speed motor. If the vanes are left wide open (or, if there are no vanes), what speed would be required to produce 50,000 cfm? What would the power be?
e. In the "real world," variable inlet vane control is usually preferred over variable-speed control. What reasons can you offer for this? What is the strongest point in favor of variable-speed control?

2.37. An ID fan has variable inlet vanes for flow control. This fan is guaranteed to deliver 750,000 cfm with a (total) pressure rise of 17.0 in. wg and requires 2375 hp when operating at a speed of 709 rpm (a 10-pole motor with 1.5% slip) and handling gas with density 0.060 lb/ft^3. In order to verify the guarantee, a series of field tests are run to measure fan performance. (Recall that the variable inlet vanes will be adjusted in an attempt to "hit" the guarantee point.) The results of the tests are all "close" to the guarantee point but none is "right on." The three test results are as follows:

Test Number	Speed (rpm)	Density (lb/ft^3)	Pressure Rise (in. H_2O)	Flow (cfm)	Power (hp)
1	701	0.057	16.25	722,000	2174
2	701	0.059	15.76	702,000	2050
3	701	0.062	16.87	771,000	2419

a. For each test point, adjust the flow, pressure, and power to the correct values of speed and density, so that they may be compared to the guarantee point.

b. Plot the three test points, together with the guarantee point on a "standard" $\Delta p \sim Q$ plane.

c. Suggest a way that the three test points may be used to determine a value for the power at the guarantee point ($\Delta p = 17$; $Q = 750,000$; $P =???$).

2.38. A centrifugal compressor compresses air from 14.7 psia, 59°F to 40 psia, 260°F. Compute the isentropic efficiency. A compressor of the same design is to be used to compress air from 14.7 psia, 59°F to 80 psia. Compute the discharge temperature and isentropic efficiency for the second compressor.

2.39. A fan has been tested at various speeds. The data that have been accumulated are as follows:

Speed (rpm)	Flow (cfm)	Total Pressure Rise (in. H_2O)	Power (hp)
1745	0	14.85	48.2
1720	10,040	14.38	54.1
1700	19,790	11.62	55.8
1660	27,190	9.35	55.7
1620	37,550	5.70	48.9
1620	47,100	2.06	38.2

a. Determine performance curves (total pressure rise, power, and efficiency) for this fan at a speed of 1750 rpm. Both graphical and "least-squares" curve fits are required.

b. What is the "BEP" for this fan (at 1750 rpm)?

c. Calculate the specific speed and estimate the impeller diameter. The fan is operated at 1750 rpm and installed in a system characterized by $\Delta p_T(\text{in. wg}) = 0.019\,(Q\,[\text{cfm}]/1000)^{1.82}$. Find the flow, pressure rise, and power.

d. At what speed must the fan be operated to deliver 25,000 cfm in this system?

2.40. A pump has the performance curves shown in Figure P2.40 when it is operated at 2450 rpm. The pump is connected to a system that contains a length of 12.5 cm i.d. pipe and a valve. When the valve is fully open, the system curve is $H_{sys} = 10 + 28V^2/2g$. The flow in the system can be controlled by either closing the valve or reducing the pump speed. The valve characteristics are as follows:

Position	Open	1/4	1/2	3/4
Loss coefficient	7	11	17	40

a. Find the flow and pump power at 2450 rpm with the valve wide open.
b. Find the flow and pump power at 2450 rpm with the valve half open.
c. Find the flow and pump power at 2000 rpm with the valve wide open.

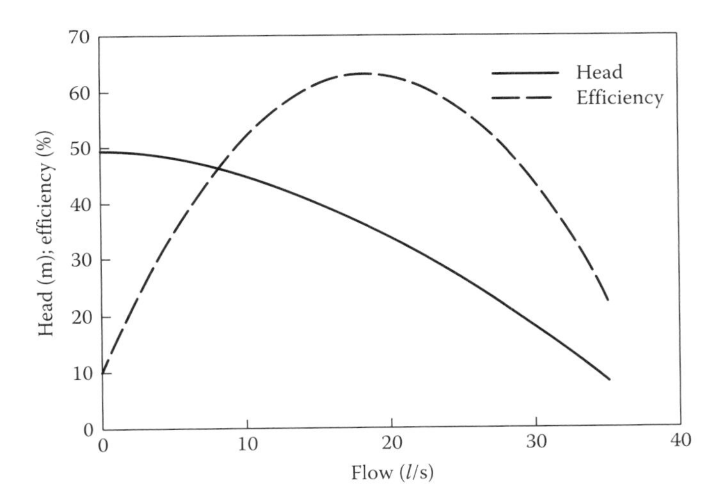

FIGURE P2.40 Pump performance at 2450 rpm.

3

Scaling Laws, Limitations, and Cavitation

3.1 Scaling of Performance

The examples and development thus far have shown how to use ϕ, ψ, ξ, and/or η to calculate any change in the level of performance of a given design caused by changes in speed, diameter, or density. The common procedure for scaling involves rearranging these nondimensional parameters into ratios of the physical (dimensional) parameters by remembering that, with a change that maintains geometric similarity, ϕ, ψ, ξ, and η will be invariant (subject to restrictions on changes in *Re* and *Ma*, which will be discussed shortly). For example, when a model pump of known dimensions is tested to determine performance curves in a given fluid, one obtains a complete description of the model pump's performance in terms of ρ_m, N_m, D_m, Q_m, P_m, H_m, and η_m. (The "m" subscript identifies the variables of the model.) From the model test results, one can predict the performance of a pump that may be of significantly different size from the model, may handle a different fluid, and may run at a different speed. Let the prototype operating and performance parameters be represented by ρ_p, N_p, D_p and H_p, Q_p, and P_p. If the model and prototype are geometrically similar, then each has a performance point such that

$$\phi_m = \frac{Q_m}{N_m D_m^3} = \phi_p = \frac{Q_p}{N_p D_p^3}. \tag{3.1}$$

This equation can be rearranged as

$$\frac{Q_p}{Q_m} = \frac{N_p D_p^3}{N_m D_m^3} = \left(\frac{N_p}{N_m}\right)\left(\frac{D_p}{D_m}\right)^3 \tag{3.2}$$

or

$$Q_p = Q_m \left(\frac{N_p}{N_m}\right)\left(\frac{D_p}{D_m}\right)^3. \tag{3.3}$$

This equation shows that Q changes linearly with speed and as the third power of the size (diameter). This relation is called a *scaling rule* or a *fan law* or a *pump law* and can be used directly for quick calculation of scaling effects on performance. For example, if one doubles speed, one doubles the flow. If one halves the speed, one halves the flow. Double the diameter, and obtain eight times the flow; halve the diameter, and obtain one-eighth of the flow; and so on.

A scaling law for pressure or head can be derived from ψ, which is also invariant when geometric similitude is maintained:

$$\psi_m = \frac{\Delta p_{Tm}}{\rho_m N_m^2 D_m^2} = \psi_p = \frac{\Delta p_{Tp}}{\rho_p N_p^2 D_p^2} \tag{3.4}$$

or

$$\frac{\Delta p_{Tp}}{\Delta p_{Tm}} = \frac{\rho_p N_p^2 D_p^2}{\rho_m N_m^2 D_m^2}, \tag{3.5}$$

and

$$\Delta p_{Tp} = \Delta p_{Tm} \frac{\rho_p N_p^2 D_p^2}{\rho_m N_m^2 D_m^2}. \tag{3.6}$$

This states that pressure rise varies linearly with density and parabolically with speed and diameter. In terms of head (constant g is assumed)

$$H_p = H_m \left(\frac{N_p}{N_m}\right)^2 \left(\frac{D_p}{D_m}\right)^2. \tag{3.7}$$

For the power, from $\xi_p = \xi_m$,

$$P_p = P_m \left(\frac{\rho_p}{\rho_m}\right) \left(\frac{N_p}{N_m}\right)^3 \left(\frac{D_p}{D_m}\right)^5. \tag{3.8}$$

Finally, recall that efficiency is invariant so that

$$\eta_{Tp} = \eta_{Tm} \tag{3.9}$$

and

$$\eta_{sp} = \eta_{sm}. \tag{3.10}$$

Equations 3.3 and 3.6 through 3.10 are the *simple scaling laws* for low-speed fans, liquid pumps, and hydraulic turbines. It should be noted that, strictly speaking, these laws assume that effects of changes in Reynolds number and Mach number are absent.

Example: Hydraulic Turbine Scaling

Consider a full-scale turbine with a head drop of 1.5 m, an efficiency of 0.696, and a volume flow rate of 55 m^3/s. The turbine is intended to operate in warm seawater ($\rho = 1030\ kg/m^3$) while moored in the Gulf Stream south of Florida. It runs at a speed of 98 rpm and has a diameter of 4 m. Devise a scale model test with a power extraction of 10 kW and a model running speed of 1000 rpm in freshwater ($\rho = 998\ kg/m^3$). Determine the proper model diameter and calculate the required head and flow to the model turbine to maintain similarity.

The turbine can be scaled to model size using the similarity rule for power (Equation 3.8)

$$P_p = P_m \left(\frac{\rho_p}{\rho_m}\right)\left(\frac{N_p}{N_m}\right)^3\left(\frac{D_p}{D_m}\right)^5.$$

Solving for diameter ratio

$$\frac{D_m}{D_p} = \left[\left(\frac{\rho_p}{\rho_m}\right)\left(\frac{P_m}{P_p}\right)\left(\frac{N_p}{N_m}\right)^3\right]^{1/5}.$$

The full-scale power, P_p, is given by

$$P_p = \eta_T \rho g H Q = 0.696 \times 1030 \times 9.807 \times 55 \times 1.5 = 580012\ \mathrm{W} = 580\ \mathrm{kW}.$$

Then the diameter ratio for the model is

$$\frac{D_m}{D_p} = \left[\left(\frac{1030}{998}\right)\left(\frac{98}{1000}\right)^3\left(\frac{10}{580}\right)\right]^{1/5} = 0.111.$$

Therefore, the proper diameter is

$$D_m = 0.111 \times 4 = 0.444\ \mathrm{m}.$$

Note that both the Reynolds numbers are very large, with $Re_p \approx 1 \times 10^9$ and $Re_m \approx 1 \times 10^8$. The assumption of negligible Reynolds number effects is very reasonable in this instance.

It remains to calculate the head and flow requirements for the model. These are determined from the simple scaling laws as

$$H_m = H_p \left(\frac{N_m}{N_p}\right)^2\left(\frac{D_m}{D_p}\right)^2 = 1.5\ \mathrm{m} \times \left(\frac{1000}{98}\right)^2 \times \left(\frac{0.444}{4.0}\right)^2 = 1.92\ \mathrm{m}$$

and

$$Q_m = Q_p \left(\frac{N_m}{N_p}\right)\left(\frac{D_m}{D_p}\right)^3 = 55\ \mathrm{m^3/s} \times \left(\frac{1000}{98}\right) \times \left(\frac{0.444}{4.0}\right)^3 = 0.768\ \mathrm{m^3/s}.$$

3.2 Limitations and Corrections for Reynolds Number and Surface Roughness

The scaling laws are very useful, but they have important limitations. They assume that there are no Reynolds number effects, no Mach number (compressibility) effects, and complete geometric similarity. Generally, whenever one changes size, speed, or fluid properties, changes in Reynolds number and Mach number are unavoidable. Fortunately, if Reynolds number is *large* enough and Mach number is *small* enough, the effect of changes in their value is negligible. Rather obviously, one would not plan to violate geometric similarity; it would be unlikely that data on a small axial fan would predict the performance of a large centrifugal fan! A more subtle effect is that of surface roughness, which usually does not scale with machine size (in fact, the absolute surface roughness, ε, tends to remain constant as machine size changes, causing variations in the relative roughness, ε/D). In this section, Reynolds number and roughness effects will be considered; Mach number and compressibility effects will be considered in the next section.

There are three approaches to handling Reynolds number and roughness effects: (1) ignore them, (2) prohibit them, and (3) apply a correction factor. Generally speaking, codes and standards for testing turbomachines tend to prohibit corrections. For fans, the Air Movement and Control Association (AMCA) and the American Society of Heating, Refrigeration and Air-Conditioning Engineers (ASHRAE) state in their jointly issued laboratory testing standard (AMCA, 1999) that the influence of *Re* is to be neglected. This is reiterated in AMCA's report on model testing (AMCA, 2002). The "fan laws" [i.e., scaling laws (Equations 3.3 and 3.6 through 3.10)] are to be used without modification subject to the following restrictions: the Reynolds number [$Re_D = (ND^2)/\nu$] for the model must be $>3 \times 10^6$ and the model diameter must be at least 35 in. (0.900 m) or 1/5 the size of the full-scale prototype, whichever is larger. For smaller full-scale equipment, the model is the prototype (i.e., a full-scale fan is the test article). Similarly, the American Society of Mechanical Engineers (ASME) Performance Test Codes for Fans (ASME, 2008) and Centrifugal Pumps (ASME, 1990) do not provide corrections for changes in *Re*. These codes are written for full-size tests and specify the use of the simple scaling laws for changes in speed and density discussed in the previous section.

Next, consider the application of correction factors. The typical "scaling law" for the effects of Reynolds number has the general form (Kittredge, 1967):

$$\frac{1-\eta_p}{1-\eta_m}\left(\frac{\eta_m}{\eta_p}\right) = a + (1-a)\left(\frac{Re_m}{Re_p}\right)^n \quad (0 \le a \le 1). \tag{3.11}$$

The term $1-\eta$ is the (dimensionless) hydraulic loss, which would be expected to depend on *Re*. One of the oldest and simplest scaling laws is that of Moody (1925):

$$\frac{1-\eta_p}{1-\eta_m} = \left(\frac{D_m}{D_p}\right)^n, \tag{3.12}$$

which is a simplification of Equation 3.11 in which it is assumed that speed and viscosity are unchanged between model and prototype. Moody's "law" was originally proposed for hydraulic turbines and is often applied to pumps. A more general form, popular for pumps, is

$$\frac{1-\eta_p}{1-\eta_m} = \left(\frac{Re_m}{Re_p}\right)^n. \tag{3.13}$$

Recommended values of the exponent n vary between 0.1 and 0.5.

A model for estimating the effects of Reynolds number on centrifugal pumps was suggested by Csanady (1964). The model uses *Reynolds number factors* according to

$$\frac{\eta_p}{\eta_m} = \frac{f_\eta(Re_p)}{f_\eta(Re_m)} \tag{3.14}$$

and

$$\frac{\psi_p}{\psi_m} = \frac{f_H(Re_p)}{f_H(Re_m)}. \tag{3.15}$$

Csanady presents graphs for f_η and f_H; they can be approximated (for $Re > 10^3$) by

$$f_\eta \approx 0.20 + 0.80\tanh[0.10\log_{10}^2(Re) - 0.30\log_{10}(Re)], \tag{3.16}$$

$$f_H \approx 0.55 + 0.45\tanh\left[0.75\log_{10}(Re)\right]. \tag{3.17}$$

Csanady's graphs and these equations suggest that there is negligible Reynolds number effect on head beyond $Re = 10^6$ and negligible effect on efficiency beyond $Re = 10^7$.

A review and analysis of compressor and fan efficiency scaling was made by Wright (1989). In that work, a method is suggested for efficiency scaling according to

$$\frac{1-\eta_p}{1-\eta_m} = 0.3 + 0.7\left(\frac{f_p}{f_m}\right), \tag{3.18}$$

where f is the Darcy friction factor, evaluated from the Moody chart or the Colebrook formula

$$f = \frac{0.25}{\left[\log_{10}\left(((\varepsilon/D)/3.7) + \left(2.51/Re_b\sqrt{f}\right)\right)\right]^2}. \tag{3.19}$$

The influence of both Reynolds number and surface relative roughness (ε/D) is included in estimating the losses. The Reynolds number, Re_b, is based on the hydraulic diameter (approximately two times the blade tip width b) and blade tip speed ($ND/2$) for centrifugal machines [$Re_b = (NDb)/\nu$]. The method has been validated by comparing with test data for centrifugal compressors and fans, but not for axial machines. Nevertheless, the method should be reasonably accurate if the blade speed and chord length near the tip are used to define the Reynolds number. Likewise, there is no particular reason to believe that the method would not apply reasonably well to pumps.

The ASME Performance Test Code for Compressors and Exhausters (ASME, 1997) permits efficiency to be corrected for the effects of Reynolds number for axial flow compressors and both Reynolds number and roughness for centrifugal compressors. The ASME PTC 10 formula for axial compressors is of the "Moody–Kittredge" type:

$$\frac{1-\eta_p}{1-\eta_m} = \left(\frac{Re_{bm}}{Re_{bp}}\right)^{0.2}; \quad \text{Axial flow compressors;} \tag{3.20}$$

$$Re_b = \frac{NDb}{\nu_{01}}; \quad Re \geq 9 \times 10^4.$$

The PTC 10 scaling law for centrifugal compressors is similar to that of Wright (Equation 3.14). The polytropic efficiency is used in the PTC 10 compressor scaling laws. No scaling law for ψ is given for either type of compressor. Likewise, no correction is recommended for ϕ or ξ.

Another method that appears to be very similar to the Wright method is the European ICAAMC (International Compressed Air and Allied Machinery Committee) method (Casey, 1985), which has an efficiency correction given by

$$\frac{1-\eta_p}{1-\eta_m} = \frac{0.3 + (0.7f_p)/f_{fr}}{0.3 + (0.7f_m)/f_{fr}}. \tag{3.21}$$

The subscript "fr" refers to "fully rough" conditions at high Reynolds number for a given relative roughness. The ICAAMC method is not clearly related to the physics of the flow, particularly in the use of f_{fr}. The ICAAMC also recommends adjustments in ϕ and ψ according to

$$\frac{\psi_p}{\psi_m} = 0.5 + 0.5\left(\frac{\eta_p}{\eta_m}\right), \tag{3.22}$$

$$\frac{\phi_p}{\phi_m} = \left(\frac{\psi_p}{\psi_m}\right)^{1/2}. \tag{3.23}$$

These relationships are heavily evaluated against experimental compressor data (Strub et al., 1984; Casey, 1985). The ICAAMC equations are also limited to centrifugal machines but show considerable promise that they can be generalized to cover axial fans, compressors, and pumps, as well as centrifugal fans and pumps through the choice of an appropriate form of Reynolds number.

The scaling laws for Reynolds number and roughness discussed here are but a few of the many that have been proposed over the years. Kittredge, writing in 1967, quotes some 17 different scaling rules dating back as far as 1909. Firms that design and manufacture turbomachinery typically have proprietary "in-house" rules; often these are minor variations on the rules presented here. Scaling of performance for Reynolds number (and roughness) is a subject of ongoing controversy and is not fully agreed upon in the international turbomachinery community at this time.

An example of efficiency scaling might be helpful in understanding what the controversy is all about. Suppose one tests a pump model in a size that has a Reynolds number of 10^5. Further suppose that the full-scale machine will operate at a Reynolds number of 10^6, and one wants to estimate efficiency for the full scale from the measured model efficiency of $\eta_m = 0.80$. Using the Moody rule as

$$\frac{1-\eta_p}{1-\eta_m} = \left(\frac{Re_m}{Re_p}\right)^n,$$

if $n = 0.25$, $\eta_p = 0.887$, which is a very substantial improvement. On the other hand, if one uses a conservative scaling exponent, $n = 0.1$, efficiency is predicted as $\eta_p = 0.841$. Using $n = 0.45$, $\eta_p = 0.929$. The results indicate the potential gains in efficiency from about 4% up to nearly 13%. This broad range of results is unacceptable, particularly since there is great temptation to take commercial advantage of using the larger exponents to predict really superior performance. A long history of commercial manipulation of efficiency scaling or "size effect" has led to a widespread feeling of healthy skepticism about Reynolds number and roughness scaling and to the adoption of excessively restrictive rules, including the AMCA statements on total restriction, that is, no scaling at all.

3.3 Compressibility (Mach Number) Limitations and Corrections

When the fluid handled by a turbomachine is a gas or vapor, it is possible that the density may change as the fluid passes through the machine, giving a so-called compressibility effect on the performance. If the machine is a high-speed compressor or turbine, it is usually necessary to explicitly account for the compressibility effect by matching the Mach number between a model test and prototype performance; that is, $Ma_m = Ma_p$ where $Ma = ND/(\gamma RT_{01})^{1/2}$.

The test standards allow some leeway in this requirement; for example, ASME PTC 10 *Performance Test Code on Compressors and Exhausters* (ASME, 1997) provides charts showing allowable Mach number "departure" ($Ma_p - Ma_m$) in terms of the nominal operating Mach number. This allowable departure varies from a high of about 0.2 at a Mach number of 0.2 to a low of 0.03 for Mach numbers above 0.6.

What happens when a compressible fluid like air is handled by a low-speed machine like a fan? Traditionally, test data and scaled performance are considered to be incompressible as long as $\Delta p_T \leq 4$ in.wg (about 21 lb/ft^2, 1 kPa). This is about 1% of typical atmospheric pressure, and either an isothermal or isentropic calculation indicates a negligible change in density (0.7–1%) for this pressure change. Earlier editions of ASME and AMCA standards specified this limit; newer versions do not state explicit upper and lower limits.

Above the 4 in.wg level, density change begins to have a significant effect on Q, Δp_T, and P. For pressure *ratios* up to about 1.2 (Δp about 80 in. wg, 420 lb/ft^2, 20 kPa), it is customary to apply a correction factor for the effects of compressibility. The standards provide an "approved" method of both correcting test data to equivalent incompressible model performance and predicting full-scale prototype performance from the incompressible data (AMCA, 1999; ASME, 2008). The procedure uses a polytropic compression model as outlined in Section 1.5. The details of the derivation are given in the AMCA/ASHRAE standard (AMCA, 1999). In general, a compressibility coefficient, K_p, is calculated, and test data can be converted to "equivalent" incompressible values according to

$$Q_i = K_p Q, \tag{3.24}$$

$$\Delta p_{Ti} = K_p \Delta p_T, \tag{3.25}$$

where the "i" subscript identifies the equivalent incompressible variable. K_p is given by the following equation:

$$K_p = \left(\frac{n}{n-1}\right) \frac{\left[(p_b + p_2)/(p_b + p_1)\right]^{(n-1)/n} - 1}{\left[(p_b + p_2)/(p_b + p_1)\right] - 1}. \tag{3.26}$$

Here, p_b is the barometric (or ambient) pressure, such that $p_b + p_2$ and $p_b + p_1$ are absolute pressures, $p_2 - p_1 = \Delta p_T$, and n is the polytropic exponent. Assuming that for the small pressure changes involved, $\eta_p \approx \eta_s \approx \eta_T$, the polytropic exponent is calculated from Equation 1.8 according to

$$\frac{n}{n-1} = \eta_T \left(\frac{\gamma}{\gamma - 1}\right). \tag{3.27}$$

When scaling model performance directly from test data to the conditions for a prototype, one must calculate the compressibility coefficient for both model and prototype conditions. As always, the "m" subscript indicates the model and the "p" subscript the (full-scale) prototype. Then,

$$Q_p = Q_m\left(\frac{N_p}{N_m}\right)\left(\frac{D_p}{D_m}\right)^3\left(\frac{K_{pm}}{K_{pp}}\right), \tag{3.28}$$

$$\Delta p_{Tp} = \Delta p_{Tm}\left(\frac{\rho_p}{\rho_m}\right)\left(\frac{N_p}{N_m}\right)^2\left(\frac{D_p}{D_m}\right)^2\left(\frac{K_{pm}}{K_{pp}}\right), \tag{3.29}$$

$$P_p = P_m\left(\frac{\rho_p}{\rho_m}\right)\left(\frac{N_p}{N_m}\right)^3\left(\frac{D_p}{D_m}\right)^5\left(\frac{K_{pm}}{K_{pp}}\right). \tag{3.30}$$

The total efficiency is unchanged (no effect of compressibility):

$$\eta_{Tp} = \eta_{Tm}. \tag{3.31}$$

Note that if the model flow is essentially incompressible such that $K_{pm} = 1.0$, the compressibility scaling relations imply that Q_p, Δp_{Tp}, and P_p are all proportional to $(1/K_{pp})$, which is larger than 1. If the prototype operates at high enough Mach number/pressure/speed, then the machine performance parameters (Q, Δp_T) will be higher than expected from the simple incompressible scaling rules. Practically speaking, a smaller than expected fan might do the job.

Example: Compressibility Effects in Scaling

Consider a model fan test in air ($\gamma = 1.4$) such that

$$\rho_m = 0.0023\ \text{slug/ft}^3, \quad P_m = 37.1\ \text{hp}$$
$$Q_m = 10{,}000\ \text{cfm}, \quad \eta_{Tm} = 0.85,$$
$$\Delta p_{Tm} = 20\ \text{in.wg}, \quad p_b = 2116\ \text{lbf/ft}^2.$$

Estimate the performance of a prototype for which

$$\rho_p = 0.0023\ \text{slug/ft}^3, \quad \frac{D_m}{D_p} = 2$$
$$p_b = 2116\ \text{lbf/ft}^2, \quad \frac{N_p}{N_m} = 0.75.$$

Since $\Delta p_{Tm} > 4$ in. wg, the model performance is affected by compressibility. Assume that $p_1 = 0$ (gage). Then,

$$\frac{n_m}{n_m - 1} = 0.85\left(\frac{1.4}{1.4-1}\right) = 2.975; \quad \frac{n_m - 1}{n_m} = 0.3361$$

$$K_{p_m} = 2.975\frac{[(2116 + 20 \times 5.204)/2116]^{0.3361} - 1}{[(2116 + 20 \times 5.204)/2116] - 1} = 0.9840$$

and the "incompressible" model performance can be calculated:

$$Q_i = 0.9840 \times 10{,}000\,\text{cfm} = 9840\,\text{cfm},$$
$$\Delta p_{Ti} = 0.9840 \times 20\,\text{in. wg} = 19.68\,\text{in. wg},$$
$$P_i = 0.9840 \times 37.1\,\text{hp} = 36.5\,\text{hp}.$$

To calculate prototype performance, a new value of K_p is needed for the prototype, K_{pp}.

According to Equation 3.26, the value of p_{2p} will be needed to calculate K_{pp}. Formally, an iteration will be required. To begin, Δp_{Tp} is estimated using incompressible scaling:

$$\Delta p_{Tp} = \Delta p_{Tm}\left(\frac{\rho_p}{\rho_m}\right)\left(\frac{D_p}{D_m}\right)^2\left(\frac{N_p}{N_m}\right)^2,$$
$$\Delta p_{Tp} = 20\,\text{in. wg} \times (1) \times (2)^2(0.75)^2 = 45\,\text{in. wg}.$$

Assuming $p_{1p} = 0$ so that $p_{2p} = 0 + \Delta p_{Tp}$,

$$p_b + p_2 = (2116 + 45 \times 5.204)\,\text{lb/ft}^2 = 2350.2\,\text{lb/ft}^2,$$
$$p_b + p_1 = 2116\,\text{lbf/ft}^2.$$

Then (recall that $\eta_{Tp} = \eta_{Tm}$, $n_p = n_m$),

$$K_{pp} = 2.975\frac{[(2350.2)/(2116)]^{0.3361} - 1}{[(2350.2)/(2116)] - 1} = 0.9653,$$

so that

$$Q_p = 10{,}000\,\text{cfm} \times (0.75)(2)^3\left(\frac{0.9840}{0.9653}\right) = 61{,}803\,\text{cfm},$$
$$\Delta p_{Tp} = 20\,\text{in. wg} \times (1)(0.75)^2(2)^2\left(\frac{0.9840}{0.9653}\right) = 45.87\,\text{in. wg},$$
$$P_p = 37.1\,\text{hp} \times (1)(0.75)^3(2)^5\left(\frac{0.9840}{0.9653}\right) = 510.7\,\text{hp},$$

and, of course,

$$\eta_{Tp} = 0.85.$$

Formally, the calculations should be repeated with the new value of Δp_{Tp} and iterated to convergence. It is left as an exercise for the reader to show that this will make very little difference.

If no compressibility corrections were made, the predicted values (and error) would be:

(Value)	(Error)
$Q = 60{,}000$ cfm	2%
$\Delta p_T = 45.0$ in. wg	2%
$P_p = 501$ hp	2%
$\eta_T = 0.85$	0%

The difference might seem insignificant, but it is larger than the allowable tolerance on performance in many equipment purchase contracts. As an exercise, calculate the errors if, for a different prototype, $\rho_p/\rho_m = 0.75$, $D_p/D_m = 1.5$, and $N_p/N_m = 1.5$.

3.4 Cavitation Avoidance in Pumps (and Turbines)

When working with liquids in turbomachines, one does not need to be concerned with Mach number effects or the influence of compressibility on performance. However, one does have to worry about another equally important problem called *cavitation*. In turbomachines, cavitation is a localized vaporization (boiling) and subsequent recondensation of a liquid, which occurs inside the impeller passages of a machine.

Normally, we think of boiling as occurring as the temperature of a liquid is raised, such as in a pot of water on a stove, but vaporization can also be caused by lowering the pressure of a liquid. Vaporization will occur when the liquid pressure is less than or equal to the liquid's vapor pressure, which primarily depends on the fluid temperature. Water, for example, has a vapor pressure of 12.49 lb/ft^2 at 32°F, 100.5 lb/ft^2 at 100°F, and 2116 lb/ft^2 (which is equal to standard sea level atmospheric pressure) at 212°F. Stated differently, 212°F water boils when its pressure is 2116 lb/ft^2. At lower temperatures, the fluid boils if the local pressure drops to the level of the vapor pressure at that particular temperature. The relationship between vapor pressure and temperature for water is shown in Figure A.3 in Appendix A. The relationship depends on the identity of the liquid. The $p_v = f(T)$ relationship for selected petroleum products is shown in Figure A.4.

To illustrate, suppose that a pump is generating 30 m of head with 6 m of suction (i.e., the pressure at the inlet flange is 6 m of water below atmospheric) and 24 m of water discharge pressure. The absolute pressure at the suction flange is

$$p_{inlet} = p_{atm} + p_{gage} = 101.3\,\text{kPa} + \left(9.810\,\text{kN/m}^3 \times [-6\,\text{m}]\right) = 42.4\,\text{kPa}.$$

The vaporization temperature at this pressure, from Figure A.3, is 78.3°C. Vaporization will occur at the suction flange if the water temperature is 78.3°C.

Now suppose that the water is actually at "room temperature," 25°C. Then the vapor pressure is 3.17 kPa and the suction head can be increased to [(101.33–3.17)/9.81] 10.0 m before vaporization occurs at the suction flange.

Mass boiling at or ahead of the pump suction flange is called "massive suction line cavitation" and leads to a "vapor lock" of the pump, where the impeller is trying to operate with a vapor whose density is several orders of magnitude lower than the intended liquid density. The result is a virtually complete loss of head generation, which may lead to a backflow surge of water into the pump that may well destroy the pump. Obviously, this situation must be avoided.

More typically, cavitation occurs in smaller, localized regions inside the impeller of a pump or turbine. These regions typically occur near the entry of the pump—at the inlet rim of the pump shroud or along the leading edges of the blades (sometimes along the juncture of the hub and blade or shroud and blade)—and near the outlet of turbines.

Generally, cavitation causes three different types of problems. The first is loss of performance. If regions of the rotor flow passages are filled with vapor rather than liquid, then the flow rate (in liquid volume terms) is reduced as is the head and, correspondingly, the efficiency. The second problem associated with cavitation is noise and vibration caused by irregular growth and collapse of vapor regions within the impeller. Closely associated with this is possible pulsation of the flow.

The third and most serious problem caused by cavitation is erosion and corrosion of the working surfaces of the rotor (and possibly the casing) of pumps and the downstream piping of hydraulic turbines. Ironically, these problems are not caused by the vaporization of the liquid in the lowest pressure regions but rather by the recondensation of the vapor into liquid. As bubbles of vapor collapse, a phenomenon similar to an intense water hammer occurs locally and at high frequency. Typically, at frequencies reaching 25 kHz, shock waves induce temperature rises in the range of 500–800°C and pressures of 4000 atm. The subsequent intense hammering of machine surfaces in the collapse region leads to local fracturing and breakdown of the surface structure, followed by progressive pitting and erosion. Prolonged exposure to even very limited cavitation can lead to structural failure of the machine, preceded by progressive deterioration of efficiency and performance. Figure 3.1 is a photograph of a pump impeller showing long-term cavitation damage. Note that the damage is concentrated in regions where the rising pressure causes the vapor bubbles to recondense (at the leading edges and impact surfaces of the blades).

Clearly, good design or selection and installation of a pump or turbine require that the machine operates as far away as possible from conditions that will lead to cavitation. In the simplest terms, all one must do is to assure that the lowest pressure occurring within the machine is sufficiently larger than the liquid's vapor pressure. In practice, this situation is slightly more complicated than it would seem. When designing, selecting, or operating a pump, one works with the pressure at the suction flange, the interface between the system

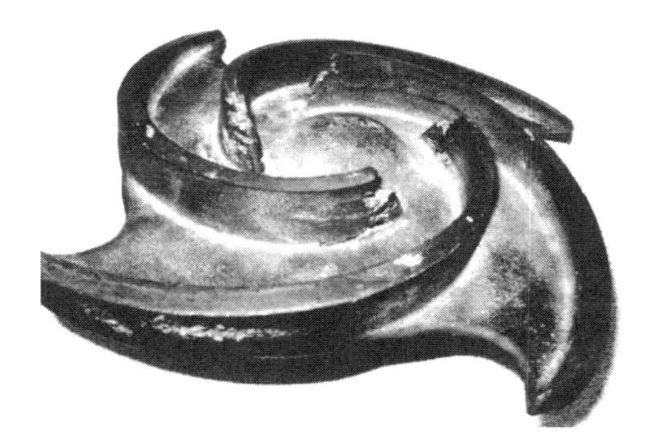

FIGURE 3.1 Pump impeller showing cavitation damage. Note material loss at blade leading edges and impact surfaces of blades.

and the pump itself. The minimum pressure does not occur there (as may be implied by the example above), but at some location within the machine, near the entry onto the impeller, following pressure drops associated with the entrance nozzle and turning of the flow onto the impeller. What is required is a parameter that can be evaluated *at the pump inlet* that indicates the possibility of cavitation *inside the pump*. In practice, that parameter is the *Net Positive Suction Head* (typically, if awkwardly, written as NPSH). In simple terms, NPSH is the difference between the total energy available in the liquid at the pump inlet and the energy associated with the liquid vapor pressure. Mathematically:

$$\text{NPSH} \equiv h_{\text{T1}} - h_{\text{v1}} = \frac{p_1}{\rho g} + \frac{V_1^2}{2g} + z_1 - \frac{p_{\text{v1}}}{\rho g}. \tag{3.32}$$

In this definition, "1" represents the pump inlet and *the pressures must be expressed in absolute units*. The elevation, z_1, is measured relative to the probable location of minimum pressure, typically, but not always, the center of the impeller inlet eye. If the pump inlet lies below the impeller eye, z_1 is negative.

One way to think about NPSH is in terms of a pressure or energy budget. The total head represents the energy available to the fluid as it enters the pump. As the fluid flows from the inlet to the point of minimum pressure, some of this energy is "spent" on losses. Also, the pressure will change (usually, decrease) because of Bernoulli (acceleration) effects and hydrostatic (elevation) effects. The resulting minimum pressure must exceed the vapor pressure in order to avoid cavitation. In a sense, the vapor pressure represents a "minimum balance" that must be maintained in the energy account to avoid a troublesome overdraft!

The value of NPSH at which cavitation occurs within a pump is determined by testing. Such tests can be made using the same facility as a standard (head vs. flow) performance test. Referring to Figure 1.10, the value of the vapor pressure can be controlled by varying the liquid temperature with the heating/cooling coils. The total head at the pump inlet can be varied by changing the air pressure or liquid level in the reservoir or by using a throttling valve (not shown) in the suction line. The test is performed by setting a high value of NPSH and establishing a particular value of Q, and hence H and η. The NPSH is then reduced and the system (discharge-line valve) is manipulated to re-establish the same flow. The head and efficiency are again determined. The process is repeated until the head and efficiency drop. By convention, a head or efficiency drop of a few percent (typically 2–3%) is interpreted as the inception of cavitation. The NPSH at which cavitation begins is referred to as the "NPSH Required" and is usually written as NPSHR. For a particular pump, the value of NPSHR depends on speed and flow rate

$$\text{NPSHR} = f(Q, N).$$

Pump performance curves typically include information on NPSHR; Figure 3.2 is an example of such curves (see also Figure P1.41).

For a given installation (existing or proposed), the actual value of NPSH at the pump inlet depends on the suction system. This existing value is called the "NPSH Available" and is written NPSHA. In order to avoid cavitation, it is required that

$$\text{NPSHA} \geq \text{NPSHR}. \tag{3.33}$$

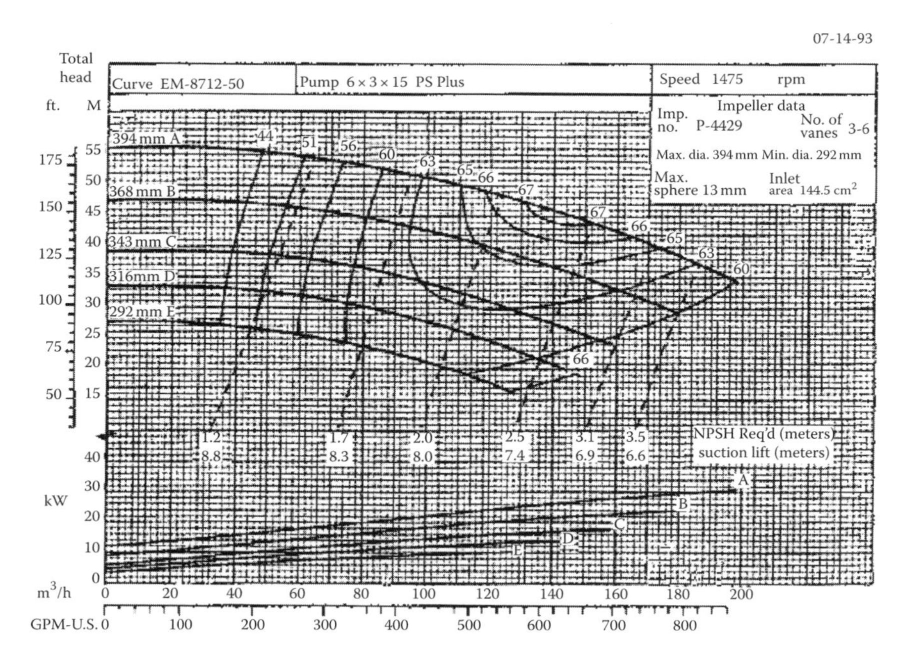

FIGURE 3.2 Performance curves for a single-suction pump, showing NPSHR information. (From ITT Fluid Technology. With permission.)

This requirement is another pump-system matching situation as NPSHR is a property of the pump and NPSHA is a property of the (suction) system.

How does one determine NPSHA? The defining equation is essentially identical to Equation 3.32:

$$\text{NPSHA} = \frac{p_1}{\rho g} + \frac{V_1^2}{2g} + z_1 - \frac{p_{v1}}{\rho g}. \tag{3.34}$$

If the pump in question is already installed and operating, and if suitable instruments and connections are available, then NPSHA can be determined from measurements of inlet pressure, temperature (to determine p_v), and flow rate (to determine V). It is important to remember that p_1 and p_v are absolute pressures.

More frequently, NPSHA must be determined by analyzing the inlet system. Figure 3.3 shows a pump drawing water from a reservoir through a rather simple intake system. To evaluate NPSHA for this installation, apply the mechanical energy equation to a streamline running from the free surface of the reservoir to the pump inlet:

$$\frac{p_{atm}}{\rho g} + z_0 = \frac{p_1}{\rho g} + \frac{V_1^2}{2g} + z_1 + h_L,$$

where p_{atm} is the absolute pressure of the atmosphere at the free surface, z_0 is the elevation of the free surface relative to the pump impeller inlet, and h_L is the total head loss in the intake system due to pipe friction and minor losses. Then

$$\text{NPSHA} = \frac{p_{atm}}{\rho g} - h_z - h_L - \frac{p_v}{\rho g}, \tag{3.35}$$

where $h_z = z_1 - z_0$. This equation shows the four primary methods to control NPSHA if cavitation is likely to be a problem:

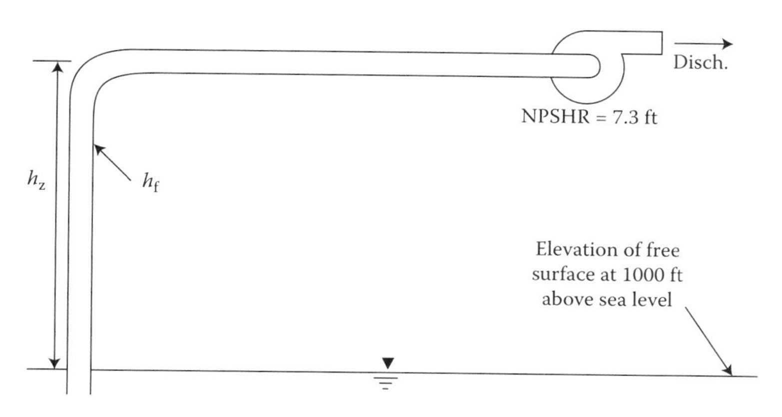

FIGURE 3.3 A pump with intake (suction) system.

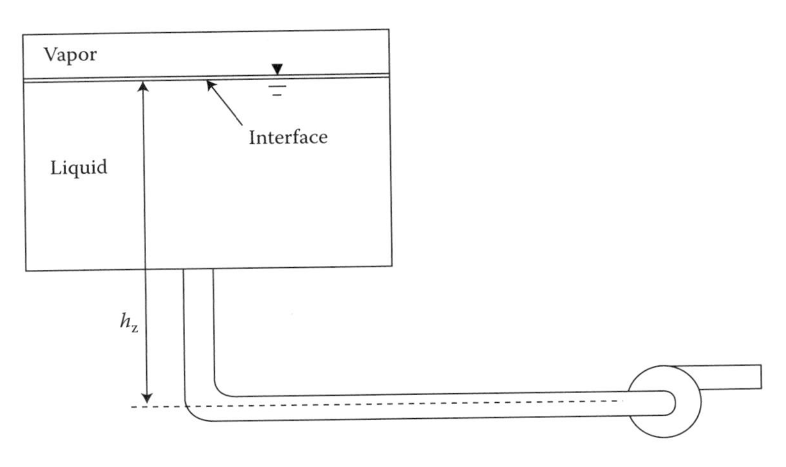

FIGURE 3.4 A pump drawing a volatile liquid from a closed, elevated tank.

- Increase reservoir pressure
- Minimize suction lift (or install the pump below the reservoir surface)
- Minimize head losses in the intake system
- Raise vapor pressure (typically by lowering fluid temperature).

Figure 3.4 shows a different installation in which a pump handles a volatile liquid drawn from an elevated, closed tank. Applying the mechanical energy equation in the same fashion results in the following:

$$\text{NPSHA} = \frac{p_V}{\rho g} + h_Z - h_L - \frac{p_V}{\rho g} = h_Z - h_L.$$

In this case, the vapor pressure term cancels because the liquid surface is in equilibrium with the vapor above it. Also, in this case, $h_Z = z_1 - z_0$.

Example: Cavitation and Suction Lift

The pump shown in Figure 3.3 has an impeller diameter of 343 mm (performance data shown as Curve C in Figure 3.2). The reservoir surface is at 1000 ft elevation ($p_{atm} = 14.17$ psia) and the water is at 85°F ($p_V = 0.60$ psia, by converting from Figure A.3). Compute the maximum suction lift (h_Z) for a flow rate of 500 gpm. Assume that friction loss is given by $h_L = 5(V^2/2g)$ and the pipe diameter is 5.0 in.

From Figure 3.2, NPSHR = 2.0 m = 6.56 ft. The water velocity is

$$V = \frac{Q}{A} = \frac{4Q}{\pi D^2} = \frac{4 \times 500/449}{\pi(5/12)^2} = 8.17\ \text{ft/s}.$$

To just avoid cavitation NPSHA = NPSHR so,

$$\text{NPSHA} = \frac{p_{atm}}{\rho g} - h_L - h_Z - \frac{p_V}{\rho g} = \text{NPSHR} = 6.56\ \text{ft}.$$

Solving

$$h_z = \frac{p_{atm} - p_v}{\rho g} - (5)\frac{V^2}{2g} - 6.56\,\text{ft} = \frac{(14.17 - 0.60) \times 144}{62} - 5 \times \frac{8.17^2}{2 \times 32.2} - 6.56 = 18.74\,\text{ft}.$$

In practice, a "safety margin" of about 10% is recommended, so an NPSHA of about $[1.10 \times 6.56\,\text{ft} = 7.22\,\text{ft}$ (say $7.5\,\text{ft})]$ should be used and the suction lift should be no larger than about 17.9 ft.

Cavitation parameters can be formed into dimensionless groups by the usual considerations of similarity and dimensional analysis. The two most common dimensionless groups are the *Thoma cavitation parameter* $\sigma_V \equiv (\text{NPSHR}/H)$ (Thoma and Fischer, 1932) and the *suction specific speed* $S \equiv N\sqrt{Q}/(g \times \text{NPSHR})^{3/4}$. Some experts assert that S should be preferred over σ_V because cavitation in a pump depends almost exclusively on conditions near the inlet while the head H used to form σ_V depends most strongly on conditions near the outlet. In any case, the parameters are not independent parameters; they are related by $S = Ns/\sigma_V^{3/4}$.

Typical values of σ_V at the BEP, based on collected experimental data (Thoma and Fischer, 1932; Shepherd, 1956; Stepanoff, 1948), have been presented as curve fits:

$$\sigma_{V,BEP} \approx 0.241\,Ns^{4/3} \quad \text{(single-suction pumps)}, \tag{3.36}$$

$$\sigma_{V,BEP} \approx 0.153\,Ns^{4/3} \quad \text{(double-suction pumps)}. \tag{3.37}$$

Simpler, and equally satisfactory, are the equivalent correlations (the reader is invited to demonstrate their equivalence to Equations 3.36 and 3.37):

$$S_{BEP} \approx 2.9 \text{ (single suction) and } S_{BEP} \approx 3.7 \text{ (double suction)}. \tag{3.38}$$

For hydraulic turbines (where cavitation occurs near the outlet),

$$S_{BEP} \approx 4.4. \tag{3.39}$$

Example: Cavitation Correlations

Investigate the accuracy of Equations 3.36 and 3.38 for the 394-mm diameter pump whose performance is shown in Figure 3.2.

For the pump in question, the BEP is at $Q = 140\,\text{m}^3/\text{h}$ $(0.0389\,\text{m}^3/\text{s})$; $H = 47\,\text{m}$; and $\eta_T = 0.68$. Also $N = 1475\,\text{rpm} = 153.8\,\text{s}^{-1}$. Thus $Ns = \left(153.8 \times \sqrt{0.0389}\right)/(9.81 \times 47)^{3/4} = 0.305$. Since the pump is single suction:

$$\sigma_{V,BEP} \approx 0.241(0.305)^{4/3} = 0.0495,$$

$$\text{NPSHR} = \sigma_V H = 0.0495 \times 47\,\text{m} = 2.33\,\text{m}.$$

The curves specify about 2.3 m, so agreement is very good in this case.
Now turning to Equation 3.38

$$S_{BEP} = \frac{N\sqrt{Q}}{(g \times \text{NPSHR})^{3/4}} \approx 2.9 \quad \text{gives}$$

$$\text{NPSHR} \approx \left(\frac{153.8 \times \sqrt{0.0389}}{2.9}\right)^{4/3} \times \frac{1}{9.81} = 2.34\,\text{m};$$

an equally accurate value.

3.5 Summary

The concepts of scaling turbomachinery performance according to a fixed set of rules were introduced. The previously developed rules of similitude were reformulated to generate a set of fan laws and pump affinity rules for machinery with essentially incompressible flows, and examples were presented for illustration.

The major restrictions on scaling of performance with changes in speed, size, and fluid properties were identified in terms of the matching of Reynolds and Mach numbers between a model and a prototype. This precise condition for similitude and permissible scaling was extended to embrace the idea of a "regime" of Reynolds and Mach numbers for which scaling will be sufficiently accurate. Several algorithms by which the simple scaling rules could be extended to account for the influence of significant changes in Reynolds number and surface roughness on efficiency and performance of fans pumps and compressors were presented. The influence of Mach number and allowable disparities between values for a model and a prototype were reviewed for fans and compressors, with clear limitations governing the accuracy of scaled performance. Corrections for compressibility effects on scaling between model and prototype were developed and illustrated with an example.

For liquid pumps and turbines, the phenomenon of cavitation was introduced in terms of pressure in the flow and the temperature-dependent vapor pressure of the fluid. The influence of low-pressure regions near the suction or entry side of a pump and discharge side of a turbine was related to the onset of conditions favorable to cavitation or local boiling and recondensation of the flowing fluid. The governing parameters of NPSH or NPSHA at the pump impeller inlet and the value of NPSH needed to preclude the development of cavitation (NPSHR) were presented. The semiempirical prediction of NPSHR for a pump, in terms of the specific speed of the pump, was presented and illustrated with examples to provide a uniform approach to analysis of cavitation problems.

EXERCISE PROBLEMS

3.1. The pump of Figure P1.1 is to be sized requiring geometric similarity such that at its BEP, it will deliver 300 gpm at a speed of 850 rpm. Show that the diameter should be 16.7 in. and the head will be 58.8 ft.

3.2. Calculate the specific speed and specific diameter for the pump of Problem 3.1 and compare the results to the Cordier diagram. Also, compare the efficiency to the Cordier estimate and comment.

3.3. The characteristic curve of a forward-curved bladed furnace fan (Figure P3.3) can be approximately described by the simple curve fit:

$$\Delta p_s = \frac{\Delta p_{s,BEP}\left[1 - (Q^2/Q_{max}^2)\right]}{1 - (Q_{BEP}^2/Q_{max}^2)}.$$

a. Derive the equation by assuming a parabolic curve fit of the form $\Delta p_s = a + bQ^2$.

b. For a particular fan, $\Delta p_{s,BEP} = 1$ in. wg, $Q_{BEP} = 2000$ cfm, and $Q_{max} = 3500$ cfm. The fan supplies air ($\rho g = 0.075\,\text{lbf/ft}^3$) to a duct system with $D = 1$ ft, $L = 100$ ft, and $f = 0.025$. What flow rate will the fan provide?

c. If the duct diameter in part (b) were $D = 1.5$ ft, what flow rate would result?

3.4. If the fan of Problem 3.3 is pumping air at 250°F and 2116 lbf/ft^2 inlet conditions, what volume flow rate and static pressure rise will it move through the 1 ft duct?

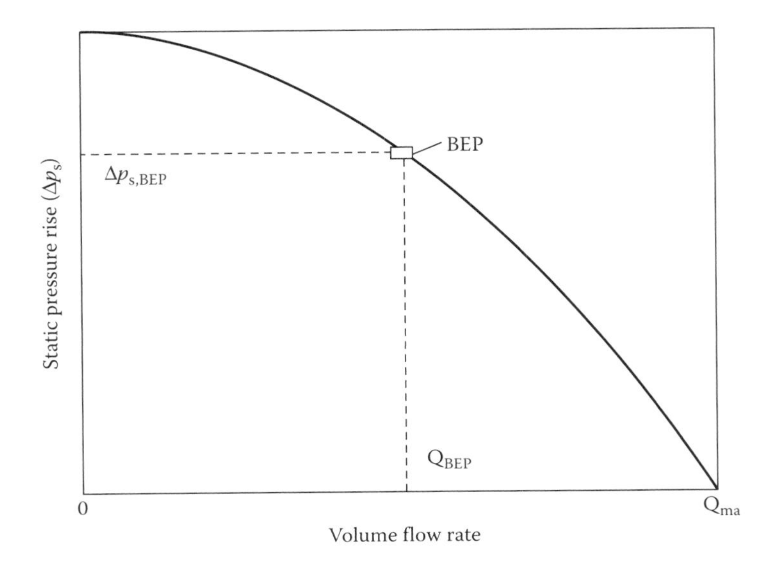

FIGURE P3.3 Characteristic curve for a forward-curved blade furnace fan.

3.5. A model fan is tested at 1200 rpm in a 36.5-in. diameter size to meet AMCA rating standards. This fan delivers 13,223 cfm at a total pressure rise of 2.67 in. wg and a total efficiency of 89% when tested on standard air.

   a. What type of fan is this?

   b. If the model is scaled to a prototype speed and diameter of 600 rpm and 110 in., respectively, what flow rate and pressure rise will result?

   c. If both model and prototype are run in standard air, what is the estimated efficiency for the prototype fan?

3.6. The fan design described in Problem 3.5 is to be modified and used in a low-noise, reduced-cost application. The diameter is to be increased slightly to 40 in., and the speed is to be reduced to 885 rpm. Recalculate the performance. Include the effects of Reynolds number on efficiency and required power.

3.7. In a large irrigation system, an individual pump is required to supply 200 ft^3/s of water at a total head of 450 ft. Determine the type of pump required if the impeller runs at 450 rpm. Determine the impeller diameter. Estimate the required motor power.

3.8. Verify in detail the curve fits used for the Cordier N_S–D_S and efficiency relations.

3.9. A centrifugal pump must deliver 250 ft^3/s at 410 ft of head running at 350 rpm. A laboratory model is to be tested; it is limited to 5 ft^3/s and 300 hp. The test fluid is water with $\rho g = 62.4\,lbf/ft^2$. Assuming identical model and full-scale efficiency, find the model speed and size ratio.

3.10. A hydraulic turbine must yield 30,000 hp at 80 rpm when supplied with 125 ft of head. A testing lab can accommodate 50-hp models at a head of 20 ft. Determine the model speed, the size ratio, and the model flow rate.

3.11. A pump that was designed for agricultural irrigation is to be scaled down for use in recirculation of water in a small aquarium for pet fish. The irrigation pump supplies 600 gpm at 100 ft of head and is a well-designed pump, with an efficiency of 75% at 1200 rpm. The aquarium pump must deliver 6 gpm at 5 ft of head.

   a. Find the proper speed and diameter for the aquarium pump.

   b. Estimate the efficiency of the aquarium pump.

   c. How do you expect the small pump to deviate from the simple scaling laws?

3.12. If the fan of Problem 3.5 is scaled to a prototype speed and diameter of 1800 rpm and 72.0 in., respectively, what will the flow and total pressure rise be? Estimate η_T.

3.13. Shown in Figure P3.13 are the performance curves of an Allis-Chalmers single-suction pump.

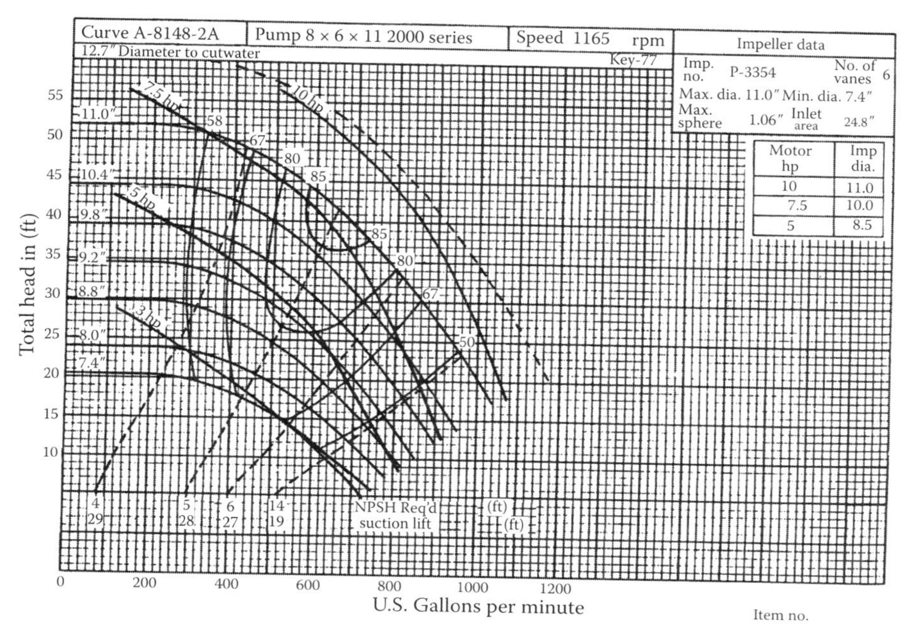

FIGURE P3.13 Performance curves for a single-suction Allis-Chalmers pump. (From ITT Fluid Technology. With permission.)

a. Estimate the head-flow curve for a 10-in. diameter impeller and plot the result on the curve.

b. Calculate the N_S and D_S values at the BEP of the 9-, 9.6-, and 10-in. curves. Compare the experimental efficiencies to the Cordier values.

c. Can you use a 10-in. impeller with the 5-hp motor? Support your answer with suitable calculations.

3.14. A small centrifugal pump is tested at $N = 2875$ rpm in water. It delivers 252 gpm at 138 ft of head at its BEP ($\eta = 0.76$).

a. Determine the specific speed of the pump.

b. Compute the required input power.

c. Estimate the impeller diameter.

d. Sketch two views of the impeller shape.

3.15. Two low-speed, mixed flow blowers are installed to operate in parallel, pumping chilled air (5°C) through a refrigerated truck trailer filled with grapes in plastic bags. The resistance to flow through the fairly tightly packed cargo yields a K-factor of 150.0 with an effective cross-sectional area of 5 m^2.

a. If the required flow rate for initial cool-down is 10.0 m^3/s, calculate the net pressure rise required.

b. After initial cool-down, the flow rate requirement can be supplied by only one of the two fans. Estimate the new flow and pressure rise.

c. Select a suitable fan size with $N = 650\,\text{rpm}$ for this installation.

3.16. If one uses the pump of Problem 2.16 to pump jet engine fuel, one would expect a reduction in efficiency due to higher viscosity. Assuming a specific gravity for JP-4 at 25°F of 0.78 and a kinematic viscosity 14 times that of water, determine whether the 10-hp motor will be adequate to drive the pump at 1750 rpm and 250 gpm, within an overload factor of 15%.

3.17. If one can run the model test of Problem 2.9 in a closed-loop, variable-pressure test duct, what pressure should be maintained (at the same temperature) to achieve complete dynamic similitude?

3.18. A model pump delivers 20 gpm at 40 ft of head running at 3600 rpm in 60°F water. If the pump is to operate at 1800 rpm, determine the proper water temperature to maintain dynamic similarity. Find the flow rate and head under these conditions.

3.19. A shrouded wind turbine operates in a 30 mph wind at standard air density. The shroud allows the 30-ft diameter turbine to capture a volume flow rate of twice its area times the freestream velocity, V_{fs}. According to the "Betz limit" (see Dixon, 1998), the optimum head recovered by the turbine should be 59% of the freestream velocity pressure.

a. Calculate the fluid power captured by the turbine in kW.

b. With the turbine running at 90 rpm, estimate its efficiency and net power output.

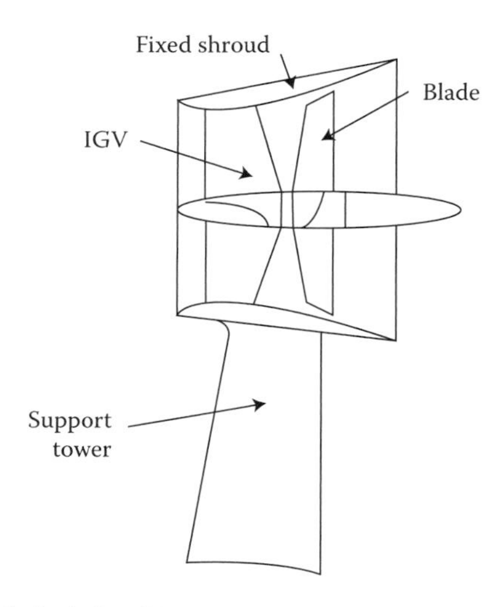

FIGURE P3.19 A shrouded wind turbine.

3.20. For the model and prototype fans of Problem 3.5, use the ASME, ICAAMC, and Wright models to estimate the efficiency for the full-scale fan. The model efficiency is $\eta_T = 0.89$. Compare and discuss the various results for the full-scale prototype fan. Assume hydraulically smooth surfaces.

3.21. A 40-in. diameter axial fan operates at 965 rpm. The fan produces 15,000 cfm at 2.0 in. wg. Estimate the total efficiency of the fan for a range of surfaces roughness values from $\varepsilon/D = 0.001$ to 0.01. Develop a graph of η_T and η_s versus ε/D. Can a value of $\eta_T = 0.80$ be achieved? Select a combination of parameters to achieve this efficiency and defend your choice on a design decision basis. (Hint: Begin your efficiency calculations based on the Cordier efficiency and modify using *Re*, assuming initially that roughness is negligibly small.)

3.22. The very large pump described in Problem 3.7 is to be tested in air to verify the manufacturer's performance claims. Predict the static pressure rise (in in. wg), the flow rate (in cfm), and the static efficiency running in standard air.

3.23. A pump is designed to provide 223 gpm at 98 ft of head in 85°F water. The 9.7-in. impeller operates at 1750 rpm. Estimate the NPSHR and maximum suction lift assuming no suction system losses.

3.24. A double-suction version of the pump in Problem 3.23, providing 450 gpm at 95 ft of head, has suction system head losses of 5 ft. Estimate the NPSHR and maximum suction lift for 85°F water. Repeat for 185°F water.

3.25. The pump of Problem 3.23 must pump gasoline (S.G. = 0.73, p_{vp} = 11.5 psi) from an evacuated tank. Find the NPSHR and maximum suction lift.

3.26. The influence of water temperature on cavitation in pumps is related to the behavior of h_{vp}, the vapor pressure head, with temperature. NPSHR for a pump must then be rated by the manufacturer at some stated temperature. For example, the Allis-Chalmers pump of Figure P2.12 is rated at 85°F. For any other temperature, this value must be corrected. For moderate temperature excursions (35°F $<$ $T <$ 160°F), the correction is estimated by NPSHR = NPSHR$_{85}$ + Δh_{vpT}, where NPSHR$_{85}$ is the quoted value and Δh_{vpT} is the difference in vapor pressure at 85°F and the value of T being considered [$\Delta h_{vpT} = h_{vp}(T) + h_{vp}(85°)$]. This is approximately equal (in water) to $\Delta h_{vp} = 5\ (T/100)^2 - 3.6$ (ft).

The NPSHR$_{85}$ values for the Allis-Chalmers pump vary, of course, with flow rate and can be approximated by NPSHR$_{85}$ = 1.125 + $6.875(Q/1000)^2$ with Q in gpm, for the 12.1-in. diameter impeller.

a. Verify the equation for Δh_{vpT}.

b. Verify the equation for NPSHR$_{85}$.

c. Develop an equation for NPSHR$_{85}$ for the 7.0-in. diameter impeller.

3.27. Using the information in Problem 3.26, do the following.

a. Develop a curve for the maximum allowable flow rate without cavitation for the 12.1-in. impeller as a function of temperature, if NPSHA = 34, 24, and 14 ft.

b. Comment on the allowable suction lift for these conditions.

3.28. For the information in Problem 3.27, discuss the influence of $T < 85°F$ water on the results in (a) and compare to the effects of water at $T > 85°F$.

3.29. For the pump of Problem 3.26, the values of NPSHR at a given flow rate are much higher for the small pump (the 7-in. diameter impeller) than for the large pump (the 12.1-in. diameter impeller). Explain this in terms of Thoma correlation for NPSHR. Does the variation for NPSHR for either impeller make sense? Show all work.

3.30. Rework Problem 2.4 considering compressibility.

3.31. A small double-suction pump has a design point rating of $H = 60$ ft at $Q = 75$ gpm, running at 3600 rpm. It will be installed at some height above an open pond to pump cold water into an irrigation ditch. Atmospheric head is 33 ft, vapor pressure head is 1.4 ft, and the suction line head loss is 4 ft. Estimate the maximum safe height of the pump above the pond.

3.32. Correct the results of Problem 2.5 for compressibility using both isentropic and polytropic pressure–density laws. Comment on the comparison of these results with each other and with the incompressible calculations. What are the economic implications in using an incompressible result?

3.33. A high-pressure blower supplies 2000 cfm of air at a pressure rise of 42 in. wg, drawing the ambient air through a filtering system whose upstream pressure drop is 32 in. wg (i.e., the fan discharge pressure is 10 in. wg). Assume an isothermal pressure drop upstream of the fan inlet to estimate inlet density from ambient conditions. The ambient air properties are $T = 80°F$ and $p_{amb} = 30$ in. Hg. Select a single-stage, narrow centrifugal fan for this application, and estimate diameter, speed, and efficiency. (Hint: Size a fan using inlet density and correct the required flow and pressure for compressibility.)

3.34. For the pump described in Problem 1.1, develop an estimated curve of NPSHR versus Q for:

a. A single-suction pump.

b. A double-suction pump.

3.35. A pump must deliver 1 ft^3/s of water in the system shown in Figure P3.35. This pump is to be a single-suction type, direct-connected to an AC induction motor ($N = 600, 900, 1200, 1800$, or 3600 rpm).

a. Find the direct-connected speed for which the pump will not cavitate. Use Thoma's criterion to size NPSHR $= 1.1\sigma_V H$, where 1.1 is a "safety factor."

b. Determine the pump diameter, pump type, and probable efficiency from Cordier.

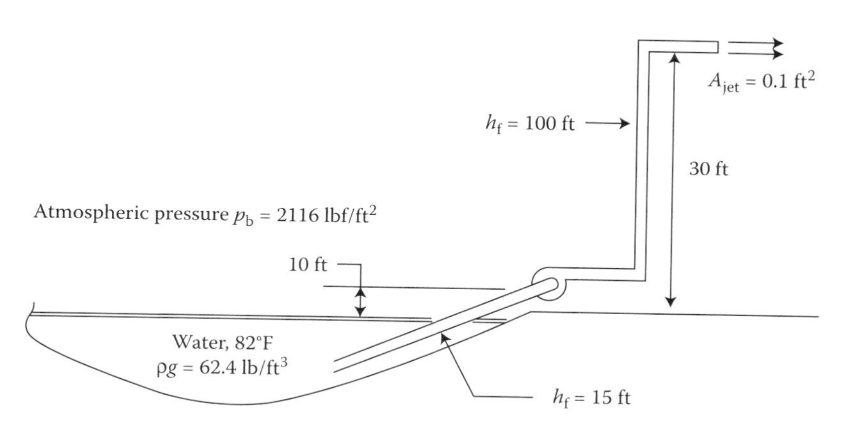

FIGURE P3.35 A pump and system configuration.

3.36. Warm gasoline must be pumped from an underground tank using a submerged centrifugal pump. The tank is effectively vented to atmospheric pressure (although for environmental protection, a carbon filter is installed in the vent). If the NPSHR of the pump is 8 m, the vapor pressure of the gasoline is 40.0 kPa, and the suction line friction head losses are 0.8 m, estimate the required submersion depth of the pump to avoid cavitation. The specific gravity of the gasoline is 0.841, and the barometric pressure is 100 kPa.

3.37. A large centrifugal fan is designed to operate at 600 rpm, with an impeller diameter of 60 in. The fan uses 40 hp at 60,000 cfm and 3.0 in. wg *static* pressure rise. The fan is driven by a variable-frequency motor that can operate at any speed between 600 and 7200 rpm.

a. Estimate the flow rate, static pressure rise, and power required at 1200 rpm.

b. Estimate the change in efficiency at 1200 rpm compared to the value at 600 rpm.

c. At what speed (in rpm) would the scaled performance need to be corrected for compressibility effects?

3.38. A 1.25-m diameter model fan is tested for certification of performance. The model operates at 985 rpm. The fan achieves BEP performance of 6.0 m^3/s at 700 Pa and a total efficiency of 0.875.

a. If the model is scaled to prototype speed and diameter of 600 rpm and 2.5 m, what will be the flow rate and pressure at standard density?

b. Estimate the prototype efficiency.

3.39. A three-speed fan, when operating at its highest speed of 3550 rpm, supplies a flow rate of 5999 cfm and a total pressure rise of 30 lb/ft^2

with $\rho = 0.00233\ \text{slug/ft}^3$. Lower speeds of 1750 and 1175 rpm are also available.

a. Estimate the size of the fan and its efficiency at 3550 rpm.

b. Estimate the flow rate and pressure rise at the low speed of 1175 rpm.

c. Use the Reynolds numbers of the fan at 1750 and 1175 rpm to estimate the efficiency of the fan at these lower speeds.

3.40. The low-pressure section of a steam turbine has nine stages. Steam enters the section at 380 lbm/s (173 kg/s), 160 psia (1.30 MPa), 755°F (401°C). The turbine spins at 3600 rpm. The steam leaves the section (after the ninth stage) at 4 psia (27.5 kPa).

a. Assuming that each of the nine stages has the same pressure ratio, find the pressure ratio for the first stage.

b. Find the first stage volume flow (use average density for the first stage).

c. Determine an expected diameter for the first stage.

d. Estimate the efficiency of the first stage.

e. Assuming that all nine stages have the same efficiency, calculate the efficiency for the entire section.

f. Calculate the specific work in the first stage.

3.41. A gas-turbine engine has an 8-stage axial flow compressor. Each stage has a pressure ratio of 1.24 and a polytropic efficiency of 87.2%. The compressor's rated air flow is 30 lb/s.

a. Calculate the overall pressure ratio of the compressor and its isentropic efficiency.

b. The engine is rated with inlet air at 59°F and 14.696 psia. Calculate the power consumed at rated conditions.

c. The compressor is operated at a location where the inlet air conditions are 90°F and 13.8 psia. Calculate the new values of mass flow and input power.

3.42. A turbocharger is used to increase the power output of a reciprocating engine by compressing the air/fuel mixture prior to induction into the engine. The power to run the compressor is provided by a turbine in the exhaust gas stream. The compressor and the turbine share a common shaft and hence rotate at the same speed. Consider a turbocharger with the following specifications: Compressor inlet air—14.0 psia, 70°F, $c_p = 0.24\ \text{Btu/lbm °R}$, $\gamma = 1.4$, $\dot{m} = 0.5\ \text{lbm/s}$; Compressor discharge pressure = 28 psia; Turbine exhaust gas properties are 1100°F, $c_p = 0.27\ \text{Btu/lbm °R}$, and $\gamma = 1.3$; the turbine exhausts to atmospheric pressure. Neglect the fuel mass added between the compressor discharge and the turbine inlet.

a. Assuming an optimum choice for the compressor, estimate its speed and size.

b. What power must the turbine supply to run the compressor?

c. Since turbines are naturally more efficient than compressors, assume that ($\eta_{t,s} = \eta_{c,s} + 0.05$) and estimate the pressure at the turbine inlet.

d. What type of turbine would be expected to yield this efficiency? What would be its size?

3.43. A centrifugal compressor is being designed to handle high-temperature air ($T_{inlet} = 800°F$, $p_{inlet} = 14.7$ psia). The design pressure ratio is to be 5:1, the design speed is to be 25,000 rpm, and the design mass flow is 25 lbm/s.

a. Estimate the size of the compressor and sketch the impeller.

b. A rating test is to be conducted with cold air (59°F). At what speed, pressure ratio, and flow rate should the test be conducted?

c. Assume that the test yields an *isentropic* efficiency of 84.2%. Determine the efficiency at design conditions. Include effects of Reynolds number, if any.

3.44. Use the pump from Problem 2.3 operating at 1200 rpm. Assume that the pump's NPSHR is given by

$$\text{NPSHR} = 5 + 20\left(\frac{Q}{100}\right)^{1.8}$$

and determine how far downstream from the system inlet that the pump can safely be located. Assume 90°F water.

3.45. Consider once again the Taco Model 4013 family (series) of pumps in Figure P1.41 (Problem 1.41).

a. Calculate the specific speed for this type pump. Use the 13.0-in. impeller.

b. Compare the actual efficiency of this pump with the expected efficiency from the $\eta - N_S$ correlation in Figure 2.9.

c. Compare the shape of the performance curve for this pump to that expected for a pump with this specific speed.

d. Make a sketch of the probable shape of the impeller.

e. Calculate the specific speed using the 10.6-in. impeller. Compare with your answer from part "a" and discuss any differences.

3.46. Consider using a number of the Taco pumps (in Figure P1.41) to deliver 500 gpm of water through a system that has a 300 ft elevation rise and a (minimum) head loss of (approximately) 50 ft. In order to save capital costs, it is contemplated to use a single 13.0-in. pump of this family and to use either a 1750-rpm (4-pole) or a 3550-rpm (2-pole) motor. The flow requirement is still 500 gpm; the head loss can be increased if necessary by closing a valve.

a. Can a single pump deliver 500 gpm through the system at 1750 rpm? At 3550 rpm? Which speed would you recommend? Why? (Hint: Plot performance curves for both speeds.)

b. What is the maximum distance above the lower reservoir that the pump can be located to avoid cavitation. Assume that pipe length in the suction system is 1.10 times the elevation and the water temperature is 80°F. The pipe is 4-in. schedule 40 and $f = 0.018$.

c. The cheapest (smallest) pump–motor combination will have a 2-pole motor and run at 3550 rpm. Specify a single pump (likely NOT a Taco Model 4013) to do the required pumping job and estimate size and power required.

3.47. The dimensionless performance curves for a certain family of pumps can be represented by

$$\psi = 0.15 - 79.0\phi^2; \quad \eta = \eta_{BEP}(59.6\phi - 3.1 \times 10^4\phi^3),$$

where η_{BEP} is the maximum attainable efficiency.

a. What is the specific speed for this family of pumps? What is the most likely value for η_{BEP}?

b. Sketch the impeller shape (two views).

c. A pump from this family is to pump 1000 gpm of cold water (60°F) against a head of 60 ft. Estimate the size and speed required and also the power and the NPSHR.

d. The pump from part (c) is to be used to move oil with a kinematic viscosity of 1.0×10^{-4} ft^2/s. Estimate the (BEP) head, flow, and efficiency with the oil.

4

Turbomachinery Noise

4.1 Introductory Remarks

In recent years, it has become necessary for designers and users of turbomachinery to understand and predict the acoustic as well as aerodynamic performance of turbomachines. A great deal has been written on the subject of acoustics, including acoustics in turbomachinery. This chapter introduces some general concepts and then examines sound production and noise prediction in turbomachines. This chapter focuses on the acoustics of fans and blowers, with the understanding that the acoustic phenomena in other types of turbomachines behave in a similar fashion. Interested readers are encouraged to consult the extensive reference list of recommended books and papers on acoustics in turbomachinery given at the end of this book.

4.2 Sound and Noise

In a physical sense, sound is a small amplitude vibration of particles in a gas, solid, or liquid (Broch, 1971). Undesirable sound is often called *noise*. Sound vibration can be thought of as a transfer of momentum from particle to particle in air. Due to the finite spacing between air molecules, sound waves travel at a finite propagation speed. In air at 20°C, the propagation speed, or "speed of sound," is 343 m/s (1126 ft/s). The propagation of the vibration occurs as a density/pressure variation in a longitudinal fashion, that is, a longitudinal wave. The variation is a sinusoidal function, and it is periodic and of constant amplitude. The corresponding "frequency spectrum" associated with the sine waves is often a useful data item in acoustic analysis. The spectrum is an illustration of the frequencies and their corresponding magnitude that make up a signal. Spectral analysis is useful in understanding the effect of discrete frequency phenomena as well as the bandwidth of background noise.

One common type of wave formation found in acoustics is the spherical wave. This type of wave can be thought of as a pulsing sphere emanating from a point source and can be used to illustrate the relationship between the sound emitted from a source and the sound pressure at a distance, x, from a point source. Assume that a sound of power E is emitted from a point source. The power will radiate in a spherical fashion. The power intensity, I, is given by

$$I = \frac{E}{(4\pi x^2)} = \frac{\text{Power}}{\text{Area}}. \tag{4.1}$$

At a sufficient distance away from a source, I becomes proportional to the square of the sound pressure p at x. Therefore,

$$p = \frac{k}{x}. \tag{4.2}$$

This relationship is known as the "inverse-distance" law for free sound in an acoustic far field.

Acoustic literature often refers to near fields and far fields (Beranek and Ver, 1992). The nature of a field (i.e., near or far) is important in acoustic measurements. If a source is assumed to be in the far field, it can be treated as a point source. The magnitude of an emission sensed or measured from a point source is solely based on the distance from the source. On the other hand, the magnitude of an emission measured in the near field will be affected by the three-dimensional (3D) location of the transducer relative to the source. To be certain that a given measurement is in the far field, acoustic measurements should be made 2–5 characteristic diameters from the source.

One must keep reflection of waves in mind when working with acoustics. Sound waves will reflect from a surface and affect sound pressure levels. The sound pressure at a point in a field is determined not only by the pressure of the original wave but also by an additional effect of the reflected wave.

The physical scale of sound pressures covers a dynamic range of approximately 1–1,000,000. Therefore, despite the small magnitude of pressure variations (of the order of a few Pascals), the relative range of pressure variation is large. It is then convenient to utilize a decibel scale to report or characterize sound pressures. The decibel scale is a relative measure based on a ratio of power. The typical relationship utilized for sound pressure is given by

$$L_W = 10 \log_{10}\left(\frac{p^2}{p_0^2}\right) = 20 \log_{10}\left(\frac{p}{p_0}\right). \tag{4.3}$$

Although this is a ratio of pressures, its use is acceptable due to the relationship between the sound power and the square of the sound pressure (in the far field). p_0 is a reference pressure, with a value of 2×10^{-5} Pa.

The decibel relationship can also be used with reference to sound power levels. The relationship is given by

$$L_w = 10 \log_{10}\left(\frac{P}{P_0}\right). \tag{4.4}$$

In this equation, P_0 is a reference power level with a value of 10^{-12} W (1 picowatt). The sound power level is, due to its relative nature, essentially the same relationship as the sound pressure level (unless the near field is being considered). Table 4.1a gives examples of the relationship between sound pressures and sound pressure (noise) levels of some commonly occurring phenomena. Table 4.1b shows allowable limits for human exposure to certain levels of sound.

The reader should try to gain a feel for the decibel scale, noting that 20 dB accounts for an order of magnitude difference in pressure. Also note that doubling the sound pressure levels (doubling the sound) results in a change of 6 dB. Therefore, a reduction of 6 dB would reduce the sound by one-half.

TABLE 4.1a

Typical Noise Levels

Noise Type	SPL (Pa)	Noise Level (dBA)
Hearing threshold	2×10^{-5}	0
Recording studio	10^{-4}	20
Normal sleep	10^{-3}	30–35
Living room	10^{-3}	40
Speech interference, 4 ft	10^{-2}	65
Residential limit	10^{-2}	68
Commercial limit	10^{-1}	72
Air compressor, 50 ft	10^{-1}	75–85
OSHA 8-h limit	1	90
Pneumatic hammer (at Operator)	3	100
Airplane (Boeing 707)	8	112
Concorde SST	40	123
Threshold of pain	110	140

Source: Thumann, A., 1990, *Fundamentals of Noise Control Engineering*, 2nd ed., Fairmont Press, Atlanta.

TABLE 4.1b

Detailed OSHA Limits

Duration per Day (h)	Sound Level (dBA)
8	90
6	92
4	95
3	97
2	100
1.5	102
1	105
0.5	110
0.25 or less	115

Source: Beranek, L. L. and Ver, I., 1992, *Noise and Vibration Control Engineering Principles and Applications*, Wiley, New York.

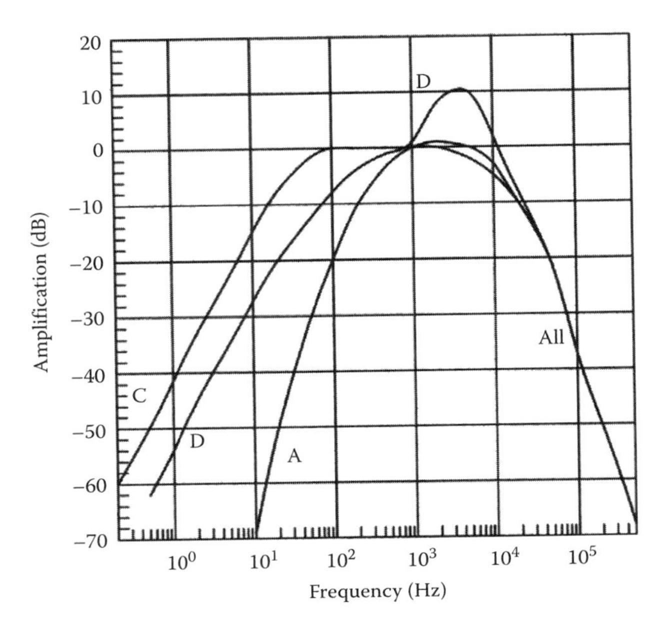

FIGURE 4.1 Common "filter" scales for frequency weighting. (Adapted from Broch, J., 1971, *Acoustic Noise Measurement, The Application of Bruel & Kjaer Measuring Systems*, 2nd ed., Bruel & Kjaer, Denmark. With permission.)

A simple measurement of sound pressure level at a particular frequency, although useful, is not completely definitive due to the lack of knowledge of the frequency distribution of the entire sound emission (Broch, 1971; Crocker, 2007). To make acoustic measurements more useful, a system of frequency weighting utilizing tone-specific weighting curves has been developed. The curves were developed with the idea of equal loudness (loudness is defined based on the average human physiological response to an acoustic event) levels at all frequencies by considering the effect of frequency as well as magnitude on the level of loudness of a signal. "A-scale" weighting has become a standard in acoustical measurements of sound pressure, with the resulting sound pressures having units of dB(A). Generally, in a sound measurement activity, spectrum analysis techniques are utilized, but the weighting system allows for ready comparison of independent acoustic fields, in a single quantity. The A-scale and its associated distribution of reduction values are shown in Figure 4.1, along with similar scales—the C-, and D-scales.

4.3 Fan Noise

Axial flow fans are a good example to study with respect to turbomachine noise due to the wide spectrum of sound associated with such fans. There are

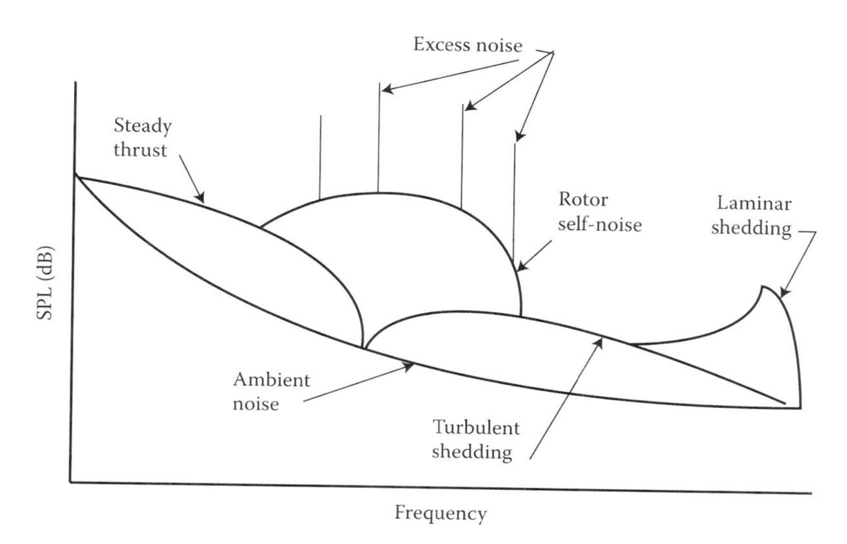

FIGURE 4.2 Acoustic spectrum for axial flow turbomachines. (From Wright, S. E., 1976, *Journal of Sound and Vibration*, 85. With permission.)

two general types of noise associated with fans. *Broadband noise* and *discrete frequency noise* are shown in the frequency spectrum illustrated in Figure 4.2.

The broadband is illustrated as the smooth curve, whereas the discrete frequency noise is illustrated as spikes centered at the blade pass frequency (bpf) and its harmonics. The contribution of each type of noise is related to the performance characteristics of the fan. Broadband noise arises from sources that are random in nature and can be of two types (Wright, 1976). First source arises from the fluctuations of the aerodynamic forces on the blade, whereas the second source arises from turbulent flow in the blade wakes. Generally, the force fluctuation source dominates the spectrum.

There are two sources of force fluctuations. The first source arises from the shedding of vorticity at the blade trailing edge in smooth inflow, which introduces surface pressure fluctuations on the blade. The second source is generated when the blades move through turbulent flow. The turbulence causes random variations in the angle of incidence of the flow, relative to the blade, thus creating random pressure loadings and force fluctuations. If a body is in nonturbulent flow, the vortex shedding is periodic (the well-known Karman Vortex Street). This produces regularly fluctuating lift. As the Reynolds number is increased, the fluctuations become more random in nature as the flow transitions to turbulence. Also, the effect of turbulent inflow is influenced by the scale of turbulence relative to the chord length.

Since broadband noise is affected by the blade pressure distribution, the incidence angle of the blade will affect the noise levels due to the change in blade load and pressure distribution with the change in incidence angle. In this light, it has been found that fans exhibit minimal noise characteristics when the fan is operated near the design point. An increment of about 5–10 dB usually accompanies the fan operating at or near the stall point, or near free delivery. Historically, early investigators tried to model noise solely on the basis of gross performance parameters. Although these methods were a large step forward, they are now being supplanted by schemes that more nearly relate to the aerodynamic and geometric parameters of a machine.

A large number of correlations have been developed to predict noise in various situations. The empirical expressions generally attempt to relate, as simply as possible, the sound output of a rotor to its physical and operating characteristics, with most of the correlations limited to a particular situation.

A general method for the prediction of fan noise (axial and centrifugal) was developed by J.B. Graham beginning in the early 1970s (Graham, 1972, 1991). The method was based solely on performance parameters and was formulated for fans operating at or near the design point. Graham tabulated specific sound power levels for various types of fans at octave band center frequencies (Table 4.2). The specific sound power level was developed to allow for comparison of different types of fans and is defined as the sound power level generated by a fan operating at 1 cfm and 1 in. wg static pressure rise. Table 4.2 is used with the following correlation:

$$L_W = K_W + 10\log_{10} Q + 20\log_{10} \Delta p_S, \tag{4.5}$$

where L_W is the estimated sound power level (dB relative to 10^{-12} W), K_W is the specific sound power level, Q is the volume flow rate (cfm), and Δp_S is the static pressure rise (in. wg). In addition, a "blade frequency increment" is

TABLE 4.2

Specific Sound Power Level Data

Fan type	63	125	250	500	1000	2000	4000	8000	bpi
Centrifugal airfoil blade	35	35	34	32	31	26	18	10	3
Centrifugal backwardly curved blade	35	35	34	32	31	26	18	10	3
Centrifugal forward curved blade	40	38	38	34	28	24	21	15	2
Centrifugal radial blade	48	45	43	43	38	33	30	29	5–8
Tubular centrifugal	46	43	43	38	37	32	28	25	4–6
Vane axial	42	39	41	42	40	37	35	25	6–8
Tube axial	44	42	46	44	42	40	37	30	6–8
Propeller	51	48	49	47	45	45	43	31	5–7

Source: Specific Sound Power Level Data for Fans and Blowers from Graham, J. B., 1991, *ASHRAE Handbook: HVAC Applications*, American Society of Heating, Refrigerating and Air-Conditioning Engineers, Inc. With permission.

Note: Octave Center Band Frequencies, Hz.

added to the octave band into which the bpf falls. This frequency is given by

$$\text{bpf} = \frac{N_B \times \text{rpm}}{60}. \tag{4.6}$$

Therefore, the method is as follows:

1. Find the specific K_w values from Table 4.2.
2. Calculate $10 \log_{10} Q + 20 \log_{10} \Delta p_s$ and add to the K_w values.
3. Calculate bpf and bpi and add to the proper bandwidth.

This method is very general, is a good way to estimate noise levels for comparison and selection, and provides an estimate of the spectral distribution.

Although the methods for predicting sound power have historically encompassed a wide range of theoretical, semitheoretical, and empirical approaches, most of these methods can be considered too complex for routine analysis (Baade, 1982, 1986). In light of Baade's comments, Graham's relatively simple and well-known method for sound power estimation can be re-examined. K_w is the vehicle for capturing the intrinsic difference between the various fan types—centrifugal, forward curved, radial bladed, radial tipped, vane axial, tube axial, and propeller. For these groupings of fans, Graham defines expected levels of specific noise K_w in terms of type. Coincidentally, these types of fans can also be grouped, as noted earlier, in the appropriate "zones" on the Cordier N_s–D_s diagram. To use Graham's data directly, it is necessary to assign values of D_s to the fan groupings as delineated by Graham. Using his classifications by type and estimating reasonable specific diameters for them, the overall K_w values can be estimated as well. In addition, a large group of axial fans described by Wright (1982), a comparison of centrifugal fans (Beranek and Ver, 1992), and additional data from industrial catalogs (Zurn, 1981; Chicago Blower Company, 1998; Industrial Air, 1986) were analyzed in terms of static pressure, flow rate, and sound power level to generate additional values of K_w across the complete range of D_s. The results of these studies are shown in Figure 4.3 as K_w versus D_s. Shown with the data is a linear regression curve fit in the form $(\log K_w) = a + b \log D_s$. The resulting fit to the data yields

$$K_w \approx \frac{72}{D_s^{0.8}}, \tag{4.7}$$

with a correlation coefficient of -0.90 as reflected in the scatter of the data. The results are reasonably consistent and are within the level of accuracy requirements appropriate to simple fan selection and preliminary design layout. The data also suggest a leveling off of values near the larger specific diameters. A separate estimate of K_w above $D_s \approx 2.0$ is provided by

$$K_w \approx 40, \tag{4.8}$$

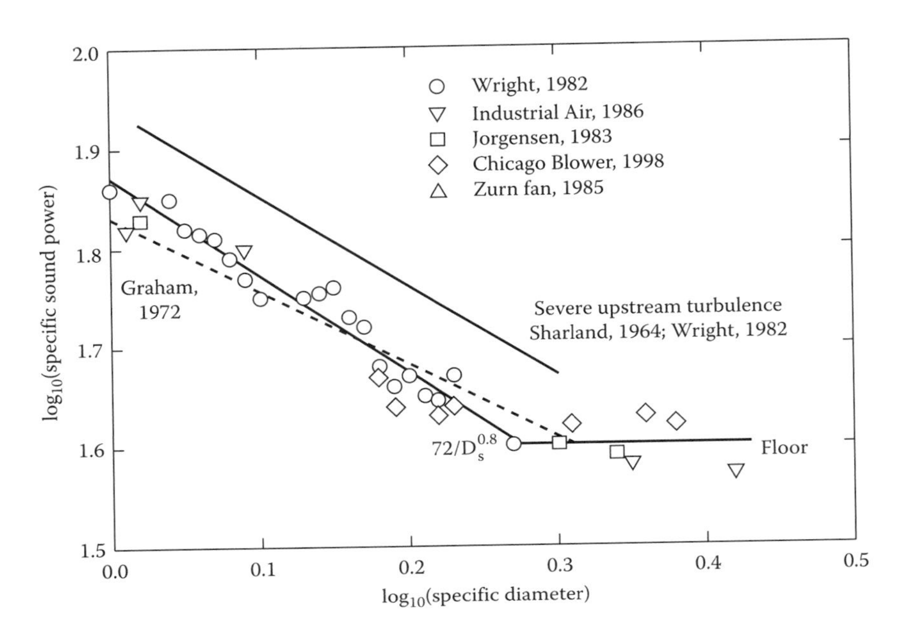

FIGURE 4.3 K_W results for axial and centrifugal fans.

based on the apparent "floor level" indicated in Figure 4.3. Note that the scatter of the data suggests a probable error in these estimates of perhaps ±4 dB.

Fans with inlet guide vanes (IGVs) and with other forms of extreme inflow disturbances were not included in the database for this correlation. Such fans show substantially higher noise levels for the same values of Q and Δp_S, resulting in levels of K_W far above the rest of the data. If an estimate for fans with IGVs or strong upstream turbulence or disturbances is needed, then the equation for K_W can be modified to

$$K_W \approx \frac{84}{D_S^{0.8}}, \tag{4.9}$$

resulting in an additional 12 dB (Wright, 1982). This is clearly a punitive result but is relevant to selection or preliminary design when dealing with preswirl vanes, severe obstructions, or generally rough inflow.

4.4 Sound Power and Sound Pressure

Frequently, the safety and legal constraints applied to noise or sound are written in terms of sound pressure rather than power. For example, the

Occupational Safety and Health Administration (OSHA) health and safety standards are given in terms of the sound pressure level that exists at the workstation of an employee. Local regulations may establish maximum values of sound pressure level at the property or boundary line of a plant or other facility. Exposure times per day for a worker are limited to a fixed number of hours per day for a given sound pressure level. Table 4.1b shows these permissible exposure values.

In the development of estimation methods for the noise from fans and blowers, several algorithms for predicting the sound power level, L_w, have been discussed. For purposes of controlling exposure in the workplace, one often needs to be able to estimate sound pressure level (L_p) in the vicinity of a fan or blower to meet design or selection criteria. The relation one seeks is based on the diminishing of the signal or wave strength with distance from the noise source.

4.5 Outdoor Propagation

For a source operating in a "free field," with no reflections from surfaces near the source, the decay rate is given by (Moreland, 1989)

$$L_p = L_W + 10\log_{10}\left(\frac{1}{4\pi x^2}\right) + c, \tag{4.10}$$

where $c = 0.1$ for x in meters and $c = 10.4$ for x in feet. The distance x is the distance from the source and should be at least 3 times the characteristic length of the sound source. Although the theory considers the source to exist at a point (a point source), in the case of turbomachinery, one can take the impeller diameter, machine length, casing size, or a similar physical length of the machine as the characteristic length. Then it is necessary that the point where pressure is to be predicted be at least three or four times than the length away from the machine to allow use of the free field equation.

For example, consider a 1-m-diameter fan that generates a sound power level of 85 dB. To estimate the free field sound pressure level 5 m away from the fan, one calculates

$$L_p = 85 + 10\log_{10}\left[\frac{1}{4\pi 5^2}\right] + 0.1 = 60.1\,\text{dB},$$

where dB is the sound pressure level referenced to $p_{ref} = 2 \times 10^{-5}$ Pa. At a distance of 10 m away from the fan, the same calculation will give an L_p value of 54.1 dB. At 20 m away, the sound pressure level would be 48.1 dB and so on. Note that with each doubling of the distance from the source, the sound pressure level is reduced by 6 dB. This is a general result and is an accurate "rule of thumb" for free field noise decay.

Machinery is frequently located in areas where nearby surfaces cause non-negligible reflection of sound waves, so the decay rate of sound pressure in a free field environment must be modified if accurate estimates are to be made. Two factors dominate this effect. First, the presence of reflective walls or a floor near the source can focus the sound waves, yielding an effective increase in pressure. This focusing or concentration is characterized by a *directivity factor*, Λ, where Λ is based on the number of adjacent walls (reflective surfaces).

For example, if the sound source, a fan, is sitting on a hard reflective plane (e.g., a large, smooth concrete pad), then the plane will reflect all of the sound waves that (in a free field) would go down and send them upward. That is, all of the energy will be concentrated into half the original space, thus doubling the sound pressure above the plane. This is expressed as $\Lambda = 2$. A very soft plane, such as grass, sod, or heavy carpet, can absorb much of the downward directed energy rather than reflect it so that the resulting pressure more nearly resembles a free field than a hard single-plane field. (This will be accounted for in Section 4.6.)

If the source is located near two walls or planes, as near a hard floor and a hard, vertical wall, then the pressure waves will not only be reflected up, but they will not be allowed to move either left or right. Instead, the leftward waves will be reflected to the right, causing another doubling of the sound pressure in the field. This is quantified by a directivity factor, Λ, of 4.0. Three walls, as in a corner formed by two vertical, perpendicular walls on a horizontal floor, will again double the pressure level for a value of $\Lambda = 8.0$.

These Λ values are used to modify the free field decay equation as

$$L_{\mathrm{p}} = L_{\mathrm{w}} + 10\log_{10}\left(\frac{\Lambda}{4\pi x^2}\right) + c. \tag{4.11}$$

If the fan with $D = 1\,\mathrm{m}$ and $L_w = 85\,\mathrm{dB}$ were mounted on a large totally reflective, hard plane, the sound pressure 5 m away would be

$$L_{\mathrm{p}} = 85 + 10\log_{10}\left[\frac{2}{4\pi 5^2}\right] + 0.1 = 63.1\,\mathrm{dB}.$$

Note that the difference from the free field result is simply $10\log_{10}(2) \approx 3$. For a two-wall situation, the increment is given by $10\log_{10}(4) \approx 6$, and for the three-wall arrangement, $10\log_{10}(8) \approx 9$. Clearly, these hard, totally reflective walls have a substantial effect on the local values of sound pressure level and the health and safety risks and annoyance associated with fan noise.

If the nearby walls are not really hard (i.e., if they are not totally reflective to sound waves) and some of the pressure energy striking the surfaces is absorbed, then the reflective form of $\Lambda/(4\pi x^2)$ must be modified accordingly. Let α represent the fraction of energy that is absorbed. Since a totally absorbing wall or walls would revert the field to a nonreflective environment, the limit of total absorption ($\alpha = 1.0$) would be equivalent to $\Lambda = 1.0$. Similarly,

TABLE 4.3

Modified Directivity Factors with Similar Walls

Number of Walls	$\alpha = 0$	$\alpha = 0.5$	$\alpha = 1.0$
0	1	1.0	1
1	2	1.5	1
2	4	2.5	1
3	8	4.5	1

the total reflection value $\alpha = 0.0$ uses the Λ values of 2, 4, and 8. When α is "in between," the Λ values must be reduced to a modified form. The Λ value for a partially absorbing wall can be written as $(2 - \alpha)$. For a two-wall configuration, the factor becomes $\Lambda = (2 - \alpha_1)(2 - \alpha_2)$, where the subscript refers to the properties of each wall. For a three-wall layout, $\Lambda = (2 - \alpha_1)(2 - \alpha_2)(2 - \alpha_3)$ if all three α values are distinct. For walls with the same absorption coefficients, the results simplify accordingly. Examples are given for similar walls in Table 4.3.

If, for example, the three walls have distinct α values of 0.2, 0.4, and 0.6, respectively, then the directivity factor would be $\Lambda = (2 - 0.2)(2 - 0.4)(2 - 0.6) = 4.03$. Values of α depend on the wall surface properties and the frequency content of the sound waves striking the surface. Low frequencies may be absorbed to a lesser or greater degree than high frequencies by a given surface. A thorough treatment of absorption or reflection thus requires that one should not only know the L_W values but also know the details of spectral distribution, as well as the detailed frequency-dependent values of α. For a careful treatment of the details, see *Fundamentals of Noise Control* (Thumann, 1990) or *Noise and Vibration Control Engineering* (Beranek and Ver, 1992).

Rough estimates can be made somewhat cautiously and will be used here to approximate the trends and overall influence of absorption. Thumann presents a range of α values for a variety of surfaces; for example, glazed brick and carpeted flooring are characterized by the values shown in Table 4.4.

One can use an averaged value or the value near the mid-frequency range, say at 1000 Hz, to characterize the material. Thus, one might assign brick the value $\alpha = 0.04$ or carpet the value $\alpha = 0.37$. Similar nonrigorous values of α for several materials are given in Table 4.5.

TABLE 4.4

Sample α Values at Octave Band Center Frequencies

Frequency	125	250	500	1000	2000	4000
Brick, α	0.03	0.03	0.03	0.04	0.05	0.07
Carpet, α	0.02	0.06	0.14	0.37	0.60	0.65

Source: Thumann, A., 1990, *Fundamentals of Noise Control Engineering*, 2nd ed., Fairmont Press, Atlanta.

TABLE 4.5

Approximate α Values

Material	Overall α value
Brick	0.04
Smooth concrete	0.07
Rough concrete	0.30
Asphalt	0.03
Wood	0.10
Heavy carpet	0.37
Carpet with foam rubber pad	0.69
Plywood panels	0.10
Plaster	0.04

As a more complex example, consider the noise level from a vane axial fan in terms of the sound pressure level at a distance of 50 ft from the fan. The fan is resting on a large rough-surfaced concrete pad. Fan diameter is 7 ft, speed is 875 rpm, flow rate is 100,000 cfm, pressure rise is 3.6 in. wg (static; 4.0 in. wg total), and the fan has 11 blades. Using the simple methods, one can estimate the sound power of the source and the decay to sound pressure level 50 ft away. Sound power is given by

$$L_W = \frac{72}{Ds^{0.8}} + 10\log_{10}(Q) + 20\log_{10}(\Delta p_s). \quad (4.12)$$

For a vane axial fan, D_s is about 1.7. The resulting sound power level for the fan is 108 dB. Using the approximate value of $\alpha = 0.3$ for rough concrete gives $\Lambda_{mod} = 1.7$. Then the L_p calculation yields the approximate value of $L_p = 76$ dB at 50 ft from the fan.

Thumann gives the octave bandwidth values for α for rough concrete, and one can estimate the octave bandwidth power levels of the fan using the full Graham method. Table 4.6 summarizes the details of the calculation. $L_W - K_W$ is simply $10\log_{10} Q + 20\log_{10} \Delta p_s$, and bpf is the blade pass frequency increment in the octave containing the frequency (NB × rpm/60). The octave

TABLE 4.6

Detailed Bandwidth Calculations

Frequency (Hz)	63	125	250	500	1000	2000	4000	8000
α	0.25	0.36	0.44	0.31	0.29	0.39	0.25	0.12
K_W	42	39	41	42	40	37	35	25
$L_W - K_W$	61	61	61	61	61	61	61	61
bpf	0	6	0	0	0	0	0	0
Λ_{mod}	1.75	1.64	1.56	1.69	1.71	1.61	1.79	1.88
L_{p-oct}	60.3	63.0	59.3	59.3	58.4	55.1	53.6	43.8

increments are summed according to $L_p = 10\log_{10}(\Sigma 10^{L_{p-\mathrm{oct}/10}})$ to yield $L_p = 78\,\mathrm{dB}$. The approximate result of 76 dB compares favorably with this more detailed, more definitive result.

4.6 Indoor Propagation

When considering the propagation or decay of noise inside a room, the question of the influence of numerous wave reflections and reverberation arises. One can account for these effects through the use of a reflective "room constant" factor, R. (One could call it a "wall factor," but room constant is the accepted term.) R is based on the total surface area of the walls, S, and a coefficient of absorption as considered earlier. The same complexities of frequency dependence pertain as before, and the different walls can be made of different materials as well. Keeping things relatively simple for now, and assuming the surfaces to be at least very similar, R is defined as

$$R = \frac{S\alpha}{1-\alpha}. \tag{4.13}$$

The pressure level equation is subsequently modified to the form

$$L_p = L_w + 10\log_{10}\left[\frac{1}{4\pi x^2} + \frac{4}{R}\right] + c. \tag{4.14}$$

If, for example, the walls are totally absorptive, R approaches infinity and the "correction" disappears from the equation as expected. Propagation or decay is calculated exactly as in a free field with $\Lambda = 1$. For perfectly reflective walls, the value of α is 0.0 and R becomes 0, implying that the sound pressure will build indefinitely in the room. Fortunately, no materials are totally reflective, so the infinite limit is not seen in practice.

Consider the problem of a large room $50 \times 100 \times 300$ ft having a 6-ft-diameter fan mounted on one of the 50×100 ft walls. The fan has a sound power level of $L_w = 105\,\mathrm{dB}$. It is necessary to estimate the sound pressure at the center of the floor of the room where there is a workstation. The walls and floor are made of wood and one can use an average value of α for plywood panels of 0.10. Surface area S calculates to $S = 100{,}000\,\mathrm{ft}^2$, so that

$$R = \frac{S\alpha}{1-\alpha} = 11,111\,\mathrm{ft}^2.$$

With the fan mounted in an end wall far from the adjacent walls, one approximates, as described earlier, $\Lambda = 2 - \alpha = 1.9$. Then, calculate L_p as

$$L_p = 105 + 10\log_{10}\left(\frac{1.9}{4\pi \times 152^2} + \frac{4}{11{,}111}\right) + 10.4 = 81\,\mathrm{dB}.$$

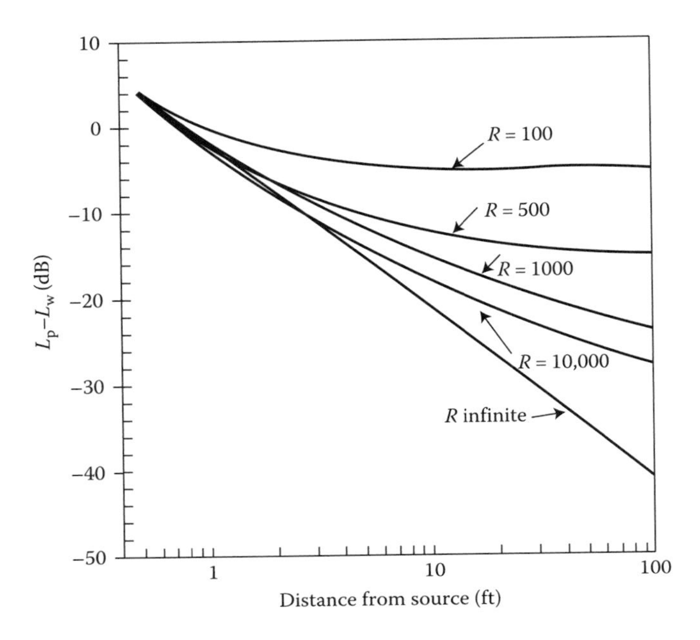

FIGURE 4.4 Regions of free field and far field. (From Thumann, A., 1990, *Fundamentals of Noise Control Engineering*, 2nd ed., Fairmont Press, Atlanta. With permission.)

Harder walls made of finished plaster over concrete with $\alpha = 0.04$ yield $R = 4167$ and $L_p = 85$ dB. For comparison, fully absorptive surfaces would yield $L_p = 61$. A smaller room $30 \times 20 \times 10$ ft with the same fan installed would yield $L_p = 101$ dB with plywood walls and $L_p = 108$ dB with plastered concrete walls. Clearly, the reverberation effects can strongly influence the sound pressure results for a given fan or blower.

In general, if the dimensions of a room and the values for the room surfaces are known, a simple calculation of R allows a convenient calculation of the influence of the installation of a fan or blower. Figure 4.4, taken from Thumann (1990), allows a quick estimation of the decay of L_p with distance from the source as $(L_p - L_w)$ versus x (in feet) with R as the parameter in ft^2. As seen in Figure 4.4, for small values of R, the room is largely reverberant and not much decay is evident. That is, the acoustic results are dominated by the high reflectivity of the room surfaces (low absorptivity). For example, with $R = 100$ ft^2, any position greater than about 10 ft from the source will see very little additional decay of the signal. For $R = 500$ ft^2 (decreased reflectivity), a gradual decay continues until about $x = 100$ ft, where the decay has reached about 15 dB. These numbers can be used to determine required room properties for a given constraint on L_p if L_w and size are known.

Consider an example in which a room has dimensions of $12 \times 36 \times 120$ ft, so that the total surface area is $S = 10,440$ ft^2. With a sound source in the center of the ceiling (an exhaust fan) of strength $L_W = 86$ dB, the constraint on noise requires that the L_P value near an end wall be less than 60 dB. This constraint is for intelligible normal speech at 3 ft. Since the fan is exposing only its inlet side to the room, one assumes that only half of the total sound power is propagated into the room so that $L_W = 83$ dB, and the requirement is for $(L_P - L_W) = -23$ dB. Given that $S = 10,440\ \text{ft}^2$ and the distance from the source is about 60 ft, a quick check on the free field noise shows that the maximum attenuation is given by $L_P = 83 + 10\log_{10}(1/(4\pi \times 3600)) + 10.4 = 46$ dB. This value is below the allowable value of 60 dB. For stronger noise sources, greater than 97 dB, the lowest achievable levels—even in an anechoic room ($\alpha = 1.0$)—will be greater than the required 60 dB. The criterion could not be met, and the choices would be to select a fan with L_W less than 97 dB or to install a muffler or silencer on the fan as a first step toward meeting the design criterion.

For the 83-dB fan, the $(L_P - L_W)$ requirement is -23 dB at $x = 60$ ft. Figure 4.4 shows that the room constant must be about $10{,}000\ \text{ft}^2$ or better. Since $\alpha = (R/S)/(1 + R/S)$, one requires a room whose surface properties yield a value of nearly $\alpha = 0.5$. This probably implies the need for extensive "acoustical treatment" of all room surfaces, but—at some cost—the goal can be achieved. As implied, the criterion for the example might best be interpreted as a limitation on the allowable sound power of the fan (or other device), particularly if the room structure is given. For rough concrete walls ($\alpha = 0.3$), a paneled ceiling ($\alpha = 0.1$), and a carpeted floor ($\alpha = 0.37$), then a weighted absorption value can be estimated according to

$$\alpha = \sum \frac{\alpha_{surf} S_{surf}}{S}, \tag{4.15}$$

where the "surf" subscript implies summing the values for each of the six surfaces of the room. For this example, the calculation is

$$\alpha = \frac{0.3(2 \times 12 \times 36) + (2 \times 12 \times 120) + 0.37(36 \times 120) + 0.1(36 \times 120)}{10{,}440}$$
$$= 0.30.$$

The room constant becomes $R = 0.30 \times 10{,}440/0.7 = 4474$. Figure 4.4 indicates that the attenuation at 60 ft from the fan will be about 20 dB. The allowable fan source noise in the room is thus limited to about 80 dB. One must select a fan with $L_W = 83$ dB or less.

If dealing with a fan that has a sound power level above 83 dB, then as mentioned before, one has to absorb or attenuate some of the sound power at the source itself, that is, the fan. An example of a fan inlet/outlet silencer for a series of high-pressure blowers is shown in Figure 4.5. Table 4.7 shows the attenuation values for each octave band. Using an estimate of the sound pressure broken down in octave bands allows these silencer values to be subtracted from the fan values to arrive at a new estimate of the source sound

FIGURE 4.5 Pressure blower and silencer. (From New York Blower, 1986, *Catalog on Fans*, New York Blower Co. With permission.)

power level. Table 4.8 summarizes the calculations necessary to estimate the noise reduction achievable using a silencer from Table 4.7 on a pressure blower providing 3450 cfm (inlet density) at 56 in. wg pressure rise. One can estimate the specific diameter at about 3, with a specific speed of 0.7. The corresponding efficiency is approximately 0.90 based on the Cordier line. Using the high specific diameter floor to predict specific sound power level gives $K_W \approx 40$ dB. The contributions of flow and pressure rise give another 70 dB, yielding $L_W \approx 110$ dB. However, to use the silencer attenuation values, the octave band breakdown for the sound power levels is required. These are shown in Table 4.8 along with the Size-10 silencer values. (Note that the specific diameter provides an estimate on a tip diameter of 26 in.; with $d/D = 0.4$, the inlet diameter should be about 10 in. or a nominal Size-10.) The overall value

TABLE 4.7

Silencer Attenuation Values for Several Sizes

Size	63 Hz	125 Hz	250 Hz	500 Hz	1000 Hz	2000 Hz	4000 Hz	8000 Hz
4	4	18	26	34	37	30	23	21
6	2	14	23	32	34	29	25	23
8	1	11	21	30	31	29	26	25
10	2	14	23	32	31	28	25	24
12	1	11	24	33	32	28	25	24

Source: From New York Blower, 1986, *Catalog on Fans*, New York Blower Co. With permission.

TABLE 4.8

Calculations for Silencer with High-Pressure Blower

Frequency (Hz)	63	125	250	500	1000	2000	4000	8000
K_W	35	35	34	32	31	26	18	10
$L_W - K_W$	70	70	70	70	70	70	70	70
bpf	0	0	0	0	3	0	0	0
$L_{W-\text{oct}}$	105	105	104	102	104	96	88	80
$I.L.$	2	14	23	32	31	28	25	24
L_W net	103	91	81	70	73	68	63	56

for the fan sound power level (row 4, Table 4.8) is estimated as $L_W = 111$ dB. From the net L_W values, the overall sound power level is 103 dB, a reduction of about 8 dB. This reduction could be nearly doubled by adding another Size-10 silencer in series with the first, yielding a source noise of about 95 dB. The process of adding more silencers becomes self-defeating in terms of size and bulk, cost, added flow resistance, and the flow-induced self-noise of the silencers. Figure 4.5 shows the pressure blower and the silencer used in this example.

4.7 A Note on Pump Noise

Noise in pumps falls into several categories (Karassik et al., 2008). The first is the fluid-borne noise as it propagates in the pump environment; it is considered primarily a problem similar to the propagation of fan noise mentioned earlier in the chapter. Given a suitable estimate or manufacturer's rating for a given pump, one can treat this noise as if it were fan noise. Potential liquid-generated noise sources include flow separation, turbulence, and impeller interaction with the cutwater of the volute, flashing, and cavitation. The broadband noise of a pump generally arises from the turbulent flow over blades and volute surfaces and is worst at high velocities. Discrete frequency noise is generally associated with the blade passage frequency and the cutwater interaction and is worst under conditions of very high head requirement. Noise associated with cavitation in the eye or impeller will be perceived as intermittent high-intensity noise at relatively high frequency, typically heard as a "snapping" or "crackling" noise. Its appearance in the spectrum of a pump may be used as a means of monitoring the flow conditions and the absence or presence of harmful degradation of the pump impeller and potential loss of performance, or even the pump itself.

A typical pump noise spectrum is shown in Figure 4.6, in which the tonal spikes are clearly seen. As in all turbomachines, the sound power will be a strong function of the impeller speed and the correct selection of pump at the proper design performance. Noise mitigation techniques, as discussed by Karassik et al. (2008), include isolation, interdiction of propagation or

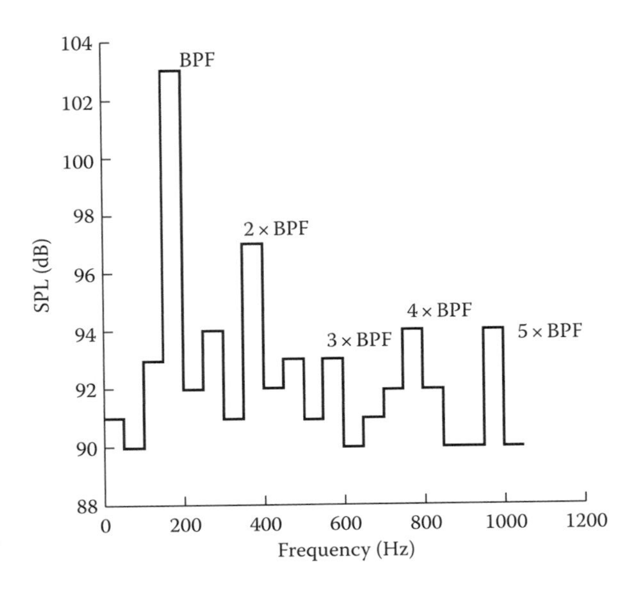

FIGURE 4.6 Pump noise spectrum. (From Karassik, I. J. et al., 2008, *Pump Handbook*, 4th ed., McGraw-Hill, New York. With permission.)

transmission path, reduction of speed or change of speed to avoid resonance conditions, and avoiding cavitation. This can be achieved by an increase in backpressure or injection of air or another gas (perhaps argon) into the eye of the pump, if possible. Where the hydrodynamic interaction of the cutwater and the pump blades is a major noise source, the level of the noise generated can frequently be reduced by modifying the shape or location of the cutwater (Karassik et al., 2008). The distance of the blade tips from the cutwater can be increased, or the cutwater can be made more rounded or shaped to avoid being parallel with the blade trailing edges. Such changes can also reduce the performance level of the pump while alleviating the noise problem.

4.8 Compressor and Turbine Noise

A modest review of gas turbine engine noise, with emphasis on the high-speed axial compressor component, is available in Richards and Mead (1968) and in various AIAA and ASME journals. The fundamental acoustic properties of these machines are characterized by blade pass frequencies and pure tones at much higher frequencies than are typical of fans and blowers, due to both high rotating speeds and large numbers of rotating blades. The older turbo-jet engines were clearly characterized by strong radiation of compressor noise from the inlet and very strong turbulent jet noise from the discharge of the engine. A frequency spectrum typical of the older design practice is given

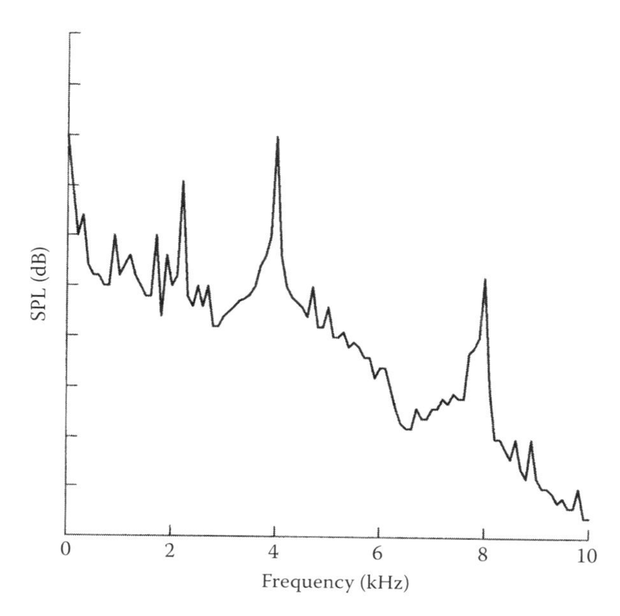

FIGURE 4.7 Noise spectrum of a turbo-jet engine. (From Cumpsty, N. A., 1977, *ASME Journal of Fluids Engineering*. With permission.)

in Figure 4.7 (taken from Cumpsty, 1977). Note the very strong magnitudes concentrated in the blade pass and higher harmonic spikes associated with the compressor. Excellent review papers are available (Cumpsty, 1977; Verdon, 1993) and serve as the primary source for this brief section on engine noise.

Considerable work during the 1960s and 1970s went into the reduction and control of these high noise levels as the basic form of the engines continued to evolve. As engines moved to higher transonic Mach numbers in the compressor rotors, the basic acoustic signatures changed to include the very characteristic multiple pure tone (mpt) noise at frequencies below blade pass or even the rotating speed of the compressor. This behavior, which is related to formation and propagation of shock waves from the compressor rotor, is quite distinctive, and a more contemporary spectrum for a gas turbine engine is shown in Figure 4.8. This mpt component of noise, also known as "buzz-saw" noise, is amenable to duct-lining treatment in the engine intake, and suppression of the component has been achieved (Morfey and Fisher, 1970).

The broadband noise components associated with compressors are more closely related to fan and blower noise as treated earlier in the chapter. Cumpsty shows the clear relation between the diffusion level in a blade row (see Chapters 6 and 8) and the magnitude of the broadband noise component, as shown in Figure 4.9 (as developed by Burdsall and Urban, 1973). While the "self-noise" of compressors is said to depend on locally unsteady flow on the blade surfaces (Wright, 1976; Cumpsty, 1977), it may also be strongly affected by the ingestion of turbulence into a moving blade row (Sharland,

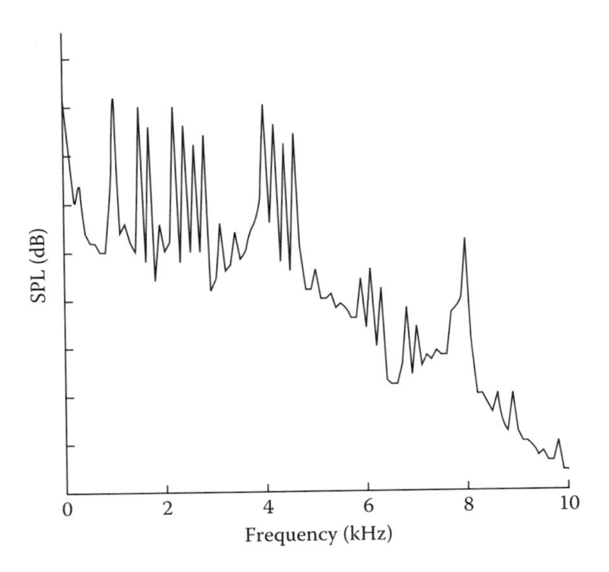

FIGURE 4.8 Noise spectrum at higher Mach number with mpt noise. (From Cumpsty, N. A., 1977, *ASME Journal of Fluids Engineering*. With permission.)

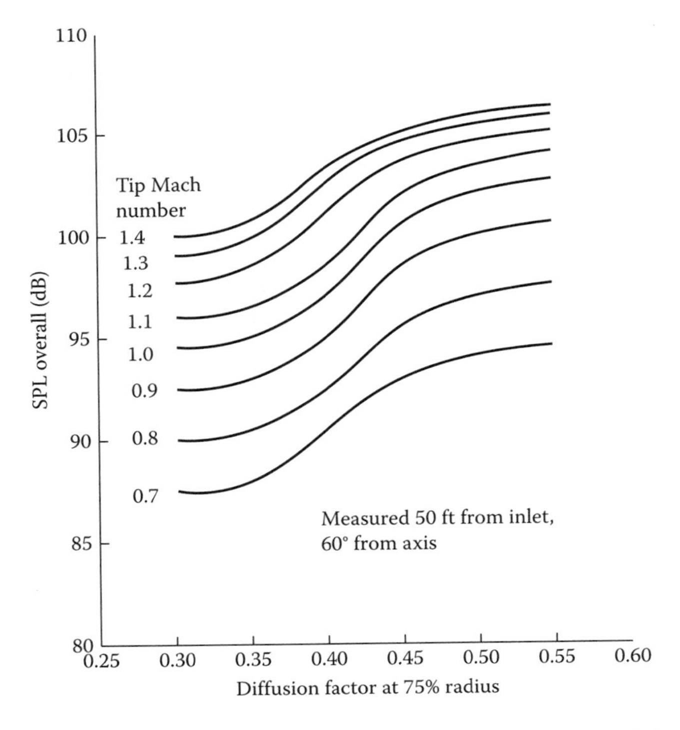

FIGURE 4.9 Broadband noise and blade row diffusion level. (From Cumpsty, N. A., 1977, *ASME Journal of Fluids Engineering*. With permission.)

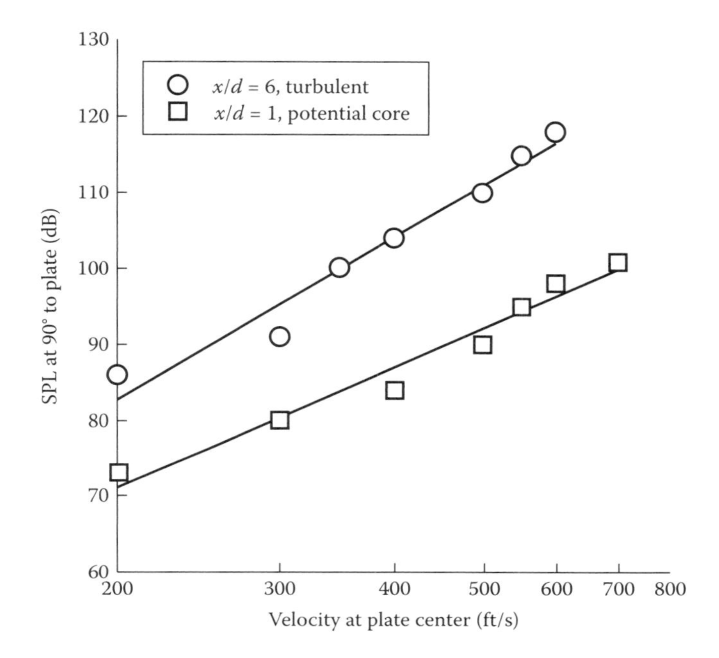

FIGURE 4.10 Turbulence ingestion influence on compressor noise. (From Sharland, C. J., 1964, *Journal of Sound and Vibration*, 1(3). With permission.)

1964; Hanson, 1974). Such influence is illustrated in Figure 4.10, in which an upstream flat plate disturbance was used to introduce turbulence systematically into a compressor blade row (Sharland, 1964). The results show a dramatic increase in the noise level.

The noise contribution associated with the turbine section of an engine has not been investigated or understood as well as the more dominant compressor and jet contributions to the spectra. Noise levels associated with the relatively low blade relative velocities in the turbine are usually masked by strong jet noise (in the older turbo-jet engines) and by aft propagation of fan and compressor noise in high bypass ratio engines. In addition, the downstream turbulence and flow masking have led to a "broadening" of the pure tone spikes of the turbine into less distinguishable "semibroadband" noise.

Along with the modification to the noise signatures of the engines caused by the high bypass turbo-fan developments, intensive efforts to reduce the high-speed jet noise component had been effective in reducing downstream propagated noise. An example of the performance of a noise suppression nozzle design is shown in Figure 4.11, with reductions of 10 dB or more when compared to a conventional nozzle.

Trends in the 1980s and 1990s developed along the lines of numerical computation and theoretical prediction of aeroacoustic performance of turbomachines. These methods (Verdon, 1993) seek to accurately cover a wide

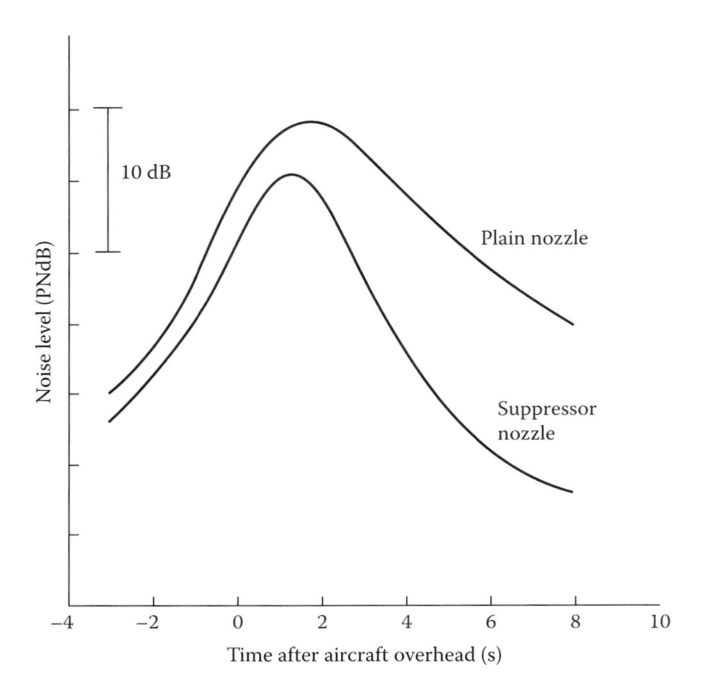

FIGURE 4.11 Nozzle noise suppression performance. (From Richards, E. J. and Mead, D. J., 1968, *Noise and Acoustic Fatigue in Aeronautics*, Wiley, New York. With permission.)

range of geometries and operating conditions, leading to a broad and very difficult task, demanding great efforts toward minimization of computer storage requirements and highly time-efficient solution algorithms. Verdon characterized this work (in 1993) as being developmental, with actual design relying on the experimental information and the earlier analytical methods of the 1970s and 1980s.

The fully viscous, unsteady methods for numerical solutions of the aerodynamics and aeroacoustics, which attempt to drive an unsteady solution to a converged periodic result, offer the greatest rigor but require tremendous computational resources as well. The inviscid linear or linearized analyses are still being developed and are capable of close agreement with the more complete approach to such solutions. Of course, the simpler methods remain somewhat more limited in applicability (Verdon, 1993).

There appears to be no real question as to the continued development and application of the modern numerical methods to the problems of turbomachinery acoustics. However, the improved state of experimental measurement will surely lead to an improved and needed experimental database to complement the advances in numerical predictive capabilities.

4.9 Summary

The topic of noise associated with turbomachines was introduced in a fairly basic, conceptual form. The most fundamental concepts of sound wave propagation and the idea of a spectrum or frequency distribution of sound or noise were discussed. These ideas were extended to the free field propagation and diminution of the sound pressure level or intensity of the sound at increasing distances from the noise source. Propagation in the free field was conceptually extended to include the influence of reflective surfaces near the noise source.

The quantitative physical scales of acoustic power and pressure were introduced in terms of reference quantities and decibel levels in power and pressure. Illustrations were provided to introduce decibel levels that are typical of familiar environments in an attempt to develop a "feel" or sense of the loudness of a given noise level.

Fan and blower noise were used as the primary vehicles to discuss and quantify the noise characteristics of turbomachines. The concepts of broadband (wide spectrum) and pure tone (narrow spectrum or spike) noise components were developed in terms of the physical behavior and characteristics of the machines. The Graham method was introduced as a semiempirical procedure for estimating the overall sound power level of both fans and blowers and means for estimating the spectral distribution of the sound power as well. This well-established model was modified to simplify calculation and to unify the concept of specific sound power of a machine as a simple function of specific diameter. Subject to inherent limits in accuracy and some loss in detail, this "modified Graham method" was used in examples to illustrate the process of designing for low noise levels.

Brief sections were provided in the chapter to extend the discussion of turbomachinery noise to pumps, compressors, and turbines. The references provided in the sections may serve as a beginning point for more thorough treatment of these topics as needed by the reader. Finally, the subject of numerical analysis of the fluid mechanics and acoustics in turbomachinery was mentioned briefly and is mentioned again in Chapter 10, along with a few references on the subject.

EXERCISE PROBLEMS

4.1. Estimate the sound power level for the fans described in Problem 3.5—both the model and the prototype. Develop the octave bandwidth data for both.

4.2. Estimate L_W for the fan of Problem 3.3 using the modified or simplified Graham method. Use the Cordier approach to size this fan and re-estimate the noise using the very simple tip speed model, $L_W = 55 \log_{10} U_t - 24$. Compare with the result using the Graham method.

4.3. From the information given in Problem 2.4, calculate the noise level of the wind tunnel fan (L_W). Compare this result to the sound pressure levels shown in Table 4.2. Using appropriate techniques, try to minimize the value of L_P at a distance of 60 ft from the fan inlet (assume a free field).

4.4. Use the results from Problem 2.5 to estimate the noise level of the compressor. Use the simplified Graham method for overall L_W.

4.5. From your solution of Problem 2.6, estimate the L_W values at the original speed. Use the tip speed algorithm from Problem 4.2 to approximate the value of L_P at a distance of 30 ft from the fans (assume Λ=4). Compare these results to Table 4.2 values and discuss what measures (if any) should be taken for workers operating in the environment of these fans at distances greater than 30 ft.

4.6. A ducted axial fan is tested at 1750 rpm and delivers 6000 cfm of air at 4 in. wg (static).

a. Calculate the expected size and efficiency for this fan.

b. Estimate the power required for the fan motor.

c. Calculate the sound power level for this fan using $K_W = 72/D_s^{0.8}$ in the simplified Graham method. Compare this result with an estimate using the Onset Velocity method, $L_W = 55 \log 10\, W_{1t} - 32$, where $W_{1t} = [U_t^2 + (Q/A)^2]^{1/2}$ is the inlet blade relative velocity at the blade tip. Also compute and compare a sound power level calculation based on the fan shaft power, P_{sh} (in hp), according to $L_W = 20 \log_{10} P_{sh} + 81$, and using a "speed-corrected" method based on $L_W = 20 \log_{10} P_{sh} + 81 + 4 \log_{10}(U_t/800)$ (U_t is the tip speed in ft/s). State all assumptions and simplifications.

4.7.. A laboratory room 16 × 10 × 8 ft high must be ventilated so that the air in the room is exchanged twice per minute. The resistance to the flow is a short exhaust duct with a set of screened louvers over a vent ($K = 1$). Choose a fan whose level of sound power is less than about 70 dB in order to allow for low speech interference levels within the lab.

4.8.. The laboratory room of Problem 4.7 has walls, ceiling, and floor treated with an absorbent padding roughly equivalent to heavy carpeting. Estimate the sound pressure level at the room center ($x = 8$ ft) if the fan is in an end wall. Compare the results to the criterion for speech interference in Table 4.2.

4.9.. A large forced draft fan ($D = 10$ ft) supplies cooling air to a steam condenser in an oil refinery. The inlet of the fan is open and faces the boundary of the plant property, where the sound pressure level cannot exceed 55 dB. The fan supplies 100,000 cfm at 8 in. wg static pressure rise and is sited against a smooth concrete wall on a smooth concrete pad. The inlet is 250 ft from the boundary. Will the fan inlet need a silencer?

4.10. A large mechanical draft fan supplies cooling air to a primary power distribution transformer in a suburban environment. The fan is a wide, double-width centrifugal type with the inlet of the fan open and facing the boundary of the utility property. At the boundary line, the fan sound pressure level must not exceed 58 dB. The fan supplies 400,000 cfm at 21 in. wg static pressure rise and is sited against a smooth concrete wall with deep, soft sod on the ground between the fan and the boundary line which is 65 ft away. Will the fan inlet need a silencer? If so, how much attenuation is required? Could an acoustic barrier or berm be used? (See Baranek or Thumann.) (Hint: Assume $\alpha = 0$ for the wall and $\alpha = 1.0$ for the sod. Also, size the double-width fan as a single-width fan with half the flow to establish D_S. Then double K_W, logarithmically of course.)

4.11. There have been many semiempirical scaling rules proposed for the estimation of sound power level for fans and blowers, with many of them based on impeller tip speed. They usually take the form

$$L_W = C_1 \log_{10} U_t + C_2.$$

If the speed (rpm) is increased, then the change in the sound power level is given by

$$L_W = L_{W2} - L_{W1} = C_1 \log_{10} \left(\frac{U_{t2}}{U_{t1}} \right).$$

Use the Graham method to show that the C_1 value consistent with his correlations is $C_1 = 50$ for N-scaling. That is, restrict the study to variable N and constant D.

4.12. Pursue the same exploration as was proposed in Problem 4.11, but examine the influence of diameter variation with fixed speed (rpm). What value of C_1 makes sense in the context of Graham's model? How can you reconcile this result and the result of Problem 4.11 with the typical empirical values of C_1 in the range $50 < C_1 < 60$?

4.13. Select a fan that can supply air according to the following specifications: $Q = 6\,\text{m}^3/\text{s}$, $\Delta p_S = 1\,\text{kPa}$, and $\rho = 1.21\,\text{kg/m}^3$ with the design constraint that $L_W \leq 95\,\text{dB}$. Choose the smallest fan that can meet the specifications and the constraint on sound power level. Use a Cordier analysis, not a catalog search.

4.14. On large locomotives, axial fans are used to supply cooling air to a bank of heat exchangers. Two of these fans draw ambient air through louvers in the side panels of the locomotive. Each fan draws $5.75\,\text{m}^3/\text{s}$ of air at 28°C and 98.0 kPa and must supply a total pressure ratio of $p_{02}/p_{01} = 1.050$ ($\Delta p_T = 5\,\text{kPa}$).

a. If locomotive design requirements impose a diameter limit on the fans of $D = 0.6\,\text{m}$, calculate the minimum speed required and estimate the sound power level (L_W).

b. If the allowable level of sound pressure level (L_p) is 55 dB 15 m to the side of the locomotive, is noise attenuation treatment needed? How much?

c. Evaluate the need to include compressibility effects on the fan design and selection.

4.15. Calculate the reduction in noise available for the locomotive fans of Problem 4.14 if the constraint on size can be relaxed. Examine the range from 1 m up to 2 m diameter for the stated performance requirements. Develop a curve of L_W and L_P for the stated conditions.

4.16. A small axial flow fan runs at 1480 rpm to supply a flow rate of $3\,m^3/s$ of air with a density of $1.21\,kg/m^3$ and a static pressure rise of 100 mm of water.

a. Determine a proper fan size, efficiency, and power requirement. Adjust the Cordier efficiency estimate for the Reynolds number.

b. Estimate the value of L_W for the fan using the modified Graham method.

c. Estimate the sound pressure level, L_P, for an open inlet configuration propagating over water. Develop a curve of L_P versus x up to 100 m.

4.17. An axial flow cooling tower fan has a diameter of 32 ft and is a tube axial type. It provides 1,250,000 cfm at a total pressure rise of 1.0 in. wg with an air weight density of $0.075\,lbf/ft^3$.

a. Estimate the sound power level, L_W, of the fan using the modified Graham method and the tip speed method (see Problems 4.1 and 4.11).

b. Calculate the sound pressure level, L_P, at a distance of 110 ft from the fan inlet (assume free field propagation).

4.18. A large cooling tower fan has a diameter of 10 m and is a tube axial type. It provides $600\,m^3/s$ at a total pressure rise of 500 Pa with an air weight density of $10.0\,N/m^3$.

a. Estimate the sound power level, L_W, of the fan using the modified Graham method and the tip speed method.

b. Calculate the sound pressure level, L_P, at a distance of 50 m from the fan inlet (assume free field propagation).

4.19. Railway locomotives require air blowers to assist the heat transfer from the coolant "radiators" of the very large diesel engines powering these vehicles. Typical performance numbers for these blowers (there are two for each of the engines) are $Q = 3\,m^3/s$ and $\Delta p_S = 15\,kPa$ with $\rho = 1.2\,kg/m^3$.

a. Estimate the probable unattenuated sound power level of one of these blowers. Assume a centrifugal single-width blower of a fairly high specific diameter, say $D_S = 2.4$.

b. Estimate the combined sound power level of all four blowers.

c. Calculate the sound pressure level at 15 m from the side of the locomotive, over soft grassy earth, with only two of these radiator blowers in operation (i.e., assume that the two radiators on one side of the locomotive are the dominant sound power source).

4.20. To reduce the space requirements of the fans for the locomotive engine cooling application studied in Problem 4.19, compact three-stage axial fans can replace the centrifugal blowers used earlier.

a. Using a specific speed near the high range for axial fans, say $N_S = 3$, determine the speed and diameter needed.
b. Estimate the sound power level for one stage.
c. Estimate the combined power level for the multistage fan.
d. Finally, estimate the overall sound pressure level of all four of these multistage fans at the 15-m evaluation distance of Problem 4.19 and compare it with the specified maximum level of Problem 4.14 (55 dB).

4.21. Cooling tower fans such as the one studied in Problem 4.18 are frequently used in much smaller sizes in adjacent “cells” of several fan-tower units operating in parallel. Typical size will be of the order of 3-m diameter with groups of 5, 10, 20, or more of the smaller fans and tower units. Consider a grouping of 20 fans set in a straight line. Each fan supplies 25 m^3/s of air with a static pressure rise of 150 Pa at $\rho = 1.02\ kg/m^3$.

a. Estimate the sound power level of one such fan.
b. Estimate the combined sound power level of all 20 fans.

4.22. For the fans described in Problem 4.21, with a center-to-center spacing of the fans of 4 m

a. Estimate the sound pressure level from the line of fans, at a position situated 20 m from the 10th fan, measured perpendicular to the line.
b. Use line source theory for the arrangement of fans to provide an estimate of the sound pressure of part (a). (Hint: Refer to other references on noise propagation such as Beranek and Ver, 1992; Thumann, 1990.)
c. From a distance of 10 m from the line, approximately how much error would be incurred in using the line source approximation on the group of 20 fans?

4.23. For fan or blower performance stated in SI units (m^3/s and Pa), show that the Graham equation for sound power level can be modified to

$$L_W = K_W + 10 \log_{10} Q + 20 \log_{10} \Delta p_S - 14.6,$$

with $K_W = f(D_S)$ as before.

4.24. An axial cooling tower fan has a diameter of 18 m and is a tube axial type. It provides a volume flow rate of 4000 m^3/s at a total pressure rise of 350 Pa with an air density of 1.10 kg/m^3. Estimate the sound power level of the fan using the Graham method and the Onset Velocity method (Problem 4.6). Estimate the sound pressure level 120 m from the fan. Assume a ground plane absorption coefficient of 0.25.

4.25. A small woodworking shop has a floor plan area 6 m by 9 m with a 3 m ceiling. A recirculation fan and filter system is placed at one end of the shop for removing dust and volatile vapors. To control the sound pressure level in the work areas, the reverberant properties of the room must be examined. Calculate the reduction in the fan noise level at the room center. That is, calculate $(L_P - L_W)$ for the following construction options:

   a. The floor is smooth concrete and the walls and ceiling are plywood paneling.

   b. The floor is smooth concrete, the ceiling is treated with an absorptive batting with $\alpha = 0.4$, and the walls are plywood paneling.

   c. The floor is smooth concrete, the ceiling is treated with an absorptive batting with $\alpha = 0.4$, and the walls are covered in heavy carpet.

   d. The concrete floor is covered with a heavy carpet, the ceiling is treated with an absorptive batting with $\alpha = 0.4$, and the walls are covered in heavy carpet with foam rubber pad.

4.26. The fan and filter system of Problem 4.25 must be able to recirculate and clean the entire volume of air in the shop in 5 min against a pressure drop in the filter of 2 kPa.

   a. Choose a suitable centrifugal fan for this performance (use $2 < D_S < 3$) and estimate the sound pressure level at the room center for the various conditions given in Problem 4.25.

   b. Repeat part (a) using an axial fan ($1.5 < D_S < 2$).

   c. Compare these results for the two types of fans to the limitations given in Table 4.1 for the OSHA 8 h limit, the Commercial Limit, and the 4 ft Speech Interference level.

4.27. A low specific speed fan is used to provide tertiary combustion air to provide control of the formation of oxides of nitrogen in a coal-fired boiler (using air-over control). Typical performance requirements for the forced draft fan will be $Q = 25\ \text{m}^3/\text{s}$ at $\Delta p_S = 10\ \text{kPa}$ with a density of $\rho = 0.57\ \text{kg/m}^3$ ($T = 265°\text{C}$). The fan will run at 1500 rpm. Use the simple Graham method to estimate the sound power level of such a fan. (Hint: the value of D_S should lie well beyond the correlation of Figure 4.3. Use a "floor value" of $K_W = 40\ \text{dB}$ for a conservative estimate of L_W.)

4.28. For the fan of Problem 4.27, develop the octave bandwidth sound power distribution using the specific noise from Graham's data in Table 4.6. Assume that the fan has to be constructed with 16 backwardly inclined nonairfoil blades. Use this distribution to calculate the overall L_W value for the fan. Compare this with the approximate result from Problem 4.27.

4.29. The induced draft fans used in cement plants to filter hot, dust-laden air drawn from a lime kiln must operate in severe conditions. At the inlet side of these fans, the hot, low-density air enters the fan at 320°F and barometric pressure (giving $\rho = 0.048\,\text{lbm/ft}^3 = 0.0015\,\text{slug/ft}^3$). Required flow is $Q = 250{,}000$ cfm at a pressure rise of $\Delta p_S = 25$ in. wg to overcome the resistance of the downstream baghouse filters. Because of the harsh operating conditions, the fan speed is limited to $N = 880$ rpm $= 92.1\,\text{s}^{-1}$. Estimate the sound power level, L_W, of the fan using the modified Graham method.

4.30. The induced draft fan of Problem 4.29 must meet a property boundary noise requirement of 59 dB (sound pressure level). Although the fan inlet draws air through the kiln, we assume that there is no attenuation through the kiln. The ground between the kiln and the property boundary is grassy turf with a mean absorptivity of $\alpha = 0.75$. How far must the fan and kiln be placed from the boundary to meet the noise constraint?

5

Performance Estimation, Machine Selection, and Preliminary Design

5.1 Preliminary Remarks

Thus far, emphasis has been placed on characterizing the performance of existing machines, including forms for performance curves and the extrapolation ("scaling") of available performance information for changes in size, speed, or fluid properties. Now suppose that what is known is a set of performance requirements for some application. For example, a heat exchanger may require a water flow of 10,000 gpm and an available head of 8 ft to overcome losses or a fan may be needed to supply 50 m^3/s of air with a noise level of 95 dB or less. In these cases, one wants to select a machine to carry out the required task, typically subject to one or more constraints (usually economic). In other cases, the task is to design a machine rather than select one for purchase from a supplier.

The beginning point in either selection or design is to determine the type of machine best suited for the task and to make preliminary estimates of speed, size, and perhaps other performance characteristics such as noise and cavitation characteristics. In almost all cases, estimates of efficiency (or power) are required so that an economical choice can be made. Because the design and manufacture of turbomachinery is a mature technology, the starting point for these investigations is almost always found in the data accumulated over the past several decades.

5.2 Cordier Diagram and Machine Type

In attempts to organize historical performance data, Cordier (1955), as well as many others (e.g., Balje, 1981), has found that machines of different types—axial flow, mixed flow, or radial flow—naturally group into different regions on the N_s–D_s diagram. On the left side (low D_s, high N_s), machines with low head rise and high flow rate are grouped. These are predominantly axial flow machines of various types. In the middle range of the chart, machines of

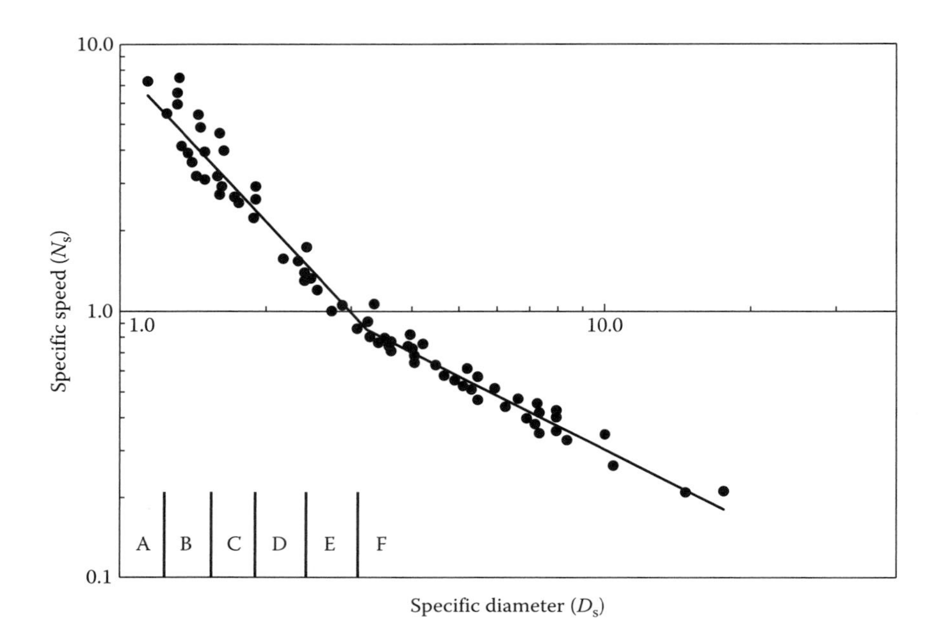

FIGURE 5.1 The Cordier diagram showing regions for different machine types.

moderate head and flow with a mixed flow path fall together. Those machines with small flow and large head rise group to the right side of the diagram, at low specific speed and high specific diameter. In Figure 5.1, this breakdown of regions of the diagram is refined somewhat beyond the initial observation to depict six regions of the Cordier diagram; each region corresponds more or less to a particular type of machine. In truth, the boundaries of each region are indistinct and there is a natural tendency for them to overlap. Nevertheless, this breakdown will help identify the application range, in flow and pressure, for different types of machines, whether they are called pumps, fans, blowers, compressors, or turbines.

In Region A, in a roughly defined range with $6 < N_s < 10$ and $0.95 < D_s < 1.25$, one finds propeller-type machines used for moving large quantities of fluid with very little change in specific energy. Ship and aircraft propellers, horizontal axis wind turbines, and light ventilation equipment such as floor, desk, and ceiling fans fall into this range. They are usually of open or unshrouded configuration. Figure 5.2 illustrates this type of machine. Note from the efficiency curve (Figure 2.9) that "Region A" machines also lie in a range of low efficiency, with $0.50 < \eta_T < 0.70$ for pumping machinery.

In Region B, the range is roughly $3 < N_s < 6$ and $1.25 < D_s < 1.65$. Here, one finds ducted axial flow equipment such as tube axial fans, vane axial fans with small hubs, axial flow pumps, or "inducers," (used to improve cavitation resistance) and shrouded horizontal axis wind turbines capable of somewhat

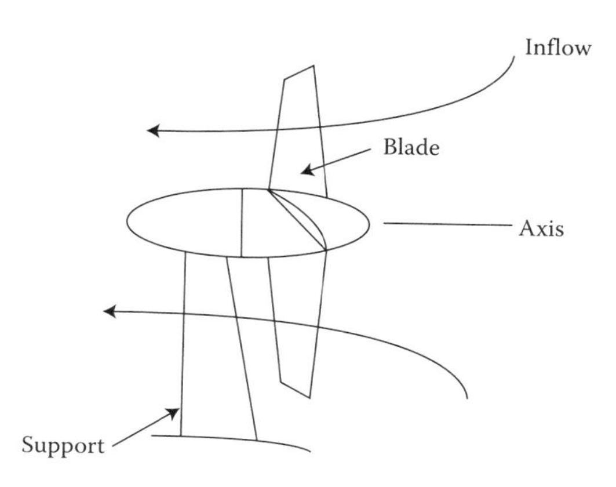

FIGURE 5.2 Open-flow axial fan, pump, or wind turbine (Region A; $6 < N_S < 10$ and $0.95 < D_S < 1.25$).

higher specific energy change (for a given diameter and speed). Ventilation applications such as attic fans, window fans, and air conditioner condenser fans are typical machines in this range. Axial flow pumps and shrouded propellers (air or water) also fall into this category. Figure 5.3 illustrates a machine of this type.

For Region C, the range is approximately $1.8 < N_s < 3$ and $1.65 < D_s < 2.2$. This is a region of "full-stage" axial equipment, as illustrated in Figure 5.4. The blade row operates enclosed in a duct or shroud. A pump, fan, or compressor usually has a set of diffuser vanes downstream of the blade row that straightens and recovers the energy in the swirling flow before the fluid leaves the machine. A turbine will have a set of nozzle vanes upstream of the rotor to accelerate and preswirl the flow. Typical applications include vane axial fans with fairly large hubs, bulb-type hydraulic turbines, and single-stage or multistage axial compressors, pumps, and vapor or gas turbines.

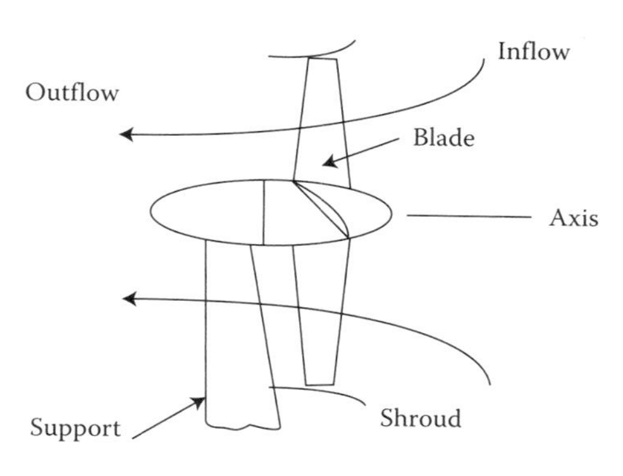

FIGURE 5.3 Ducted axial flow machine (Region B; $3 < N_S < 6$ and $1.25 < D_S < 1.65$).

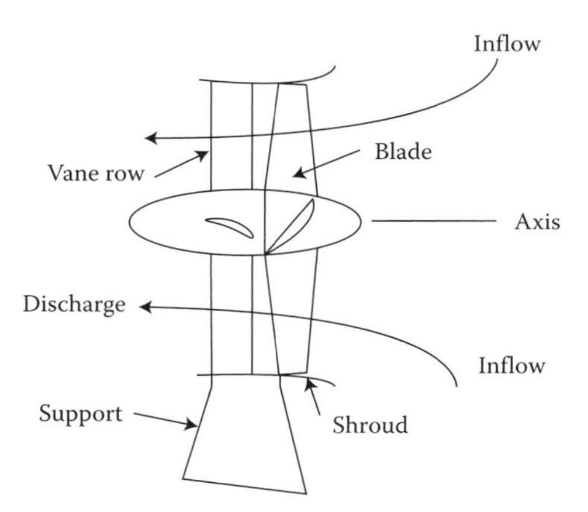

FIGURE 5.4 Full-stage axial flow machine (Region C; $1.8 < N_S < 3$ and $1.65 < D_S < 2.2$). Note stationary vane row.

Region D, for $1.0 < N_S < 1.8$ and $2.2 < D_S < 2.80$, is a "mixed flow" region. Figure 5.5a shows a "Region D" machine of primarily axial flow type, while Figure 5.5b shows a "Region D" machine of primarily radial flow type. Machines in this region include multistage quasi-axial flow equipment, mixed flow pumps and blowers, Francis-type mixed flow hydraulic turbines and pumps as well as centrifugal/radial flow pumps, compressors, or blowers with unusually wide blading at the exit of the impeller. Applications requiring higher pressure rise with a large flow rate will use this equipment. Heat exchanger applications in HVAC equipment, moderately low-head liquid pumping, and mechanical draft for combustion processes are typical.

Note the interface between Regions D and E; $0.7 < N_S < 2.0$ and $2.2 < D_S < 4.0$ is the region of inherently highest efficiency.

In Region E, with $0.70 < N_S < 1.0$ and $2.80 < D_S < 4$, radial discharge equipment dominates. Centrifugal fans of moderate width and centrifugal liquid pumps lie in this range. Heavy-duty blowers, compressors, and high-head turbines running at fairly high rpm are also typical. Blading will typically be of the backwardly inclined or backwardly curved type. Figure 5.6 illustrates equipment in this region.

Region F is characterized by fairly narrow centrifugal or radial flow equipment, with $N_S < 0.70$ and $D_S > 4$ (Figure 5.7). High-pressure blowers, centrifugal compressors, and high-head liquid pumps are typical. For small flow and very high head, one begins to see narrow radially oriented blades and even forward-curved blading. Efficiencies for machines with $N_S < 0.20$, $D_S > 10$ are quite low. Generally, positive displacement machines are preferable beyond these limits.

The association of a specific machine type with a region in "N_S–D_S space" is not completely rigorous. To complicate matters, there are some commercially available "hybrid" machines for special-purpose applications. These

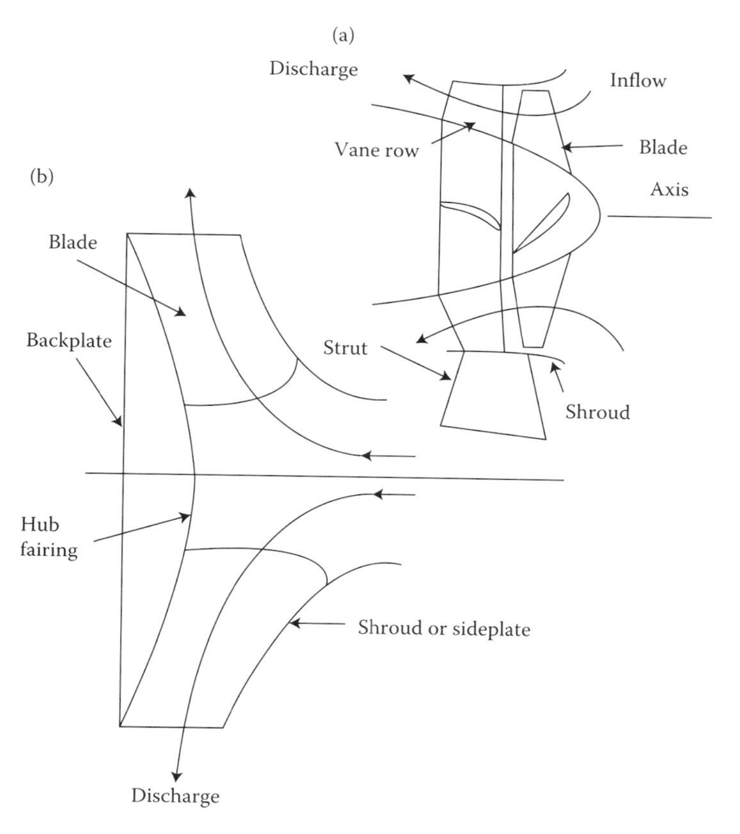

FIGURE 5.5 (a) Mixed flow (quasi-axial) machine (Region D; $1.0 < N_S < 1.8$ and $2.2 < D_S < 2.80$). (b) Mixed flow (quasi-radial) machine (Region D; $1.0 < N_S < 1.8$ and $2.2 < D_S < 2.80$).

include very wide centrifugal blowers with forward-curved blading, often called "squirrel-cage blowers," that are used as furnace fans or air conditioner evaporator fans. Radial discharge centrifugal machines, which operate in a cylindrical duct, an essentially axial flow path, also exist. These fans are often referred to as in-line or plug fans. Axial machines with a large ratio of hub to tip diameter ($d/D \sim 75$–90%) are capable of crowding into the "centrifugal range" of the diagram with relatively small flow rate and high pressure.

It is extremely important to note that the Cordier-type information given here applies only to single-stage and single-width configurations. Many centrifugal flow machines are available in the so-called double-inlet, double-width, or double-suction configurations and are found to operate well into the nominally axial range of the Cordier diagram when the total ("two-sided") flow is used to define N_S and D_S. Similarly, any number of axial-flow stages can be arranged in series in order to achieve an arbitrarily high specific energy change for a given flow (e.g., a multistage (20+) steam turbine or an axial compressor (15+ stages) for a jet engine).

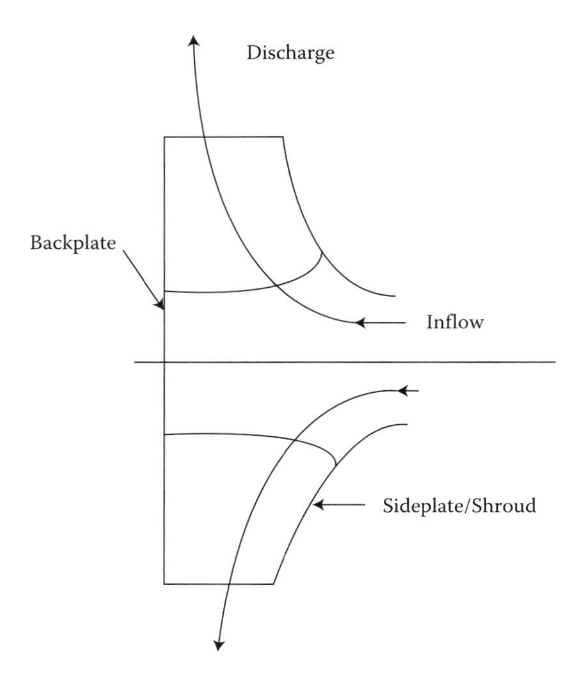

FIGURE 5.6 Radial discharge centrifugal machine (Region E; $0.70 < N_s < 1.0$ and $2.80 < D_s < 4$).

When it is necessary to select or design a machine to meet certain requirements in flow rate and specific energy (e.g., pressure) change, one can follow the overall trend prediction of the Cordier diagram to help decide what kind of machine will perform the job most efficiently. For an initial selection of speed and with known fluid density, one can estimate the specific speed. The N_s–D_s diagram (Figure 2.8 or 5.1) can be used to choose a value of D_s on or near the "Cordier line." This value of D_s can be used together with the specified flow and specific energy to determine a workable value for the machine diameter. This procedure will indicate both the size and the type of machine one should be considering and an upper bound on the efficiency one can reasonably expect to achieve. Correlations from Chapters 3 and 4 can yield estimates of noise generated and/or cavitation parameters as needed. It must be emphasized that all of these estimates pertain to the BEP.

To illustrate, suppose that one needs a fan to supply $Q = 5\,\text{m}^3/\text{s}$ with $\Delta p_T = 1250\,\text{Pa}$ when handling air with $\rho = 1.20\,\text{kg/m}^3$. To facilitate the process, calculate

$$\frac{Q^{1/2}}{(\Delta p_T/\rho)^{3/4}} = \frac{5^{1/2}}{(1250/1.20)^{3/4}} = 0.0122\,\text{s}$$

and

$$\frac{Q^{1/2}}{(\Delta p_T/\rho)^{1/4}} = \frac{5^{1/2}}{(1250/1.20)^{1/4}} = 0.394\,\text{m}.$$

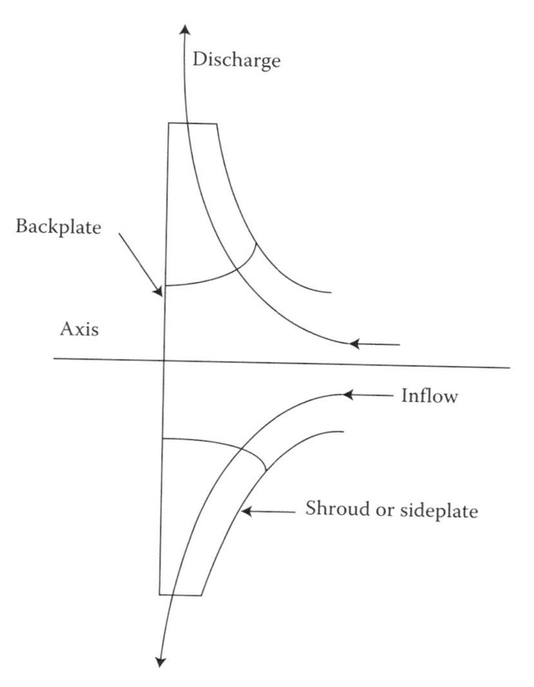

FIGURE 5.7 Narrow centrifugal/radial flow machine (Region F; $N_s < 0.70$ and $D_s > 4$).

If one chooses (arbitrarily, at this point) $N = 1800\,\text{rpm} = 188.5\,\text{s}^{-1}$, then $N_s = 0.0122 \times 188.5 = 2.30$. The Cordier diagram (or Equation 2.36) indicates $D_s \approx 1.94$, from which

$$D = D_s \left[\frac{Q^{1/2}}{(\Delta p_T/\rho)^{1/4}} \right] \approx 1.94 \times 0.394 = 0.763\,\text{m}.$$

From Figure 2.9, the maximum efficiency is no greater than 0.89 or 0.90 (in truth, very likely less). This combination of N_s and D_s lies in Region D (to the left edge) and suggests that the fan should probably be a mixed flow or a vane axial fan.

What if a lower speed is chosen? Let $N = 900\,\text{rpm}$. Now $N_s = 1.15$, with a corresponding D_s of approximately 2.77. This indicates a Region E fan; a radial flow or moderate centrifugal fan with a diameter of about (2.77×0.394) 1.1 m and an efficiency perhaps as high as 0.90 or 0.91. This fan would be (marginally) more efficient, but the diameter is quite large and the fan would be expensive to manufacture. What would happen if a higher speed, say 3600 rpm, were chosen?

The point is this: Using the Cordier diagram for guidance, one can very quickly examine some alternatives in machine type, size, and efficiency for a given performance requirement and begin to make decisions about selecting or perhaps designing a machine to perform the task.

5.3 Estimating the Efficiency

The actual selection of a machine from a manufacturer/vendor or, alternately, the selection of the exact design that will be put into production is nearly always based on factors in addition to the machine's ability to meet performance specifications. In the vast majority of situations, the choice is based strongly on economic considerations, including costs of acquisition, operation, and maintenance and costs associated with lost production if the machine fails to function. In some circumstances, considerations such as weight (in air/spacecraft applications) or noise are also important. In making such evaluations, reliable estimates of the efficiency are important, as a more efficient machine will be cheaper to operate and may well be lighter and quieter—although it may be more expensive to manufacture or purchase. Even before the selection or design decision phase, an accurate estimate of efficiency is required for fluid-dynamic design calculations, as will be shown in later chapters.

The expected (BEP) efficiency of turbomachines has been expressed as a function of specific speed (or alternately, specific diameter) in many ways over the course of many years. The Cordier efficiency diagram (Figure 2.9) provides correlations that apply to "good" (i.e., large and well-built) machines. This diagram provides an estimate of a reasonable upper limit on expected efficiency.

In practice, actual efficiency depends on many other things beside specific speed/specific diameter. A broadly defined list of other factors that are known to affect efficiency follows:

- Size (often measured by the flow rate)
- Cost (which is related to the level of manufacturing quality)
- Fluid properties (primarily viscosity)
- Direction of work transfer (is the machine a turbine or a pumping machine?).

One widely known correlation of efficiency that shows the effect of additional factors is a chart showing the expected efficiency of centrifugal pumps as a function of specific speed *and* size (flow rate). Such a chart is shown here as Figure 5.8. Note that the specific speed is given in "U.S. Customary" (quasi-dimensional) terms,

$$n_s = \frac{N\ (\text{rpm})\sqrt{Q\ (\text{gpm})}}{[H\ (\text{ft})]^{3/4}}.$$

Looking at a more "technical" list of factors influencing efficiency, which goes more to the detailed mechanisms leading to losses, one might consider

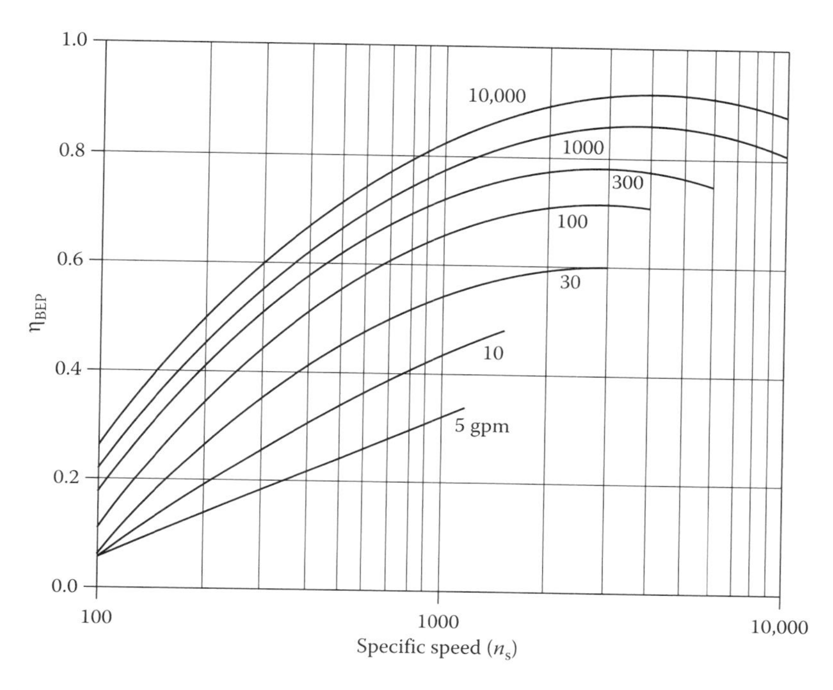

FIGURE 5.8 Expected (BEP) efficiency of centrifugal pumps as a function of specific speed and size. (After White, F. M. (2008), *Fluid Mechanics*, 6th ed., McGraw-Hill, New York; Karassik et al., (2008), *Pump Handbook*, 4th ed., McGraw-Hill, New York.)

- Reynolds number (incorporates size, speed, and viscosity)
- Surface roughness (related to cost/manufacturing quality)
- Running clearance between rotor and casing (cost/manufacturing)
- Manufacturing tolerances (rotor symmetry, blade uniformity)
- Fluid diffusion versus fluid acceleration.

Unfortunately, information on the quantitative effects of these factors is limited and of nonuniform quality. One possible approach would be to estimate efficiency by combining an estimate from the Cordier efficiency curve with a series of *derating factors* according to the following equation:

$$(1 - \eta_T) = (1 - \eta_{T,C}) \times F_{\eta,1} \times F_{\eta,2} \times F_{\eta,3} \times \cdots, \tag{5.1}$$

where $\eta_{T,C}$ is the "Cordier Efficiency" (from Figure 2.9 or Equations 2.38 and 2.39) and the F_ηs are the derating factors. Usually, one would expect the actual efficiency to be less than the Cordier value (i.e., the losses would be greater), so that $F_{\eta,i} \geq 1$; but it is possible that a very well-designed, very well-built machine could achieve a higher efficiency.

The issue is, of course, determining what to use for the F_ηs. A (very) limited list of possible choices is

- *Reynolds number derating factor*:

$$F_\eta \approx \left(\frac{10^7}{Re_D}\right)^{0.17}; \quad Re_D \leq 10^7 \text{ (Koch and Smith, 1976).} \tag{5.2}$$

Note that this is basically a "Moody-type" scaling law. Data indicate that effects of Reynolds number are negligible above $Re_D = 10^7$, so F_η is taken as 1.0 in that range.

- *Combined Reynolds number and surface roughness derating factor*:

$$F_\eta \approx 0.3 + 0.7\,\frac{f(Re_b, \epsilon/2b)}{f_{\text{Cordier}}}, \tag{5.3}$$

where b is the blade tip-width and f is the Darcy friction factor. This derating factor is obtained from Equation 3.18 what is being done is to assume that the "Cordier machine" is the model and the machine whose efficiency is to be estimated is the prototype. The obvious difficulty is on deciding what value to use for f_{Cordier}.

- *Fan radial clearance gap derating factor*:

$$F_\eta \approx 1 + 2.5\tanh(0.3[((c/D)/0.001) - 1]) \text{ (Wright, 1974, 1984c),} \tag{5.4}$$

where the clearance gap, c, is defined for axial fans and for radial fans in Figures 5.9a and 5.9b, respectively. Note that this derating factor has been validated only for fans.

A strong warning is in order. While the form of Equation 5.1 is plausible, the stated estimates for the derating factors should be used with great care mixed with a healthy degree of skepticism. One should not mix factors that represent essentially the same (broadly defined) effect—for example, for a pump, either the Reynolds number factor may be used *or* the size (flow) effect might be estimated from Figure 5.8, but not both. This list of factors is far from exhaustive and each has been validated against only limited data. In general, these factors should only be used if no better information is available. Often, an equally valid approach would be to simply lower the Cordier efficiency estimate by a value chosen from experience (e.g., $\eta_T \approx \eta_{T,C} - 0.20$ for an inexpensive fan). Alternatively, one might base selection and preliminary design decisions on the relative merits of the Cordier efficiencies, without putting much confidence in the numerical values.

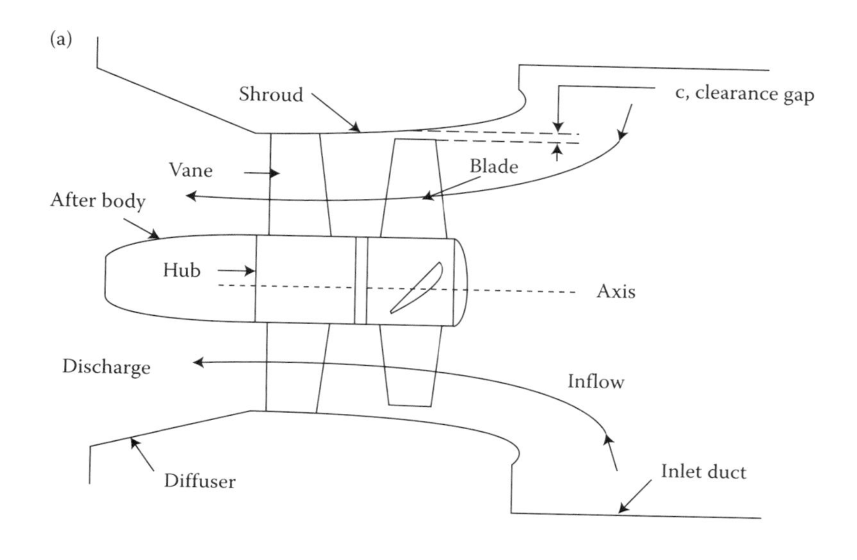

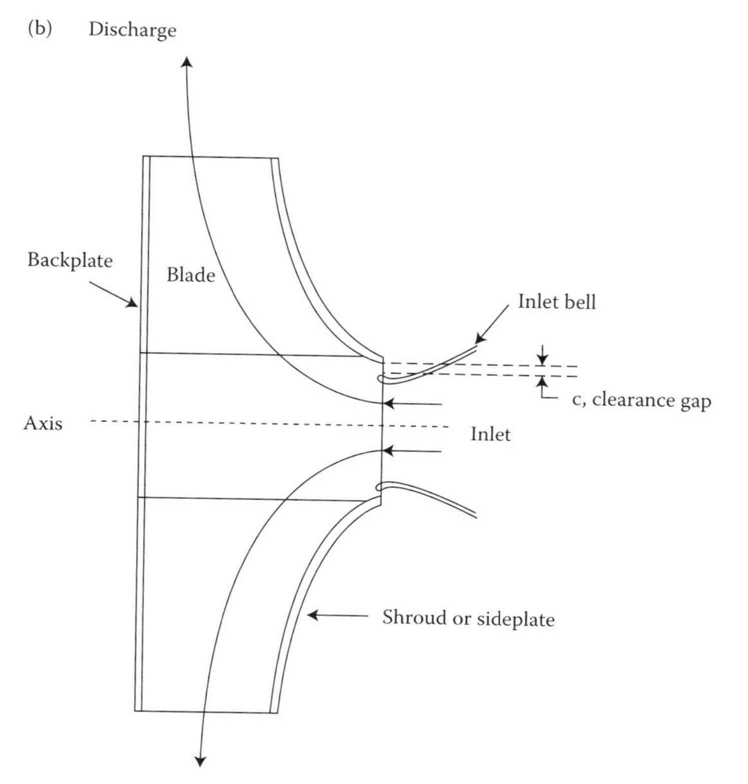

FIGURE 5.9 (a) Axial fan geometry showing radial clearance gap. (b) Centrifugal fan geometry showing radial clearance gap.

Example: Efficiency Estimation

Compare various efficiency estimates for the 394 mm diameter pump whose performance is described by curve A in Figure 3.2.

For the pump in question, the BEP is at $Q = 140\,\text{m}^3/\text{h}$ ($0.0389\,\text{m}^3/\text{s}$) and $H = 47\,\text{m}$. Also $N = 1475\,\text{rpm} = 153.8\,\text{s}^{-1}$. Thus $N_S = 153.8 \times (0.0389)^{1/2}/(9.81 \times 47)^{3/4} = 0.305$ and $D_S = 0.394 \times (9.81 \times 47)^{1/4}/\sqrt{0.0389} = 9.25$. These values define a point near, but not precisely on, the Cordier line [$D_S(N_S) = (2.5/0.305)^{1.092} = 9.95$ and $N_S(D_S) = 2.5 \times 9.25^{-0.916} = 0.326$].

The Cordier efficiency is estimated by

$$\eta_{T,C} = 0.898 + 0.0511(\ln[0.305]) - 0.0645\left(\ln[0.305]\right)^2 + 0.0046\left(\ln[0.305]\right)^3$$
$$= 0.739.$$

To apply Equation 5.2, the Reynolds number must be estimated. If the pump handles room temperature water, $\nu = 9 \times 10^{-7}\,\text{m}^2/\text{s}$ and $Re_D = ND^2/\nu = 2.6 \times 10^7$ and there is no derating for Reynolds number because Re_D is $>10^7$.

To apply Equation 5.3, the tip-width and surface roughness must be estimated. The tip-width, b, is assumed to be about one-tenth of the diameter (this is a "Region F" pump), giving a tip-width Reynolds number (NDb/ν) of about 2.6×10^6. The roughness is assumed to be about 0.05 mm (the impeller is probably made of cast steel or bronze). Then $\varepsilon/2b = 0.05/79 = 0.0006$. The value of the friction factor, f, for this Reynolds number and roughness is about 0.018. Assume that the Cordier friction factor, $f_{Cordier}$, is about 0.015 (implying a Reynolds number of about 10^6 and a relative roughness of about 0.0002). Then

$$1 - \eta_T \approx (1 - 0.739)\left(0.3 + 0.7 \times \frac{0.018}{0.015}\right) \quad \text{giving } \eta_T = 0.702.$$

Finally, since the machine is a pump, Figure 5.8 can be read directly, if imprecisely, using $n_S = 1435\,\text{rpm} \times \sqrt{620\,\text{gpm}}/(154\,\text{ft})^{3/4} = 834$ and $Q = 620\,\text{gpm}$ to give $\eta \approx 0.70$.

The estimates vary from about 0.74 to 0.70. The best *estimate* is probably that taken from Figure 5.8, because that figure applies specifically to liquid pumps and one can deal explicitly with the effect of size (flow). A more important question is—What is the correct value for the efficiency? Of course, it is $\eta \approx 0.68$, taken directly from the pump's own performance curve! *Estimates of efficiency (or any other performance parameter) must only be used if the actual performance data are unavailable.*

5.4 Preliminary Machine Selection

This section describes how to use the Cordier relationships to make a preliminary selection of a turbomachine to meet given performance specifications. The procedure is as follows: The performance requirements and the type of fluid will be known; that is, ρ (and other fluid properties such as μ, γ, and p_v as needed), Δp_T (or H), and Q are specified. Recall that the total pressure change must be used in the Cordier analysis; however, pressure requirements for fans, in particular, are often stated in terms of static pressure (recall that "fan static pressure rise" is actually the discharge static pressure minus the inlet total pressure). In this case, one must convert the static values to total values and, if necessary, iterate the velocity pressure calculation as machine size changes at constant mass or volume rate. This will be illustrated in an example.

To determine the recommended machine type, one explores combinations of size (D) and speed (N) in relation to points on or near the N_s–D_s locus of good equipment as indicated by the Cordier diagram (Figure 2.8). Given ρ, Δp_T (H), and Q, one can choose D and find N—or choose N and calculate D—using the definitions of N_s and D_s. The resulting region on the N_s–D_s diagram will imply the type of machine to use, and the efficiency can be estimated according to the methods described above. If necessary, noise estimates can be made using the methods of Chapter 4 and the required NPSH can be estimated using the correlations from Section 3.4.

The beginning point is the selection of N or D. In most situations, this selection is not completely arbitrary. It may be necessary to select a fan diameter to closely match existing or planned ductwork, in which case, D is specified and N determined by the selection process. It is more common that the possible values of speed are limited. Most turbomachines are driven by AC electric motors. The ideal or "no load" speed of such motors is given by

$$N_{\text{no load}} = \frac{120 \times \text{line frequency}}{\text{number of poles } (n_p)}. \tag{5.5}$$

The line frequency is 60 Hz in the United States (60 cycle AC current) or 50 Hz for European applications. The number of poles is an even number, $n_p = 2, 4, 6, 8$, and so on. The actual speed of a synchronous motor will be close to the no-load value. An induction motor, which has lower purchase cost, has actual running speeds that are slightly lower as the motor develops full-load torque (Nasar, 1987). The full-load or design condition for an induction motor usually requires a "slip" ranging from 15 rpm to as much as 150 rpm, depending on motor quality and rated power. The smaller values of slip are associated with high-quality and large-rated power (hundreds of hp). Higher values of slip will generally be seen in fractional horsepower, low-cost motors.

A turbomachine that is direct-coupled to an AC motor will run at the motor speed, which has a finite number of possible values. In order to obtain other speeds, a transmission or belt-driven system must be placed between the motor and the machine. This adds cost and the drive system will have some energy losses. Sometimes, the cost can be justified in savings in motor cost or by permitting the machine speed to be adjusted to operate more efficiently at off-design flow rates.

Large machines (those that exchange large amounts of power) such as those used in power plants, oil refineries, and chemical processing facilities sometimes use steam turbines or expensive variable-frequency electric motor drives in order to achieve speeds in excess of 3600 rpm or the ability to vary speed for efficient flow control.

Example: Preliminary Pump Selection

A water slide in an amusement park requires a pump to deliver 0.5 m^3/s of water against a head of 50 m. For best economy, an AC induction motor with either two poles or four poles will be direct-coupled to the pump. Specify which motor to use and the type of pump required and estimate the pump size, power requirement, and required NPSH.

Assume motor slip of 50 rpm, so that the pump speed, N, will be either 3550 rpm with a 2-pole motor or 1750 rpm with a 4-pole motor. The specific speeds are

$$N_{s2p} = \frac{3550 \times (2\pi/60) \times (0.5)^{1/2}}{(9.81 \times 50)^{3/4}} = 2.52;$$

$$N_{s4p} = \frac{1750 \times (2\pi/60) \times (0.5)^{1/2}}{(9.81 \times 50)^{3/4}} = 1.24.$$

According to Figure 5.1, the "2p" pump falls in Region C (full-stage axial) and the "4p" pump falls in Region D (mixed flow). In reality, neither of these is a likely choice. Noting that this application requires a fairly large flow rate, a double-suction pump would be a likely candidate. Recalling that the Cordier correlations apply to single-stage, single-suction machines, one-half of the flow (0.25 m^3/s) is used to recalculate the specific speeds:

$$N_{s2p} = \frac{3550 \times (2\pi/60) \times (0.25)^{1/2}}{(9.81 \times 50)^{3/4}} = 1.78;$$

$$N_{s4p} = \frac{1750 \times (2\pi/60) \times (0.25)^{1/2}}{(9.81 \times 50)^{3/4}} = 0.88.$$

The "2p" double-suction pump lies in the mixed flow Region D, and the "4p" double-suction pump lies in Region E, implying a radial discharge centrifugal machine.

The specific diameters are calculated as

$$D_{s2p} = \left(\frac{8.26}{1.78}\right)^{0.517} = 2.21; \quad D_{s4p} = \left(\frac{8.26}{0.88}\right)^{0.517} = 3.18.$$

Calculating the corresponding pump diameters

$$D_{2p} = D_s \frac{Q^{1/2}}{(gH)^{1/4}} = 2.21 \times \frac{0.25^{1/2}}{(9.81 \times 50)^{1/4}} = 0.235\ \text{m};$$

$$D_{4p} = 3.18 \times \frac{0.25^{1/2}}{(9.81 \times 50)^{1/4}} = 0.338\ \text{m}.$$

There are two tools available for estimating the efficiency: the Cordier curve or the pump efficiency chart, as shown in Figure 5.8. The Cordier curve will be used here. The efficiencies are read from Figure 2.9 or calculated from Equation 2.38 and are $\eta_{2p} \approx 0.91$ and $\eta_{4p} \approx 0.89$. Considering the accuracy of the Cordier efficiency correlation, these efficiencies are essentially equal; so take $\eta \approx 0.90$ and estimate the power for either pump as (remembering to use the total flow):

$$P = \frac{\rho g Q H}{\eta} \approx \frac{1000 \times 9.81 \times 0.5 \times 50}{0.9} = 272{,}000\,\text{W} = 272\,\text{kW}.$$

The required NPSH is estimated from Equation 3.37 (the single-suction equation will be used together with the single-suction specific speeds):

$$\text{NPSHR}_{2p} \approx 0.241\,N_{s2p}^{4/3} \times H = 0.241 \times 1.78^{4/3} \times 50 \approx 26\,\text{m},$$

$$\text{NPSHR}_{4p} \approx 0.241\,N_{s4p}^{4/3} \times H = 0.241 \times 0.88^{4/3} \times 50 \approx 10\,\text{m}.$$

The pump driven by the 4-pole motor will be larger and perhaps consumes slightly more power, but it has two advantages. First, it is much more resistant to cavitation. Second, the radial discharge design is more common and probably less expensive for a given size, so it will likely be used. The final (preliminary) selection is as follows:

- Double-suction, radial discharge centrifugal pump (N_s[one side] $= 0.88$)
- Driver: A direct-coupled 4-pole induction motor
- Impeller diameter: About 340 mm
- Power requirement (allowing for a motor efficiency of 0.95): 290 kW
- Net positive suction head required: 10 m.

Example: Preliminary Fan Selection

Select the basic size and speed for a fan that must move $4.80\,\text{m}^3/\text{s}$ with a static pressure rise of 500 Pa at standard sea-level air density ($\rho = 1.21\,\text{kg/m}^3$).

In order to use the Cordier correlations, total pressure must be used. Assume an air velocity of 20 m/s (to be checked later). Then the velocity pressure will be about 240 Pa and the total pressure rise about 740 Pa. This is a lot of flow and not much pressure rise, so the best selection will likely have a fairly high N_s. First calculate:

$$\frac{Q^{0.5}}{(\Delta p_T/\rho)^{0.75}} = \frac{4.80^{0.5}}{(740/1.21)^{0.75}} = 0.01782\,\text{s} \quad \text{and}$$

$$\frac{Q^{0.5}}{(\Delta p_T/\rho)^{0.25}} = \frac{4.80^{0.5}}{(740/1.21)^{0.25}} = 0.4406\,\text{m}.$$

Arbitrarily choosing $N_s = 5$ yields $N = 5/0.01782\,\text{s} = 280.6\,\text{s}^{-1} = 2679\,\text{rpm}$. From the Cordier correlation $D_s = (8.26/N_s)^{0.517} = 1.296$, so $D = D_s \times 0.4406 = 0.571\,\text{m}$. The estimated efficiency is read from Figure 2.9, that is, $\eta_T \approx 0.83$. The noise (sound power) generated by this fan is estimated by $K_W \approx 72/Ds^{0.8} = 58.5\,\text{dB}$ and $L_W = K_W + 10\log_{10} Q + 20\log_{10} \Delta p_s = 58.5 + 10\log_{10}(2119 \times 4.8) + 20\log_{10}(500/249.1) = 104.6\,\text{dB}$. This is not an intolerable selection, but is it optimum? Other values of N_s should be checked; some calculations are summarized in Table 5.1.

Some general observations on these results can be made:

- At the highest range of N_s, the efficiency is poor (the actual efficiency may be as much as 15 points lower than the Cordier estimate) but the fan is small and probably inexpensive. At 4000 rpm ($N_s \approx 7.5$), it is very fast and is probably a "screamer." Note the very high sound power level of 115 dB.
- At the lowest range of N_s, the efficiency is dropping off and the fan is getting outlandishly large and expensive, and at 100–200 rpm, the speed is ridiculously slow. It is, of course, very quiet at about 65 dB.
- In the middle range, efficiency is good, both size and speed are moderate, sound power level is also moderate, and one may have a good starting point for preliminary design or selection.

One can start the search for an actual fan (from a vendor) with, say, $N_s = 2.5$ such that $N = 1340\,\text{rpm}$, $D = 0.82\,\text{m}$, and $\eta_T \approx 0.89$; this would be a vane axial fan with an acceptable sound power level of about 90 dB. The actual fan would probably have total efficiency of about 0.75.

Recall that the actual specifications were on static pressure. With $D \approx 0.82\,\text{m}$, the average discharge velocity is about $V_d = 4.80\,(\text{m}^3/\text{s})/((\pi/4)(0.82\,\text{m})^2) = 9.1\,\text{m/s}$. This gives a velocity pressure of only about 50 Pa with the corresponding total pressure of only 550 Pa. The assumption of 242 Pa velocity pressure (742 Pa total pressure) was a little conservative and should be corrected in a second iteration (it would not be necessary to construct an entire table, however).

As an exercise, assume a velocity pressure of 500 Pa, making $\Delta p_T = 1$ kPa, and rework the example, choosing a suitable initial fan. One should get a high specific

TABLE 5.1

Cordier Calculations for Fan Selection Example

N_s	D_s	N (rpm)	D (m)	L_W (dB)	η_T	Region	Fan Type
7.5	1.05	4019	0.463	115.3	0.78	B	Propeller
5	1.30	2679	0.571	104.6	0.83	B	Tube axial
4	1.45	2144	0.641	99.5	0.86	C	Vane axial
3	1.69	1608	0.744	93.5	0.88	C	Vane axial
2	2.08	1072	0.917	86.2	0.90	D	Full-stage axial
1	2.98	536	1.313	76.2	0.90	E	Mixed/radial discharge
0.5	4.26	268	1.878	68.7	0.83	F	Narrow centrifugal
0.25	6.10	134	2.688	63.1	0.69	F	Narrow centrifugal

speed vane axial fan with a diameter of approximately 70 cm running at around 1800 rpm (mean discharge velocity pressure will be about 450 Pa). Would organizing the calculations differently, say by starting with a selected diameter rather than specific speed, eliminate the need for iteration to account for velocity pressure?

5.5 Fan Selection from Vendor Data

The process described above uses the Cordier diagrams to estimate the type of machine required to meet certain performance specifications and to estimate the size, speed, efficiency, power requirement, and noise or cavitation limitations. The results are very general and often quite approximate. The next step is to select an actual machine from a manufacturer/vendor. Typically, this process requires consulting catalogs supplied by vendors in order to make the choice. For fans (and pumps too), there are many possible vendors and dozens of catalogs, often comprising a few hundred pages each! Fortunately, most of this information is now available electronically, either on the Internet or on CD-ROM. One may even find the selection process quite automated. In many cases, vendor's representatives are only too happy to assist. The selection process described here follows the basic Cordier concept to establish a base of candidate size and speed values, which can then be compared to available vendor data.

For fans, vendor catalog data are usually available in one of the two forms: (graphical) performance curves or *multirating tables*. In general, performance information in either of these forms can be reduced to *dimensionless* performance curves. Bear in mind that, although the tables are very limited in the examples here, a broad search could involve entire catalogs from many vendors.

Table 5.2 is a portion of a (rather generic) multirating table for a family of vane axial fans, assumed to be from a specific vendor. The fans listed are belt-driven fans rather than direct-connected or direct-drive fans. A typical vane axial fan is shown in Figure 5.10.

The key to understanding multirating tables is to recall that the primary purpose of a fan is to move air (flow) and that fans are available in several sizes and several speeds (because of the belt drive). Take a moment to examine the structure of Table 5.2. The flow, Q (in cfm), controls the leftmost column entry. Since the fan (and duct) size is fixed for each table, a unique value of velocity pressure $\left(\frac{1}{2}\rho V_d^2\right)$ is associated with each flow rate. Increasing values of static pressure rise (Δp_s, labeled S.P. in Table 5.2) head a series of columns to the right of the first two columns. Note that the static pressure rise ranges from 0.0 to 1.5 in. wg but the highest pressures are not available for the lower volume flow rates. Finally, the corresponding values of speed (rpm) and power input (hp) required to produce the particular combination of flow and static pressure

TABLE 5.2

Multirating Tables for a Vane Axial Fan

Volume (cfm)	V.P. (in. wg)	S.P. = 0 (in. wg) (rpm/hp)	0.25 (rpm/hp)	0.50 (rpm/hp)	0.75 (rpm/hp)	1.00 (rpm/hp)	1.25 (rpm/hp)	1.50 (rpm/hp)
24-in. Fan								
Fan Diameter = 24 in.; Casing Diameter = 24.3 in.								
3850	0.09	545/0.20	663/0.35	790/0.56				
4500	0.12	640/0.30	740/0.50	840/0.71				
5150	0.16	730/0.44	810/0.66	900/0.90	1000/1.15			
5800	0.20	820/0.65	900/0.90	965/1.11	1050/1.44	1150/1.7		
6400	0.25	900/0.88	975/1.12	1050/1.40	1120/1.75	1200/2.00		
7000	0.30	1000/1.16	1060/1.45	1130/1.70	1190/2.00	1250/2.40	1350/2.80	1400/3.10
27-in. Fan								
Fan Diameter = 27 in.; Casing Diameter = 27.3 in.								
4875	0.09	486/0.24	590/0.45	700/0.70				
5676	0.12	560/0.40	650/0.60	750/0.90				
6500	0.16	650/0.55	725/0.80	800/1.10	885/1.45	945/1.23		
7300	0.20	730/0.80	800/1.10	865/1.41	950/1.80	1000/2.2		
8100	0.25	800/1.11	875/1.40	950/1.80	1000/2.15	1000/3.0		
9000	0.30	900/1.50	950/1.80	1010/2.25	1050/2.60	1125/3.05	1190/3.50	1250/4.00

FIGURE 5.10 Typical belt-driven vane axial fan. (From New York Blower (1986), *Catalog on Fans*, New York Blower Co. With permission.)

are shown in Table 5.2 at the intersection of the "flow" row and "pressure" column. Note that, if Table 5.2 were extensive enough, it would be possible to plot "standard" performance curves of pressure and power versus flow at a variety of speeds using data from the table (possibly requiring some interpolation).

Example: Fan Selection Using a Multirating Table

Use the multirating table, Table 5.2, to select a fan to meet the following specifications: $Q = 3.0\,\mathrm{m^3/s}$, $\Delta p_S = 250\,\mathrm{Pa}$, and $\rho = 1.21\,\mathrm{kg/m^3}$.

Since Table 5.2 is very limited, the fan selection could be made very quickly from it; however, it must be remembered that the actual table would be very much larger, encompassing many more fan sizes, each with several more performance points. To illustrate the process of selecting from a larger table, Cordier analysis will first be used to "home in" on a specific portion of the tables.

The first issue is that both the specification and the multirating tables employ static pressure rise while Cordier requires total pressure rise. To get around this, one can carry out the initial analysis using D as the independent variable. If it is assumed that d/D is 0.50 (d = hub diameter), one can calculate the annulus area and the velocity pressure, so that the total pressure is known for each choice of D. A Cordier analysis table (Table 5.3) is constructed using the defining equations for D_S and N_S together with the usual correlations for efficiency and sound power level.

Since a vane axial fan from Table 5.2 is to be selected, look at the upper range of the Cordier table (Table 5.3). The diameter of such a choice must lie between about 0.6 and 0.8 m, or roughly around 0.7 m. The spread of data underlying the Cordier curve suggests a degree of flexibility in the exact value of the parameters. Thus, one should be able to find a candidate selection with a specific speed of about 3.23, implying a true speed of about $N = 1160\,\mathrm{rpm}$.

Having estimated the parameters of the required fan, next one locates the appropriate region of the multirating table (in this case, Table 5.2). Since Table 5.2 is in U.S. Customary units, temporarily convert the requirements to $Q = 6350\,\mathrm{cfm}$,

TABLE 5.3

Cordier Analysis

D (m)	p_V (Pa)	Δp_T (Pa)	D_s	N_s	N (rpm)	L_W (dB)	η_T	P (kW)	Region	Fan Type
0.4	613	863	1.19	5.87	4463	100.6	0.81	3.19	A/B	Propeller
0.6	121	371	1.45	4.03	1626	91.6	0.86	1.30	B	Tube axial
0.8	38	288	1.81	2.61	871	82.8	0.89	0.97	C	Vane axial
1	16	266	2.22	1.76	553	76.1	0.91	0.88	C	Axial/mixed
1.2	8	258	2.65	1.26	386	71.1	0.91	0.85	D	Axial/mixed
1.4	4	254	3.08	0.94	285	67.4	0.89	0.85	E	Mixed/radial discharge
1.6	2	252	3.51	0.73	220	64.4	0.87	0.87	F	Radial discharge

$\Delta p_s = 1$ in. wg with $D \approx 28$ in. at 1160 rpm, requiring about 1.1 kW. Checking Table 5.2 for the 27-in. size fan, the closest match to the specifications appears to be a fan with 6500 cfm at 1.00 in. wg running at 945 rpm requiring 0.92 kW. Another option is just below this one in Table 5.3. It gives 7300 cfm at 1.0 in. wg at 1000 rpm requiring 1.64 kW. The next smaller size, the 24-in. fan, yields two choices. At 6400 cfm, a 1.0 in. wg static pressure rise is available at 1200 rpm and 1.5 kW. At 5800 cfm, 1 in. wg is available at 1150 rpm, requiring 1.27 kW. Examination of additional tables for larger and smaller versions of this family of vane axial fans would provide more choices for fans that can operate close to (but probably not exactly at) the desired performance point.

Although this may appear somewhat confusing, the target is surrounded, so to speak, with fan parameters reasonably close to the specifications derived from the Cordier analysis. The notable exception is the required power, which is high in every case. For the actual fans, both the Reynolds number and the radial clearances are relatively far from the excellent conditions underlying the Cordier efficiency estimate. It is possible to derate this fan using information from Section 5.3. The Reynolds number is estimated as $Re = ND^2/\nu \approx 0.4 \times 10^7$. From Table 5.2, the 27-in. fan has a blade diameter of 27.0 in. and the casing or inlet diameter is 27.3 in. Thus, the radial clearance is $c = 0.15$ in. and $c/D = 0.0056$. Using Equations 5.2 and 5.4 to estimate derating factors, the total efficiency is reduced to about 0.61 from a Cordier value of 0.88. This raises the nominal value of required power to about 1.6 kW, fairly close to the various multirating values. An additional explanation for the difference in efficiency and power is that the Cordier estimate applies only at the BEP, while the multirating table entry may not represent the BEP.

What fan should be ultimately chosen? The best choice appears to be the 27-in. fan at 945 rpm and 0.92 kW power required. The 24-in. fan at 6400 cfm and 1.0 in. wg could provide the performance, but the power is increased by more than 60%. Estimating the sound power for the two fans shows the 24-in. fan at 91.1 dB and the 27-in. fan at 87.6 dB. The choice between the two could be made on the basis of sound power level, but the difference is not very great. Table 5.4 summarizes the parameters of these two fans.

The two fans are very similar and the final choice would depend on the user's preference either for smaller size and perhaps slightly lower initial cost or a larger fan at increased first cost, but a reduced power requirement running at reduced noise. The preference may well depend on the usage of the fan. If the fan is going to be used more or less continuously over a period of many years, the 27-in. fan with decreased operating cost will be chosen. If the fan is for occasional or seasonal use and noise is not very important, the smaller fan will be chosen.

The same kind of Cordier analysis could be used to select a tube axial fan but is deferred to the problem assignments at the end of the chapter.

An alternative to using (extensive) multirating tables is to reduce the tables to a set of dimensionless performance curves to characterize an entire family of fans, which may be available in different sizes and speeds. For some variety, consider a family of single-width, single-inlet (SWSI) centrifugal backwardly inclined blade fans. The fans in question are available in a variety of sizes, in this case ranging between 0.46 and 1.85 m.

Centrifugal fans are divided into "Classes," which relate to structural strength and ruggedness. A Class I fan is limited to blade tip speeds <55 m/s, a Class II fan to tip speeds of 68 m/s, and a Class III fan to tip speeds of 86 m/s. As a general rule, fan pressure rise is proportional to the square of tip speed, so higher-pressure applications require fans of a higher class-rating. Of course, cost increases with class-rating. The fans in question are available in all three class ratings.

The multirating tables for this family of fans would fill a sizeable catalog. By selecting several performance points (flow, static pressure, and power) from the tables for a variety of combinations of speed and size, one can generate a number of dimensionless performance points from the usual definitions

$$\phi = \frac{Q}{ND^3}; \quad \psi_S = \frac{\Delta p_S}{\rho N^2 D^2}; \quad \eta_S = \frac{Q \Delta p_S}{P}.$$

The equivalent total performance variables can be generated from the relations given in Problem 2.22

$$\psi_T = \psi_S + \frac{8}{\pi^2}\phi, \tag{5.6}$$

$$\eta_T = \eta_S \frac{\psi_T}{\psi_S}. \tag{5.7}$$

Note that Equation 5.6 is only an approximation for a centrifugal fan whose diameter, in general, is not the same as the duct diameter. (In fact, the duct may be rectangular instead of circular.)

TABLE 5.4

Fan Options

Fan	D (m)	N (rpm)	Q (m^3/s)	Δp_S (Pa)	P (kW)	L_W (dB)
24-in.	0.610	1200	3.03	250	1.50	91.1
27-in.	0.686	1000	3.07	250	0.92	87.6

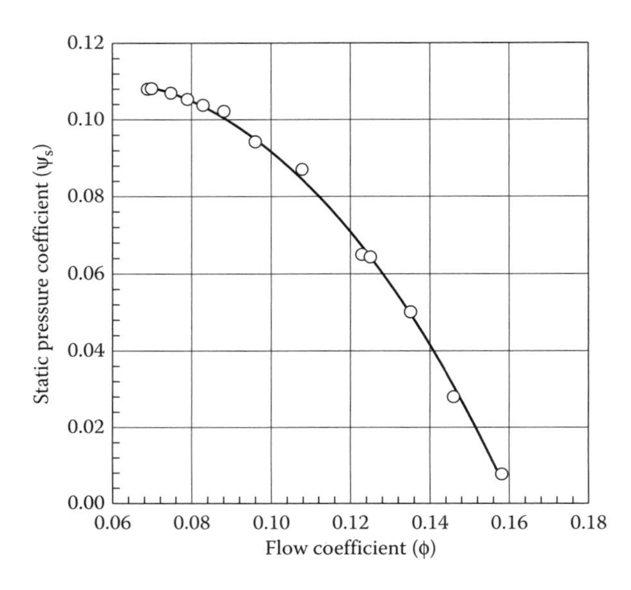

FIGURE 5.11a Nondimensional pressure curve for a family of SWSI centrifugal fans. Available sizes are 0.46, 0.61, 0.91, 1.02, 1.25, 1.375, 1.50, 1.67, and 1.85-m diameter. Class I, Class II, and Class III models are available.

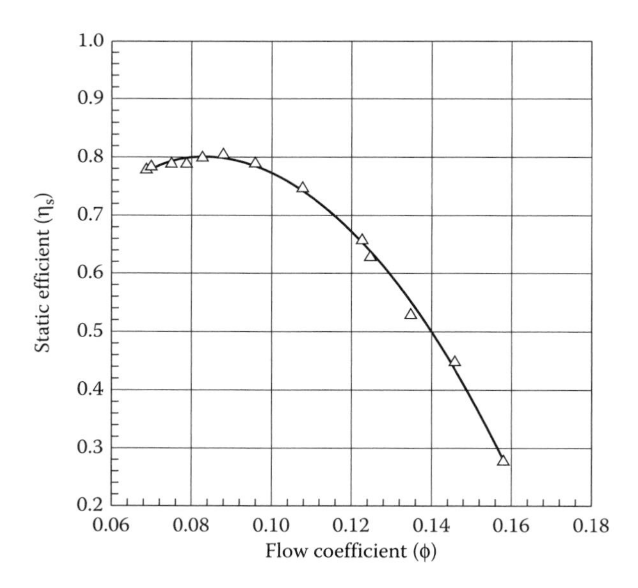

FIGURE 5.11b Nondimensional efficiency curve for a family of SWSI centrifugal fans. Available sizes are 0.46, 0.61, 0.91, 1.02, 1.25, 1.375, 1.50, 1.67, and 1.85-m diameter. Class I, Class II, and Class III models are available.

FIGURE 5.12 Impeller for a large DWDI fan. (From Process Barron. With permission.)

The points so generated can be plotted as performance curves. Figures 5.11a and 5.11b are typical curves of (static) pressure coefficient as a function of flow coefficient and (static) efficiency as a function of flow coefficient. Total pressure and efficiency curves (not shown) would be similar with somewhat higher values on the ordinate. These curves could be transformed into those for a double-width, double-inlet (DWDI) version of the same fan (Figure 5.12) by replotting with $\phi_{DWDI} = 2\phi_{SWSI}$ while leaving ψ_s and η_s unchanged.

The following example illustrates how to use these dimensionless performance curves to select a specific fan, after Cordier analysis helps identify the general type of fan to seek.

Example: Selecting a Fan with a Noise Constraint

Select a fan to meet the following performance specifications: $Q = 15{,}000\,\text{cfm}$, $\Delta p_S = 6.0\,\text{in. wg}$, $\rho = 0.0023\,\text{slug/ft}^3$, and subject to the constraint $L_W < 100\,\text{dB}$.

Begin by using the noise constraint to estimate the minimum value for D_S.

$$L_W \approx \frac{72}{D_S^{0.8}} + 10\log(Q\ [\text{cfm}]) + 20\log(\Delta p_S\ [\text{in. wg}])$$

using $L_W = 100\,\text{dB}$ and the specified values for flow and pressure give $D_S > 1.92$. Using the Cordier correlation then gives $N_S < 2.33$. Considering the Cordier efficiency curve, Figure 2.9, a high value of efficiency can be realized with $0.8 < N_S < 2$. This implies a fan from Cordier Region D (mixed flow) or E (radial discharge). The fan whose performance is shown in Figures 5.10a and 5.10b is representative of this region, so the selection will be made from this line of machines.

To obtain good efficiency, investigate a fan with $\phi \approx 0.085$, which is near the BEP from Figure 5.10b. At this point, $\psi_S = 0.102$. Using

$$\phi = \frac{Q}{ND^3} = \frac{15{,}000/60}{ND^3} \quad \text{and} \quad \psi_S = \frac{\Delta p_S}{\rho N^2 D^2} = \frac{6 \times 5.204}{0.0023 \times N^2 D^2}$$

one can solve for D (Equation 2.28); $D = 2.84$ ft (=0.865 m). The closest available size is 0.91 m (2.99 ft, 35.8 in.). What about the speed? To use the (most economical) Class I construction, the maximum tip speed is 55 m/s. The corresponding fan speed is $N = 2 \times 55$ (m/s)/0.91 m $= 120.88\,\text{s}^{-1} = 1154$ rpm. *If* $N = 1154$ rpm and $D = 0.91$ m are specified, the desired dimensionless performance point is

$$\phi = \frac{15{,}000/60}{120.88 \times 2.99^3} = 0.0774 \quad \text{and} \quad \psi_S = \frac{6 \times 5.204}{0.0023 \times 120.88^2 \times 2.99^2} = 0.104.$$

This point *is not* on the performance curve—but it is close. The final specification will be

- Use the 0.91 m diameter, Class I Fan. Operate at 1150 rpm.

The actual performance point will, as always, be determined by the match between the fan curve and the system resistance curve. The process would be as follows:

- Obtain dimensional performance curves for the fan using $\Delta p_S = \psi_S(\rho N^2 D^2)$; $Q = \phi(ND^3)$, with values of ψ_S and ϕ selected from Figure 5.10a.
- Assume a parabolic system resistance curve that passes through the specified operating point $\Delta p_{S,\text{system}} = 6$ [in. wg](Q [cfm]/15,000)^2.
- Determine the operating point as the intersection of the two curves.

The details are left to the reader as an exercise.

5.6 Pump Selection from Vendor Data

Generally speaking, the overall process for selecting a pump is similar to the process for selecting a fan. There are dozens of pump manufacturers/vendors and seemingly limitless pages of catalog data available. Electronic tools are abundantly available to assist in the process, either on the Internet or on CD-ROM. It is possible to find Internet sites that combine data from many manufacturers and with a more or less completely automated selection process. Of course, manufacturer's/vendor's representatives are usually more than pleased to assist an engineer in selecting one of their pumps. Often, pump selection is influenced by nonhydraulic factors such as corrosion resistance or suitability for handling suspended solids in the liquid stream.

The selection process described here parallels the process for fans in that Cordier analysis is first employed to identify the general type of pump required and to estimate its size and performance. After this, vendor data, most usually in the form of performance curves (multirating tables are not typically used for pumps), are used to make a final selection. Typically, in selecting a pump, cavitation performance (NPSH) is an important criterion but noise is not.

Typical vendor-catalog pump performance curves are shown in Figures 5.13a–5.13c. These particular curves represent single-suction (sometimes called end-suction) pump data from one manufacturer, Allis-Chalmers pumps (by ITT Industrial Pump Group, Cincinnati, OH). These pumps are directly coupled to the motor, which is a common practice in the pump industry, so available speeds are limited. Pump vendors often supply a number of different impeller sizes in the same casing. This provides cost savings in manufacturing and purchasing, but it does affect the efficiency for the smaller impeller sizes, as shown.

Note the wealth of information provided with the curves. There are separate curves for head, power, efficiency, and NPSHR. Geometric data and motor data are given in the legend. The somewhat vague item "maximum sphere" tells the largest size sphere that could pass through the impeller unhindered. The following example illustrates the use of curves such as these in pump selection.

Example: Pump Selection, Including a Cavitation Constraint

Select a pump to deliver 275 gpm against a head of 150 ft. The fluid is water at 85°F, and the system imposes the requirement that NPSHR must be <10 ft. For economy, consider only direct-driven pumps.

A Cordier analysis table is constructed using the number of motor poles as the independent variable (Table 5.5). The motor slip is assumed to be 50 rpm. Equation 3.37 is used to estimate the Thoma cavitation parameter and hence the pump's NPSHR.

It is clear that a pump driven by a 2-pole motor cannot meet the NPSHR requirement. Also, pumps driven by 6-pole (and higher) motors will be very inefficient; therefore, a pump driven by a 4-pole motor, with an impeller diameter of the order

TABLE 5.5

Cordier Analysis for Direct-Connected Pump

Motor Poles (n_p)	Motor Speed	N_s	D_s	D (in.)	η_T	σ_V	NPSHR (ft)
2	3550	0.50	5.77	6.50	0.83	0.097	14.6
4	1750	0.25	12.49	14.07	0.69	0.038	5.7
6	1150	0.16	19.75	22.26	0.57	0.022	3.2

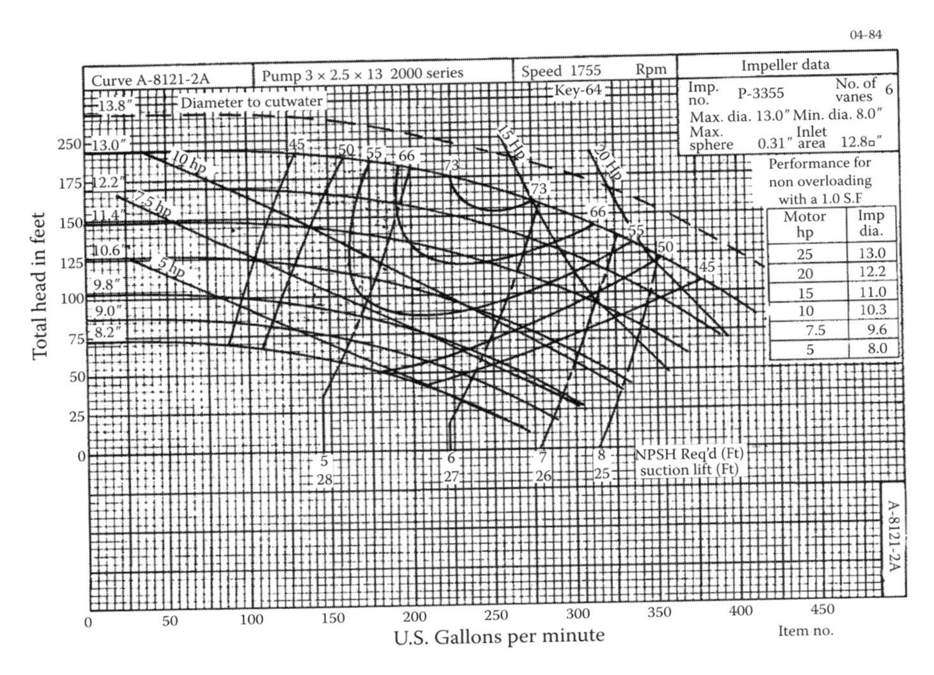

FIGURE 5.13a Performance curves for a line of Allis-Chalmers pumps. (From ITT Fluid Technology. With permission.)

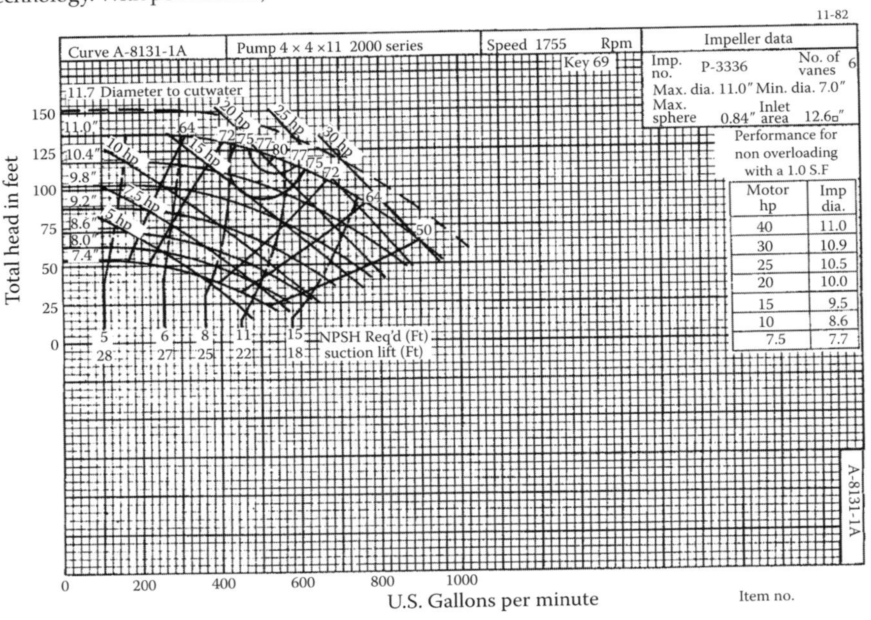

FIGURE 5.13b Performance curves for another line of Allis-Chalmers pumps. (From ITT Fluid Technology. With permission.)

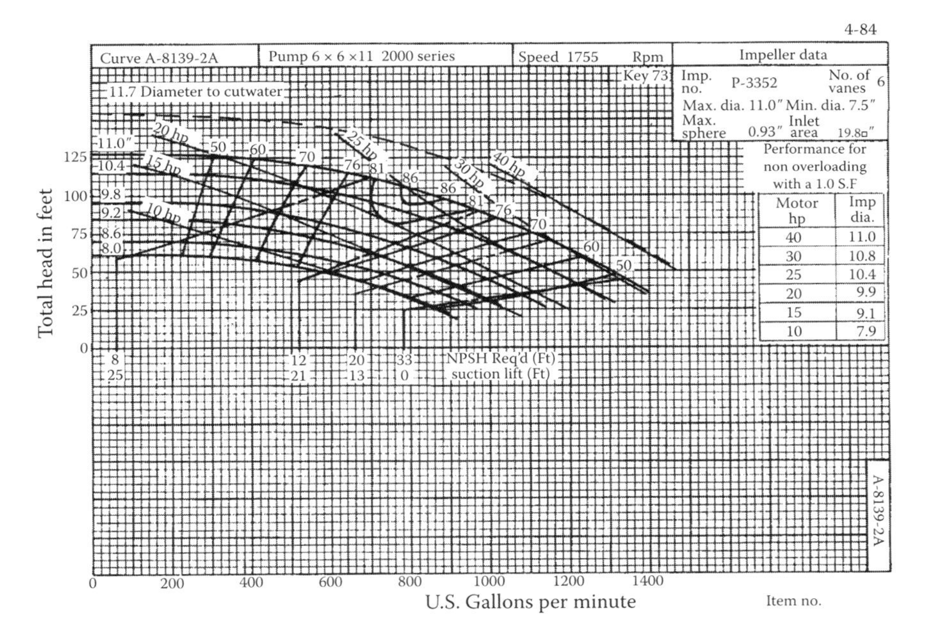

FIGURE 5.13c Performance curves for another line of Allis-Chalmers pumps. (From ITT Fluid Technology. With permission.)

of 1 ft, will be examined. The low specific speed suggests that a narrow-bladed centrifugal pump will be the machine of choice. The pumps whose performance is described in Figures 5.13a through c meet these requirements and they will be investigated for the best choice. (Of course, pumps from numerous other manufacturers will also meet these requirements.)

Consider a few specific pumps:

- Curve A-8139 shows an 11-in. impeller at 1755 rpm with 275 gpm at 124 ft of head. NPSHR = 6 ft and $\eta_T \approx 0.50$. Note that the pump would operate far from the zone of best efficiency and requires a 20-hp motor.
- Curve A-8131 also shows an 11-in. impeller at 1755 rpm that delivers 275 gpm at 150 ft of head. NPSHR = 8 ft and $\eta_T \approx 0.68$. This is better, but it still operates to the left of the zone of best efficiencies. The head is closer to the desired value but the pump still has a 20-hp motor.
- Curve A-8121 shows a somewhat larger pump with $D = 13$ in. at $N = 1755$ rpm, which delivers 275 gpm at 154 ft of head. NPSHR = 6 ft and $\eta_T \approx 0.73$. This is the best candidate thus far. Some efficiency is gained, but it is at the expense of a moderate increase in size and a small increase in NPSHR (which is still <10 ft). More importantly, there has been a drop-down to a motor requirement of 15 hp, which will reduce the cost substantially, even though the pump is larger. This appears to be a good choice based on the review of very limited catalog data.

The final selection is

$$\text{Choose pump } 3 \times 2.5 \times 13 \text{ Model 2000}$$

$$\text{with 12.2-in. diameter and 20-hp motor}$$

Note that this pump will operate very near its BEP and the Cordier estimates of efficiency and NPSHR are quite accurate, although the actual diameter is somewhat smaller than the Cordier estimate.

5.7 Selection of Variable Pitch and Variable Inlet Vane Fans

Thus far, the selection process has been focused on obtaining a single performance point. There are many applications in which mass flow or volume flow rate must vary over fairly wide ranges or must be very precisely controlled. One way to accomplish this goal is to use dampers (for fans) or valves (for pumps) to vary the system resistance and hence shift the operating point. This requires little capital investment but is usually quite wasteful of energy (and hence operating cost dollars) because extra system losses are introduced and the machine is forced to operate at less than peak efficiency. A second option would be to vary machine speed, keeping operation near the BEP. This would require a costly variable speed transmission, steam turbine drive, or perhaps a variable frequency drive motor. Generally, variable speed systems are economically justified only for very large or very high energy-consuming systems.

Two available alternates for fans are *variable blade-pitch vane axial fans* and *variable inlet vane centrifugal fans*. These devices can provide flow variability and control over very wide ranges of performance. Variable pitch vane axial fans are constructed with blades mounted in bearing assemblies that allow the blades to rotate around their radial axes so that the blade or cascade pitch angle can be continuously varied to change the flow and pressure rise while the fan is in operation (Figures 5.14 and 5.15). Variable inlet vane centrifugal fans (Figure 5.16) have a set of vanes mounted on radial pivot rods installed in the fan's inlet cone. These vanes can be pivoted in such a way as to impart a spin (or swirl) to the air stream entering the impeller. This changes the flow-pressure characteristic of the fan, with a different curve for each setting angle of the vane assembly (Problem 2.36). The bearing assemblies, linking mechanisms, and feedback control systems necessary to adjust the blades or vanes may add somewhat to the complexity and cost of a given machine, but the resulting configuration is capable of adjustment and control to meet very exacting requirements. As their overall performance characteristics are similar, only variable pitch vane axial fans will be discussed further here.

FIGURE 5.14 Variable pitch axial fan with diffuser. (Howden Denmark.)

The Variax Series of fans from Howden Denmark, for example, range in size from 794 to 1884 mm in diameter and operate at 1470 rpm (50 Hz line frequency), with some smaller sizes available at 2950 rpm. Flow ranges from about 4 to 110 m^3/s, at total pressure rises from 250 to 3000 Pa. Figure 5.17 is a fan selection chart for these fans, showing the available range of flow and pressure, as well as the specific fan model recommended for various subranges. The particular fan illustrated in Figures 5.14 and 5.15, model ASV 1000/630-10, has a diameter of 1000 mm (1.0 m). The fan hub diameter is 630 mm; there are 10 blades, and the fan runs at 1470 rpm.

In essence, each different setting of the blades represents an (aerodynamically) different fan. A unique performance curve exists for each blade setting. The combination of several of these curves on Figure 5.18 is called a *performance map*. An example performance map (for ASV 1000/630-10) is shown in Figure 5.18. The pressure rise versus flow rate graph gives total pressure in Pa and flow rate in m^3/s for a machine without the optional downstream diffuser (catalog information is also available for fans with diffusers). Also given is a graph of shaft power required, in kW. Efficiency contours are superposed on

FIGURE 5.15 Detail of a large variable pitch axial fan showing circular blade mounts for controllable blades. (From Howden Denmark. With permission.)

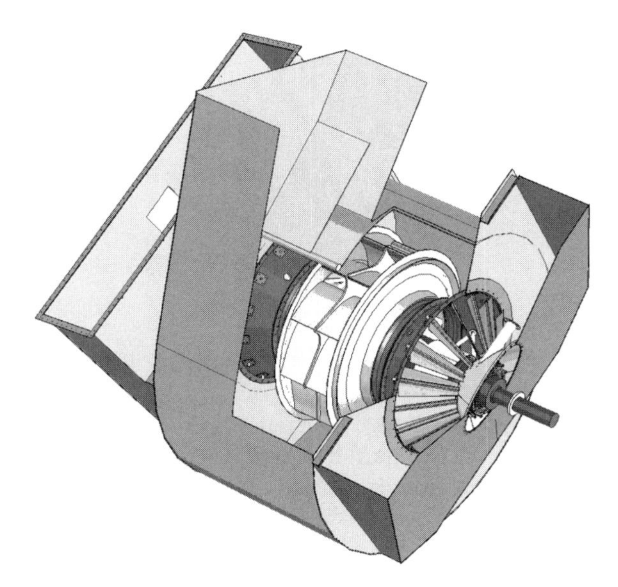

FIGURE 5.16 DWDI centrifugal fan assembly showing variable inlet vanes. (From FlaktWoods Americas. With permission.)

the pressure-flow curves, with total efficiencies shown. One pair of curves is shown for each pitch angle setting (represented by α_g, the geometric pitch). The angle is generally measured from the plane of rotation to the blade chord line for the section located at the hub or base of the blade. For this fan, α_g ranges from 10° to 55°, and flow rates can vary from about 0.7 to 28 m^3/s, with total

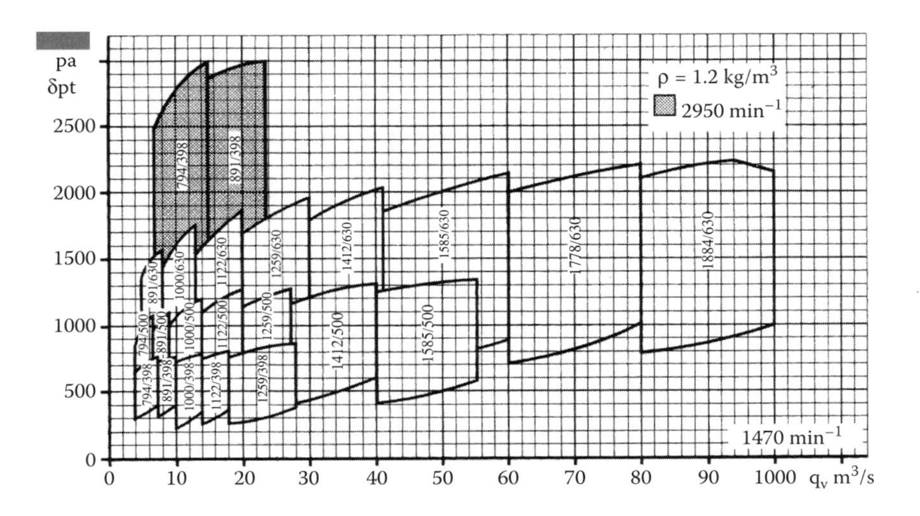

FIGURE 5.17 Fan selection chart for variable pitch axial fans. (From Howden Denmark. With permission.)

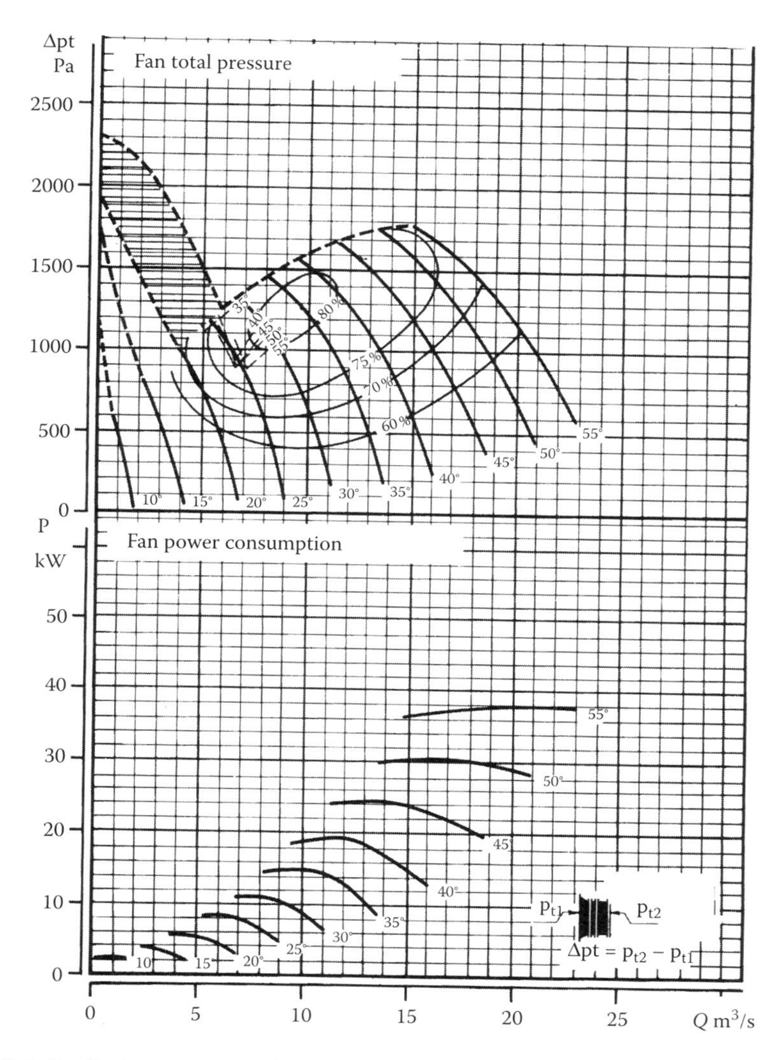

FIGURE 5.18 Performance map for a variable pitch fan. (From Howden Denmark. With permission.)

pressure rises up to about 1800 Pa. The zone of best efficiency (replacing the BEP) lies in an oval-shaped region within the 80% total efficiency contour, roughly between 1000 and 1500 Pa and 7–11 m^3/s.

When selecting a variable pitch fan, the performance point at which the system operates most of the time should be near the basic BEP of the machine and certainly must lie within the best efficiency contour. If possible, the identifiable alternative points of operation should fall within this zone as well. In no case should an operating point be allowed to lie outside of the dashed curve that defines the locus of *stall entry*. Attempts to operate in this region can result in pulsating flow and pressure, excessive noise, and mechanical vibration, which can lead to serious damage and structural failure. The various performance curves are not defined in the stall region as operation is unstable. Any choice of operation within the unstalled region of the fan map should provide an adequate margin of operation away from the stall boundary. A satisfactory margin is usually defined as being at a pressure rise no greater than 90% of the stall pressure at that flow rate. For example, if the ASV 1000/630-10 fan of Figure 5.18 operates at 15 m^3/s, the allowable pressure rise is about 1620 Pa (0.9 × 1800 Pa), so the blade pitch would be about 51°.

Example: Operating Points for a Variable Pitch Axial Fan

A fan is required to meet the following three operating points: 9 m^3/s at 1100 Pa; 13 m^3/s at 1500 Pa; and 6 m^3/s at 1000 Pa. Show that the ASV 1000/630-10 fan is a reasonable choice.

By reading the performance map (Figure 5.18), one obtains the following:

- At 9 m^3/s and 1100 Pa, set $\alpha_g = 32°$ and obtain $\eta_T = 0.81$, operating pressure is at 73% of stall.
- At 13 m^3/s and 1500 Pa, set $\alpha_g = 45°$ and obtain $\eta_T = 0.78$, operating pressure is at 85% of stall.
- At 6 m^3/s and 1000 Pa, set $\alpha_g = 24°$ and obtain $\eta_T = 0.76$, operating pressure is at 77% of stall.

Clearly, these points work with the chosen fan and exhibit good efficiencies and stall margins, so the fan is at least a reasonable choice.

Example: Using a Fan-Selection Chart

Select a fan to meet the following three performance points: 50 m^3/s at 1200 Pa; 60 m^3/s at 2000 Pa; and 20 m^3/s at 1100 Pa.

With three desired points of operation, it is assumed that the "middle point," 50 m^3/s at 1200 Pa, controls the selection of fan type. Assuming operation at 1470 rpm, the specific speed is 6.12. This implies a vane axial or tube axial fan. From Cordier analysis, the specific diameter would be about 1.2 and the actual diameter about 1.5 m. Assuming that the economics justify the expense, the decision is to use a variable pitch fan. Using the fan selection chart (Figure 5.17) for a nominal operating point of 50 m^3/s at 1200 Pa indicates that ASV 1585/630 would be adequate as a starting selection for this fan. Note that "1585" implies a fan diameter of 1.585 m, reasonably close to the Cordier estimate.

5.8 Summary

The chapter began with a discussion of the subdivision of the Cordier diagram into zones of performance capability. These zones or regions were identified with particular kinds of turbomachines, ranging from a region on the left side of the diagram (low values of specific diameter/high values of specific speed) to a region at the right edge (high values of specific diameter/low values of specific speed). The leftmost region was associated with lightly loaded axial flow machines, and the rightmost region with heavily loaded radial flow machines. The intervening zones were subsequently assigned to the more heavily loaded axial flow equipment, mixed flow machines, and moderately loaded centrifugals. The hybrid or crossover turbomachines were discussed, and the practice of multistaging both axial flow and centrifugal impellers was introduced.

Next, methods for (roughly) estimating the BEP efficiency of a turbomachine were discussed. The concept of combining "derating" factors was introduced, and a very limited amount of derating information was presented.

The concept of classification zones and efficiency estimation was then extended to develop a logical method for determining the appropriate machine configuration for a given set of performance specifications. The process was quantified in a set of useful but approximate procedures developed in terms of the variables of the Cordier diagram, the specific speed, specific diameter, the total efficiency, and sound power level and Thoma cavitation parameter, if necessary. Illustrations of the technique were used to show the procedure as well as to emphasize the approximate nature of the algorithms and the need for flexibility in the selection and determination of efficiency.

Several rather extensive selection exercises, using manufacturers' performance and geometric information, illustrated the procedure with fans, blowers, and pumps. The procedure was shown to produce an optimal choice of equipment under a variety of constraints.

EXERCISE PROBLEMS

5.1. A ventilating fan is required to provide 15,000 cfm of standard density air with a total pressure rise of 2 in. wg. The fan is to be operated for 1200 h per year for 5 years. Use Cordier analysis to select the cheapest fan/motor combination possible with the following constraints:

- Use a total efficiency (including the motor) 10 points lower than Cordier
- Motor cost is estimated in dollars/hp based on the number of poles [$/hp = 70 + 5 (n_p - 4)^2$]
- Consider direct-connected motors of 2, 4, 6, 8, and 10 poles
- Assume that the initial cost is twice the motor cost, and power costs $0.08/kWh
- Neglect the time value of money.

5.2. Use a Cordier analysis to evaluate candidate fan configurations for a flow rate of 10,000 cfm with a total pressure rise of 5 in. wg (standard air). Assume a belt-driven configuration and cover a wide range of equipment. Choose a best candidate based on total efficiency. If discharge velocity pressure is not recovered (base V_d on impeller outlet area), choose a best fan based on static efficiency.
5.3. A chemical process fan must deliver 5000 cfm of air ($\rho = 0.0023\,\text{slug/ft}^3$) at 32 in. wg total pressure rise. Select the proper single-stage centrifugal fan for this application, and compute the diameter, speed, efficiency, and power. Also select a simple two-stage series centrifugal fan arrangement and compute the diameter, speed, efficiency, and power. Compare the two selections as quantitatively as possible and discuss the advantages and disadvantages of both configurations. Neglect compressibility effects.
5.4. Rework Problem 5.3 using a multistage axial configuration. Justify the number of stages selected.
5.5. Reconstrain the analysis of Problem 5.4 to include a minimization of the overall fan noise.
5.6. Use Table 5.2 with a Cordier analysis to select a vane axial fan with requirements of $Q = 12{,}000$ cfm and $\Delta p_s = 0.50$ in. wg at standard density. Assume initially that the velocity pressure is 0.75 in. wg and iterate the solution once only.
5.7. Select a tube axial fan that delivers $2\,\text{m}^3/\text{s}$ at a static pressure rise of 70 Pa using a Cordier analysis aimed at Figures P5.7a and P5.7b.

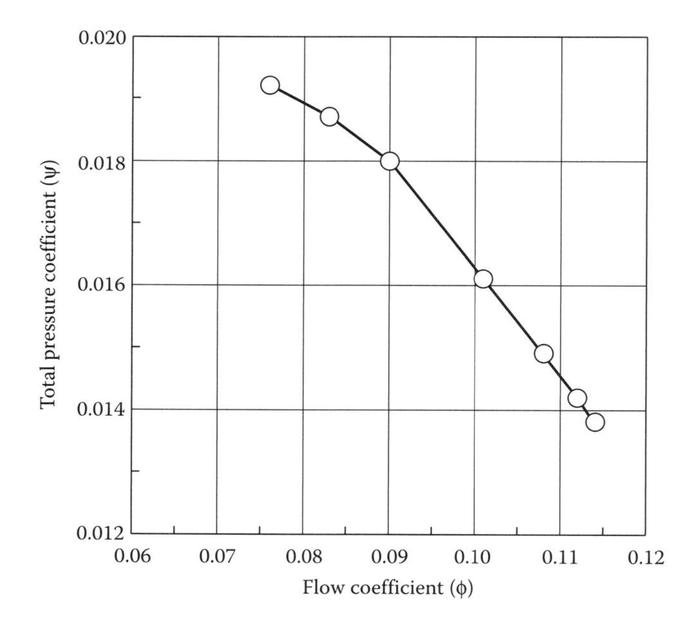

FIGURE P5.7a Dimensionless performance of a tube axial fan. Sizes available (diameter, in.): 12, 15, 18, 21, 24, 27, 32, 36, 42, 48, 54, 60, 72, 84, 96. Maximum tip speed, $U_t \leq 590$ ft/s.

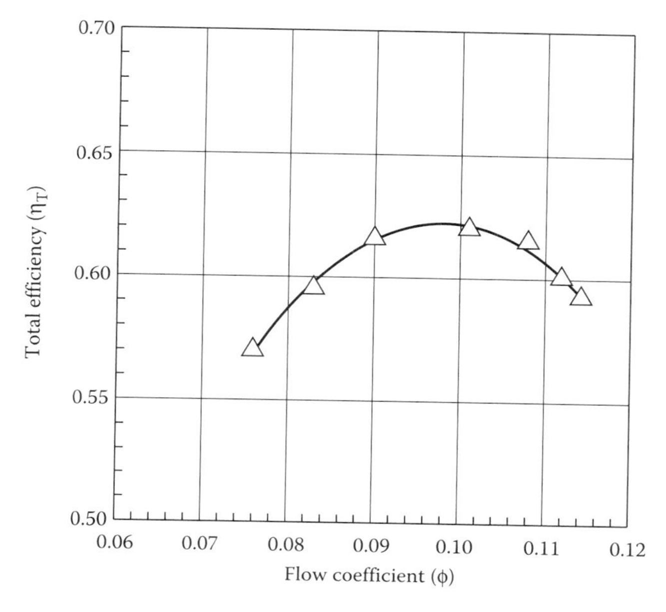

FIGURE P5.7b Total efficiency of a tube axial fan.

Analyze based on available diameters and the speed limitations. Select based on best fit to Figure P5.7 and minimum power required.

5.8. Begin with a Cordier analysis, then use Figure P5.7 to choose a tube axial fan suitable for $Q = 200{,}000$ cfm at $\Delta p_s = 3.75$ in. wg, with a low noise as a selection criterion.

5.9. Use Table 5.2 to search for a fan to provide $Q = 6250$ cfm at $\Delta p_T = 0.75$ in. wg, in standard air.

5.10. A farmer needs to pump 900 gpm of 85°F water at 100 ft of head from an open pond. A suction lift of 8 ft is required. Select a pump that can perform to this specification with an adequate cavitation margin. Try for the best efficiency attainable.

5.11. Select a fan to deliver 15,000 cfm of air (68°F at 1 atm) that provides a static pressure rise of 4.0 in. wg. To keep costs down, try to avoid a need for more than 12 hp, and choose a comparatively quiet fan.

5.12. A pump is required to supply 50 gpm of 85°F water at 50 ft of head. NPSHA is only 10 ft. Which is the smallest pump available?

5.13. Find a fan that can supply 100,000 cfm of 134°F air with $p_b = 28.7$ in. Hg (barometric pressure) at a pressure rise of 1.0 in. wg. Use minimum noise as a criterion for the best choice.

5.14. Identify a minimum sound power level blower that can deliver 25 m^3/s of standard density air ($\rho = 1.2\ kg/m^3$) at 500 Pa static pressure rise.

5.15. A small-parts assembly line involves the use of solvents requiring a high level of ventilation. To achieve removal of these toxic gases, a mean velocity of 8 ft/s must be maintained across the cross-section of a 12 ft high by 30 ft wide room. The toxicants are subsequently removed in a wet scrubber or filter, which causes a pressure drop of 30 lb/ft^2. Using a Cordier analysis, select the design parameters for a fan that will provide the required flow and pressure rise, while operating at or near 90% total efficiency, and that minimizes the diameter of the fan. Give dimensions and speed of the fan and describe the type of fan chosen.

5.16. For the performance requirements stated in Problem 5.15, constrain your selection of a fan to less than $L_W = 100$ dB. Relax the size and efficiency requirements in favor of this noise requirement.

5.17. A water pump must supply 160 gpm at 50 ft of head, and the pump is provided with 4.0 ft of NPSHA. For a single-suction pump direct-connected to its motor, determine the smallest pump one can use without cavitation (define D, N, and η).

5.18. A small axial flow fan has its performance at BEP defined by $Q = 2000$ cfm, $\Delta p_S = 0.10$ in. wg, and $\rho = 0.00233$ slug/ft^3 with a diameter of 20 in.

a. Estimate the specific speed and required rpm.

b. What type of fan should one use?

c. How much motor (shaft) hp is needed for the fan?

d. Estimate the sound power level in dB.

5.19. An aircraft cabin must be equipped with a recirculation fan to move the air through the heaters, coolers, filters, and other air conditioning equipment. A typical recirculation fan for a large aircraft must handle about 0.6 m^3/s of air at 1.10 kg/m^3 (at the fan inlet) with a total pressure rise requirement of 5 kPa. The electrical system aboard the aircraft has an alternating current supply at a line frequency of 400 Hz, and the fan must be configured as a direct-connected impeller for reliability. Size the fan impeller to achieve the best possible static efficiency.

5.20. If one constrains the fan of Problem 5.16 to generate a sound power level of < 85 dB, how are design size and efficiency affected?

5.21. For the recirculation fan of Problem 5.19, if one must keep the sound power level below 90 dB, what are the design options?

5.22. The performance requirements for an air mover are stated as flow rate, $Q = 60$ m^3/min; total pressure rise, $\Delta p_T = 500$ Pa. Ambient air density, $\rho = 1.175$ kg/m^3. The machine is constrained not to exceed the 90 dB sound power level.

a. Determine a suitable speed and diameter for the fan and estimate the total efficiency.

b. Correct the total efficiency for the effects of Reynolds number, a running clearance of 2.5 × Ideal, and the influence of a belt-drive system (assume a drive efficiency of 95%).

5.23. Use the dimensionless performance curves of this chapter to select three candidates for the requirements of Problem 5.22, which closely match with the results of your preliminary design decision in that problem.

5.24. To evacuate a stone quarry (to preclude the pleasure of unauthorized swimming), we need a pump that can move at least 0.5 m^3/s against a head of 80 m. The pump must be floated on a barge with its inlet 1.125 m above the surface of the 12°C water. The atmospheric pressure during pumping may fall as low as 98 kPa. Select a noncavitating pump for this application and match your modeling of size and speed with the best available pump from Figure P5.24.

5.25. Repeat the fan selection exercise of Problem 5.7 using the flow rate of 2 m^3/s but changing the static pressure rise to 140 Pa and then again with 35 Pa. Compare the results with each other and with the results of Problem 5.7. Use the dimensionless curves of Figure P5.7 for the fans.

5.26. A water pump, which is required to provide 21 m^3/h with 16 m of head, must operate with 1.125 m of NPSHA. For a single-suction machine, directly connected to its motor, determine the smallest pump that can be used, defining the D, N, and η. What would the minimum size be for an NPSHA of 0.5 m?

5.27. The FD fan of Problem 4.27 was required to provide 25 m^3/s of air with a static pressure rise of 10 kPa subject to an inlet density of 0.57 kg/m^3. At the imposed speed of 1500 rpm, estimate the fan size, its static and total efficiencies, and the required input power.

5.28. The inlet air temperature for the fan of Problem 5.27 was $T = 265°C$. High temperature fans frequently require very large radial clearances to accommodate the thermal distortion of the inlet cone

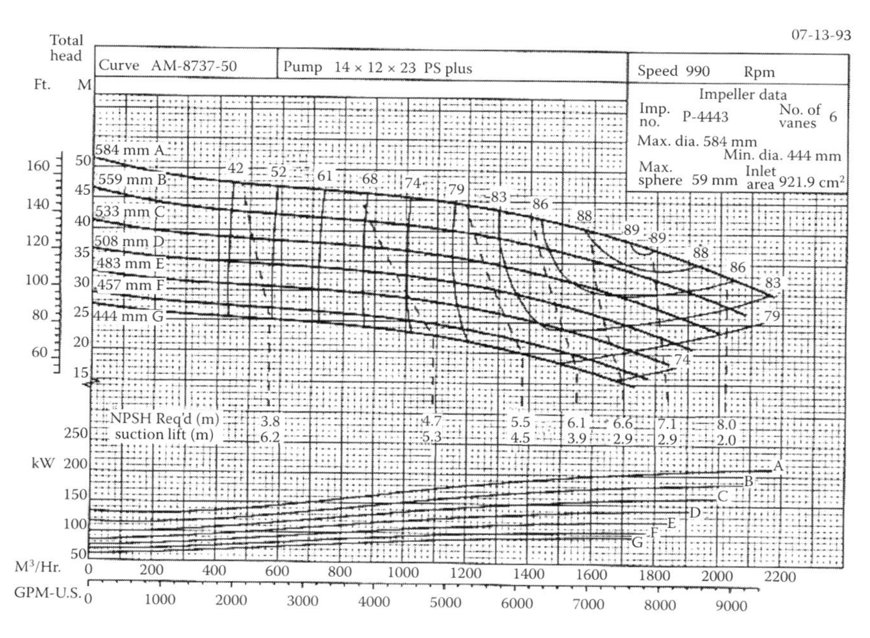

FIGURE P5.24a Pump performance curves (990 rpm) (ITT Fluid Systems. With permission).

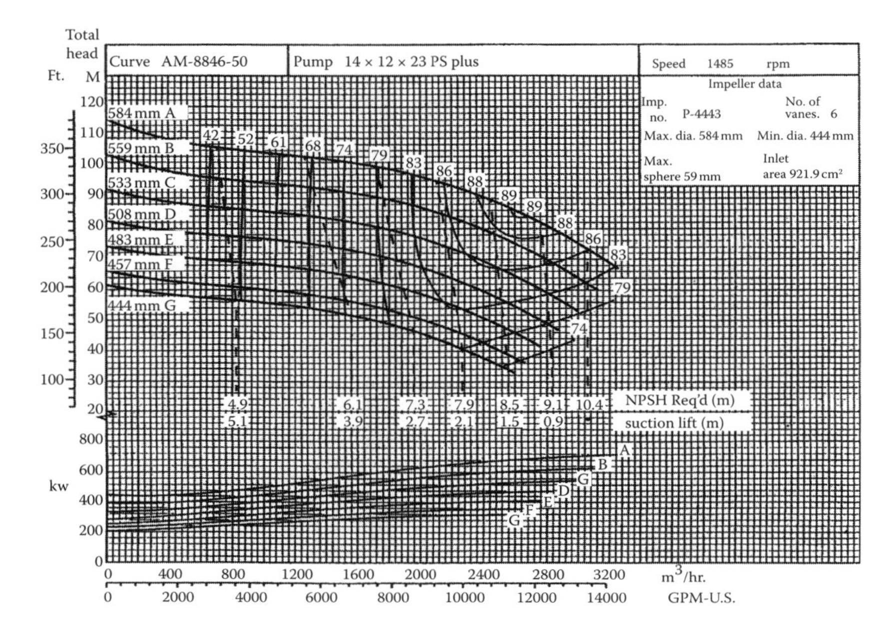

FIGURE P5.24b Pump performance curves (1485 rpm) (ITT Fluid Systems).

relative to the impeller. Use a clearance ratio of C/D = 8 and determine the actual radial clearance of this fan in mm. Use $T = 265°C$ to modify the kinematic viscosity of the air and correct the total efficiency for both Reynolds number and the generous clearance.

5.29. A lime-kiln fan similar to the one described in Problem 4.29 has performance requirements of $Q = 250{,}000$ cfm and $\Delta p_S = 45$ in. wg with an inlet density of 0.037 lbm/ft^3. The fan is assumed to be a double-width impeller configuration and is to be sized for high total efficiency. Determine size and power requirements for a fan to meet this specification. Correct the total efficiency and required power for the Reynolds number (set C/D = 1).

5.30. Assume that the variable pitch fan of Figure 5.15 can be arranged in a multistage configuration to achieve very high-pressure rise performance. For an FD application requiring $Q = 100{,}000$ cfm and $\Delta p_T =$ 28 in. wg (at $\rho = 0.072$ lbm/ft^3), select a variable pitch fan and specify size and number of stages from the data of Figure 5.17. Use $N = 1470$ rpm.

5.31. In the discussion of the flow regions in Section 5.2, the "hybrid" machines are mentioned. One example is the Airfoil Plenum Fan shown, along with its performance chart in Figure P5.31. These fans may offer a practical alternative for air handling or conditioning, or in retrofit applications. Elimination of the conventional scroll can allow acceptable performance in a smaller space. Estimate

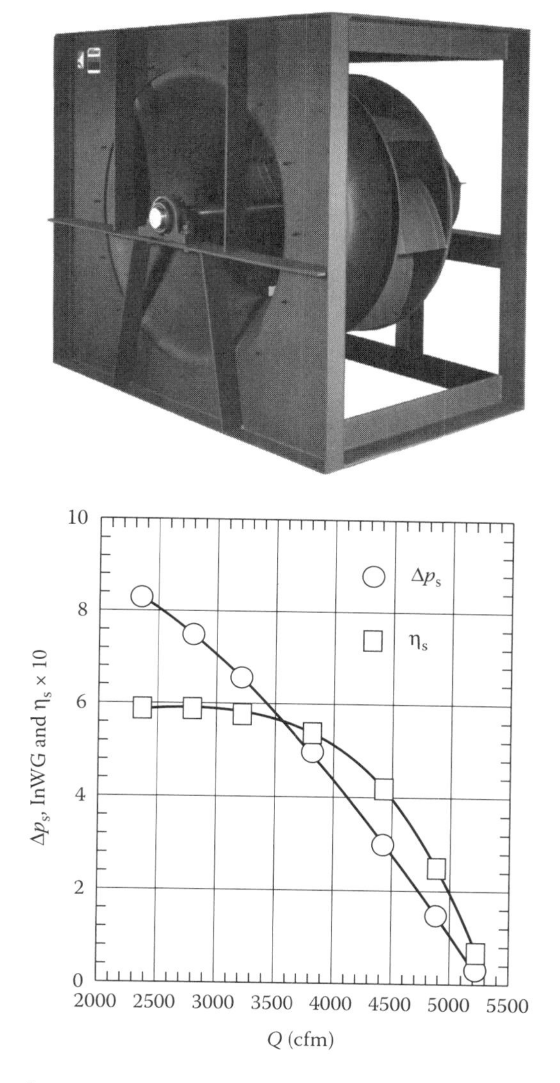

FIGURE P5.31 Configuration and static performance of a centrifugal Airfoil Plenum Fan (no scroll installed). $D = 15$ in., $N = 3600$ rpm, $\rho g = 0.075$ lbf/ft^3. (From Chicago Blower Corporation. (1998) *Fan Catalogs.* With permission.)

the specific speed and diameter of the fan and compare with the expected region for a centrifugal fan.

5.32. A small woodworking shop (similar to the one considered in Problem 4.25) has a ventilation requirement of 1.5 m^3/s with a static pressure rise of 1.5 kPa. Because of very tight space constraints on the fan installation, it would be advantageous to use a compact plenum fan as illustrated in Figure P5.31. Is it possible to use a Class I version of the fan or must the more robust—and more expensive—Class II version of the fan be selected? [Hint: Work from the specific speed and diameter values of the given plenum fan in the region where $\eta_s = 0.58$.]

5.33. The small plenum fan of Problem 5.31 is rated at a sound pressure level of 79 dBA at a distance of 10 ft from the open inlet of the fan (based on a free field environment). Use the modified Graham method of Chapter 4 to estimate L_P 10 ft from the inlet of the 15-in., 3600-rpm fan and compare the estimate with the rated value. The best total efficiency was found to occur at $Q = 3,200$ cfm in Problem 5.31.

5.34. Electric power generating stations use many different pumps (one expert estimates that at least 100 pumps may be used in a 500 MW unit). The three main types used in the plant cycle are boiler feed pumps, condensate pumps, and (condenser) circulating water pumps. The following paragraphs describe the performance requirements for these three types of pumps as applied in a 500-MW generating unit.

Boiler feed pump. This pump delivers water through the high-pressure feedwater heaters and into the boiler. The pump takes suction from the deaerating feedwater heater (essentially an open tank where liquid water and steam are mixed at approximately 200 psia) and delivers water at approximately 3200 psia. As the hot water has a specific weight of about 58 lb/ft^3, the pump's head is about 7500 ft. The flow rate is about 4,200,000 lb/h or 9000 gpm.

Condensate pump. This pump takes suction from the condenser hotwell, where the just-condensed water has a pressure of about 1.5 psia, and delivers it through the low-pressure feedwater heaters into the deaerating feedwater heater. The deaerator is at about 200 psia and is typically located about 175 ft above the pump. Accounting for friction losses, and with a water density of about 61 lb/ft^3, the condensate pump must develop a head of about 700 ft. The flow is about 6900 gpm.

Circulating water pump. This pump supplies cooling water to the condenser. Water is drawn from a lake or river (or from a cooling tower), pumped through the tubes in the condenser, and returned to the source. The head is only about 30 ft and is due only to friction losses. The flow rate is in the neighborhood of 200,000 gpm.

a. If all pumps have efficiency of 85%, estimate the power required for each. [Use efficiency of 85% only for part (a).]

b. Which pump is the largest (impeller diameter)? Which is the smallest? Explain.

c. Which pump is the fastest (rpm)? Which is the slowest? Explain.

d. Which pump is more likely employed in parallel with other identical pumps? Why?

e. Which pump is almost certainly built in several stages? Why? (A multistage pump has several impellers in series—all contained in the same casing and mounted on a common shaft—so a multistage pump is really the same as several pumps in series.)

f. Which pump(s) is likely driven by a steam turbine? Which are likely driven by electric motors? Explain.

g. To save initial cost, the condensate pump is built as a single, one-stage pump and driven at about 1800 rpm (so the specific speed of the condensate pump is around 0.4). Of the remaining pumps, one has impeller(s) designed for a specific speed of 0.6 and the other has impeller(s) designed for a specific speed of 3.0. Which pump (boiler feed or circulating water) has which specific speed (0.6 or 3.0)? Sketch the impeller shape for each of the three pumps.

h. For each pumping system, using the appropriate specific speed for each, specify a reasonable rotating speed, a reasonable number of stages if multistage, or a reasonable number of identical pumps in parallel if applied in parallel. (Hint: 300 rpm $<$reasonable speed $<$6000 rpm; reasonable number of stages or machines in parallel $<$8.)

5.35. A plant has a water pump that has been in service for a few years. The pump is driven by an AC electric motor. A check of plant monitoring instruments indicates the following performance data: Flow $\approx$4300 gpm; Head $\approx$50 ft; and Motor power input $\approx$52 kW. A check of the motor nameplate reveals that it operates at 1750 rpm and has a maximum output capacity of 60 kW. A change in plant operating strategy now requires that the pump operates at 50% capacity for 12 h per day and full capacity for the other 12 h per day. Modifications must be made to the pumping system to accommodate these changes. Options to be considered are (1) install a throttling valve and (2) install a variable speed drive between the pump and the motor. Unfortunately, all manufacturers' information on the pump has been lost.

a. Assuming that the pump was originally optimally matched with the system (i.e., it operated at BEP), sketch a set of performance curves (Head–Flow and Power–Flow) for the pump. The curves must be as realistic as possible.

Now consider the expected pump performance at 50% flow.

b. Estimate the pump head for the throttling option, the required pump speed for the variable-speed option, and the power for both options (assume that the variable-speed drive has an efficiency of 95%).

c. Some costs associated with both options are as follows:

Throttling valve	\$800
Variable speed drive	(\$) = 850 (hp/10)$^{0.87}$ [hp = Input horsepower]
Electricity	\$0.07/kWh.

Company procedures require payback of a capital investment in 1 year and do not assign a time value to money. Calculate the annual cost for both options and recommend the most economical.

5.36. Coal-fired power plants frequently employ a "balanced draft" condition in the steam generator furnace. The furnace is kept at a pressure slightly below atmospheric (typically, −0.5 in. wg). These units then employ large FD fans to supply air to the furnace and ID fans to draw the combustion products from the furnace and through the air heater and flue gas cleanup equipment. A 650 MW unit is expected to generate 6.5×10^6 lb/h of flue gas. The total pressure requirement is estimated to be 23 in. wg. In order to allow some "margin" to accommodate air in-leakage and system blockage, the design point for the ID fan is set 10% higher on both flow and pressure. The gas temperature is 300°F and it can be modeled as air with density 0.050 lbm/ft^3 and kinematic viscosity 3.1×10^{-4} ft^2/s. If a centrifugal fan design is selected, then two DWDI fans will be employed (equivalent to four fans operating in parallel). If an axial fan design is selected, then four separate fans operating in parallel will be used. The fans will be driven by large, high-quality electric motors (assume 0.5% slip and 98.5% motor efficiency) that are direct-coupled to the fans.

a. What are the design requirements (flow and pressure) for a single fan (i.e., one of the four axials or one side of the two DWDI centrifugals). Neglect any compressibility effects.

b. Using the number of motor poles (2, 4, 6, 8, 10, 12 ...) as the independent variable, specify a fan (speed, size, and type) to minimize the electrical power consumption by the fans. What is the total electrical power consumption by the fans?

A model of the selected fan is to be laboratory tested to verify the performance and to develop the performance curve. The model fan is to be 36 in. in diameter, driven by a 1750 rpm motor, and handle standard air. Neglect any compressibility effects.

c. What values of flow and total pressure rise would be expected to verify the design (BEP) point?

d. What would be the expected efficiency of the test fan? (Correct for Reynolds number and roughness effects [if any]. Both model and full-scale fans will be fabricated from steel with roughness 0.00015 ft.)

f. What power would be required by the test fan?

5.37. Select a pumping system to deliver 100,000 gpm against a head of 25 ft (these specifications would be typical for condenser circulating water pumping system for a 250-MW generation unit). Specify a number of identical pumps, between one and five, operated in parallel. Specify the speed (i.e., number of motor poles and slip); estimate the total power required; and sketch the impeller shape (including size). The pump(s) will be driven by synchronous AC electric motor(s).

6

Fundamentals of Flow in Turbomachinery

6.1 Preliminary Remarks

In the previous chapters, attention was focused on understanding and estimating the performance of turbomachines. In a sense, the emphasis was on what various machines do, as opposed to how they do it. This chapter begins with the consideration of the details of the flow processes within turbomachines. The essential features are the interaction of the flowing fluid with a moving blade row and the accompanying interchanges of momentum, energy, and pressure within the fluid. First, consideration is given to the geometry of blading and the vector relationships between fluid and blade velocities. Then, the fundamental equations describing rotor/fluid energy transfer are developed. Next, certain limitations on fluid diffusion are considered. Finally, preliminary application of the principles to design and performance analysis is illustrated. Later chapters then add both breadth and depth to design and performance prediction methods.

6.2 Blade and Cascade Geometry

Figure 6.1 shows a single blade in a fluid stream. Such a two-dimensional shape is commonly called an *airfoil*. In a turbomachine rotor, both the blade and the fluid are moving. Here it is assumed that the blade is stationary and the fluid is moving, so that the velocity of the fluid relative to the blade is considered.

The blade length is measured by the *chord* (c), a straight line between the trailing edge and the leading edge. The *camber line* is a (generally) curved line connecting the leading and trailing edges and lying halfway between the upper and lower surfaces. If the camber line and the chord line are coincidental straight lines, the blade is said to be straight; otherwise the blade is curved. The thickness of the blade is the distance between the upper and lower surfaces, perpendicular to the camber line. Typically, the "upper," more highly curved surface, is called the *suction surface* because the pressure is, on average, lower

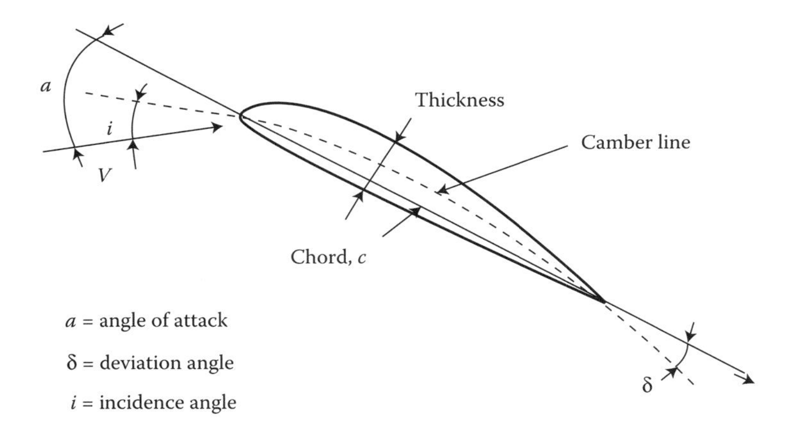

FIGURE 6.1 A single (airfoil) blade in a fluid stream showing geometry and nomenclature.

than the pressure in the approach stream and the "lower," less curved surface, is called the *pressure surface* because the pressure is, on average, greater than the pressure in the approach stream. A line tangent to the camber line at the leading and trailing edges defines the so-called blade direction at these edges.

The angle between the fluid relative velocity vector and the chord line at the leading edge is called the *angle of attack* (a). The angle of attack is considered positive as shown, when approaching from "below." The angle between the fluid relative velocity vector and the (tangent to the) camber line at the leading edge is called the *angle of incidence* or simply, the *incidence* (i). Incidence is positive as shown and, for a curved blade, is usually smaller than the angle of attack. The angle between the fluid relative velocity vector leaving the blade and the (tangent to the) camber line at the trailing edge is called the *deviation* (δ). Typically, the fluid relative velocity vector lies "above" the trailing edge direction, that is, the blade fails to turn the fluid stream sufficiently so that it perfectly follows the blade at the leading edge.

As shown in Figure 6.1, blades/airfoils typically have a rounded leading edge, which allows them to perform well with a finite incidence angle, and a sharp trailing edge, which fixes flow separation from the surface at a definite point. A particularly simple model of a blade and the flow over it is to assume that

- The blade has zero thickness, so that it collapses to its camber line (think of a thin sheet of metal bent into the shape of the camber line)
- The incidence is zero (the approach flow is aligned with the leading edge)
- The deviation is zero (the leaving flow follows the blade perfectly).

This model will be used throughout this chapter and the following chapter, although, at times, a blade will be sketched with a more realistic profile shape.

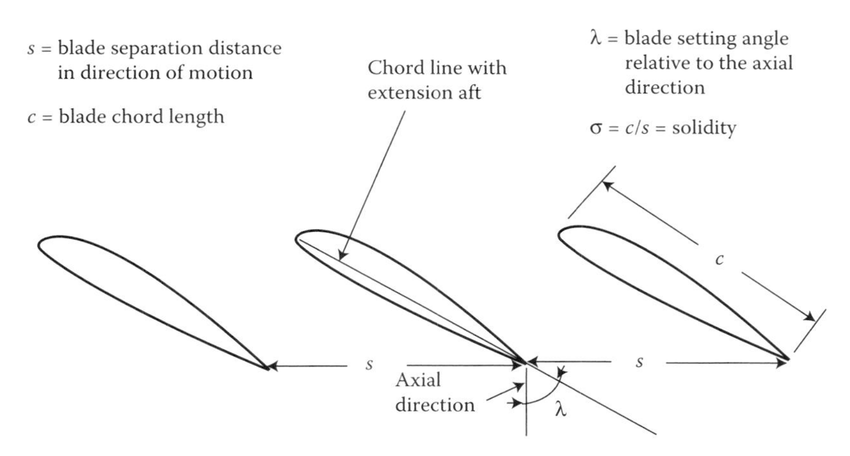

FIGURE 6.2 A blade cascade.

Turbomachines, of course, never consist of a single, isolated, two-dimensional blade. A number of blades (never less than two) are arranged in repetitive fashion around an impeller, and the blades have finite height (or width, in the case of a radial flow impeller). A more sophisticated model than a single blade is a two-dimensional *cascade* of blades, defined as an infinitely repeating array of identical blades, as shown in Figure 6.2. In addition to the geometric properties of the individual blades as defined above, one is also concerned with the blade-to-blade spacing (s) and the *setting angle* (λ) (the complementary, angle $90° - \lambda$, is sometimes called the pitch angle) of the blades relative to the normal to the plane of the cascade itself. Detailed consideration of flow in cascades, both planar (representing axial flow machines) and radial (representing radial flow machines), will be delayed until Chapter 8. For the present, note that if the *solidity*, $\sigma \equiv c/s$, is about 1 or larger, the flow passage between adjacent blades can be modeled as a curved channel, with relative flow parallel to the centerline between the blades. This model will be used extensively in this and the next chapter. Alternatively, for very low solidity, blades may indeed behave as isolated airfoils.

6.3 Velocity Diagrams

When a fluid flows in a turbomachine rotor, there are always two relevant velocities; the fluid velocity and the velocity of the rotor associated with its rotation. Arguably, the most significant similarity parameter for turbomachinery is the flow coefficient, ϕ, which was shown in Chapter 2 to represent the ratio of these two velocities (actually, the ratio of the two speeds, without any consideration of direction). When considering the fluid mechanics of turbomachinery, one is always concerned with, seemingly, *three* velocities, namely

- The linear velocity of the (rotor) blade ($\boldsymbol{U}$)—usually written simply as U because the direction is known to be tangent to the rotor. U is related to radius and rotational speed by U (m/s) $= r$ (m) $\times N$ (Hz)
- The velocity of the fluid in an absolute frame of reference (i.e., relative to the machine casing) ($\boldsymbol{C}$)
- The velocity of the fluid relative to the rotor/blade ($\boldsymbol{W}$).

Of course, there are in fact only two actual velocities, fluid and rotor, but there are two different frames of reference.*

The relationship between these velocities is extremely important; it is

$$\boldsymbol{C} = \boldsymbol{U} + \boldsymbol{W}. \tag{6.1}$$

This relationship is typically represented in a *velocity diagram*; examples of which are shown in Figure 6.3. Note that a fixed blade and a moving blade are also shown for reference. Since a velocity diagram usually forms a plane triangle, the vector notation is dropped to give the more common

$$C = U + W. \tag{6.2}$$

In this book, diagram angles are referred to the blade (tangential) velocity U. Angle α, the angle between the absolute velocity C and the tangential velocity U, is called the absolute angle. Angle β, the angle between the relative velocity W and the tangential velocity U, is called the relative angle. As will soon be evident, the various components of the velocity vectors are often important. The axial or radial component ($C \sin \alpha$ [$= W \sin \beta$]) represents the throughflow, while the tangential component ($C_\theta = C \cos \alpha$) is important in calculating energy exchange.

A velocity diagram sketch is an indispensable aid in visualizing and analyzing flow in a turbomachine. The following pointers often assist in making these sketches:

- Because of the vector addition, W and U always meet head to tail
- Owing to the parallelogram rule of vector addition, there are always two ways to draw any diagram
- C is nearly tangent to the blade direction (camber line) of fixed blades, especially when exiting a fixed blade row
- W is nearly tangent to the blade direction (camber line) of moving blades, especially when exiting a moving blade row.

* In this and later chapters, the general velocity symbol, $\boldsymbol{V}$ or V will be used for a fluid velocity when either C or W may be implied (as in flow over a blade which could be either fixed or moving).

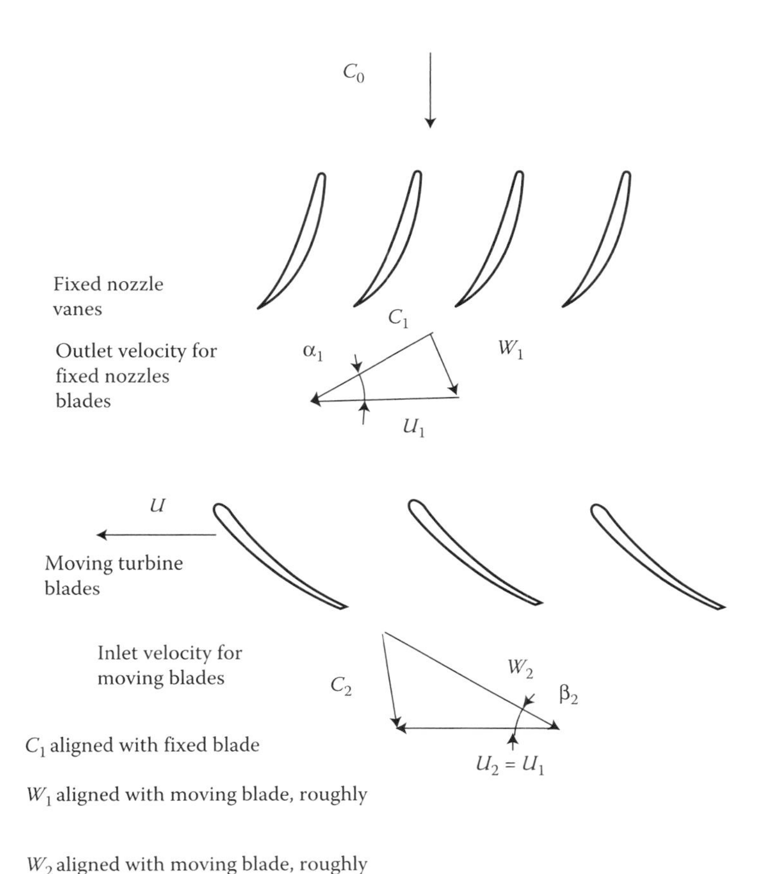

FIGURE 6.3 Velocity diagrams with fixed and moving blades for orientation.

6.4 Energy (Work) Transfer in a Rotor

An expression to evaluate the work done on/by the fluid passing through the moving rotor of a turbomachine can be developed through application of the Reynolds Transport Theorem to the angular momentum principle (Newton's second law of motion for systems with rotational motion). Recall from earlier study of fluid mechanics that the rate of change of an arbitrary extrinsic property of a system, B, as the system in question passes through a control volume, can be determined through the relation:

$$\frac{\mathrm{d}B_{\text{sys}}}{\mathrm{d}t} = \frac{\partial}{\partial t}\iiint_{\text{CV}} \rho b \,\mathrm{d}\forall + \iint_{\text{CS}} b\rho(\boldsymbol{C}\cdot\boldsymbol{n})\,\mathrm{d}A, \tag{6.3}$$

where $b = \mathrm{d}B/\mathrm{d}m$, CV is the control volume, and CS is the control surface. In this and the following equations, the symbol $\boldsymbol{C}$ is used for velocity because Newton's second law only applies in absolute (i.e., "inertial") reference frames while in this chapter, the more commonly used $\boldsymbol{V}$ can stand for either absolute or relative velocity.

To obtain the form for the angular momentum principle, note that $B_{\text{sys}} = \boldsymbol{H} = \int (\boldsymbol{r} \times \boldsymbol{C})\, \mathrm{d}m$; $b = \boldsymbol{r} \times \boldsymbol{C}$, and

$$\frac{\mathrm{d}H}{\mathrm{d}t} = \sum \boldsymbol{M} = \sum (\boldsymbol{r} \times \boldsymbol{F}). \tag{6.4}$$

Combining Equations 6.3 and 6.4 and using the terms specific to angular momentum, one obtains

$$\sum \boldsymbol{M} = \frac{\partial}{\partial t} \iiint_{\text{CV}} \rho(\boldsymbol{r} \times \boldsymbol{C})\, \mathrm{d}\forall + \iint_{\text{CS}} \rho(\boldsymbol{r} \times \boldsymbol{C})(\boldsymbol{C} \cdot \boldsymbol{n})\, \mathrm{d}A. \tag{6.5}$$

Now consider the flow through the rotor of an arbitrary turbomachine, as shown in Figure 6.4.

The following assumptions and restrictions are made:

- Steady flow is assumed ($\partial/\partial t = 0$)
- There is no "leakage" so that $\dot{m}_{\text{out}} = \dot{m}_{\text{in}} = \dot{m}$

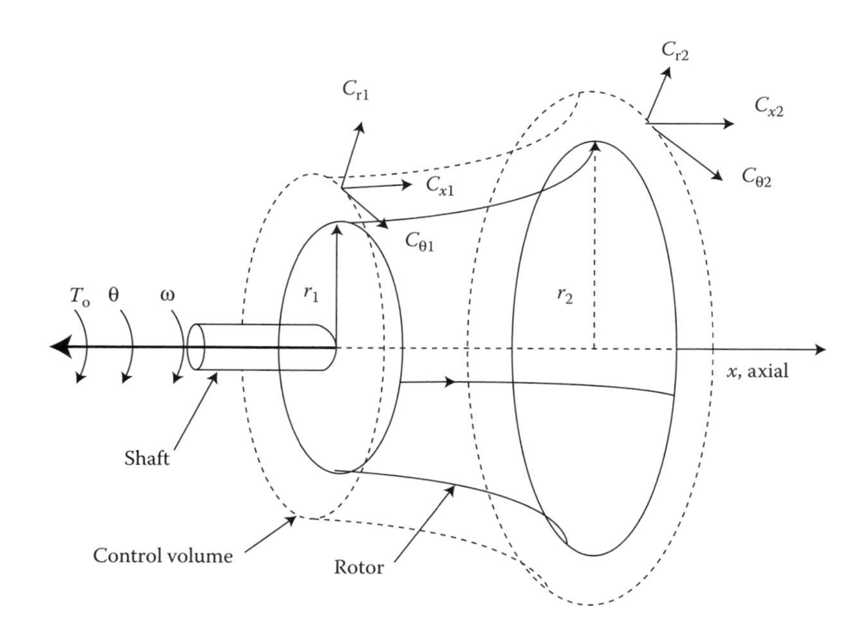

FIGURE 6.4 Control volume enclosing a turbomachine rotor.

- The moment of momentum equation, Equation 6.5, is applied to only the component in the direction of the shaft axis (i.e., only moments and momentum about the shaft axis are considered)
- The only moment about this axis is the shaft torque T_0
- The angular momentum, $\boldsymbol{r} \times \boldsymbol{C}|_{\text{axis}} = rC_\theta$, is uniform over the inlet and the outlet. (The angle θ defines the direction or rotation such that $N(\text{Hz}) = \mathrm{d}\theta/\mathrm{d}t$.)

With these conditions, Equation 6.5 gives

$$T_0 = \dot{m}_{\text{out}} r_2 C_{\theta 2} - \dot{m}_{\text{in}} r_1 C_{\theta 1} = \dot{m}(r_2 C_{\theta 2} - r_1 C_{\theta 1}). \tag{6.6}$$

From the principles of mechanics, the power transmitted in the rotating shaft is

$$P_{\text{shaft}} = T_0 N = N\dot{m}(r_2 C_{\theta 2} - r_1 C_{\theta 1}) = \dot{m}(N r_2 C_{\theta 2} - N r_1 C_{\theta 1}),$$

but $U = rN$, so this becomes

$$P_{\text{shaft}} = \dot{m}(U_2 C_{\theta 2} - U_1 C_{\theta 1}).$$

From an energy balance, the power transmitted in the shaft is the power exchanged with the flowing fluid. From the point of view of thermodynamics, work and power require an algebraic sign. The proper sign is determined as follows. If the torque and power are positive [i.e., if $(r_2 C_{\theta 2} > r_1 C_{\theta 1})$], then the torque and the direction of rotation are the same. This would mean that the shaft and rotor are driving the fluid so that the device is acting as a pump and work is being done on the fluid. In thermodynamics, this is negative work, so the proper form for the equation would be

$$-P = \dot{m}(U_2 C_{\theta 2} - U_1 C_{\theta 1}). \tag{6.7}$$

Dividing by the mass flow rate gives

$$-w = U_2 C_{\theta 2} - U_1 C_{\theta 1}, \tag{6.8}$$

where w is the work per unit mass (the specific work). This extremely important equation is known as *Euler's Pump and Turbine Equation*. Conventionally (Shepherd, 1956), the equation is written separately for turbines and for pumping machines as

$$w = U_2 C_{\theta 2} - U_1 C_{\theta 1} \quad \text{(pumps)}, \tag{6.9}$$

$$w = U_1 C_{\theta 1} - U_2 C_{\theta 2} \quad \text{(turbines)}, \tag{6.10}$$

with no sign on the specific work.

Euler's equation is without question the most widely used equation in the analysis of turbomachinery flows. It is, unfortunately, often misunderstood. Note carefully the following points:

- This is a general relation for incompressible or compressible flow and for ideal (frictionless) or viscous fluid.
- It is restricted to steady flow.

The torque in Equation 6.6 and the power in Equation 6.7 are in fact the net values transmitted to the fluid by the rotor and blades. To obtain the actual shaft torque or power, one must account for any extraneous torques associated with bearing and seal drag, or disk fluid friction.

The restriction to uniform values of fluid angular momentum across the inlet and outlet can be relaxed. Equations 6.8 through 6.10 can be considered to apply to a single streamline extending from some point in the inlet to some point in the outlet. An analysis across inlets and outlets with varying properties can be carried out by integration or summation and averaging as necessary.

Most importantly, one must realize that since Euler's equations are developed from mechanics (rather than thermodynamics), work is calculated from force and motion, without the need to assume any "process," ideal or otherwise. The quantity calculated is actually the work—no efficiency is needed.

Example: A Turbine Stage

Consider a single stage of a turbine, including fixed nozzle blades and moving rotor blades, as shown in Figure 6.5. For this example, $\dot{m} = 20\,\text{kg/s}$ and $U_1 = U_2 = U = 1047\,\text{m/s}$ (i.e., the flow does not shift radially upon passing through the stage). The flow enters axially in the turbine nozzle section and exits axially from the turbine rotor section. The nozzle approach flow is turned by the nozzle vanes through 45° to an absolute velocity of $C_1 = 500\,\text{m/s}$. The flow then approaches the moving blade row for which $U = 1047\,\text{m/s}$.

The power delivered to the rotor blades is

$$P = \dot{m}(U_1 C_{\theta 1} - U_2 C_{\theta 2}) = \dot{m}U(C_{\theta 1} - C_{\theta 2}).$$

The components of the absolute velocity are

$$C_{\theta 1} = C_1 \cos\alpha_1 = 500 \times \cos(45^\circ) = 353.6\,\text{m/s},$$

$$C_{\theta 2} = C_2 \cos(90^\circ) = 0.$$

Then

$$P = \dot{m}U(C_{\theta 1} - 0) = 20\,(\text{kg/s}) \times 1047\,(\text{m/s}) \times 353.6\,(\text{m/s})$$

$$= 7{,}404{,}400\,(\text{kg}\cdot\text{m}^2/\text{s}^3) = 7.4044 \times 10^6\,(\text{N}\cdot\text{m/s}) = 7.404\,\text{MW}.$$

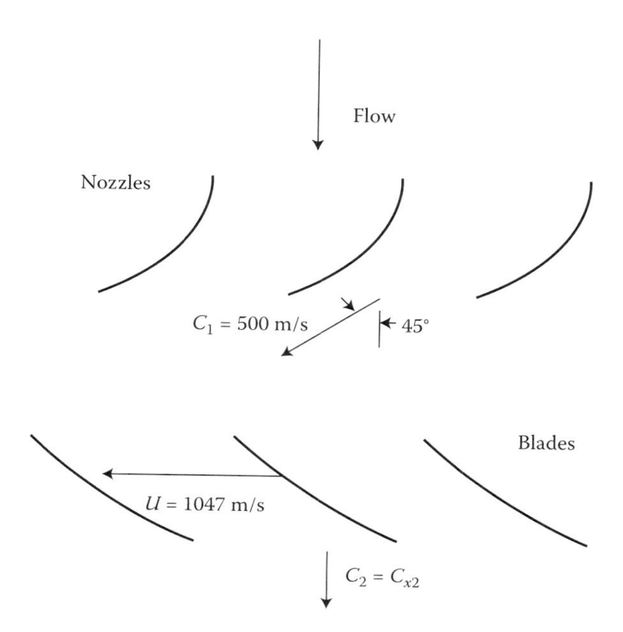

FIGURE 6.5 Cascade or planar representation of a turbine stage.

One should note the following:

- For this example, the fixed nozzle vanes and the moving blades were reduced to their camber lines
- The velocity diagrams show that C_1 is roughly parallel to the vanes at the exit from them and W_1 and W_2 are roughly parallel to the moving blade at entrance and at exit from the row
- It was not necessary to specify anything about the flowing fluid; it might have been steam (compressible), water (this is unlikely at the speeds involved), or anything else.

Actual power delivered to an external load would be reduced by mechanical losses such as bearing and seal friction so that

$$P_{\text{output}} = P - P_{\text{mechanical}} = \eta_M \times P \quad \text{(turbine)}, \tag{6.11a}$$

where η_M is the mechanical efficiency. If η_M were, say, 97%, then the power delivered to the external load would be reduced to 7.182 MW.

Note that if a pumping machine were considered, then the input power equation would be

$$P_{\text{input}} = P + P_{\text{mechanical}} = \frac{P}{\eta_M} = \frac{\dot{m}(U_2 C_{\theta 2} - U_1 C_{\theta 1})}{\eta_M} \quad \text{(pump)}. \tag{6.11b}$$

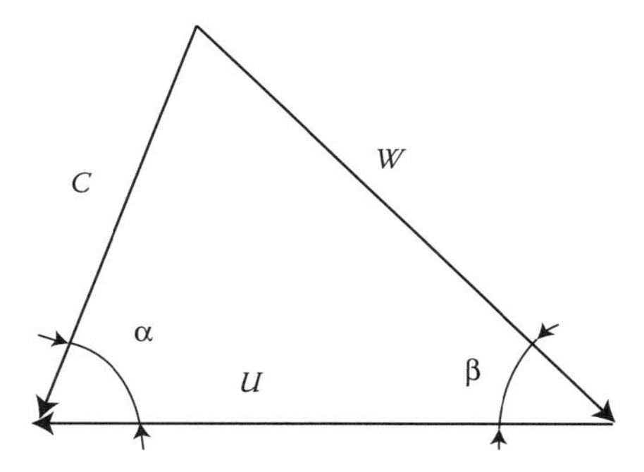

FIGURE 6.6 A velocity diagram.

There is an alternate form for Euler's equation, which can be developed from the trigonometry of the velocity diagram. Figure 6.6 shows a single velocity diagram with the three velocity vectors U, C, and W and the two angles α and β.

From the law of cosines

$$W^2 = U^2 + C^2 - 2UC\cos\alpha,$$

but $C\cos\alpha = C_U = C_\theta$ since the "U" and "θ" directions are identical. Substituting and rearranging give

$$UC_\theta = \frac{1}{2}(U^2 + C^2 - W^2).$$

From Euler's equation, $-w = U_2C_{\theta 2} - U_1C_{\theta 1}$ so that

$$-w = \frac{C_2^2 - C_1^2}{2} + \frac{U_2^2 - U_1^2}{2} - \frac{W_2^2 - W_1^2}{2}. \tag{6.12}$$

This alternate form is fully equivalent to the original Euler equation, Equation 6.8. It is sometimes more convenient. Additionally, it shows the three means by which a rotor exchanges energy (work) with a fluid stream:

- The first term, $(C_2^2 - C_1^2)/2$, is the increase in the fluid's (absolute) kinetic energy. It is sometimes called the *external term* because if a pressure rise is desired, it must occur external to the rotor, corresponding to a reduction in the absolute velocity.
- The second term, $(U_2^2 - U_1^2)/2$, is called the *centrifugal term* and can be thought of as work or potential energy associated with an (imaginary) "centrifugal force field." This term is nonzero, only if the fluid changes radius as it flows through the rotor.
- The third term, $-(W_2^2 - W_1^2)/2 = (W_1^2 - W_2^2)/2$, is called the *relative diffusion term*. It contributes to a pressure rise if the fluid decelerates relative to the rotor.

6.5 Work, Head, Pressure, and Efficiency

Euler's equation, in either form (Equation 6.8 or 6.12), calculates the actual specific work exchanged between the rotor and the fluid. Including the mechanical efficiency as in Equation 6.9a or 6.9b evaluates the shaft output or input power. Of at least equal interest is the change of useful *fluid* specific energy, that is, either the head or pressure. These are related to the work exchanged with the fluid by an internal efficiency, so in order to calculate them, this efficiency must be known.

If it is assumed that there are no hydraulic/thermodynamic losses, then the efficiency would be 100% and one would calculate the so-called *theoretical head* (Wislicenus, 1965):

$$gH_{th} = -w = U_2C_{\theta 2} - U_1C_{\theta 1}; \quad H_{th} = \frac{gH_{th}}{g}.$$

H_{th} is called the theoretical head because it corresponds to a "theoretical" machine with no losses, not because the "theoretical" Euler equation is used to calculate work. Because of this confusion over the meaning of the word "theoretical," H_{th} will not be used further in this book.

For any real machine, an appropriate efficiency is required to calculate the head or pressure change. For incompressible flow, the appropriate efficiency is the hydraulic efficiency, η_H so that

$$gH = \frac{\Delta p}{\rho} = \eta_H(-w) = \eta_H(U_2C_{\theta 2} - U_1C_{\theta 1}) \quad \text{(pump or fan)}, \tag{6.13}$$

$$gH = \frac{\Delta p}{\rho} = \frac{-w}{\eta_H} = \frac{U_2C_{\theta 2} - U_1C_{\theta 1}}{\eta_H} \quad \text{(turbine)}. \tag{6.14}$$

Alternatively, if one works with the hydraulic losses, then

$$gH = \frac{\Delta p}{\rho} = -w - \text{"Losses"} = (U_2C_{\theta 2} - U_1C_{\theta 1}) - \sum gh_L. \tag{6.15}$$

For compressible flow, adiabatic flow is assumed and the appropriate efficiency is either the isentropic efficiency, η_s, or the polytropic efficiency, η_p. The equations are

$$gH_s = \Delta h_{0s} = \eta_s(-w), \tag{6.16a}$$

$$\frac{p_{02}}{p_{01}} = \left[1 + \eta_s \frac{U_2C_{\theta 2} - U_1C_{\theta 1}}{c_pT_{01}}\right]^{\gamma/\gamma-1}, \tag{6.16b}$$

$$\frac{p_{02}}{p_{01}} = \left[1 + \frac{U_2C_{\theta 2} - U_1C_{\theta 1}}{c_pT_{01}}\right]^{\eta_p(\gamma/\gamma-1)} \quad \text{(compressors)} \tag{6.16c}$$

and

$$gH_s = \Delta h_{0s} = \frac{(-w)}{\eta_s}, \tag{6.17a}$$

$$\frac{p_{02}}{p_{01}} = \left[1 + \frac{U_2C_{\theta 2} - U_1C_{\theta 1}}{\eta_s c_p T_{01}}\right]^{\gamma/\gamma-1}, \tag{6.17b}$$

$$\frac{p_{02}}{p_{01}} = \left[1 + \frac{U_2C_{\theta 2} - U_1C_{\theta 1}}{c_p T_{01}}\right]^{(1/\eta_p)(\gamma/\gamma-1)} \quad \text{(turbines).} \tag{6.17c}$$

Of course, values for the efficiencies must be known if any of these equations are to be useful. In many situations, values can be reasonably assumed or estimated from Cordier or similar relationships. Empirical or ad hoc models for component (nozzle and diffuser) efficiencies or losses can often be used as well as more complete data from cascade experiments (see Chapter 8). Sophisticated viscous/turbulent flow models utilizing computational fluid dynamics methods are sometimes available for state-of-the art design or analysis.

Note that, for compressible flow machines

$$-w = U_2C_{\theta 2} - U_1C_{\theta 1} = h_{02} - h_{01}, \tag{6.18}$$

for either compressors or turbines, irrespective of any efficiencies.

Example: A Simple Pumping Machine

Consider a pumping (axial flow fan) example (Figure 6.7). Here, the full thickness of the blades is shown. The flow with $\dot{m} = 12$ slug/s approaches the blade row axially at $C = 100$ ft/s, and the blades are moving at $U = U_1 = U_2 = 300$ ft/s. The flow exits the blade row at a 15° angle to the axial (x) direction, that is, the blades have deflected the flow through 15° of "fluid turning." Find the power input to the fluid and estimate the total pressure rise if the fluid is air.

Assume that the blade height (perpendicular to the paper) is the same at the front and rear of the blades. Also, neglect any density change. Then, to maintain the given mass flow, the axial velocity of the fluid leaving the blade row must be the same as that entering the blade row, that is, $C_{x2} = C_{x1} = C_1 = 100$ ft/s. Using Euler's equation, the power transferred to the fluid is

$$P = \dot{m}(U_2C_{\theta 2} - U_1C_{\theta 1}) = \dot{m}U(C_{\theta 2} - C_{\theta 1}),$$

where

$$C_{\theta 2} = C_{x2}\tan(\theta_{fl}) = 100 \times \tan 15° = 26.8\ \text{(ft/s)} \quad \text{and } C_{\theta 1} = 0.$$

Thus,

$$P = 12\ \text{(slug/s)} \times 300\ \text{(ft/s)} \times 26.8\ \text{(ft/s)} = 96{,}480\ (\text{lb}\cdot\text{ft/s}) = 175.4\ \text{hp}.$$

The specific work done on the fluid is

$$-w = \frac{P}{\dot{m}} = UC_{\theta 2} = 300 \times 26.8 = 8040\ (\text{ft}^2/\text{s}^2).$$

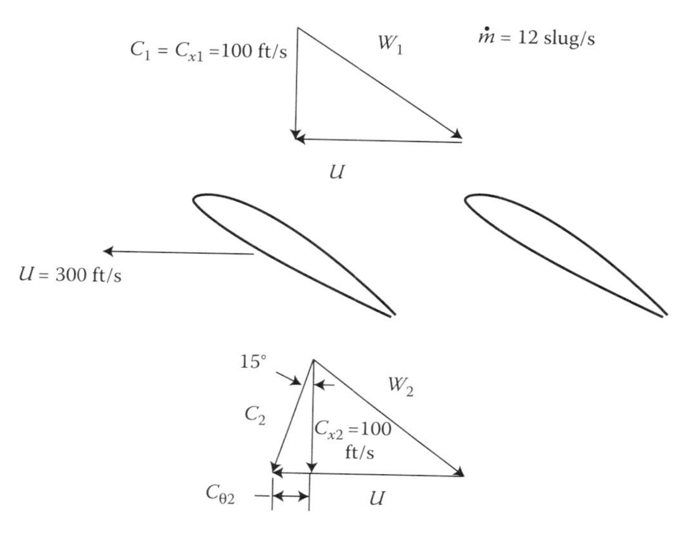

FIGURE 6.7 Blade row of an axial flow fan.

To estimate the pressure rise, the efficiency is needed. This machine is an axial fan (a fairly high-speed one with $U = 300$ ft/s). The specific speed is perhaps around 5, so Cordier would give an efficiency estimate of about 0.84. Assuming that this is a "typical" machine and that the data correspond to the design (BEP) point, use $\eta_H \approx 0.8$ to give

$$\Delta p_T = \rho g H = \rho \eta_H(-w) \approx 0.0023\ (\text{slug/ft}^3) \times 0.8 \times 8040\ (\text{ft}^2/\text{s}^2) \left(\frac{1.0 \text{ in. wg}}{5.2 \text{ lb/ft}^2} \right) = 2.84 \text{ in. wg.}$$

The amount of shaft power (which must account for the mechanical losses) that must be input depends on the mechanical efficiency of the machine. For $\eta_M \approx 0.95$, this fan would require about 185 hp (175.4/0.95).

6.6 Preliminary Design of an Axial Fan

Next, take a look at a more complete example for a typical turbomachine to examine what happens to the velocity vectors as flow moves through the machine. These vectors will be related to the performance variables and also to the shape of the (moving) blade and (stationary) vane elements. The results will be used to introduce an important limitation on pumping machinery.

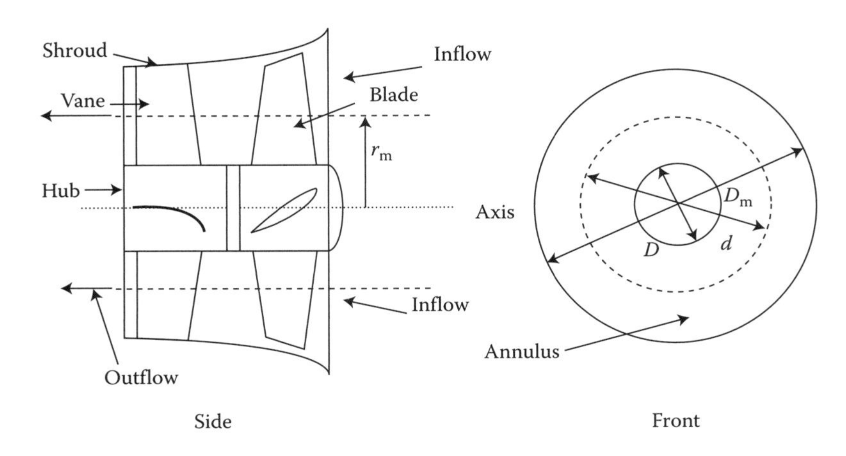

FIGURE 6.8 A full-stage vane axial fan.

Consider a full-stage vane axial fan with a single moving blade row followed by a single fixed vane row. The purpose of the fixed vanes is to "straighten" the flow, removing the swirl and converting the kinetic energy to pressure. The fan has an outer diameter of $D = 1.0$ m and an inner or hub diameter $d = 0.67$ m. We will analyze blades, vanes, work input, and pressure rise along the "mean radius" (r_m), as shown in Figure 6.8, where $r_m = 1/2(r_{tip} + r_{hub}) = (D + d)/4$.

The fan is to operate at 900 rpm, producing $\Delta p_T = 1250$ Pa and $Q = 2.0\,\text{m}^3/\text{s}$ with an air density of $\rho = 1.2\,\text{kg/m}^3$. The mean radius is $r_m = 0.417$ m and $N = (2\pi/60) \times 900$ rpm $= 94.3\ \text{s}^{-1}$. The blade velocity is $U_m = U_1 = U_2 = N \times r_m = 94.3 \times 0.417 = 39.3$ m/s. The inlet axial velocity is assumed to be equal to the average velocity of the flow over the annulus area between the hub and the tip (tip leakage and the blockage of the blades are neglected). Also, as there are no vanes ahead of the rotor to "preswirl" the fluid, C_1 has no tangential component, thus

$$C_1 = C_{x1} = \frac{Q}{A_{annulus}} = \frac{Q}{(\pi/4)(D^2 - d^2)} = \frac{4 \times 2}{\pi(1.0^2 - 0.67^2)} = 4.62\,\text{m/s}.$$

One can combine these velocities to examine the relative velocity, W_1, at the leading edge of the blade as shown in Figure 6.9. Since the inlet diagram is a right triangle

$$W_1 = \left[U_M^2 + C_1^2\right]^{1/2} = \left[39.3^2 + 4.62^2\right]^{1/2} = 39.6\ \text{m/s} \quad \text{with}$$

$$\beta_1 = \tan^{-1}\frac{C_{x1}}{U_m} = \tan^{-1}\frac{4.62}{39.3} = 6.7^\circ.$$

For widely spaced blades (low solidity), the blade chord should be aligned at its leading edge with the relative velocity W_1, such that the flow turns

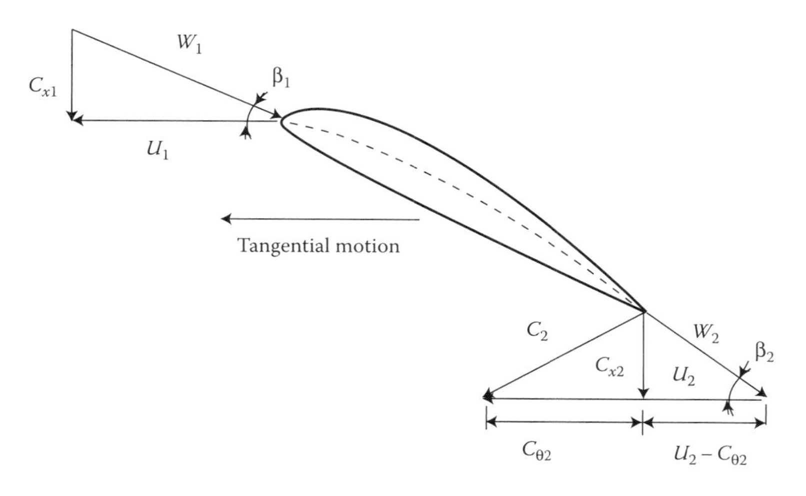

FIGURE 6.9 Axial flow fan blade with inlet and outlet velocity diagrams.

upward slightly and moves smoothly over the blade with minimum initial disturbance. For closely spaced blades (moderate to high solidity), the proper alignment moves to become tangent to the camber line at the leading edge. This alignment of the velocity vector and blade is illustrated in Figure 6.9. The fan considered here will be assumed to have moderate to high solidity and, soon, the blades will be collapsed to their chord line.

Now consider the flow at the trailing edge of the blade. First, calculate C_{x2} by imposing conservation of mass for incompressible flow so that $C_{x2} = (Q/A_{\text{annulus}}) = C_{x1}$. Next, use the specified pressure rise with the Euler equation so that

$$\Delta p_{\text{T}} = \rho\eta_{\text{T}}(-w) = \rho\eta_{\text{T}}U_{\text{m}}(C_{\theta 2} - C_{\theta 1}) = \rho\eta_{\text{T}}U_{\text{m}}C_{\theta 2}.$$

An estimate of the fan total efficiency is necessary to proceed. The point of this analysis will not be affected by assuming that $\eta_{\text{T}} = 1$, so that will be done. Then

$$C_{\theta 2} = \frac{\Delta p_{\text{T}}}{\rho\eta_{\text{T}}U_{\text{m}}} = \frac{1250\ \text{N/m}^2}{1.2\ \text{kg/m}^3 \times 1.0 \times 39.3\ \text{m/s}} = 21.2\ \text{m/s}.$$

The above assumptions comprise the "simple-stage" model for an axial flow turbomachine. Next, use $C_{\theta 2}$, U_{m}, and C_{x2} to construct the velocity diagram at the exit of the blade row, also illustrated in Figure 6.9. The relative velocity W_2 can then be calculated as

$$W_2 = \left(W_{2\theta}^2 + W_{2x}^2\right)^{1/2} = \left[(U_{\text{m}} - C_{\theta 2})^2 + C_{2x}^2\right]^{1/2}$$
$$= \left[(39.3 - 21.2)^2 + 4.62^2\right]^{1/2} = 18.7\ \text{m/s}$$

and the flow angle is

$$\beta_2 = \tan^{-1}\left(\frac{W_{2x}}{W_{2\theta}}\right) = \tan^{-1}\left(\frac{C_{2x}}{U_m - C_{2\theta}}\right) = \tan^{-1}\left(\frac{4.62}{39.3 - 21.2}\right) = 14.3°.$$

The blade relative inlet flow velocity is related to the outlet relative velocity as follows:

$$\frac{W_2}{W_1} = \frac{18.7}{39.6} = 0.472; \quad \theta_{fl} = \beta_2 - \beta_1 = 14.3° - 6.7° = 7.6°.$$

The ratio of velocities relative to the blade, W_2/W_1 for a moving row and C_2/C_1 for a fixed row, is called the *de Haller ratio* and θ_{fl} is the *angle of deflection* or *flow turning angle* caused by the blade acting on the fluid. Both provide a measure of how hard the blade is working or the *blade loading*. The larger the flow turning angle, the harder the blade (or vane) is working (it is also said that the blade is more highly loaded). Conversely, a smaller de Haller ratio corresponds to a harder working, more highly loaded, blade or vane (Wilson, 1984; Bathie, 1996).

One should also examine the vane row loading and flow turning. The vane row inlet velocity is the absolute velocity leaving the blade row (Figure 6.10). Using previous results, one obtains

$$C_2 = \left[C_{x2}^2 + C_{\theta 2}^2\right]^{1/2} = \left[4.62^2 + 21.2^2\right]^{1/2} = 21.7 \text{ m/s}$$

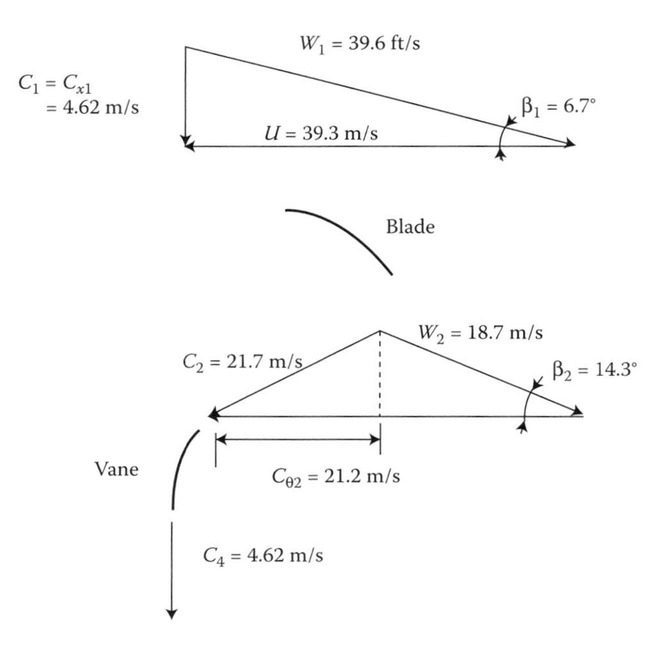

FIGURE 6.10 Complete layout of blade and vane elements for axial fan.

and

$$\alpha_2 = \tan^{-1}\left(\frac{C_{x2}}{C_{\theta 2}}\right) = \tan^{-1}\left(\frac{4.62}{21.2}\right) = 12.3^\circ.$$

Since the purpose of a vane row, at least for a single-stage fan, is to straighten the flow to a pure axial discharge, the flow must be turned through $\theta_{fl} = 77.7^\circ$ by the vanes so that $\alpha_3 = 90^\circ$. The required de Haller ratio is $(C_3/C_2) = (C_{x1}/C_2) = (4.62/21.7) = 0.213$.

In this fan, the flow is decelerated in both the blade row ($W_2/W_1 < 1$) and the vane row ($C_3/C_2 < 1$). In both cases, the fluid kinetic energy (relative to the blade) is decreased, with a resulting rise in pressure. Each blade row is acting as a diffuser. Do these values of the fluid turning and de Haller ratio seem reasonable?

6.7 Diffusion Considerations

In the previous example, the vane appears to be working considerably harder than the blade. While the flow turning angle seems reasonable at 6.7° for the blade row, the value of nearly 78° for the vanes seems rather extreme. In both rows, the de Haller ratio seems quite low, implying velocity reductions of the order of two to one and five to one. To determine whether these values are reasonable, examine the process of diffusion (fluid deceleration with a corresponding rise in pressure) in a duct or channel. Most books on fluid mechanics present a brief treatment of diffuser flow (e.g., Fox et al., 2009). An outstanding visual presentation of diffusion phenomena is included in the video program *Flow Visualization* (NCFMF, 1963). Typically, a subsonic diffuser is a diverging flow passage, as shown in Figure 6.11.

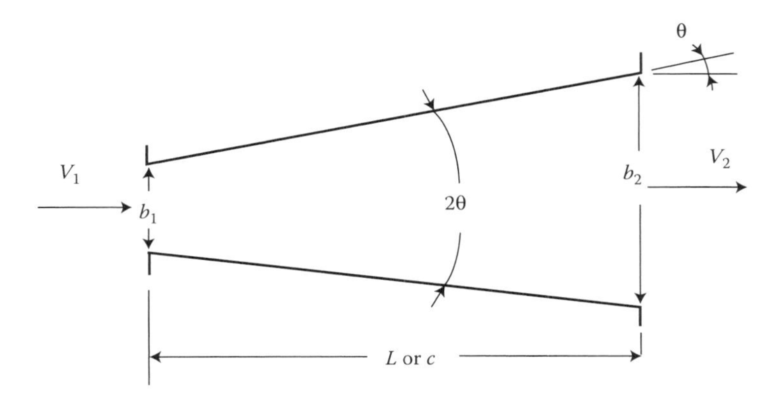

FIGURE 6.11 Typical planar (two-dimensional) diffuser showing nomenclature.

It has been well established that only a limited amount of deceleration or diffusion can occur in a flow passage. Beyond this limit, the flow undergoes high energy and momentum loss associated with boundary layer separation from the walls of the diffusing passage. The flow streamlines no longer follow the direction of solid surfaces constraining the fluid, and the flow will generally become unsteady and unstable. This breakdown is referred to as *stall*. There are actually several different stall regimes, as illustrated in the well-known (empirical) diffuser performance map illustrated in Figure 6.12. Generally, stall is a flow condition that should be carefully avoided in design, selection, and operation of any flow passage or fluid-handling machine.

How might this information be applied to a turbomachine blade row? A blade-to-blade channel is compared with a two-dimensional planar diffuser in Figure 6.13. The velocities entering and leaving the blade row are labeled V_1 and V_2, implying absolute velocity for fixed blades and relative velocity for moving blades. When excessive deceleration in either device is attempted, the results are similar—flow separation and stall. It follows that the de Haller ratio, V_2/V_1, must be above a certain limit to avoid stall. This limit is developed, at least approximately, as follows.

Applying the continuity equation to the straight-axis planar diffuser gives $V_2 b_2 h = V_1 b_1 h$, where b is the width of the opening and h is the depth, perpendicular to the paper. Then

$$\frac{V_2}{V_1} = \frac{b_1}{b_2} \approx \frac{b_1}{b_1 + 2\theta L} = \frac{1}{1 + 2\theta(L/b_1)},$$

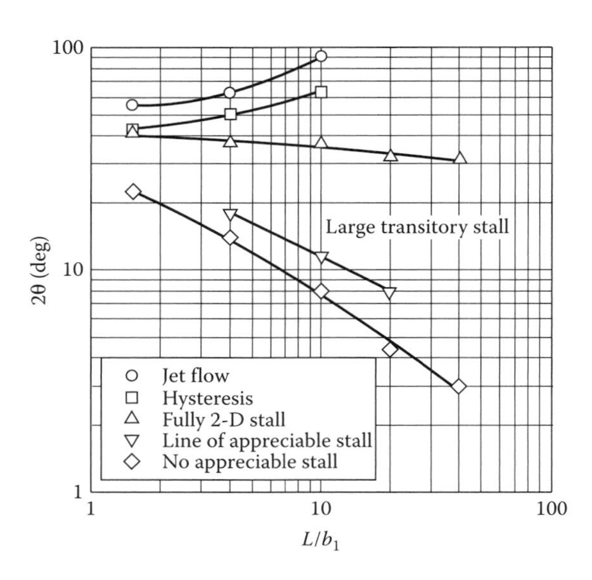

FIGURE 6.12 Planar diffuser performance map. (After Reneau, L.R., Johnson. J.P., and Kline, S.J. 1967. *ASME Journal of Basic Engineering*, 89(1). With permission.)

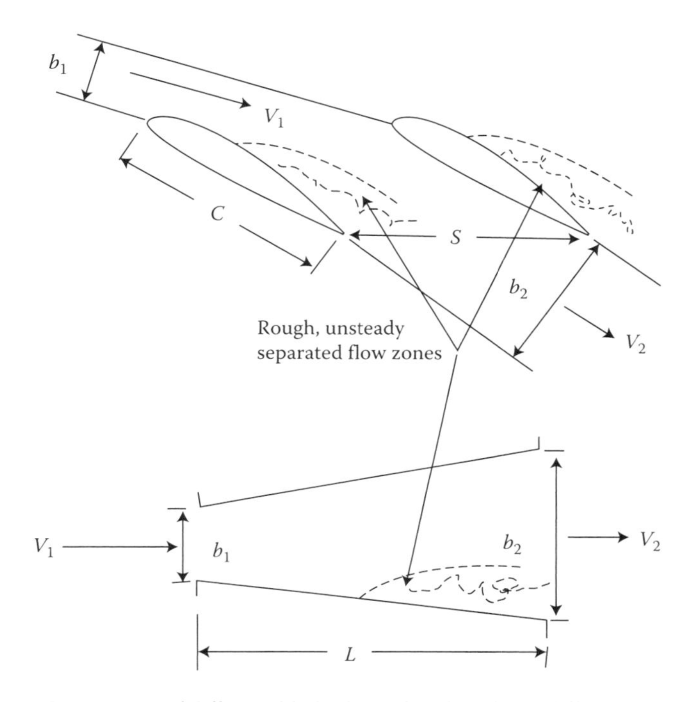

FIGURE 6.13 Comparison of diffusing blade channel with a planar diffuser.

where θ is the diffuser opening angle, in radians. In order to avoid stall, the diffuser must operate along the lower line ("No Appreciable Stall") in Figure 6.12. Consider two values near the left side of the chart:

L/b_1	2θ (°)	2θ (rad)	V_2/V_1
1	24	0.42	0.70
2	20	0.35	0.60

Recognizing that L/b_1 in the planar diffuser corresponds (at least approximately) to the solidity, σ ($=c/s$) in the blade channel, the data can be correlated in the following rule for the de Haller ratio:

$$\frac{V_2}{V_1} \geq \frac{1.05}{1+0.45\sigma} \quad \text{(to avoid stall in a diffusing blade row).} \tag{6.19}$$

The limit stated by de Haller, for $\sigma \approx 1$, matches the correlation very well; he suggested that $(V_2/V_1)_{\min}$ should be at least 0.72 for good, clean, stall-free flow over the blades. Increasing blade row solidity allows a slight increase in blade loading or diffusion.

Before leaving this topic, two additional diffuser performance parameters are introduced. The pressure coefficient for a diffuser is defined by

$$C_p \equiv \frac{p_2 - p_1}{(1/2)\rho V_1^2}. \tag{6.20}$$

From Bernoulli's equation, if there are no losses, then $C_p = 1 - (V_2/V_1)^2$. In an ideal diffuser, it would be possible to have C_p approach a value of 1; that is, all of the kinetic energy could be converted to pressure. Considering the de Haller limit, however, the actual value for C_p is limited to about 0.5–0.6.

An actual diffuser or diffusing blade row will have losses, even if there is no stall. These losses are quantified by either a *diffuser efficiency*, η_D or a *loss coefficient*, ζ_D.These are defined as follows:

$$\eta_D \equiv \frac{p_2 - p_1}{(p_2 - p_1)_{\text{ideal}}} = \frac{p_2 - p_1}{(1/2)\rho V_1^2\left[1 - (b_1/b_2)^2\right]} \tag{6.21}$$

and

$$\zeta_D \equiv \frac{gh_{L,D}}{(1/2)V_1^2}$$

so that

$$p_2 - p_1 = \frac{1}{2}\rho V_1^2\left[1 - \left(\frac{b_1}{b_2}\right)^2 - \zeta\right]. \tag{6.22}$$

Finally, it should be noted that there is *no de Haller ratio limit on nozzle (flow-accelerating) blade rows* because the flow does not stall under acceleration.

6.8 Diffusion Limits in Axial Flow Machines

In axial flow machines, a practical limit on solidity, dictated by the consideration of overall machine weight and surface area for skin frictional losses, is about two or a little more (Johnsen and Bullock, 1965). This implies (from the diffuser analogy/de Haller ratio) an overall lower limit on V_2/V_1 of about 0.55. For now, use the conservative level chosen by de Haller of $V_2/V_1 > 0.72$ as a criterion for judging acceptable blade loading.

Reconsider the axial fan "design" from Section 6.6. In that example, W_2/W_1 was 0.470 for the blade row and C_3/C_2 was 0.213 for the vane row. Thus, one concludes that the blades are much too heavily loaded, and the blade row is probably stalled, since the de Haller ratio is much less than 0.72. The vane row is in even more distress at a de Haller ratio of only 0.213, and one must conclude that the fan as designed would not perform anywhere near the specified point. One can examine this contention by comparing the fan with the Cordier diagram. For the desired performance:

$$N_s = \frac{NQ^{1/2}}{(\Delta p_T/\rho)^{3/4}} = \frac{94.3 \times (2)^{1/2}}{(1250/1.2)^{3/4}} = 0.72$$

and

$$D_s = \frac{D(\Delta p_T/\rho)^{1/4}}{Q^{1/2}} = \frac{1.0 \times (1250/1.2)^{1/4}}{2.0^{1/2}} = 4.02.$$

This places the fan near the Cordier line, but it lies in Region E. This is the upper range for radial discharge machines, whereas the axial fan as a full-stage machine should be in Region C.

There is a simple explanation for the blade overloading problem. The design is trying to achieve unrealistically high pressure rise for an axial fan, without sufficient blade speed to generate the pressure rise. Viewed from the perspective of the alternate form of Euler's equation, Equation 6.12, the specification demands too much work from the external and relative diffusion terms for the axial design, whereas a radial design that also makes use of the centrifugal effect could supply the work efficiently.

If (perhaps for cost reasons) one insists on using an axial fan, then it will be necessary to design one with a different size and speed to achieve the specified performance. From the Cordier diagram, one can choose N_s or D_s values from Region C and determine what the axial fan should look like. Suppose one chooses $D_s = 1.75$ from the middle of Region C, so that $N_s = 8.26D_s^{-1.936} = 2.8$. The proper speed and diameter are calculated as follows:

$$N = N_s \left[\frac{(\Delta p_T/\rho)^{3/4}}{Q^{1/2}}\right] = 2.8 \times \frac{(1250/1.2)^{3/2}}{2.0^{1/2}} \times \frac{60}{2\pi} = 3470\,\text{rpm}$$

and

$$D = D_s \left[\frac{Q^{1/2}}{(\Delta p_T/\rho)^{1/4}}\right] = 1.75 \times \frac{2.0^{1/2}}{(1250/1.2)^{1/2}} = 0.436\,\text{m}.$$

That is, one needs to design a much smaller fan running at a much higher speed. Now rework the mean radius blade calculations. Using $d = 0.67D$ as before, calculate the new r_m as $r_m = (0.436 + 0.291)/4 = 0.182\,\text{m}$. With $N = 3470\,\text{rpm} = 363\,\text{s}^{-1}$, $U_m = 363 \times 0.182\,\text{m} = 66.1\,\text{m/s}$. From d and D, one obtains $A_{annulus} = 0.0819\,\text{m}^2$, so $C_{x1} = Q/A_{annulus} = 2/0.0819 = 24.4\,\text{m/s}$.

Then $W_1 = [66.1^2 + 24.4^2]^{1/2} = 70.5\,\text{m/s}$ with $\beta_1 = \tan^{-1}(24.4/66.1) = 20.3°$. Using the Euler equation,

$$C_{\theta 2} = \frac{\Delta p_T}{\rho \eta_T U_m} = \frac{1250\,\text{N/m}^2}{1.2\,\text{kg/m}^3 \times 1.0 \times 66.1\,\text{m/s}} = 15.8\,\text{m/s}.$$

Then

$$W_2 = \left[(66.1 - 15.8)^2 + 24.4^2\right]^{1/2} = 55.9\,\text{m/s} \quad \text{and}$$

$$\beta_2 = \tan^{-1}\frac{24.4}{66.1 - 15.8} = 25.9°.$$

For the new blade row, $W_2/W_1 = 55.9/70.5 = 0.79$ (compared to 0.470) and $\theta_{fl} = 25.9° - 20.3° = 5.6°$ (compared to 7.6°). This is much better and indicates acceptable blade loading.

For the vane, $C_2 = [(15.8)^2 + (24.4)^2]^{1/2} = 29.1\,\text{m/s}$ and $\alpha_3 = 57.1°$. Then, $C_3 = C_{x1} = 24.4\,\text{m/s}$, yielding $C_3/C_2 = 24.4/29.1 = 0.84$ (compared to 0.213). Finally, $\theta_{fl} = 32.9°$ (compared to 77.7°). The de Haller ratio and fluid turning values for the redesigned fan are a little conservative, since 0.72 is a slightly conservative number in itself. However, one can conclude that the fan stage properly designed by using the Cordier criterion for speed and size is reasonable. If an axial fan were chosen for this duty, one would expect a fan with a diameter somewhat <0.5 m but running at high speed (e.g., a 2-pole motor, direct-coupled). Such a choice would yield a sound pressure level, estimated by the method in Chapter 4, of about 96 dB, due to the relatively high tip speed, which is rather loud for this level of performance.

For this example, the hub diameter was chosen rather arbitrarily to be two-thirds of the tip diameter. Had another value for d/D been chosen, a different value for C_{x1} and $C_{\theta 2}$ would have been obtained. All of the velocity diagrams and blade shapes would have changed along with the de Haller ratios. For example, if the hub–tip diameter ratio were $d/D = 0.5$, there would be a decrease in C_{x1} (larger annulus area) and an increase in $C_{\theta 2}$ (lower mean radius, giving lower mean blade speed), so that the (blade row) de Haller ratio decreases from 0.79 to 0.73. Conversely, if the relative hub size increases to $d/D = 0.75$, the de Haller ratio increases to 0.82. A little work along these lines, taking note of the de Haller diffusion limit, can be used to establish a constraint on d/D for a given specification of performance.

A general study of the dependence of the level of diffusion on the value of d/D can be carried out over a range of specific speed or specific diameter appropriate to axial flow machines in order to establish such a constraint. One examines the de Haller ratio calculated at the hub station of the blade, thereby examining the most strenuous diffusion conditions on the blade or vane row. At the hub station, the value of U is the smallest so that the value of $C_{\theta 2}$ required to achieve the specified value of head or pressure rise will be the largest value required. (The underlying assumption here is that the work done or the total pressure rise achieved is the same at each radial position in the fan annulus. This is not an absolute restriction on the design of axial blade rows, but it is an approximate assumption, since large variations in the work distribution across the blade will lead to serious difficulties in design. This distribution question will be considered at length in Chapter 9, for three-dimensional flow. For now, a somewhat conservative constraint on minimum hub size can be established using this assumption.)

The fan in the previous example, with $d/D = 0.67$, would have a value of U at the hub of $U_h = N \times r_h = 52.0\,\text{m/s}$. The Euler equation, using $\Delta p_T = 1250\,\text{Pa}$ and $1.2\,\text{kg/m}^3$ density and still assuming 100% efficiency, would require $C_{\theta 2,h} = 1250/(1.2 \times 1.0 \times 52.0) = 19.9\,\text{m/s}$. The de Haller ratio then becomes $W_2/W_1 = 40.6/57.7 = 0.70$, significantly smaller than the mean radial station value of 0.79. The "de Haller limit" of 0.72 can be relaxed to allow a less conservative velocity ratio criterion near the hub because the solidity is greater there. Generally, one can allow the hub station value of W_2/W_1 to decrease to around 0.60, but should continue to use the 0.70–0.72 value as a guideline for mean station analysis. For the smaller hub case considered earlier with $d/D = 0.50$, the hub station de Haller ratio would be reduced to $W_2/W_1 = 21.7/42.9 = 0.506$. Thus, a hub size that seemed to be acceptable based on the mean station calculations now appears to be too small, even allowing a diffusion level of 0.60. Clearly, the acceptable value of hub ratio for this fan is between 0.50 and 0.67. A linear interpolation suggests $d/D = 0.58$, which yields $W_2/W_1 = 0.607$. Then the smallest hub size for this fan is about 0.25 m. Recall that this (redesigned) fan has specific speed $D_s \approx 1.75$ and specific diameter $N_s \approx 2.8$.

If one carries out a systematic study (Wright, 1996) to find the smallest hub–tip ratio capable of generating a de Haller ratio (at the hub radius) of 0.6 for a range of N_s values (from 1 to 8), the results can be plotted as shown in Figure 6.14. Along with Wright's diffusion-based values, a curve

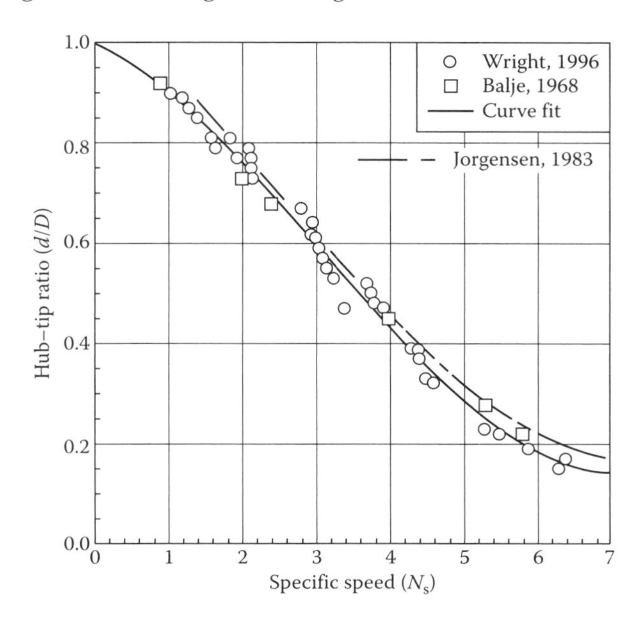

FIGURE 6.14 Hub size constraint for uniformly loaded blades.

from *Fan Engineering* (Jorgensen, 1983) is shown as well. Jorgensen's curve is somewhat more conservative. Wright's diffusion-based data points are heuristically fitted by a curve given by

$$\frac{d}{D} \geq \frac{1}{2}\left\{1 - \left(\frac{2}{\pi}\right)\tan^{-1}\left[\left(\frac{2}{\pi}\right)(N_s - 3.8)\right]\right\}. \tag{6.23}$$

Also shown in the figure are results adapted from Balje's work (Balje, 1968). Balje's estimates are based on constraining $C_2^2 \leq gH/\eta_T$ and with D_s and N_s matched from the Cordier line. Nevertheless, these results are in reasonable agreement.

It must be emphasized that Figure 6.14 and Equation 6.23 represent a design recommendation and do not necessarily apply to any specific existing axial fan. In particular, fans with larger hub ratios may be expected to yield slightly higher efficiencies, albeit at a larger size.

6.9 Preliminary Design and Diffusion Limits in Radial Flow

Now consider the preliminary design of a radial flow impeller for a pump or fan. Figure 6.15 shows a radial discharge cascade as a cutaway view of the blade row of a centrifugal impeller; in this case a centrifugal fan with backwardly inclined (i.e., opposite to the direction of rotation) airfoil-shaped blades. For preliminary design, the following assumptions are made:

- The flow entering and leaving the blades is perfectly aligned with the blade direction
- The blades are reduced to their camber lines

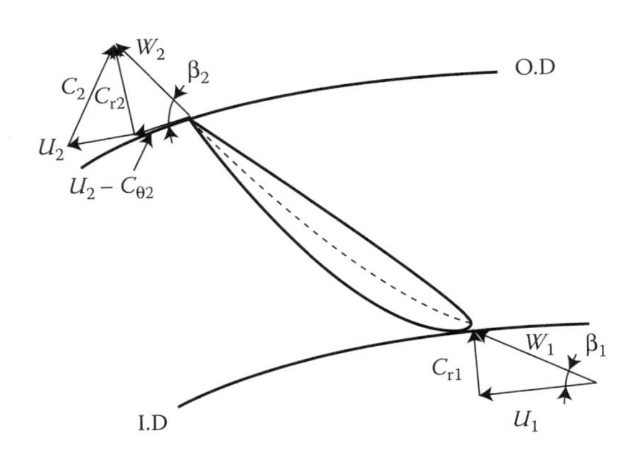

FIGURE 6.15 Radial flow fan with backwardly inclined blades.

- The flow is uniform around the circumference and across the rotor depth at inlet and outlet
- The hydraulic efficiency is 1.0 (100%)
- There are no "preswirl" vanes so that the flow enters the impeller eye in the axial direction and turns to pass onto the blades in a purely radial direction (i.e., $C_{\theta 1} = 0$).

The primary equations to relate the flow, the pressure rise, and the impeller geometry are as follows:

$$U_1 = \frac{d}{2}N, \tag{6.24}$$

$$U_2 = \frac{D}{2}N, \tag{6.25}$$

$$Q = \pi d b_1 C_{r1} = \pi d b_1 C_1 = \pi D b_2 C_{r2}, \tag{6.26}$$

so that

$$\frac{C_{r2}}{C_{r1}} = \left(\frac{d}{D}\right)\left(\frac{b_1}{b_2}\right), \tag{6.27}$$

$$\frac{\Delta p_T}{\rho} = \eta_H U C_{\theta 2}. \tag{6.28}$$

The relative velocities and blade angles are

$$W_1 = (C_{r1}^2 + U_1^2)^{1/2} = (C_1^2 + U_1^2)^{1/2}, \tag{6.29}$$

$$\beta_1 = \tan^{-1}\left(\frac{C_{r1}}{U_1}\right) = \tan^{-1}\left(\frac{C_1}{U_1}\right), \tag{6.30}$$

$$W_2 = \left(C_{r2}^2 + [U_2 - C_{\theta 2}]^2\right)^{1/2}, \tag{6.31}$$

$$\beta_2 = \tan^{-1}\left(\frac{C_{r2}}{U_2 - C_{\theta 2}}\right). \tag{6.32}$$

From these fundamental equations, one can see that the designer has the following parameters to work with in addition to the speed and impeller (outer) diameter:

- Eye diameter (d)
- Blade depth at both inlet and outlet (b_1, b_2)
- Blade angles β_1, β_2.

A particularly useful starting point is to specify that the area of the impeller eye, A_0 ($=\pi d^2/4$), be equal to the area where the fluid enters the blades,

A_1 $(=\pi d b_1)$. This will give essentially equal velocities, $C_{a0} \approx C_{r1}$. This specification gives the relationship

$$b_1 = \frac{d}{4} \quad \text{(design suggestion).} \tag{6.33}$$

The relationship between relative velocities W_1 and W_2 and the inlet and outlet blade geometry suggests an inherently high rate of diffusion when the geometry is viewed in two dimensions. However, in the design process, the rate of increase of available flow area for the relative velocity can be controlled through variation of the impeller width, b. Specifically, one can control the magnitude of C_{r1} relative to C_{r2} through choice of the blade widths b_1 and b_2. To achieve the pressure rise as governed by the Euler equation, the relative velocity through the blade channel must also be decreased. The velocity ratio C_{r2}/C_{r1} is a critical parameter in determining the diffusion ratio through the radial flow cascade; therefore, from Equation 6.25, d/D and b_1/b_2 become critical design parameters for the centrifugal blade row. d/D is analogous to the hub–tip ratio, which plays such a strong role in axial flow machine design, and b_1/b_2 provides an additional variable to maintain good flow conditions.

The diffuser analogy for the blade or vane row can be readily applied to the radial flow cascade of a centrifugal fan or pump impeller to determine an acceptable de Haller ratio. As Chapter 9 points out, in a more thorough discussion of radial flow cascades, the diffusion limit is of somewhat limited value except for blades that are backwardly inclined to yield a high efficiency, moderate pressure rise impeller design. In many cases, particularly for relatively highly loaded radial cascades, the flow is significantly separated from the blade trailing faces (suction surfaces). The flow itself is smooth and stable, forming the classic jet-wake or low-speed core flow pattern without the unsteady pulsating flow seen in axial flow machines (Balje, 1968; Johnson and Moore, 1980). The influence on overall machine performance is acceptable in light of the large amount of head rise that can be obtained in a single stage.

The Cordier diagram can suggest appropriate values for the outer diameter once speed and performance are specified, but how might the eye diameter be selected? It is generally good practice to select d (i.e., d/D) so that W_1 is minimized. This has several advantages. First, a smaller W_1 helps control the de Haller ratio, W_2/W_1. Second, in a high-speed compressor, a smaller W_1 mitigates choking problems. Similarly, in a liquid pump, a smaller W_1 reduces the likelihood of cavitation at the leading edges of the blades.

Using Equations 6.24, 6.26, 6.29, and 6.33, one can rewrite W_1 as

$$W_1 = \left[\left(\frac{4Q}{\pi d^2}\right)^2 + \left(\frac{Nd}{2}\right)^2\right]^{1/2}. \tag{6.34}$$

Note that a value of d that minimizes W_1 exists because increasing d increases U_1 while decreasing d increases C_1. Dividing both sides of this equation by

U_2 (=$ND/2$) and squaring give

$$\left(\frac{W_1}{U_2}\right)^2 = \frac{[(8/\pi)\phi]^2}{(d/D)^4} + \left(\frac{d}{D}\right)^2,$$

where $\phi = Q/ND^3$ is the flow coefficient. It is sufficient for the present purposes to seek the minimum of $(W_1/U_2)^2$ with respect to d/D by setting $d_{(W_1/U_2)^2}/d_{(d/D)} = 0$. The result is

$$\frac{d}{D} = \left(\frac{2^{7/6}}{\pi^{1/3}}\right)\phi^{1/3} \approx 1.53\phi^{1/3} \quad \text{(design suggestion)} \tag{6.35}$$

subject to the constraint that $d/D \leq 1$. This expression (Wright, 1996) supplies for centrifugal impellers the equivalent to the hub–tip ratio guideline for axial flow machines. It allows an intelligent starting point for laying out the radial blades and leaves the b_1/b_2 ratio as an additional parameter for fine-tuning the design, especially the diffusion level.

Example: Preliminary Design of a Radial Flow Fan

A centrifugal fan is required to supply 4 m^3/s of air with $\rho = 1.2$ kg/m^3. The required pressure rise is $\Delta p_T = 3$ kPa. The fan will be designed to lie in Region E of the Cordier diagram (Figures 5.1 and 5.6). Rather arbitrarily, the specific diameter is chosen as $D_s = 3.4$. The corresponding N_s is about 0.81 and, from Figure 2.9, $\eta_T \approx 0.88$. The fan diameter and speed are

$$D = D_s\frac{Q^{1/2}}{(\Delta p_T/\rho)^{1/4}} = 3.4 \times \frac{4^{1/2}}{(3000/1.2)^{1/4}} = 0.962\,\text{m}$$

and

$$N = N_s\frac{(\Delta p_T/\rho)^{3/4}}{Q^{1/2}} = 0.81 \times \frac{(3000/1.2)^{3/4}}{4^{1/2}} \times \frac{60}{2\pi} = 1370\,\text{rpm}.$$

At this point, the specific speed and the specific diameter may be adjusted if desired to achieve a particular size or speed (say a synchronous motor speed). This will not be done here.

Next, the throat size equation (Equation 6.35) is used to determine d

$$d = D \times 1.53\left(\frac{Q}{ND^3}\right)^{1/3} = 0.962 \times 1.53 \times \left(\frac{4}{143.2 \times 0.962^3}\right)^{1/3} = 0.464\,\text{m}.$$

For this design, the areas through the blade row will be made equal ($A_0 = A_1 = A_2$), which will keep the radial velocity constant. This gives

$$b_1 = \frac{d}{4} = 0.116\,\text{m} \quad \text{and} \quad b_2 = \frac{d}{D}b_1 = \left(\frac{0.464}{0.962}\right) \times 0.116 = 0.056\,\text{m}.$$

The radial velocity is

$$C_{r1} = C_{r2} = C_{a0} = \frac{4Q}{\pi d^2} = \frac{4 \times 4}{\pi \times 0.464^2} = 23.7\ \text{m/s},$$

and the blade speeds are

$$U_1 = N\frac{d}{2} = 143.2 \times \frac{0.464}{2} = 33.2\ \text{m/s}, \quad U_2 = N\frac{D}{2} = 143.2 \times \frac{0.962}{2} = 68.9\ \text{m/s}.$$

From the Euler equation

$$C_{\theta 2} = \frac{\Delta p_T}{\eta_T \rho U_2} = \frac{3000}{0.88 \times 1.2 \times 68.9} = 41.2\ \text{m/s}.$$

Note that the efficiency estimated from the Cordier diagram is used in this calculation, instead of the ideal value of 1.0, thus making the design more realistic. The relative velocities and blade angles are now calculated:

$$W_1 = (C_{r1}^2 + U_1^2)^{1/2} = (23.7^2 + 33.2^2)^{1/2} = 40.8\ \text{m/s},$$

$$\beta_1 = \tan^{-1}\left(\frac{C_{r1}}{U_1}\right) = \tan^{-1}\left(\frac{23.7}{33.2}\right) = 35.5°,$$

$$W_2 = \left(C_{r2}^2 + [U_2 - C_{\theta 2}]^2\right)^{1/2} = (23.7^2 + [68.9 - 41.2]^2)^{1/2} = 36.5\ \text{m/s},$$

$$\beta_2 = \tan^{-1}\left(\frac{C_{r2}}{U_2 - C_{\theta 2}}\right) = \tan^{-1}\left(\frac{23.7}{68.9 - 41.2}\right) = 40.6°.$$

Checking the diffusion limits, the de Haller ratio is $W_2/W_1 = 0.89$. This is a very conservative value. A more aggressive design could be obtained by reducing the speed or size (thus lowering U_2), by reducing b_2 (thus lowering C_{r2} hence W_2), by increasing the eye diameter (increasing W_1—choking and cavitation are not issues with this fan), or by some combination of these. Such optimization of the design and further details of the blade layout, such as blade count and flow-blade deviation (called *slip* in radial flow impellers), will be largely deferred to later chapters. An approximate sketch of the calculated design, based on the alignment of W_1 and W_2 with the blade mean camber line, is given in Figure 6.16.

6.10 Summary

This chapter began with a discussion of blade and cascade geometry and velocity diagrams. Next, the fundamental concept of energy transfer through the mechanism of angular momentum change was developed. The Reynolds Transport theorem was employed to develop the Euler equations for turbomachines. The resulting relations are general, subject to the steady flow

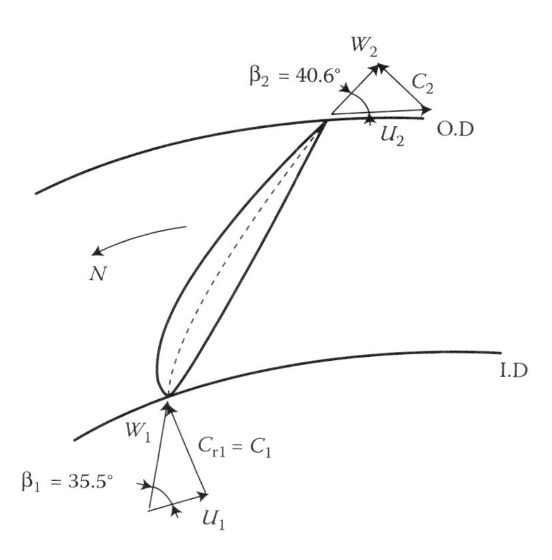

FIGURE 6.16 Radial flow fan designed in the example.

restriction and the exclusion of extraneous torques. Simple two-dimensional examples were used to illustrate the flow behavior related to the formal equations.

The need for values of efficiency or losses in order to relate rotor work and fluid specific energy was noted.

A more extensive example of energy transfer in axial flow was used to relate the energy transfer process to the geometric features and mechanical speed of axial flow machines. Through an example, the concept of blade and vane loading in a machine was developed in terms of the blade or vane relative velocities and the fluid turning angles.

The concept of loading was extended to the introduction of blade or vane row diffusion. The analogy of blade channel flow to flow in a planar diffuser was used to establish the concepts of flow separation and stall as limitations on achievable performance within a machine and to develop the de Haller limit on flow diffusion. Axial flow examples were extended in a discussion of the influence of blade row solidity on the allowable levels of diffusion. The earlier example was extended by using the guidance of the Cordier correlations to develop an acceptable flow path for an axial fan to meet both the performance requirements and the de Haller limit. The diffusion limit concept was also used to establish a correlation relating the ratio of hub diameter to tip diameter required for acceptable diffusion levels in axial flow machines.

Finally, radial flow machines were also examined to determine acceptable geometric layouts based on required performance and acceptable diffusion levels. It was noted in doing so that, for radial flow machines of very high specific diameter, the velocity ratios or diffusion levels are not as critical to good design as they are for axial flow machines. The ratio of impeller inlet diameter to exit diameter, similar to the hub–tip ratio for axial machines, was examined in terms of minimization of the relative velocity at the impeller inlet. A design recommendation was established to relate this geometric property of the impeller to the flow coefficient and hence to the specific speed and specific diameter of the radial flow machine. An example of the preliminary design of a centrifugal fan was presented.

EXERCISE PROBLEMS

6.1. A centrifugal pump is configured such that $d = 30$ cm, $D = 60$ cm, $b_1 = 8$ cm, $b_2 = 5$ cm, $\beta_1 = 20°$, and $\beta_2 = 10°$. Pumping water (20°C) at $N = 1000$ rpm, determine the ideal ($\eta = 1.0$) values of flow rate, head rise, and required power.

6.2. A pump delivers 2700 gpm at 1500 rpm. The outlet geometry is given as $b_2 = 1.5$ in., $r_2 = 8$ in., and $\beta_2 = 30°$. Estimate the head rise and ideal power required.

6.3. A centrifugal fan with a total efficiency of 85% running at 1800 rpm has $b_1 = b_2 = 6$ cm, $d = 12$ cm, $D = 40$ cm, $\beta_1 = 40°$, and $\beta_2 = 20°$. Estimate the flow rate, total pressure rise, and input hp ($\rho = 1.2\,kg/m^3$).

6.4. An axial flow water pump has a total pressure rise of 20 psig, a volume flow rate of 6800 gpm, and tip and hub diameters of 16 and 12 in., respectively. The speed is 1200 rpm.

a. Estimate the total efficiency and required power input.

b. Construct a mean station velocity diagram to achieve the pressure rise, including the influence of efficiency.

c. Lay out the blade shape at the mean station, and at the hub and tip stations. Assume zero incidence and deviation.

d. Calculate the de Haller ratio at each station.

6.5. A tube axial fan provides a static pressure rise of 1250 Pa with a volume flow rate of $120\,m^3/s$ and an air density of $1.201\,kg/m^3$. The fan has a diameter of 3.0 m with a 1.5-m hub.

a. Develop a velocity diagram at the inlet and outlet of the fan at the hub station of the blade, assuming ideal loss-free flow.

b. Estimate the total efficiency and power for the fan and correct the outlet velocity vectors for the implied losses.

6.6. Develop an appropriate vane row configuration for the tube axial fan of Problem 6.5, given that the absolute velocity leaving the blade row of that fan was $C_{x2} = 23.7$ m/s and $C_{\theta 2} = 16.3$ m/s (at the mean radius). Estimate the increase in static efficiency and static pressure rise based on the recovery of the velocity pressure associated with the swirl velocity at the hub.

6.7. Air flow ($\rho = 0.00233\,slug/ft^3$) enters the blade row of an axial fan through a set of inlet guide vanes (IGVs) and exits the blade row with pure axial velocity (Figure P6.7). Calculate the ideal power and total pressure. The exit angle from the IGVs is −32°, and the mass flow rate through the fan is 10.0 slug/s.

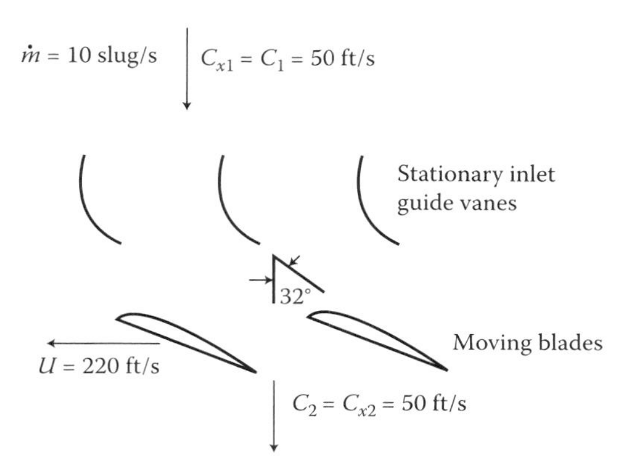

FIGURE P6.7 Fan blade cascade with IGVs.

6.8. Rework Problem 6.7, if the exit angle from the IGV row is

a. 0°
b. −10°
c. −20°
d. +20°.

6.9. For the wind turbine studied in Problem 3.19, one wants to force the turbine to have a pure axial flow discharge. Develop the required velocity diagrams for both rotor and vanes at the 70% radial station. Do the vanes make much difference? Are they really justified? What is gained, both qualitatively and quantitatively?

6.10. An axial flow fan must produce 450 Pa total pressure and provide a flow rate of 5.25 m^3/s. Choose a reasonable specific speed and diameter for a vane axial fan layout. Vary the hub–tip diameter ratio, d/D, while checking the de Haller ratio at the hub station of the blade. Using 0.6 as the lowest allowable value of W_2/W_1, determine the smallest hub–tip ratio one can use. Compare this value to the information provided in Figure 6.14.

6.11. A large axial flow pump must supply 0.45 m^3/s water flow rate with a pressure rise of 140 kPa. The hub and tip diameters for the impeller are $D = 0.4$ m and $d = 0.3$ m, respectively, and the running speed is 1200 rpm.

a. Use the Cordier diagram to estimate the pump's total efficiency, correcting the Cordier number for the influence of Reynolds number.
b. Lay out an axial flow velocity diagram for the mean radial station of the blade for this pump to achieve the required pressure rise, allowing for the influence of efficiency on the vectors.
c. Approximate the shape of the mean station of this blade based on the vector diagram of part (b).
d. Is the de Haller ratio for this station acceptable?

6.12. Develop vector diagrams, blade shapes, and de Haller ratios for the hub and tip stations of the pump examined in Problem 6.11.

6.13. Repeat the development outlined in Problem 6.5, using $d/D = 0.75$ and the "optimal" value from the curve fit in Figure 6.14. Compare these results to those of Problem 6.5 and comment on the most significant differences.

6.14. A centrifugal fan provides $0.4\,\text{m}^3/\text{s}$ of air ($\rho = 1.21\,\text{kg/m}^3$) with a pressure rise of $\Delta p_T = 1.75\,\text{kPa}$ at 1150 rpm.

a. Use $D = 0.75\,\text{m}$, $d = 0.35\,\text{m}$, and the Cordier value for total efficiency and develop the velocity vectors at the blade leading and trailing edges. Assume uniform radial velocity through the blade row.

b. Redesign the fan. Determine D, d, and widths b_1 and b_2 and develop the inlet and outlet velocity diagrams for a fan with higher speed, either 1450 or 1750 rpm. Assume that $A_1 = A_0$ for all cases.

6.15. Re-examine the velocity vectors and approximate blade layout of the fan impeller of Problem 6.14 using a d/D ratio based on the minimum W_1 analysis for centrifugal impellers ($d/D = 2^{1/6}\phi^{1/3}$).

6.16. For the fan of Problem 5.22, let 500 kPa be the static pressure requirement and

a. Estimate the change in power, total and static efficiency, and the sound power level, L_W.

b. Choose a suitable hub–tip diameter ratio for the fan and repeat the calculations of (a) and (b).

c. Calculate the de Haller ratios for the hub, mean, and tip stations for both the blades and the vanes.

6.17. A lightly loaded fan in a tube-axial configuration must deliver $1500\,\text{ft}^3/\text{s}$ of $0.0020\,\text{slug/ft}^3$ air with a static pressure rise of $5\,\text{lbf/ft}^2$. The fan has a diameter of 5 ft. Determine a reasonable speed for the fan and select a suitable hub–tip ratio.

a. Develop a vector diagram at the inlet and outlet positions on the fan blade at the 70% radial station of the blade, at the blade hub, and at the tip station.

b. Estimate the total efficiency and power and correct the outlet vectors to account for the total pressure losses implied by the total efficiency.

6.18. Further examine the velocity vectors and impeller shape of the fan of Problem 6.14 by considering blade widths that yield values of $C_{r2} = kU_2$, with $k = 0.2$, 0.3, and 0.4. These increasing k values increase the blade angles and reduce the performance sensitivity to precise accuracy requirements in blade manufacturing tolerances. Develop inlet and outlet vectors and approximate blade shapes for each case.

6.19. In Problem 6.18, narrowing the impeller to keep the blade angles fairly large results in significant acceleration of the flow from the eye to the blade leading edge. Does this matter greatly? Can one control this behavior by modifying the throat size? If so, does one gain or lose in terms of efficiency or cavitation margin?

6.20. A 0.4-m-diameter tube axial fan has a performance target of 0.8 m^3/s and a static pressure rise of 225 Pa with $\rho = 1.0\, kg/m^3$. Determine a good hub size and a reasonable speed for the fan. Develop velocity vector diagrams for the inlet and outlet stations at $2r/D = 0.75$ (the 75% blade radial station). Estimate static and total efficiencies and modify the outlet vector diagram to account for total pressure losses in the blade row (based on η_T from Cordier estimates).

6.21. A vane axial fan delivers a volume flow rate of 12 m^3/s of air with $\rho = 1.18\, kg/m^3$. The dimensions of the fan are $D = 1.0\, m$, $d = 0.7\, m$, and the rotational speed is $N = 1050$ rpm. Determine the maximum total pressure rise of this fan based on the limitation that W_2/W_1 at the blade hub station must not be <0.62. (Hint: Assume a reasonable efficiency for a vane axial fan to initiate the calculations and refine the result based on η_T as a function of N_s.)

6.22. A vane axial fan has $Q = 12\, m^3/s$, $N = 1050$ rpm, $\rho = 1.18\, kg/m^3$, $D = 1.0\, m$, and $d = 0.7\, m$. Determine the maximum attainable total pressure rise for the fan if the value of $W_2/W_1 = C_2/C_1$ at the vane hub station must be at least 0.62. Assume initially that $\eta_T = 0.75$.

6.23. For a tube axial fan with $D = 0.8\, m$, $d = 0.4\, m$, $N = 1800$ rpm, $Q = 10\, m^3/s$, and $\rho = 1.21\, kg/m^3$, determine the maximum total pressure rise achievable with a de Haller ratio of 0.65 at the hub station on the blade.

6.24. For a tube axial fan with $D = 0.8\, m$, $d = 0.4\, m$, $N = 1800$ rpm, $Q = 10\, m^3/s$, and $\rho = 1.21\, kg/m^3$, impose a total pressure rise requirement of 1200 Pa. Find the value of $r = r_{hub}$ for which the blade first meets the requirement that the de Haller ratio be >0.65. Discuss the implications for the flow quality and head loss behavior for the portion of the blade inboard of this "critical" value of radius.

6.25. A centrifugal pump impeller is designed to deliver 0.1 m^3/s of flow at 1450 rpm. If the outlet width of the impeller is 4 cm, the diameter is 20 cm, and the head rise is 25 m, determine the blade exit angle, β_2, the total efficiency of the impeller, and the required shaft power to generate this performance.

6.26. The turbine featured in Figure 1.16 has $D = 1.4$ ft, $N = 1200$ rpm and its BEP is at $Q = 3200$ gpm, $H = 33$ ft with $P_{sh} = 15.6$ kW, and $\eta = 0.78$. Make an approximate layout for the blade shape of this radial inflow–axial discharge turbine. Assume that $b_2/D = 0.05$ and that C_r is constant through the blade row. Calculate the required inlet velocity components and the inflow angle (inflow is at D). Estimate the eye diameter using Equation 6.35 and define the outlet blade angle, velocity components, and relative velocity in the plane normal to the rotational axis.

6.27. To increase the pressure rise of an axial fan with variable pitch blades, the blade stagger angle, λ, can be decreased. Using conventions for variable pitch axial fans, the "geometric pitch," the angle complementary to the stagger, is increased. Define an increment of geometric pitch as $\Delta\tau$ such that the flow angle at the trailing of the blade is perturbed from β_2 to $\beta_2' = \beta_2 + \Delta\tau$. Using the Euler equation, show that $[\psi'/\psi] = [1 - \phi((\tan\beta_2 - \tan(\beta_2 + \Delta\tau)/(1 + \tan\beta_2 \tan\Delta\tau))]/[1 - \phi\cot\beta_2]$, where ψ is associated with the unperturbed condition, ψ' is associated with the new condition, and ϕ is unchanged. {Here the definitions for ϕ and ψ are $\phi = Q/(U_h A_{ann})$; $\psi = \Delta p_T/(\rho U_h^2)$; and Δp_T is simplified to $\Delta p_T = \rho U_h C_\theta \eta_T$. U_h is the hub speed $U_h = Nd/2$; A_{ann} is the annulus area $A_{ann} = (\pi/4)(D^2 - d^2)$.}

6.28. Use the results of Problem 6.27 with the performance curve of the variable pitch fan shown in Figure 5.18. Use the one designated as 40° pitch and change the pitch angle to 45° ($\Delta\tau = 5°$). Recalculate the pressure rise for the range of flows shown for 45° and compare the calculated performance for 45° to that shown in Figure 5.18 at 45°. Use the variable forms given in Problem 6.27 and show results in the dimensionless form.

6.29. Model a centrifugal impeller with incremental increases of β_2^* so that, if β_2^* is increased by $\Delta\tau$, the pressure rise will be increased at a given flow rate. Using $\phi = C_r/U_2$ and $\psi = \Delta p_T/(\rho U_2^2)$, employ the Euler equation to show that the modified value ψ' associated with the increased flow angle $\beta_2' = \beta_2 + \Delta\tau$ compared to the unperturbed value ψ is $\psi/\psi' = [1 - ((1 + \tan\beta_2 \tan\Delta\tau)/(\tan\beta_2 - \tan\Delta\tau))]/[1 - \cot\beta_2]$.

6.30. Use the results of Problem 6.29 with the Euler equation to convert the performance curve of Figure 5.11a to a fan with $\Delta\tau = 3.0°$. Note that the results should be in terms of $\psi = \Delta p_T/\rho U_2^2$ and $\phi = Q/(\pi D^2 U_2/4)$.

6.31. The compressor impeller shown in Figure 1.20 clearly shows the inducer section of the wheel that fills the eye of the compressor. Likewise, almost all of the outer diameter of the impeller is shown in the picture. "Measure" the values of d and D. Use these approximate measurements to calculate the diameter ratio d/D. This number can be used to estimate the specific speed and diameter of the impeller (with a little help from Equation 6.35). Compare these rough values to the "regions" of the Cordier diagram. [Hint: Use $\phi = 1/(N_s Ds^3)$.]

6.32. The vane axial fan shown in Figure 1.21 (lower center) can be measured to estimate the hub–tip ratio, d/D. Use this estimate to define the specific speed of the impeller (with help from Figure 6.14) and compare the type of fan to its location in the Cordier diagram.

6.33. Repeat the "measurement" and estimates for the specific speed of the propeller-type axial fan shown in Figure 1.21 (upper left). Compare the result with the regions of the Cordier diagram.

7

Velocity Diagrams and Flow Path Layout

7.1 Preliminary Remarks

Chapter 6 examined the relationship between the energy exchange and the flow patterns within the blade and vane rows of turbomachines. Blade linear velocity (U), fluid velocity relative to the blade (W), and fluid absolute velocity (C) vectors were employed to analyze rudimentary blade layout and blade loading and to develop some basic design limitations for pumping machinery based on diffusion limits.

This chapter examines the fundamental characteristics of several distinct types of turbomachines, including both turbines and pumping machinery. Axial flow, mixed flow, and radial flow types will all be considered. Emphasis will be placed on the characteristics of the velocity diagrams; blade geometry will be derived from the simplifying assumption that the fluid velocity vectors are essentially parallel to the blade camber line, especially at the entrance and exit from blade rows. In addition, it will often be assumed that the efficiency is 100%; if not, a value for efficiency will be estimated from the Cordier relationships, which apply to the BEP of "good" machines.

The twin assumptions of high efficiency and perfect flow alignment necessarily limit the results to a very narrow region of performance near the design (or "best efficiency") point. Likewise, blade layouts are reasonably accurate only for blade rows of moderate to high solidity. It should be emphasized that these limitations apply mostly to the resulting blade layouts, not to the characteristics of the velocity diagrams themselves.

7.2 Velocity Diagram Parameters for Axial Flow Machines

In order to concisely present velocity vector information, some new non dimensional variables are introduced. First is the *work coefficient*

$$\Psi \equiv \frac{w}{U^2}. \tag{7.1}$$

From the Euler equation, $w = U_2 C_{\theta 2} - U_1 C_{\theta 1}$ and for an axial flow rotor $U_2 = U_1 = U$, so

$$\Psi = \frac{\Delta C_\theta}{U} \quad \text{(axial flow).} \tag{7.2}$$

For flow with no losses ($\eta_T = 1.0$), the work is equal to the head so that $\Psi_{\text{No losses}} = gH/U^2$. If the flow is also incompressible, $\Psi_{\text{No losses, Incompressible}} = \Delta p_T / \rho U^2$. The work coefficient is sometimes called the *loading coefficient*.

The second variable is the *axial velocity ratio* or *flow coefficient*

$$\varphi \equiv \frac{C_x}{U}. \tag{7.3}$$

As $U = ND/2$ and $Q \approx C_x(\pi/4)(D^2 - d^2)$, the new variables, Ψ and φ, are related to the head and flow coefficients defined in Chapter 2 ($\psi = gH/N^2D^2, \phi = Q/ND^3$) by

$$\phi \approx \frac{\pi}{8}\left(1 - \frac{d^2}{D^2}\right)\varphi \tag{7.4a}$$

and

$$\psi = \frac{\eta_H}{4}\Psi. \tag{7.4b}$$

In order to define the third new parameter, the concepts of impulse and reaction must be introduced. In general, when a flow stream interacts with a moving blade, the flow experiences both a change in direction and a change in speed. A blade/fluid force interaction is associated with both. When the force interaction is associated with a change in direction only, with no change in pressure, the interaction is purely impulse. A simple example of pure impulse is the pushing of a skateboard along a street by impact of a water jet from a garden hose nozzle. Conversely, when the force action is associated purely with a pressure change, the interaction is purely reaction. Examples of pure reaction are a deflating balloon propelled by the jet of air escaping from the balloon and a simple rotating lawn sprinkler that spins in reaction to jets of water emanating from "backward" bent tubes.

The third parameter, *degree of reaction* or, simply, *reaction*, characterizes the balance between impulse and reaction in a moving blade row. Many turbomachinery engineers prefer to work with a pressure-based degree of reaction defined by

$$R_p \equiv \frac{\text{Static pressure change in moving blade row}}{\text{Total pressure change}} = \frac{\Delta p}{\Delta p_T} = \frac{p_2 - p_1}{p_{T2} - p_{T1}}. \tag{7.5}$$

In order to account for both compressible and incompressible flows, as well as connect the reaction directly to the velocity diagrams, a definition based on enthalpy change rather than pressure change is preferred*:

$$R_h \equiv \frac{\text{Static enthalpy change in moving blade row}}{\text{Stagnation enthalpy change}} = \frac{\Delta h}{\Delta h_0} = \frac{h_2 - h_1}{h_{02} - h_{01}}. \quad (7.6)$$

The reaction can be related to the velocity diagrams. The first law of thermodynamics energy balance for the fluid is

$$h_1 + \frac{C_1^2}{2} + w - q = h_2 + \frac{C_2^2}{2},$$

in which potential energy has been neglected, and pumping is assumed so that w is positive. Assuming adiabatic flow ($q = 0$)

$$w = (h_2 - h_1) + \frac{C_2^2 - C_1^2}{2} = \left(h_2 + \frac{C_2^2}{2}\right) - \left(h_1 + \frac{C_1^2}{2}\right) = h_{02} - h_{01}. \quad (7.7)$$

Using Equation 7.6

$$R_h = \frac{\Delta h}{\Delta h_0} = \frac{\Delta h}{w} = \frac{w - \left[\left(C_2^2 - C_1^2\right)/2\right]}{w} = 1 - \frac{\left[C_2^2 - C_1^2\right]}{2w}.$$

Finally, using Euler's equation gives

$$R_h = 1 - \frac{\left[C_2^2 - C_1^2\right]}{2(U_2 C_{\theta 2} - U_1 C_{\theta 1})}. \quad (7.8)$$

For a "simple stage," defined by $U_2 = U_1 = U$ and $C_{x2} = C_{x1} = C_x$, note that

$$C_1^2 = C_{1x}^2 + C_{1\theta}^2 = C_x^2 + C_{1\theta}^2 \quad \text{and} \quad C_2^2 = C_{2x}^2 + C_{2\theta}^2 = C_x^2 + C_{2\theta}^2,$$

so that

$$C_2^2 - C_1^2 = C_x^2 + C_{\theta 2}^2 - C_x^2 - C_{\theta 1}^2 = C_{\theta 2}^2 - C_{\theta 1}^2 = (C_{\theta 2} + C_{\theta 1})(C_{\theta 2} - C_{\theta 1}),$$

and

$$U_2 C_{\theta 2} - U_1 C_{\theta 1} = U(C_{\theta 2} - C_{\theta 1}).$$

Then the reaction equation becomes

$$R_h = 1 - \frac{C_{\theta 2} + C_{\theta 1}}{2U} \quad \text{(simple axial stage)}. \quad (7.9)$$

* For incompressible, frictionless ($\eta_H = 1.0$) flow, $\Delta h = \Delta p/\rho$ and $R_p = R_h$.

If, in addition, there is also no swirl at the *inlet* for a pumping machine ($C_{\theta 1,\text{pump}} = 0$) or at the *outlet* for a turbine ($C_{\theta 2,\text{turbine}} = 0$), then

$$R_h = 1 - \frac{\Psi}{2} \text{ [simple stage, no inlet swirl (pump); no outlet swirl (turbine)].} \tag{7.10}$$

Thus, for this simple, but common, situation, choosing the reaction fixes the stage loading and vice versa.

7.3 Axial Flow Pumps, Fans, and Compressors

Axial flow machines have high specific speed and low specific diameter on a single-stage basis. This means that, generally, the specific energy rise is relatively small and the flow rate is relatively large. Axial flow machines that are required to accommodate large changes of specific energy, such as gas turbine engine compressors or large steam turbines, are built in several stages, such that the energy change is additive.

Mechanically, a stage comprised a single moving blade row (rotor) and, usually, a single fixed vane row (stator). In a turbine, the vane row usually precedes the rotor and acts as a nozzle, accelerating the fluid and imparting significant swirl. In a pumping machine, the vane row usually follows the rotor and acts as a diffuser, converting velocity to pressure and removing swirl. Pumping machines of several stages (e.g., axial flow compressors) are sometimes fitted with IGVs ahead of the first rotor row to set up a specific swirl pattern in the flow entering the rotor.

For a single axial flow stage, sketched in Figure 7.1, one can assume that $U_1 = U_2 = U$ and combine the blade inlet and outlet vector diagrams to form the "combined vector diagram," as shown in Figure 7.2. In this particular diagram, there is no preswirl ($C_{\theta 1} = 0$) and the axial velocity is constant ($C_{x1} = C_{x2}$). Specific numerical values shown are $U = 149$ ft/s, $C_1 = C_x = 45.1$ ft/s, and $C_{\theta 2} = 45$ ft/s. From these, we calculate $W_1 = 155.7$ ft/s, $W_2 = 113.4$ ft/s, and $C_2 = 63.6$ ft/s. The various angles are shown on the diagram.

Note that the inlet velocity to the trailing diffuser vane is C_2, 63.6 ft/s. Assuming that the diffuser vane removes all of the swirl, the flow exiting the vane has $C_3 = C_x$, $C_{\theta 3} = 0$. Such a stage is sometimes called a *normal stage*, in that the flow entering and leaving the stage is normal (perpendicular) to the plane of the rotor. The flow entering a subsequent stage is then identical to the flow entering the stage itself.

Returning to consideration of the combined velocity diagram, the critical parameters are as follows:

$$\text{Work coefficient: } \Psi = \frac{\Delta C_\theta}{U} = \frac{45 - 0}{149} = 0.302.$$

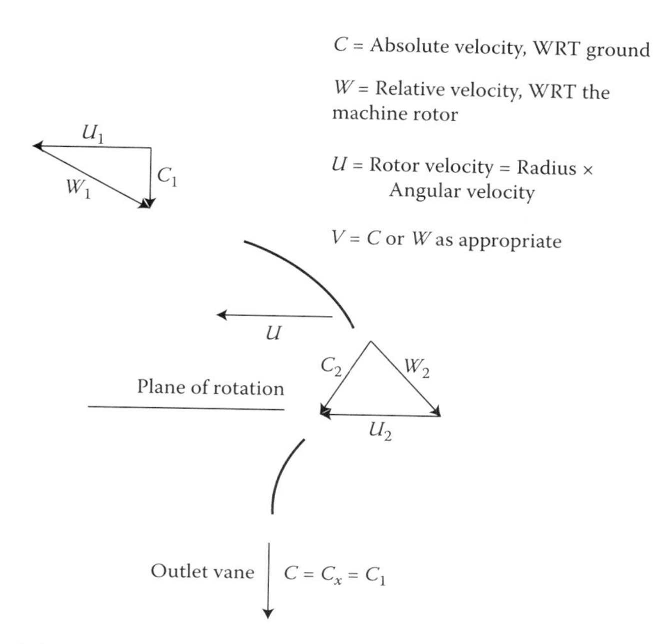

FIGURE 7.1 Velocity diagrams and blade–vane geometry for a (normal) axial flow stage.

Flow coefficient: $\varphi = \dfrac{C_x}{U} = \dfrac{45.1}{149} = 0.303.$

Reaction: $R_h = 1 - \dfrac{C_{\theta 2} + C_{\theta 1}}{2U} = 1 - \dfrac{45 + 0}{2 \times 149} = 0.849.$

Flow turning angle: $\theta_{fl} = \beta_2 - \beta_1 = 23.4° - 17.8° = 6.6°.$

de Haller ratio (blade): $\dfrac{W_2}{W_1} = \dfrac{113.2}{155.7} = 0.727.$

de Haller ratio (vane, assuming $C_3 = C_x$): $\dfrac{C_3}{C_2} = \dfrac{45.1}{63.6} = 0.709.$

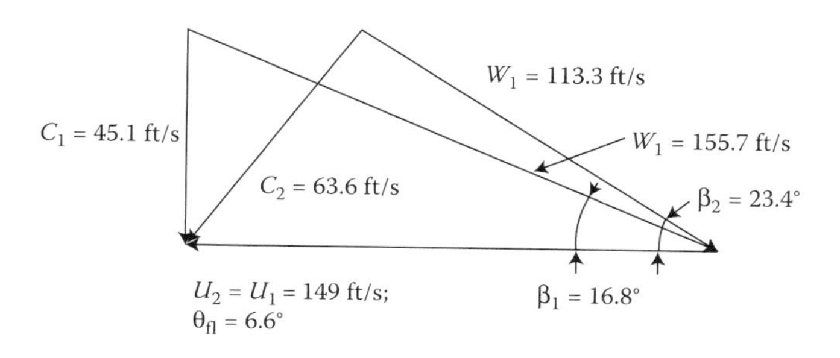

FIGURE 7.2 Combined velocity diagram for an axial flow stage.

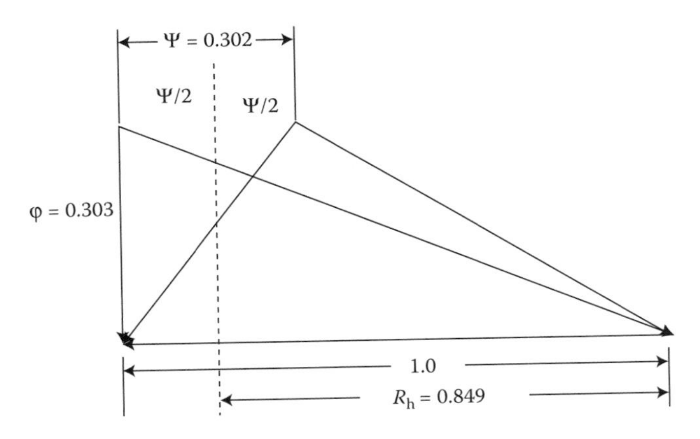

FIGURE 7.3 Dimensionless combined velocity diagram showing stage parameters.

The combined vector diagram can be normalized by dividing all velocities by U; the result is shown in Figure 7.3. This type of dimensionless diagram is a handy device because it completely describes the flow pattern entering and leaving a blade row and the entrance conditions for the vane row. In this diagram, the stage parameters appear directly, as shown. Note that

- The distance between the peaks is always the work coefficient.
- The height is always the flow coefficient.
- The reaction is the distance from the right apex to the point midway between the peaks.*
- All relative and absolute flow angles appear in the diagram, including the flow turning angle.
- The de Haller ratio is available to scale as the ratio of relative vectors for the blade and absolute vectors for the vane.

Such a dimensionless velocity diagram can be prepared for other cases that are not quite so simple. For example, a rotor in axial flow with nonaxial entry conditions typifies a pump, fan, or compressor stage with a preswirl velocity, $C_{\theta 1}$, created by a set of IGVs, as shown in Figure 7.4. The IGVs turn the flow, in this case, toward the moving blades ($C_{\theta 1} < 0$), giving a velocity vector layout as illustrated in Figure 7.5. This leads to a combined velocity diagram that looks like Figure 7.6, with the dimensionless form shown in Figure 7.7. To facilitate comparison, the values of U and C_x are kept the same as those in the previous example and the inlet swirl is set as $C_{\theta 1} = -C_{\theta 2}$. In this case, the outlet velocity diagram is identical to that in the previous example. Also,

* This property of the dimensionless diagram can be used to show that a 50% reaction design ($R_h = 0.5$) will always exhibit symmetric velocity diagrams, and for perfect fluid guidance, identical shapes (albeit inverted) for blades and vanes.

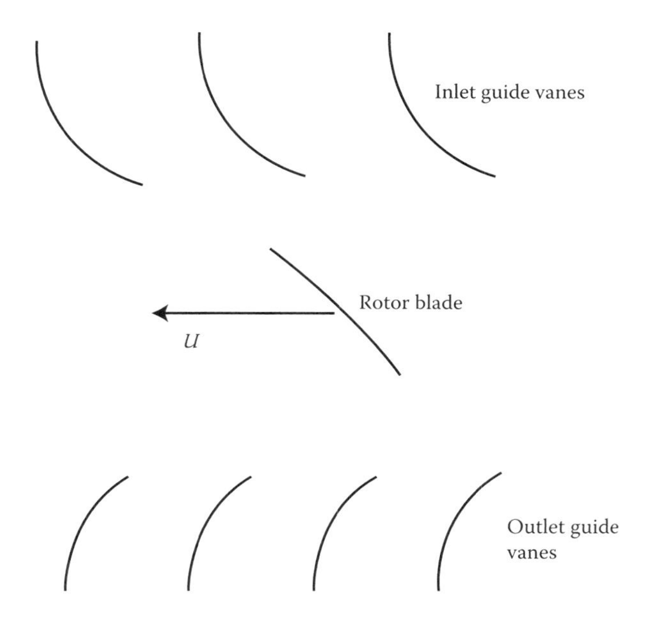

FIGURE 7.4 Axial flow stage with IGVs to generate inlet swirl.

$C_1 = C_2$, thus there is no net change in the fluid kinetic energy and the degree of reaction is 1, as verified by the dimensionless velocity diagram.

In general, negative preswirl serves to increase the work coefficient as well as the degree of reaction. The increased stage loading means that a higher specific energy rise can be obtained at a given blade speed. With the higher degree of reaction, the additional loading is placed on the blade and the blade de Haller ratio, W_2/W_1, may become excessive (in this instance, because of a significant increase in W_1). If this problem can be avoided, then using negative preswirl is useful to increase the loading of a single stage, permitting lower blade speed or avoiding the use of additional stages.

A layout that uses IGVs to produce positive preswirl ($C_{\theta 1} > 0$) is shown in Figure 7.7. For this particular case, $C_{\theta 1} = 0.25U$ and $C_{\theta 2} = 0.75U$. As shown, this gives $\Psi = 0.5$ and $R_h = 0.625$. Positive preswirl serves to reduce the work coefficient and the degree of reaction as compared to the no-pre-swirl "normal

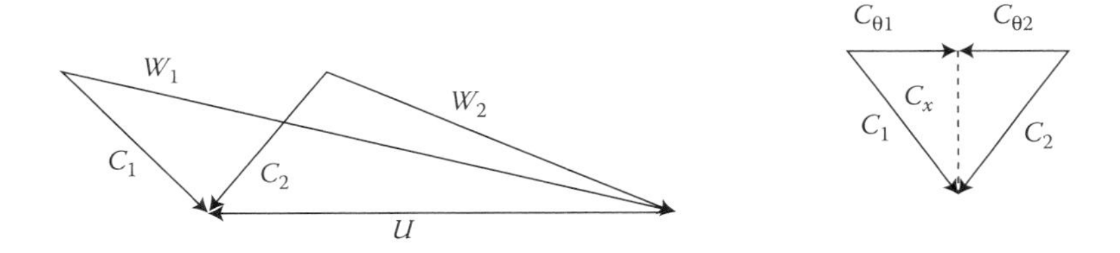

FIGURE 7.5 Combined velocity diagrams for (negative) inlet swirl.

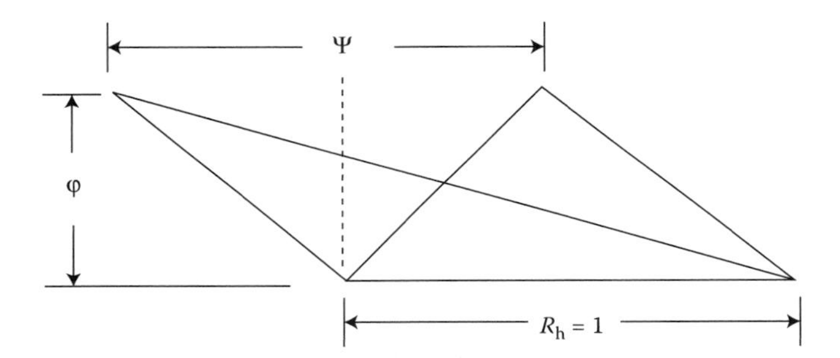

FIGURE 7.6 Dimensionless combined velocity diagram for (negative) inlet swirl.

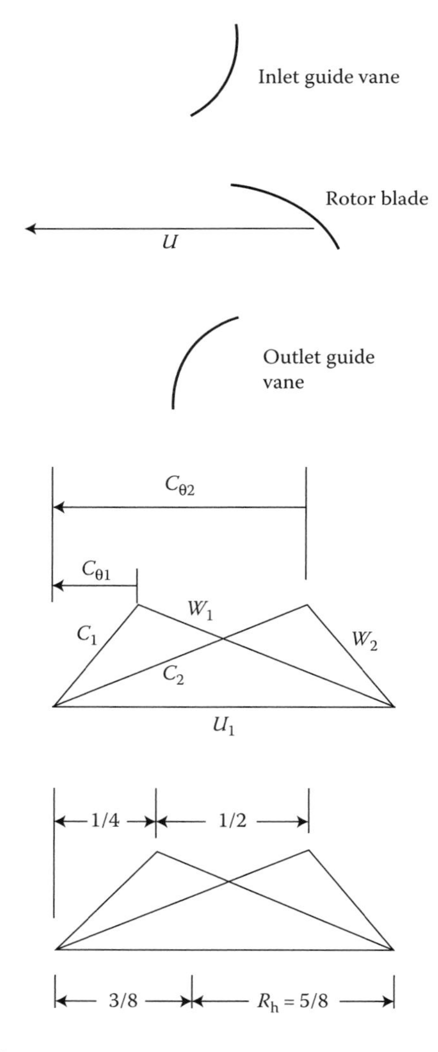

FIGURE 7.7 An axial flow stage with positive preswirl.

stage." This yields a corresponding decrease in fluid turning. Note, however, that because C_2 is the approach vector to the outlet guide vanes (OGVs), which are assumed to remove all swirl, the turning requirement imposed on the outlet vanes becomes much larger, which will lead to a greatly decreased, probably excessive, de Haller ratio. By decreasing the reaction, one is shifting loading from the blade to the vane. This can be useful up to a point but, in general, not very much is gained.

The real reason for using "positive" preswirl can be seen by using an outlet (diffuser) vane row that recovers only part of the swirl. This makes sense only for a multistage machine for which the outlet vector from the vane row is the inlet vector to the following blade row, as illustrated in Figure 7.8. In

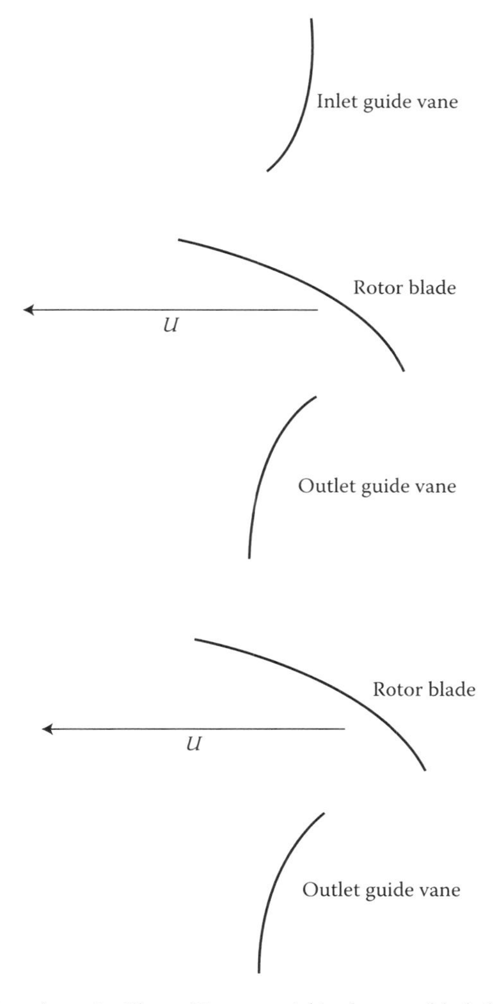

FIGURE 7.8 Multistage layout with positive preswirl to decrease blade loading.

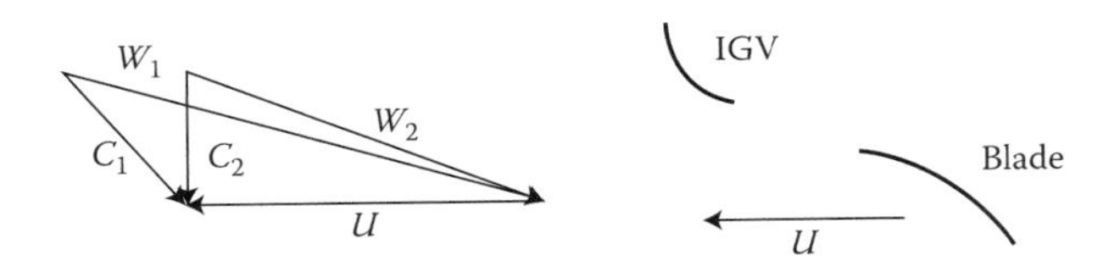

FIGURE 7.9 Axial flow stage with IGVs and no OGVs.

this configuration, there can be any number of "repeating stages" that see the same inlet–outlet conditions on blade and vane rows, and the IGVs are used only to set up the flow pattern for the first row. For the last stage, some sort of adjustment is usually necessary. If recovering all of the swirl places too much loading on the vane, the use of an extra vane row to capture the residual swirl is a possible solution, as is increasing the solidity of the final vane row.

Another approach to stage layout is to use the preswirl of an IGV to turn the flow into the rotor and add just enough opposite swirl in the blade row to yield an axial discharge from the blade. Such a configuration is shown in Figure 7.9, along with its velocity diagram. C_2 is simply equal to C_x and, for a given loading, the de Haller ratio, W_2/W_1, is no worse than the value expected for a typical normal stage with OGVs. The penalty arising from having a higher W_1 (higher friction losses) is offset by having acceleration instead of diffusion through the vane row (lower "diffuser" losses). A secondary penalty arises from having the rotor pass through a cyclic series of vane wakes from the upstream IGV elements (an issue anytime a machine has IGVs). This "slapping" of the blades through the vane wakes will yield an increase in the noise level generated by the stage (Chapter 4). In a multistage configuration, the extra noise effect may be negligible.

Example: A Multistage Axial Compressor

A seven-stage axial flow compressor is laid out as in Figure 7.8, with a single row of IGVs. The absolute velocity flow angle at the inlet to each blade row (α_1) is 65° (i.e., the flow "leans into" the blade by 25°). Each stage has 50% reaction. The rotational speed is 8000 rpm and the flow rate is 105 lbm/s. All calculations will be based on the fourth stage, which has $D = 2.10$ ft and $d = 1.70$ ft, and on the mean density, $\rho = 0.185$ lbm/ft^3. (This value could be calculated iteratively, but that would be too lengthy for this example.) Determine the velocity diagrams, the approximate blade shape, and estimate the total pressure ratio developed over the seven stages.

Calculations will be made at the mean blade radius, $r_m = (D + d)/4 = (2.10 + 1.70)/4 = 0.95$ ft. Then $U = U_m = r_m \times N = 0.95 \times (2\pi/60) \times 8000 = 796$ ft/s. The axial velocity is calculated from the flow rate

$$C_x = \frac{\dot{m}}{\rho(\pi/4)(D^2 - d^2)} = \frac{4 \times 105}{0.185 \times \pi \times (2.10^2 - 1.70^2)} = 475 \text{ ft/s}.$$

The inlet absolute velocity is $C_1 = C_x/\sin\alpha_1 = 475/\sin 65° = 524$ ft/s and $C_{\theta 1} = C_x/\tan 65° = 221$ ft/s. The inlet relative velocity is calculated as follows: $W_x = C_x = 475$ ft/s, $W_{\theta 1} = U - C_{\theta 1} = (796 - 221) = 575$ ft/s, and $W_1 = (475^2 + 575^2)^{1/2} = 745$ ft/s. Finally, the inlet relative angle is $\beta_1 = \tan^{-1}(W_x/W_{\theta 1}) = 39.6°$. Since, for 50% reaction, the velocity diagrams are symmetric, one can immediately write $C_2 = W_1 = 745$ ft/s, $W_2 = C_1 = 524$ ft/s, $\alpha_2 = \beta_1 = 39.6°$, and $\beta_2 = \alpha_1 = 65°$. Using this information, the combined velocity diagram can be drawn as shown in Figure 7.10. Dividing all of the velocities by U gives $\varphi = C_x/U = 475/796 = 0.597$. Also, $\Psi = 1 - (2C_{\theta 1}/U) = 1 - (2 \times 221)/796 = 0.445$. The fluid turning angle is $\theta_{fl} = \beta_2 - \beta_1 = 30.4°$. The dimensionless combined velocity diagram is shown in Figure 7.10, together with approximate blade and vane shapes (based on the assumption of perfect guidance and incorporating realistic, but arbitrary, thickness). The de Haller ratio for both blades and vanes is $W_2/W_1 = C_3/C_2 = C_1/C_2 = 524/745 = 0.703$, which is slightly low but could be achieved if the solidity is a little larger than 1.

The work done in a single stage is $w_{stg} = \Psi U^2 = 0.445 \times 796^2 = 2.82 \times 10^5$ ft^2/s^2. Since all stages have identical velocity diagrams, $w = n_{stg} w_{stg} = 7 \times (2.82 \times 10^5) = 1.97 \times 10^6$ ft^2/s^2. The pressure ratio achieved across the entire

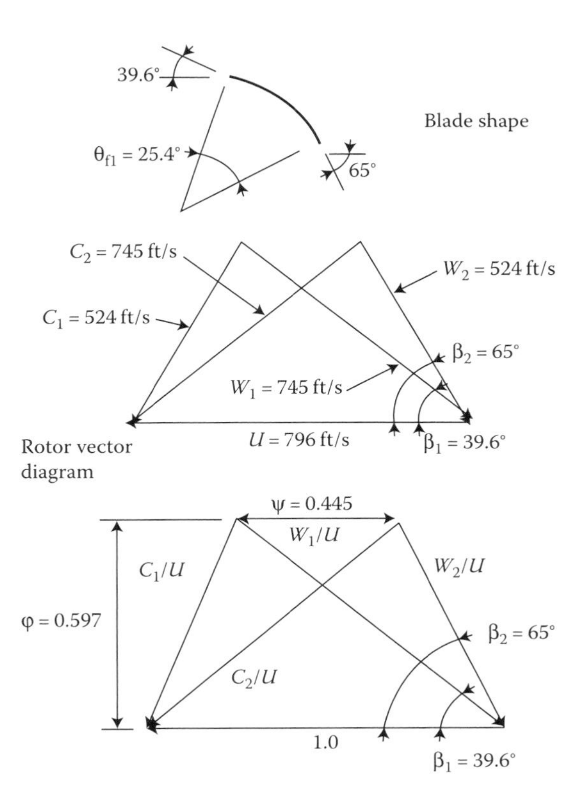

FIGURE 7.10 Velocity diagrams and blade layout (at the mean radius) for one stage of a multistage compressor.

seven stages can be calculated from

$$\frac{p_{0,o}}{p_{0,i}} = \left[1 + \frac{w}{c_p T_{0,i}}\right]^{\eta_p(\gamma/\gamma-1)},$$

where subscript "o" indicates the rotor outlet and subscript "i" indicates the rotor inlet (Equation 6.14c). Assuming that the operation is at the BEP, the polytropic efficiency can be estimated from the Cordier correlations. The specific speed is $N_s = NQ^{1/2}/(\Delta h_{0,s})^{3/4}$. Using parameters for the "representative" fourth stage, $Q = \dot{m}/\rho \approx 105/0.185 = 568\,\text{ft}^3/\text{s}$, $\Delta h_{0s} = \eta_{c,s} w \approx \eta_p w$. The calculation is clearly iterative, requiring a guess of η_p to calculate N_s, which leads to a "better" estimate of η_p from Cordier and so on. The accuracy of both the current model and Cordier does not justify more than one iteration. Starting with $\eta_p = 0.9$ gives $N_s \approx 1.77$, which gives a Cordier efficiency of about 0.9. This should probably be "derated" to about 0.85 because of the slightly low de Haller ratio and/or high friction associated with high solidity. Then, assuming that $T_{0,i} = 519°\text{R}$ (the standard sea-level value), the estimated pressure ratio is

$$\frac{p_{0,o}}{p_{0,i}} = \left[1 + \frac{w}{c_p T_{0,i}}\right]^{\eta_p(\gamma/\gamma-1)} = \left[1 + \frac{1.97 \times 10^6}{6006 \times 519}\right]^{0.85(1.4/1.4-1)} = 4.29 \approx 4.3.$$

7.4 Axial Flow Turbines

In an axial turbine stage, nearly everything in the velocity diagrams and blade–vane layout is reversed. Figure 7.11 shows a possible layout; in this case, for a normal, 50% reaction stage. Now the (fixed) vane precedes the moving blade and it increases the fluid velocity ($C_1 > C_0$). In turbines, it is more common to call the fixed vane a nozzle. One notes that $C_{\theta 1}$ is significantly large and $C_{\theta 2} < C_{\theta 1}$ (for this normal stage, $C_{\theta 2}$ is zero).

As usual for an axial flow stage, it is assumed that $U_1 = U_2 = U$. Using Euler's equation for turbines,

$$w = U_1 C_{\theta 1} - U_2 C_{\theta 2} = U(C_{\theta 1} - C_{\theta 2}) = UC_{\theta 1}. \tag{7.11}$$

(Note that the last equality only holds for a normal stage.) The turbine blade essentially recovers the swirl induced by the nozzles and discharges the flow with a purely axial outlet vector. Note that the velocity diagram for this 50% reaction turbine is symmetrical and the blade and nozzle shapes are identical, although reversed.

If one "normalizes" the diagram by dividing by U, the peaks will be separated by the distance Ψ, the height will be φ, and the degree of reaction is indicated by the distance from the right apex to a point halfway between the

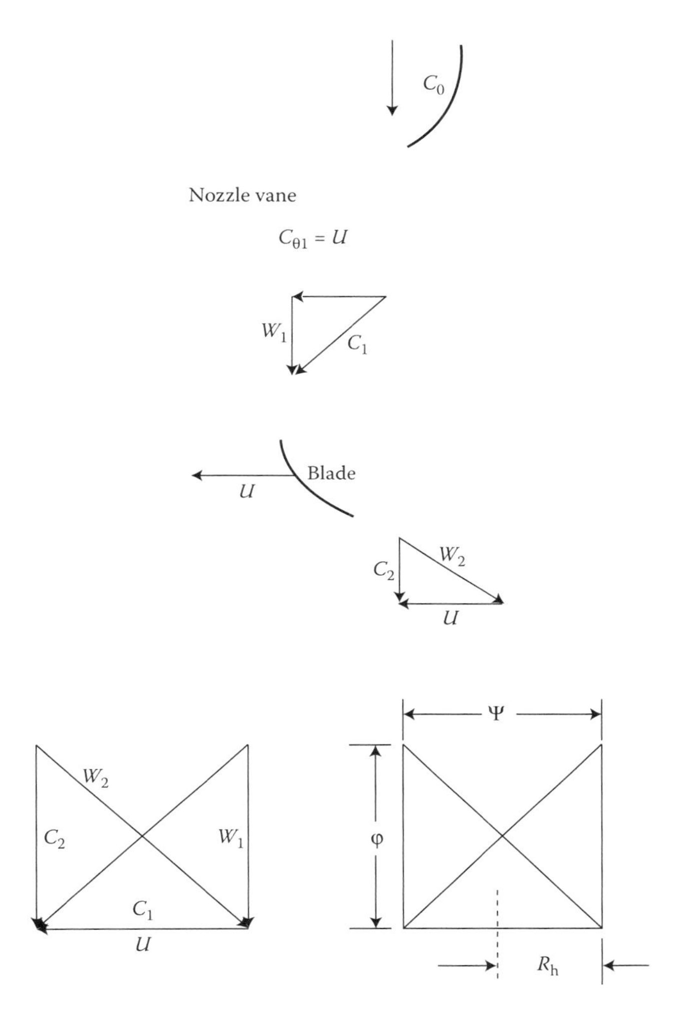

FIGURE 7.11 Velocity diagrams and blade–vane layout for a 50% reaction normal axial turbine stage.

peaks. The geometric angles, including fluid turning, are also shown as in a pumping diagram.

Note that $W_2/W_1 > 1.0$, as it generally will be in a turbine cascade. This is simply a reminder that the turbine blade accelerates the flow (blade relative) and is therefore not subject to the diffusion limitations present when dealing with pumping machines. Because this limitation is absent, it is common for turbine stages to be much more heavily loaded (w/U^2 is larger) than pumping stages.

A turbine layout that is more heavily loaded than the normal stage is sketched in Figure 7.12. In this layout, the nozzle row accelerates the flow

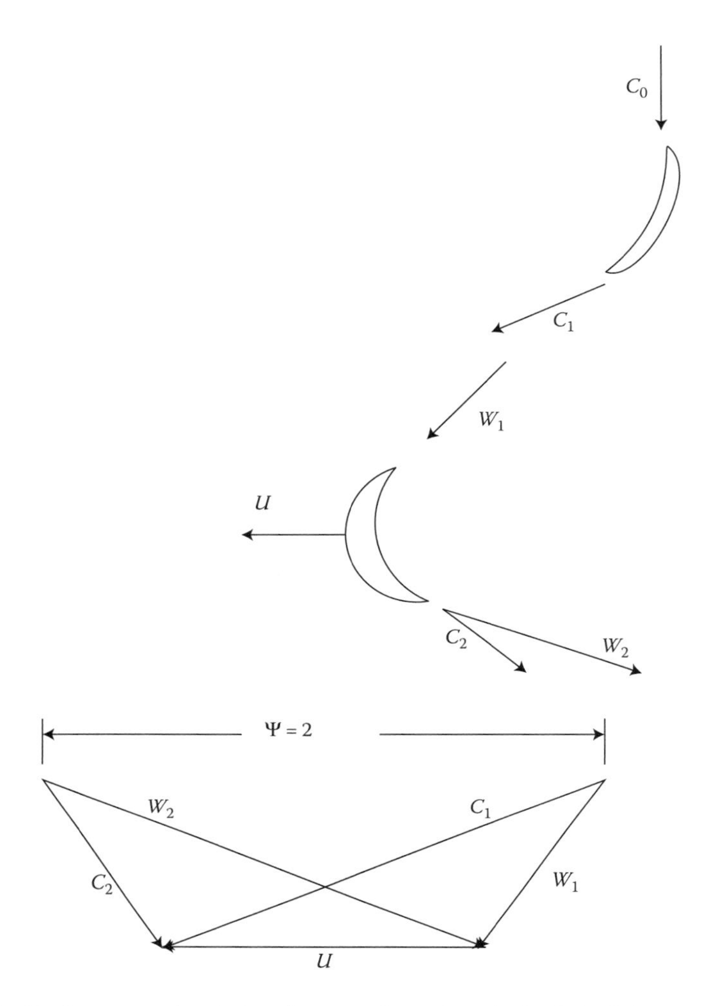

FIGURE 7.12 Layout and velocity diagrams for a heavily loaded 50% reaction turbine stage.

very sharply and relative velocities in the blade row are very high. The diagram is still symmetrical, with $R_h = 0.50$ and φ is about the same, but note that Ψ is much larger. Also note that the discharge is no longer purely axial but has a swirl component opposite to the swirl entering the blade row. This contributes part of the value of Ψ as seen from the Euler equation (here, $C_{\theta 2} < 0$). This layout has three disadvantages. If the machine has only one stage, the kinetic energy of the exiting fluid is larger than it need be, because of the residual swirl, and this energy will be lost in some downstream process. On the other hand, if the machine is a multistage configuration with identical

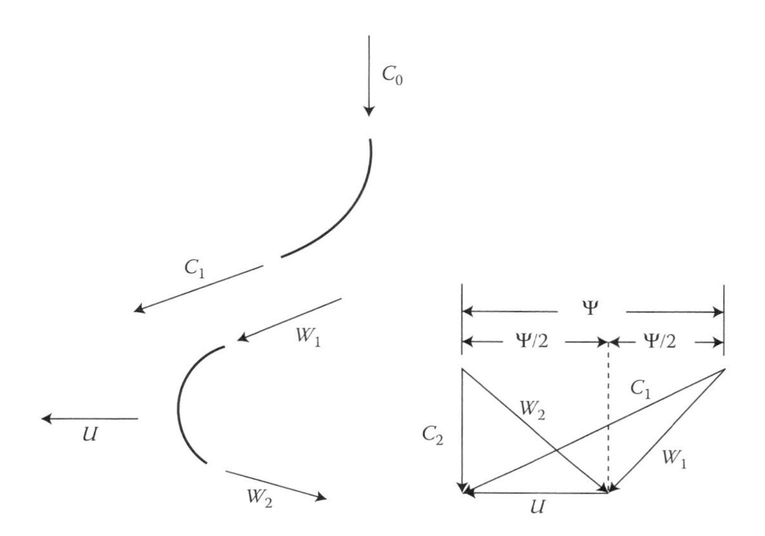

FIGURE 7.13 Normal impulse turbine stage.

velocity diagrams for each succeeding stage,* then the nozzles must impart a very large turning to the flow in order to reverse the swirl for the succeeding blade row. Lastly, the very high velocity in both nozzles and blades leads to fairly high friction losses.

Of course, not all turbines are 50% reaction designs. Other degrees of reaction are characterized by nonsymmetrical vector diagrams and nozzle/blade geometries. An important example is the "zero reaction" turbine, also known as the *impulse turbine*. Nozzle/blade layout and velocity diagrams for this stage type are shown in Figure 7.13. Note that a normal stage is shown.

In this device, all of the static enthalpy (and pressure drop) occur in the nozzles and the fluid passes through the blade row at a constant pressure. As a result, the relative velocity vectors W_1 and W_2 have equal magnitude and symmetrical direction, as shown. Because of this symmetry, the blade has geometric symmetry about its 50% chord. The characteristic impulse blade shape, often called a "bucket," immediately identifies what kind of diagram one is dealing with and that the reaction is zero. Here, there is no geometric symmetry between the blades and vanes. The velocity diagram, as shown in Figure 7.13, has $\Psi = 2.0$. The normal impulse stage has several important features including heavy blade loading, minimum "leaving loss" (stage-exit kinetic energy), and a relative minimization of friction losses because high velocities are limited to the nozzle. Another potential feature is the possibility

* To obtain identical velocity diagrams, one must neglect changes in density, which may be accommodated by increasing the annulus area of the next stage so as to maintain $\rho A =$ Constant.

of using "partial admission," where only portions of the full rotor circumference handle fluid. This is possible because the rotor operates at constant pressure. Pure impulse stages are often used in steam and gas turbines, where the desired specific energy change is very large.

Example: An Axial Flow Turbine Layout

Lay out velocity diagrams and nozzle and blade geometry for a turbine with $\Psi = 0.50$, $\varphi = 0.50$, and 50% reaction layout.

The velocity diagram is sketched in Figure 7.14. Since 50% reaction and a loading coefficient of 0.5 are desired, the velocity diagram must be symmetric and non-normal (refer to Figure 7.11 to see why $\Psi = 0.50$ requires W_1 and C_2 to lean inward). A simple nozzle/blade layout is done using circular arc camber lines to conform to the required vectors. Nozzle angles are found from C_0 (normal stage) and C_1, and blades angles are obtained from W_1 and W_2 (recall that these vectors are relative to $U = 1$). For example,

$$\alpha_1 = \tan^{-1}\left(\frac{C_{x1}}{U - C_{\theta 1}}\right) = \tan^{-1}\left(\frac{C_{x1}/U}{1 - C_{\theta 1}/U}\right) = \tan^{-1}\left(\frac{0.5}{1 - 0.25}\right) = 33.7°.$$

The nozzle/blade layout is sketched in Figure 7.14.*

Now consider a more complete design example. This example will lead to consideration of an alternative to use traditional nozzle vanes.

Example: An Axial Flow Turbine Hydraulic Turbine

Lay out a hydraulic turbine to drive a 72-pole generator ($N = 100$ rpm). The available head and flow (provided by a particular river and a dam) are 20 m and 115 m^3/s.

Begin by calculating the specific speed

$$N_s = \frac{NQ^{1/2}}{(gH)^{3/4}} = \frac{100 \times (2\pi/60) \times 115^{1/2}}{(9.81 \times 20)^{3/4}} = 2.14.$$

Then, from Cordier,

$$D_s = \left(\frac{8.26}{2.14}\right)^{0.517} = 2.01 \quad \text{and} \quad D = D_s \frac{Q^{1/2}}{(gH)^{1/4}} = 5.76\,\text{m}.$$

At this point, we have no particular way to determine d (Figure 6.14 and Equation 6.21 are not relevant because a turbine has no diffusion limits). Somewhat arbi-

* Bear in mind that the blade–vane layout given here is somewhat approximate and, for an actual design, should be corrected for the effects of imperfect guidance of the flow through a two-dimensional passage of finite geometric aspect ratio.

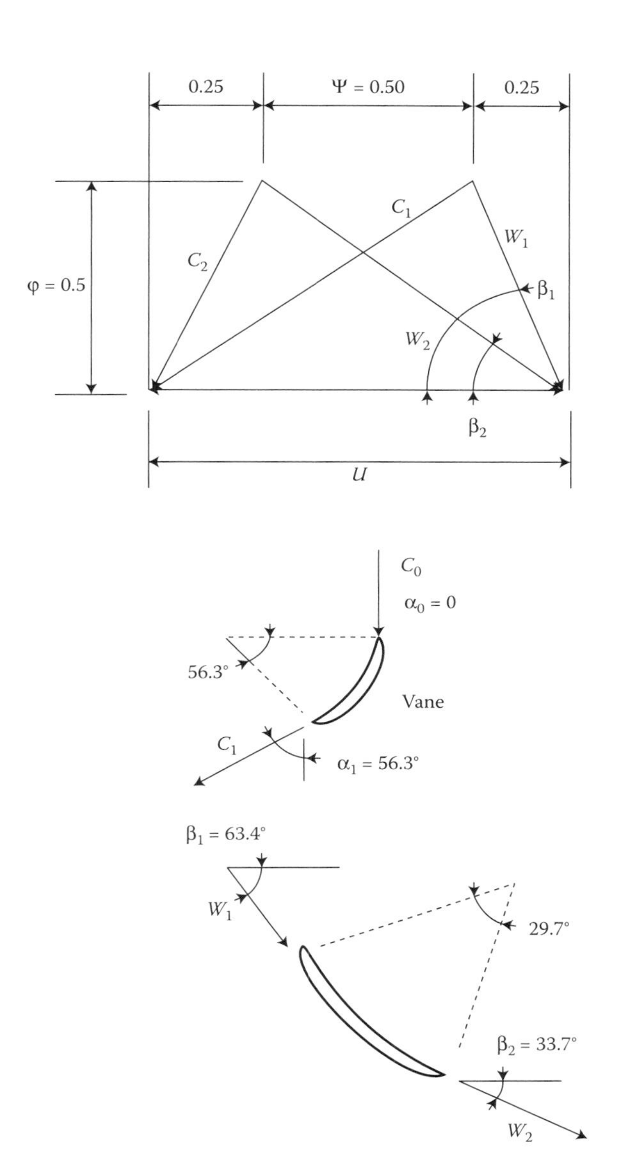

FIGURE 7.14 Velocity diagram with blade and nozzle shapes for a simple turbine stage.

trarily, choose $d/D = 1/3$. Then $d = 1.92$ m and $r_m = (D + d)/4 = 1.92$ m and $U = r_m \times N = 20.1$ m/s. The axial velocity is

$$C_x = \frac{Q}{A} = \frac{115}{(\pi/4)(5.76^2 - 1.92^2)} = 4.96 \text{ m/s},$$

giving $\varphi = 4.96/20.1 = 0.247$.

At this point, one must determine the type of turbine stage to design. A normal stage is a good choice because any swirl remaining in the water at the discharge from the blade row will be unrecoverable. Thus, $C_0 = C_x = C_2$. The loading coefficient can be determined from $\Psi = (w/U^2) = (\eta_H \times gH)/U^2$. A preliminary design could be developed by neglecting losses ($\eta_H = 1$), but it is just as easy, and certainly more realistic, to at least use the Cordier estimate for efficiency. From Figure 2.9, at $N_s = 2.14$, $\eta_H \approx 0.9$ (from the "Turbine" line). Then $\Psi = (0.9 \times 9.81 \times 20)/20.1^2 = 0.437$. From the values of φ and Ψ, together with the specification of a normal stage, the dimensionless combined velocity diagram can be sketched, as shown in Figure 7.15. From this diagram, $R_h = 1 - \Psi/2 = 0.781$. The velocity diagram angles are easily worked out; for example,

$$\beta_1 = \tan^{-1}\left(\frac{\varphi}{1-\Psi}\right) = \tan^{-1}\left(\frac{0.247}{1-0.437}\right) = 23.7°.$$

Blade and nozzle shapes can be derived as usual by assuming that the camber lines are tangent to the velocity vectors and using circular arcs as shown in Figure 7.15.

Our turbine "layout" now is as follows:

- Diameters: $D = 5.76$ m; $d = 1.92$ m.
- Blade height: $(D - d)/2 = 1.92$ m.

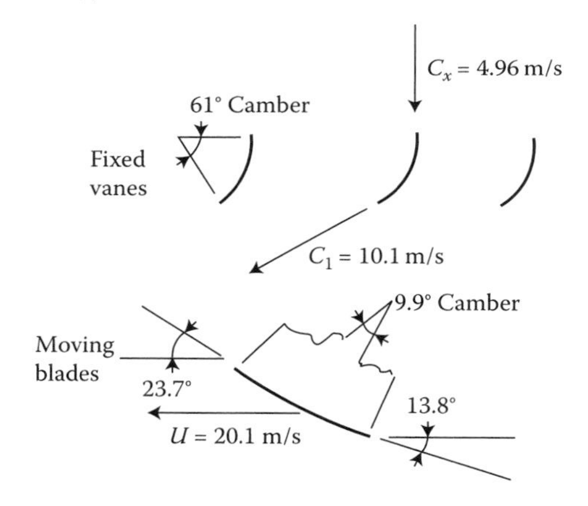

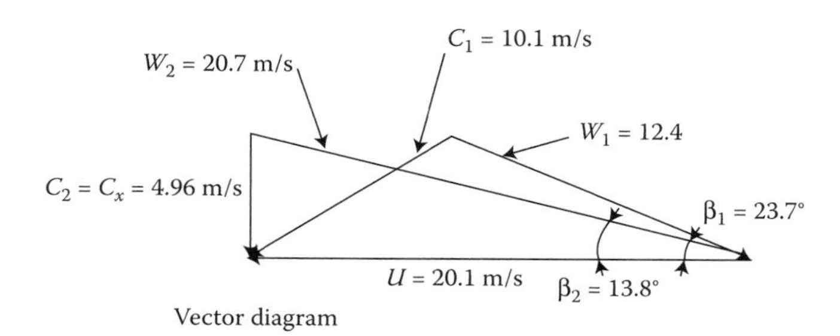

FIGURE 7.15 Velocity diagram with blade and nozzle shapes for a hydraulic turbine.

- Blade and nozzle angles and shape: As shown in Figure 7.15.
- Expected efficiency: $\eta_T \approx 0.85$ (allowing for mechanical efficiency).

At this point, we have no particular method for determining blade (chord) length or the number of blades. We do know that a solidity, σ, around 1 or larger is required for our flow model to be valid.

It should be noted that use of the dimensionless combined velocity diagram is only one way to perform the calculations for axial flow stages. An equally valid method would be to work with dimensional velocity diagrams and to draw the entrance and exit diagrams near the leading and trailing edges of the blade and vane row, rather than combining them.

The hydraulic turbine arrangement shown in Figure 7.15 is schematically correct, but it more nearly resembles an arrangement typical of steam or gas turbine nozzle and blade rows, where elements are arranged in an axial cascade configuration. In axial flow hydraulic turbines, a more typical arrangement does not use an axial flow nozzle cascade but combines the axial flow turbine rotor with a radial inflow arrangement of turning vanes or "wicket gates" upstream of the axial blade row as sketched in Figure 7.16. The incoming stream flows past the adjustable wickets, which act as turning vanes to impart a swirling motion with tangential velocity $C_{\theta 0}$. As the fluid flows

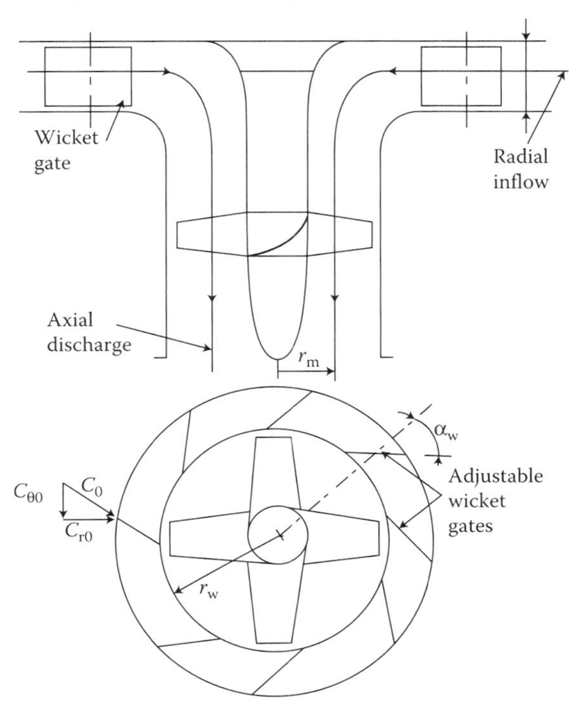

FIGURE 7.16 Wicket gate concept for an axial flow hydraulic turbine.

from the wicket gates down the duct to the turbine rotor, angular momentum is conserved. The swirl velocity seen by the turbine blade row at the radial mean station is given by

$$r_w C_{\theta 0} = r_m C_{\theta 1} \text{ so that } C_{\theta 1} = \left(\frac{r_w}{r_m}\right) C_{\theta 0}. \qquad (7.12)$$

The swirl $C_{\theta 0}$ is controlled by the wicket setting angle α_w. Assuming, as usual, perfect fluid guidance by the wicket gates, $C_{\theta 0} = C_{r0} \tan \alpha_w$, where $C_{r0} = Q/A_0$, with A_0 being the wicket cylindrical area, $A_0 = 2\pi r_w h_w$. In a sophisticated machine, both wickets and turbine blades could have adjustable angles, allowing for a broad range of power extraction levels, depending on available flow rate and power demand from the utility network.

For the previous example, the required wicket gate angle α_w can be estimated from the required swirl at the inlet to the turbine rotor,

$$\alpha_w = \tan^{-1}\left(\frac{C_{\theta 0}}{Q/2\pi r_w h_w}\right) = \tan^{-1}\left(\frac{(r_m/r_w)C_{\theta 1}}{Q/(2\pi r_w h_w)}\right).$$

From the example, $Q = 115\,\text{m}^3/\text{s}$ and the blade inlet swirl velocity is, from the velocity diagram, $C_{\theta 1} = U(1 - \Psi) = 20.1(1 - 0.437) = 11.3\,\text{m/s}$. If one specifies $r_w = D/2 = 2.88\,\text{m}$ (i.e., the rotor radius, assumed to also be the radius of the rotor inlet pipe) and $h_w = 1\,\text{m}$, then

$$\alpha_w = \tan^{-1}\left(\frac{(1.92/2.88) \times 11.3}{115/(2\pi \times 2.88 \times 1.0)}\right) = 49.8^\circ.$$

Setting the wicket gates at this angle would serve the same function as installing axial inlet nozzle vanes with 60.5° of camber (i.e., fluid turning). In general, h_w and r_w can be used as design variables (with $r_w \geq D/2$) to minimize frictional losses on the wicket gates or to control the sensitivity of power extraction to wicket gate angle or excessive incidence.

7.5 Hub–Tip Variations for Axial Flow Machines

The velocity diagrams and stage layouts discussed in the previous two sections were done for a single representative radius, namely, the arithmetic mean of the hub and tip radii. This yielded essentially two-dimensional flow patterns and blade geometries. While these geometries are certainly representative of actual turbomachines, they do not tell the whole story. Although consideration of fully 3D flow patterns will be delayed until Chapters 9 and 10, a few important points will be made here. The idea is to consider the actual 3D flow as a "stack up" of independent two-dimensional flows.

The most important effect to consider is the variation of the blade speed, U, along the span of the blade between the hub and the tip. The effect of this variation on the allowable de Haller (diffusion) ratio was discussed in Section 6.8, where it was shown that, for pumping machines, hub–tip ratios as small as 0.2 (giving a blade speed ratio as large as 5 : 1) are feasible and that ratios around 0.5 (2 : 1 blade speed) are common. Of course for turbines, there are no limits on the hub–tip ratio owing to diffusion limitations; in practice, values between about 0.2 and 0.9 may be found.

Figure 7.17 shows photographs of actual axial flow blades. Immediately, one notices two characteristics of these blades: some are twisted, and the curvature (i.e., the turning angle) is seen to vary from the hub to the tip. Consider these effects separately.

First, consider the blade twist. In most axial flow stages, the inflow is nearly normal ($C_{\theta 1} \approx 0; C_1 \approx C_x$), and C_x is nearly constant over the blade annulus area, so the relative velocity is $W_1 \approx (C_1^2 + U_1^2)^{1/2} \approx (\text{constant} + U_1^2)^{1/2}$ and the relative flow angle is $\beta = \tan^{-1}(C_x/U) \approx \tan^{-1}(\text{constant}/U)$, so that the relative velocity becomes smaller near the hub and larger near the tip, while the reverse is true for the relative flow angle and likewise the blade angle. In order to avoid flow separation and large losses near the leading edge, the blade must be twisted in order to maintain flow alignment with the leading edge (i.e., small incidence).

Now consider the curvature. In most cases, it is desirable to keep the specific work nearly constant along the blade span; otherwise, a rather serious nonequilibrium flow pattern will result as the flow leaves the blade. If w is nearly constant, then the loading coefficient Ψ ($= w/U^2$) must be larger near the hub and smaller near the tip. Larger loading nearer the hub means higher

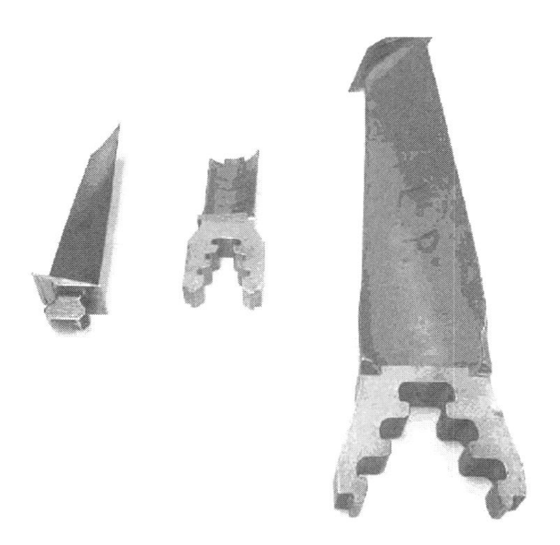

FIGURE 7.17 Axial flow blades. From left to right: a compressor blade from a gas turbine engine, a blade from a high-pressure steam turbine stage, and a low-pressure steam turbine blade.

flow turning and therefore higher blade curvature. Also, from Equation 7.9 or 7.10, the degree of reaction must be smaller nearer the hub. Long steam turbine blades (from low-pressure stages) often exhibit nearly zero reaction (the pure impulse "bucket" shape) at the root (hub) and gradually change to a nearly flat high reaction shape near the tip. It is interesting to note also that steam and gas turbine blades from "high-pressure" stages nearer the front of the turbine typically have little or no twist, with high curvature and extremely low reaction. This is because these stages feature short blades ($d/D \sim 0.8$–0.9), and smaller stage diameters, giving rise to smaller values of U and larger values of Ψ.

Example: Hub–Tip Variations in an Axial Compressor

Lay out appropriate velocity diagrams and sketch blade and vane shapes at the hub and the tip for the compressor stage considered in the example in Section 7.3. Assume that the flow angle $\alpha_1 = 65°$ all along the blade span. Also assume that specific work and axial velocity component are constant along the span.

From the previous example, $U_m = 796$ ft/s, $C_x = 475$ ft/s, and $w = 2.82 \times 10^5$ ft^2/s^2. The tip and hub diameters are $D = 2.10$ ft and $d = 1.70$ ft, and the mean diameter is 1.9 ft. The tip and hub speeds are

$$U_T = \frac{D}{d_m} U_m = \frac{2.1}{1.9} \times 796 = 879\ \text{ft/s}$$

and

$$U_H = \frac{d}{d_m} U_m = \frac{1.7}{1.9} \times 796 = 712\ \text{ft/s}.$$

To draw the dimensionless velocity diagrams, calculate

$$\varphi_T = \frac{C_x}{U_T} = \frac{475}{879} = 0.540; \qquad \Psi_T = \frac{w}{U_T^2} = \frac{2.82 \times 10^5}{879^2} = 0.365$$

and

$$\varphi_H = \frac{C_x}{U_H} = \frac{475}{712} = 0.667; \qquad \Psi_H = \frac{w}{U_H^2} = \frac{2.82 \times 10^5}{712^2} = 0.556.$$

At the tip,

$$\beta_{1,T} = \tan^{-1} \frac{\varphi_T}{1 - \varphi_T/\tan(65°)} = \tan^{-1} \frac{0.540}{1 - 0.540/\tan(65°)} = 35.8°$$

and

$$\beta_{2,T} = \tan^{-1} \frac{\varphi_T}{1 - \varphi_T/\tan(65°) - \Psi_T}$$

$$= \tan^{-1} \frac{0.540}{1 - 0.540/\tan(65°) - 0.365} = 54.6°,$$

so $\theta_{fl} = 54.6° - 35.8° = 18.8°$ (compared to 30.4° at the mean radius).

At the hub,

$$\beta_{1,H} = \tan^{-1}\frac{\varphi_H}{1 - \varphi_H/\tan(65^\circ)} = \tan^{-1}\frac{0.667}{1 - 0.667/\tan(65^\circ)} = 44.1^\circ$$

and

$$\beta_{2,H} = \tan^{-1}\frac{\varphi_H}{1 - \varphi_H/\tan(65^\circ) - \Psi_H}$$
$$= \tan^{-1}\frac{0.667}{1 - 0.667/\tan(65^\circ) - 0.556} = 78.7^\circ,$$

so $\theta_{fl} = 78.7^\circ - 44.1^\circ = 34.6^\circ$ (compared to 30.4° at the mean radius). The vane inlet angles are deduced from

$$\alpha_{2,T} = \tan^{-1}\frac{\varphi_T}{\varphi_T/\tan(65^\circ) + \Psi_T} = \tan^{-1}\frac{0.540}{0.540/\tan(65^\circ) + 0.365} = 41.2^\circ$$

and

$$\alpha_{2,H} = \tan^{-1}\frac{\varphi_H}{\varphi_H/\tan(65^\circ) + \Psi_H} = \tan^{-1}\frac{0.667}{0.667/\tan(65^\circ) + 0.556} = 37.5^\circ.$$

Figure 7.18 shows the tip, mean, and hub velocity diagrams and blade–vane shapes (assuming, as always, that the fluid follows the blade camber line).

7.6 Radial and Mixed Flow

To consider velocity diagrams in mixed flow or radial flow turbomachines in a general way, the concept of a meridional flow path or stream surface through the impeller must be introduced. The *meridional path* through a machine is the stream surface along which the mass flow occurs. In an axial flow machine, any meridional surface is cylindrical. As illustrated in Figure 7.19, all throughflow stream surfaces are described by concentric cylinders whose axis is the axis of rotation of the impeller. The flow properties (e.g., velocity or pressure) in any surface can be circumferentially averaged. The velocity vectors for the axial machine are constructed from averaged values of axial and tangential velocity components; there is no radial component of velocity.

Figure 7.20 illustrates a narrow radial flow impeller (a low specific speed machine). In this case, the meridional path, at least through the blades, can be approximated by flow along circular disks perpendicular to the axis of rotation and parallel to the backplate of the machine.

If the machine is a mixed flow type, the stream surface will be more complicated, partially axial, and partially radial. A simple example is an impeller

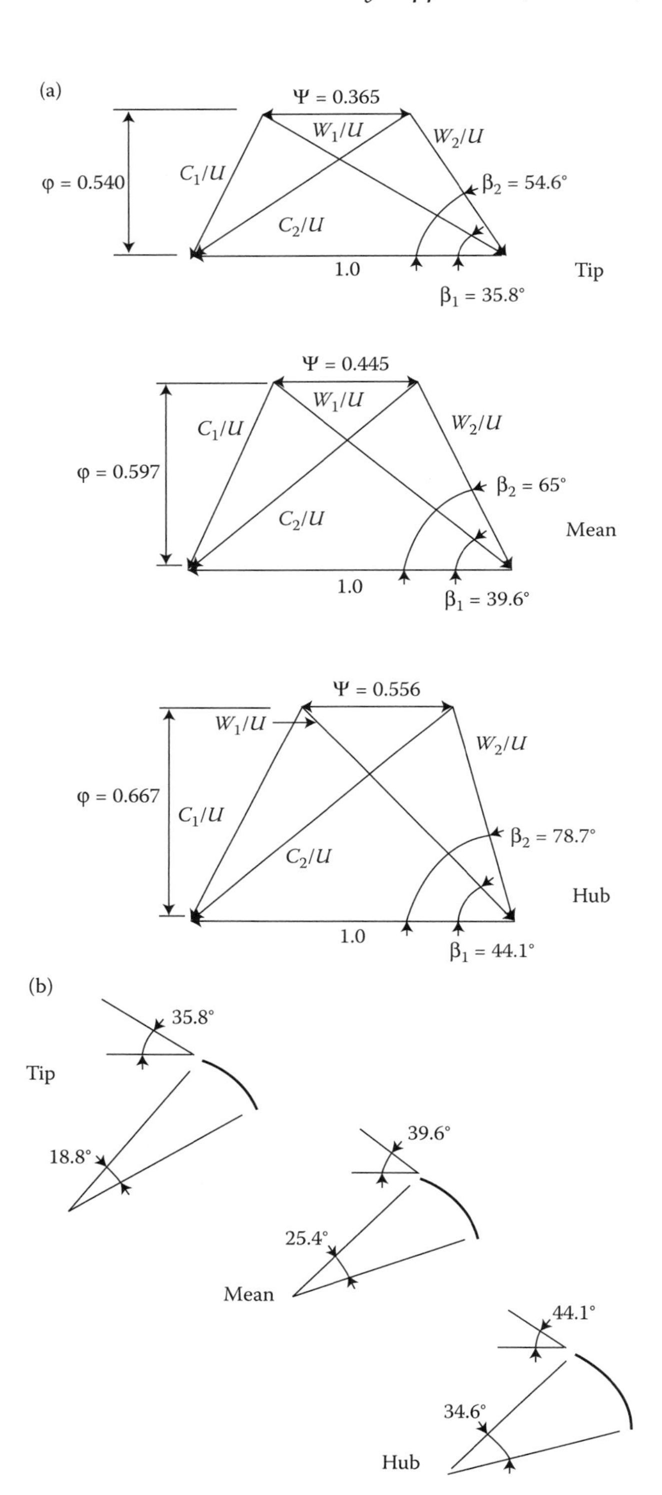

FIGURE 7.18 (a) Velocity diagrams and (b) blade layouts, and (c) vane layouts at tip, mean diameter, and hub for an axial flow compressor example.

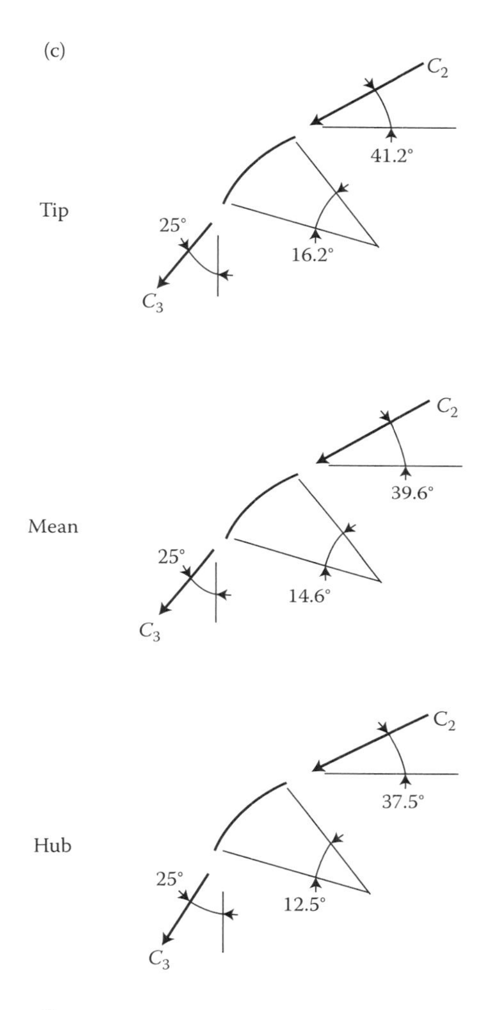

FIGURE 7.18 (Continued).

whose outer casing and centerbody consist of sections of cones concentric about the rotational axis of the impeller, as shown in Figure 7.21. In this case, the meridional surfaces can be approximated as a series of concentric cones. The most general mixed flow case must be described by concentric surfaces of revolution along curved paths that typically change from approximately axial to approximately radial, as shown in Figure 7.22.

For the limiting cases of purely axial or purely radial flow, the meridional velocity is simply the axial or radial velocity component, respectively. For both the conical and general shapes, the meridional velocity consists of both radial and axial components. In any of these cases, the tangential (swirl) component of absolute velocity can be added to form the overall absolute velocity. The

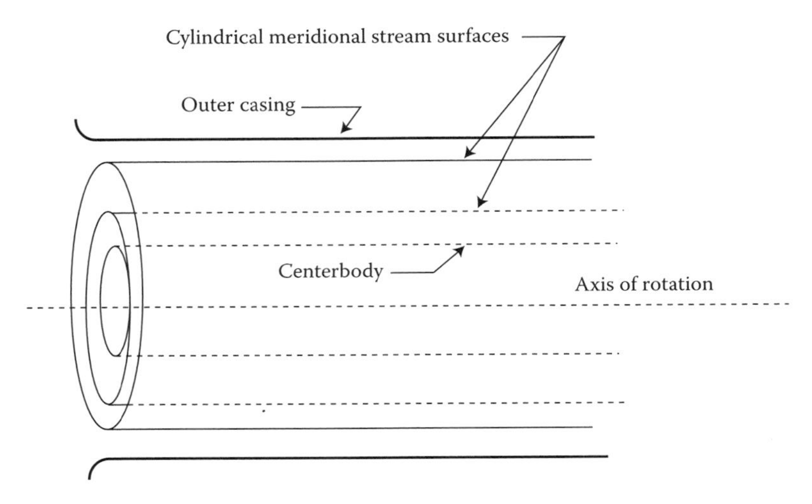

FIGURE 7.19 Meridional stream surfaces are concentric cylinders for an axial flow machine.

absolute velocity then describes a mean stream path that sweeps along and winds around the meridional surface, as illustrated in Figures 7.23 and 7.24. It is on these increasingly complex surfaces that one must describe the absolute (C) and relative (W) velocity vectors and construct the velocity diagrams. Relative velocity vectors are generated by superposing a rotation of the stream surface about its axis of revolution (the rotational axis of the machine), to give the blade linear velocity (U).

Construction of velocity diagrams for the general mixed flow case is sketched in Figure 7.25. The cross-sections of several blades, as cut by the meridional surface of revolution, are shown along with the rotational velocity U, the absolute velocity C (composed of a meridional component and a swirl component), and the resultant relative vector W. Blade "inlet" (station "1") and "outlet" (station "2") vectors are shown to illustrate the proper layout for this fairly complex situation. The blade is shown with $C_{\theta 1} = 0$ and $C_{\theta 2} > 0$ for simplicity.

7.7 Mixed Flow Example

An example will be used to illustrate some of the advantages of a mixed flow layout over a purely axial flow machine when the performance requirements are in the appropriate range. Suppose that a fan is required to provide 7000 cfm of air ($\rho = 0.00233\ \text{slug/ft}^3$) with a total pressure rise of 7 in. wg. To save cost, a 4-pole motor with a running speed (N) of 1750 rpm ($183\ \text{s}^{-1}$) is chosen. From these specifications, $N_s = 1.41$ and the corresponding D_s value from

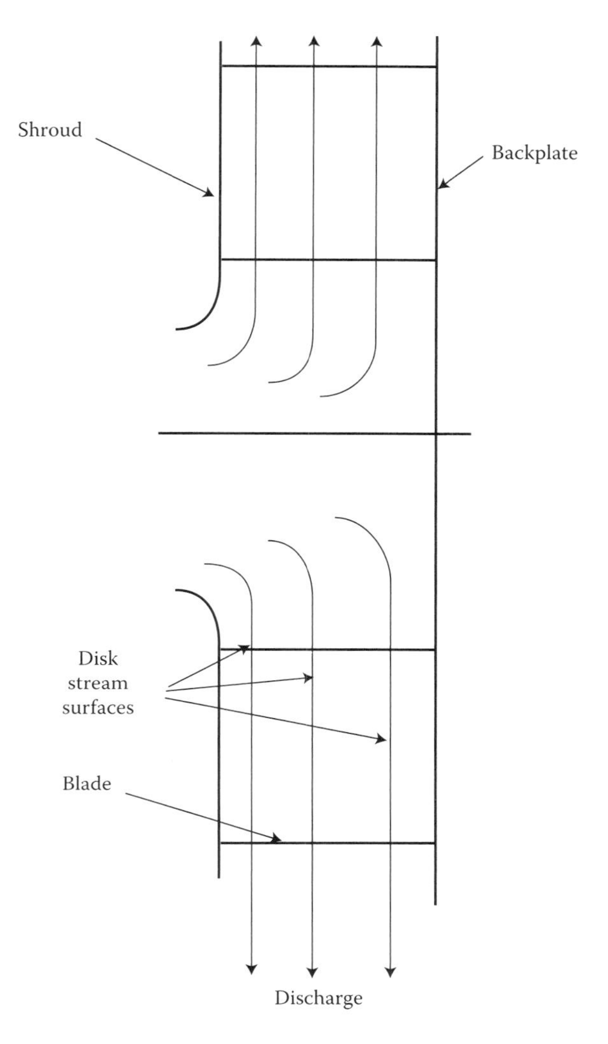

FIGURE 7.20 Meridional stream surfaces through the blades are parallel disks for a radial flow machine.

the Cordier line is 2.49. These values are in the mixed flow region (Region D). We will lay out a quasi-axial mixed flow machine. For simplicity, choose a simple conical flow path like that indicated in Figure 7.26. Use the given data to estimate the diameter:

$$D = D_{\mathrm{s}} \frac{Q^{1/2}}{(\Delta p_{\mathrm{T}}/\rho)^{1/4}} = 2.49 \times \frac{(7000/60)^{1/2}}{((7 \times 5.2)/0.00233)^{1/4}} = 2.4\,\mathrm{ft}.$$

The Cordier efficiency is about 0.9.

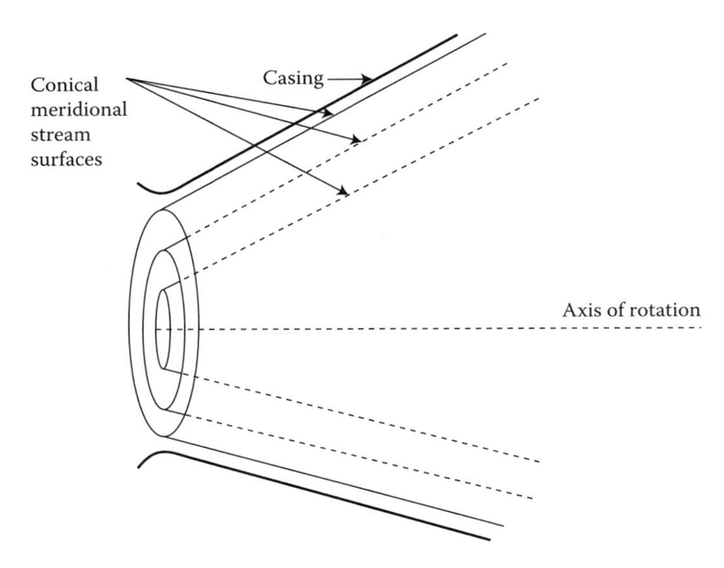

FIGURE 7.21 Meridional stream surfaces for a simplified "conical" mixed flow rotor.

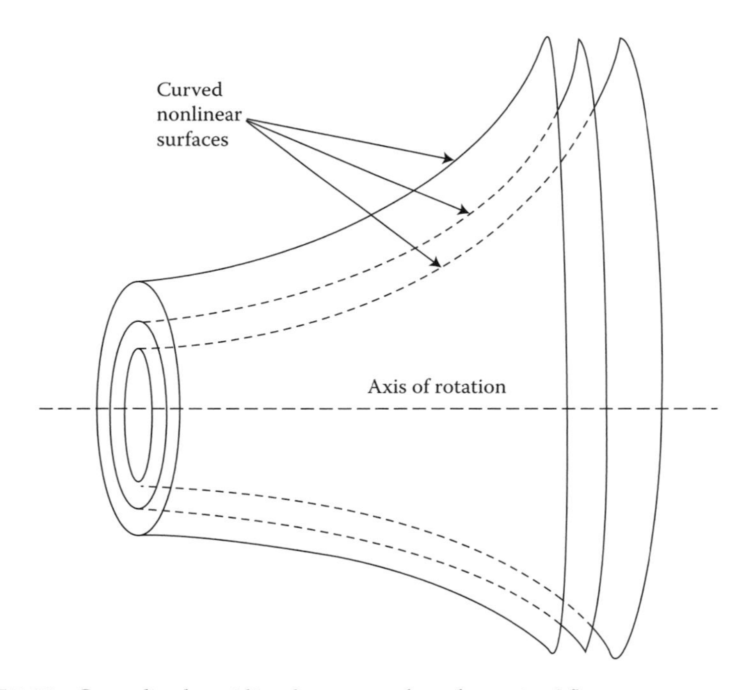

FIGURE 7.22 Generalized meridional stream surfaces for a mixed flow rotor.

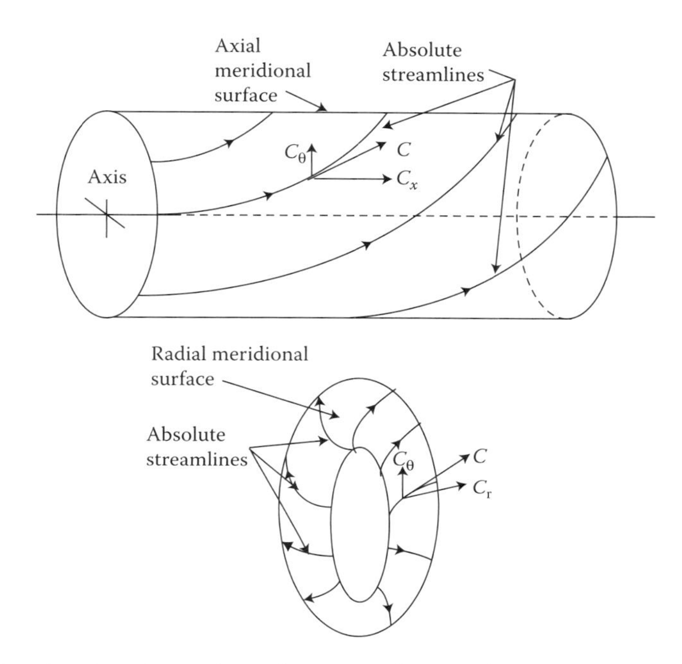

FIGURE 7.23 Resultant flow path, combining meridional and tangential velocities, for simple axial and radial machines.

A possible meridional flow path is shown in Figure 7.26. The mean radius streamline is assumed to have a slope of about 30°, so this is the path along which the blade will be laid out. Guided by Figure 6.14, choose $d = 1.9$ ft at the outlet so that $r_{\text{m,outlet}} = 1.075$ ft. At the inlet, assume $D_{\text{inlet}} = 1.8$ ft and hold $(D-d)$ constant to give $d_{\text{inlet}} = 1.3$ ft. Then $r_{\text{m,inlet}} = 0.775$ ft. At the inlet, $U_{\text{m1}} = 0.775 \times 183 = 141.8$ ft/s. The inlet area is $A_1 = 1.22\,\text{ft}^2$, so $C_{x1} = Q/A_1 = 7000/(60 \times 1.22) = 95.8$ ft/s and $C_{\text{m1}} = C_{x1}/\cos 30° = 110.7$ ft/s. Then calculate W_1 and β_1 as $W_1 = (141.8^2 + 110.7^2)^{1/2} = 179.9$ ft/s and $\beta_1 = \tan^{-1}(110.7/141.8) = 38.0°$.

At the outlet, $U_{\text{m2}} = 1.075 \times 183 = 196.7$ ft/s, $A_2 = 1.688\,\text{ft}^2$, and $C_{x2} = 69.1$ ft/s, yielding $C_{\text{m2}} = 79.8$ ft/s. $C_{\theta 2}$ is found from $w/U_2 = (\Delta p_{\text{T}}/\rho\eta_{\text{T}})/U_2 = 7 \times 5.2/(0.00233 \times 0.9)/196.7 = 88.2$ ft/s. Then $W_2 = [C_{\text{m2}}^2 + (U_2 - C_{\theta 2})^2]^{1/2} = 134.7$ ft/s and $\beta_2 = \tan^{-1}[C_{\text{m2}}/(U_2 - C_{\theta 2})] = 36.3°$. Thus, $\theta_{\text{fl}} = 2.7°$ and $W_2/W_1 = 0.75$. These are very reasonable values for the blade row. Note that the angles and the blade itself must be laid out along the meridional conical path, not on a cylinder, as shown in Figure 7.26. A more complete design may be made by also laying out blade sections at the hub and the tip, as discussed in the previous section.

A layout for a simpler purely axial flow path using $D = 2.4$ ft, $N = 1750$ rpm, and $d = 1.9$ ft (try these calculations yourself) gives a (mean radius) blade row with $W_2/W_1 = 0.62$ and $\theta_{\text{fl}} = 13.1°$, which is a blade working significantly

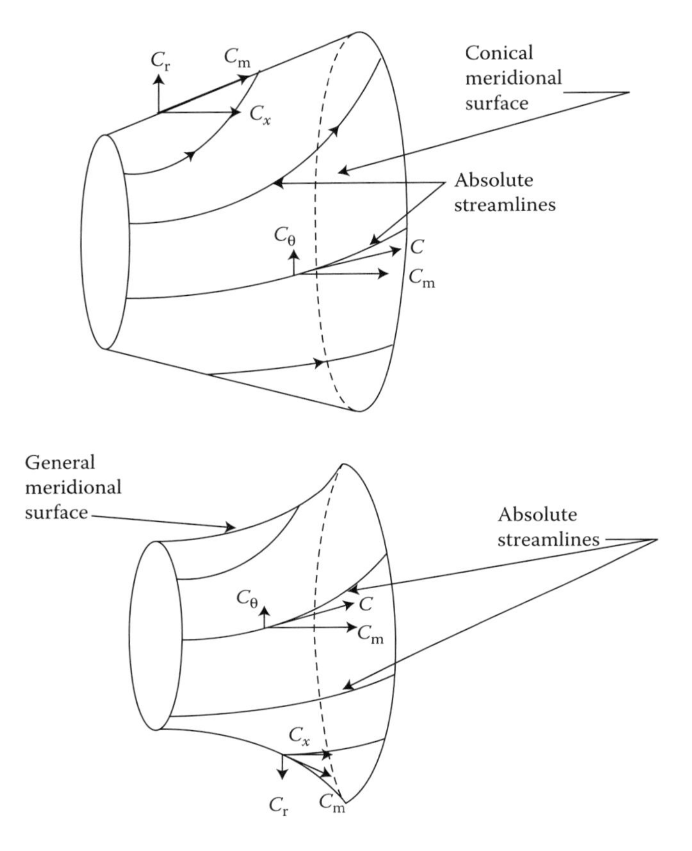

FIGURE 7.24 Resultant flow path, combining meridional and tangential velocities, for conical and general mixed flow machines.

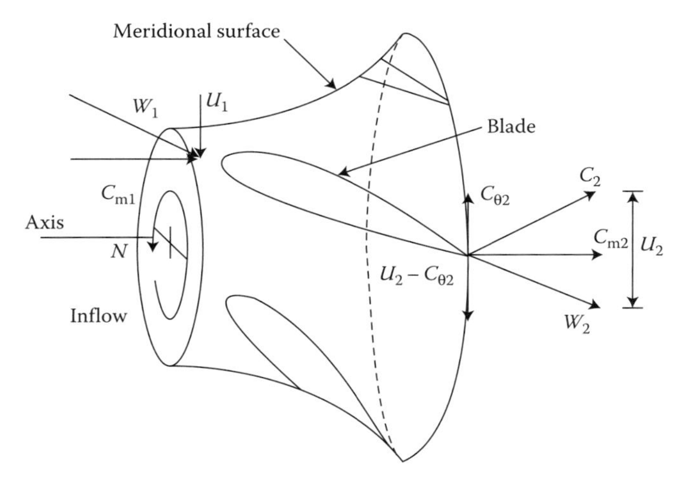

FIGURE 7.25 Construction of velocity vector diagram for a 3D mixed flow rotor.

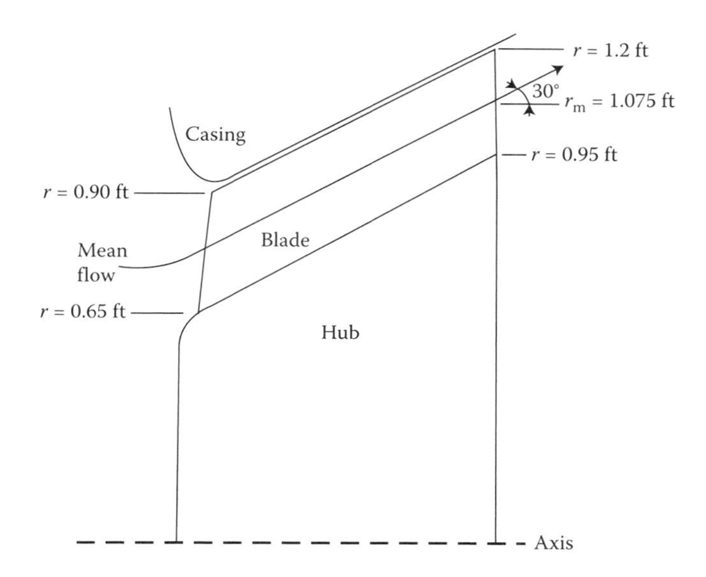

FIGURE 7.26 Quasi-axial mixed flow fan.

harder and with serious diffusion issues. Part of the problem can be overcome by using a larger hub diameter, say $d = 2.1$ ft. The resulting numbers are $W_2/W_1 = 0.702$, which is more acceptable, but now $\theta_{fl} = 14.0°$. For the purely axial flow (with $d = 2.1$), $C_x = 110$ ft/s while $C_{x2} = 69.1$ ft/s for the

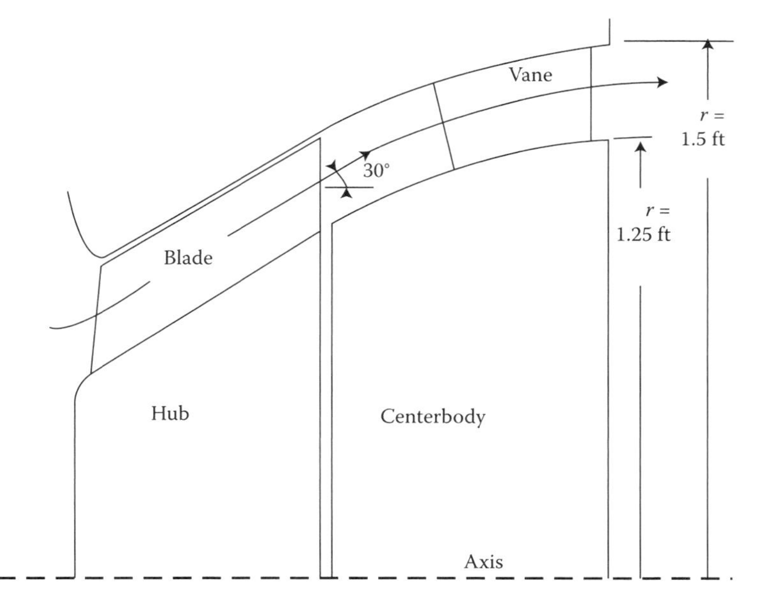

FIGURE 7.27 Meridional view of the quasi-axial mixed flow fan layout including discharge geometry.

mixed flow configuration. Axial outlet velocity pressures are 1.07 in. wg for the mixed flow and 2.71 in. wg for the axial flow. This may become a significant advantage for the mixed flow fan in terms of static efficiency. One could use a simple straightening vane row to recover the swirl for the ($d = 2.1$) axial fan with $C_x = 110\,\text{ft/s}$ and $C_{\theta 2} = 84.3\,\text{ft/s}$ obtaining $C_4/C_3 = 0.794$, which is acceptable; however, the resulting velocity of 110 ft/s would be rather high for many applications and might require further diffusion.

For the mixed flow fan, one can readily control the vane row diffusion and final velocity by continuing the conical flow path over a centerbody to an exit condition with, say, $D = 3.0\,\text{ft}$ and $d = 2.5\,\text{ft}$, followed by a turn to an axial discharge direction while simultaneously removing the swirl (Figure 7.27). This yields $C_{x3} = 54.1\,\text{ft/s}$ and, from conservation of angular momentum, $C_{\theta 3} = (r_{m2}/r_{m3})C_{\theta 2} = (1.075/1.375)88.2 = 72.2\,\text{ft/s}$. As a result, $C_4/C_3 = 0.69$, which is a bit low but acceptable level of diffusion that produces an acceptably low final velocity.

7.8 Radial Flow Layout: Centrifugal Blowers

Machines with purely radial flow are somewhat easier to examine than mixed flow geometries. We begin by considering a centrifugal blower having an essentially radial flow path along the entire blade, as shown in Figure 7.28, in which the radial flow surface is sketched next to a cutaway of the impeller. Figure 7.28 shows backwardly inclined airfoil blades, which are used if high efficiency is desired. Other frequently seen blade layouts for fans include straight radial or even forward-curved blades and blades of constant thickness

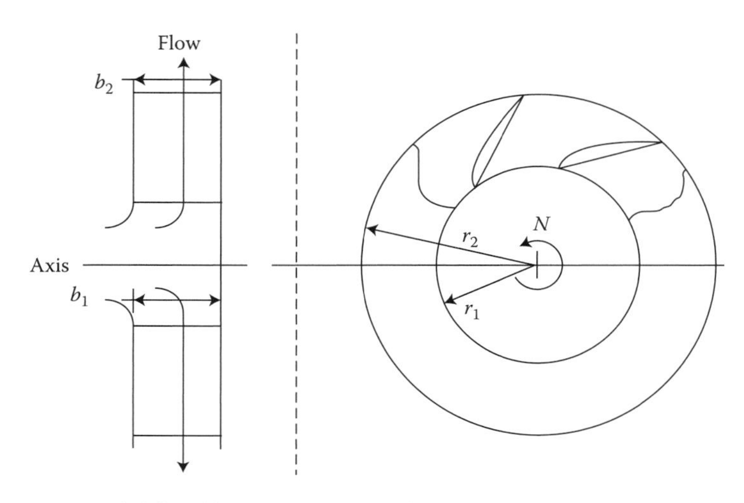

FIGURE 7.28 A radial flow blower with backward-inclined airfoil blades.

(without the airfoil shape). Of course, the approach taken in this chapter approximates all blades by their simple (zero thickness) camber line.

The usual parameters available to the designer, in addition to speed, N, and overall diameter, D, are the eye diameter, d, the blade inlet and discharge angles, and the width of the flow passage at the inlet and the outlet, b_1 and b_2, respectively. Reasonable choices for eye diameter and inlet width are discussed in Section 6.9, at least for the case of zero preswirl. This section will consider typical choices for the remaining parameters and their interaction. A final item, the number of blades, will not be considered until Chapter 8. It should be noted that the usual velocity diagram parameters from axial flow, namely, loading coefficient and flow coefficient, require redefinition because a flow streamline changes radius. The "new" definitions are

$$\Psi \equiv \frac{w}{U_2^2}, \tag{7.13}$$

$$\varphi \equiv \frac{C_{r2}}{U_2}, \tag{7.14}$$

in which the radial velocity at the rotor exit becomes the measure of the flow and work and flow are referred to the blade velocity at the rotor exit. The third parameter, degree of reaction, is seldom used because all radial flow machines have some "reaction" resulting from the radius change.

Figure 7.29 shows the details of a single (radial) blade together with the velocity diagrams at the blade inlet and outlet. Because of the change in the radial velocity, C_r, through the rotor, it is not conventional to combine the inlet and outlet vectors into a combined velocity diagram, although this can be done. If one assumes no preswirl (i.e., no IGVs), the inlet diagram is relatively simple, defined by the throughflow velocity, $C_1 = C_{r1}$ [$= Q/(2\pi r_1 b_1)$], and the blade velocity U_1 [$= Nr_1$]. The relative velocity W_1 [$= (U_1^2 + C_1^2)^{1/2}$] lines up with the blade leading edge camber line. At the outlet, W_2 aligns with the trailing edge mean camber line and is the vector sum of U_2 and C_2, as shown. For any given value of U_2, the magnitude of W_2 is "controlled" by C_{r2} and the blade trailing edge angle β_2. As can be inferred from Figure 7.29, the size of C_2 and its direction are "controlled" by the vector addition of W_2 and U_2, thus fixing the size of $C_{\theta 2}$.* The value of $C_{\theta 2}$ can be calculated from

$$\tan\beta_2 = \frac{C_{r2}}{U_2 - C_{\theta 2}}, \quad \text{giving } C_{\theta 2} = U_2 - C_{r2}\cot\beta_2. \tag{7.15}$$

* This relation can be grasped most easily by considering how the velocity diagram would be laid out graphically. First a vector of length "U_2" is drawn. Then a line is drawn from the head of the U-line at angle β_2. This line is continued until its perpendicular distance from U_2 is equal to C_{r2}. This gives the W_2 vector. Finally, a vector is drawn from the tail of U_2 to the head of W_2; this is the C_2 vector. The value of $C_{\theta 2}$ is then apparent from the diagram.

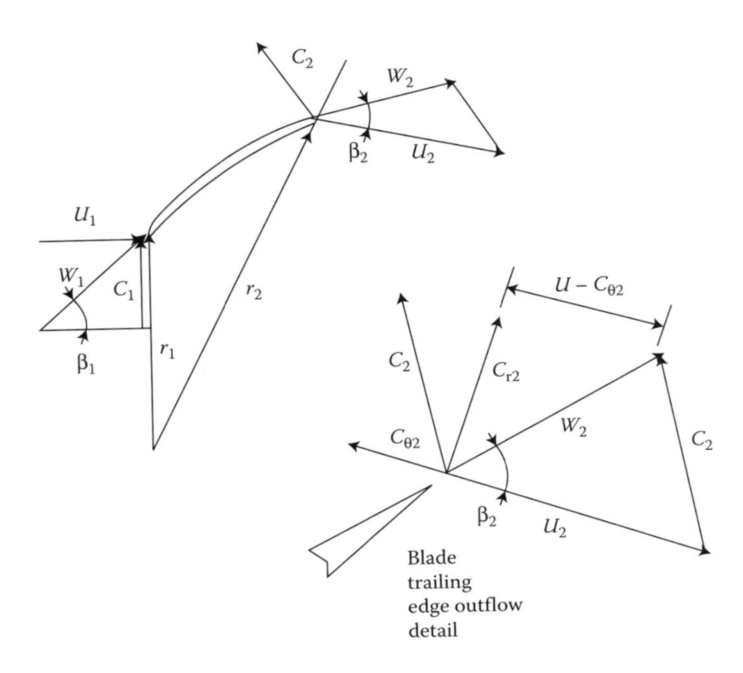

FIGURE 7.29 Radial blade, with inlet and outlet velocity diagrams.

The work coefficient and flow coefficient are related as follows. For no preswirl ($C_{\theta 1} = 0$), the work is

$$w = U_2 C_{\theta 2} - U_1 C_{\theta 1} = U_2 C_{\theta 2},$$

so that

$$\Psi = \frac{w}{U_2^2} = \frac{U_2 C_{\theta 2}}{U_2^2} = \frac{C_{\theta 2}}{U_2}.$$

Substituting from Equation 7.15

$$\Psi = \frac{U_2 - C_{r2} \cot \beta_2}{U_2} = 1 - \frac{C_{r2} \cot \beta_2}{U_2} = 1 - \varphi \cot \beta_2. \quad (7.16)$$

This is a very simplified "performance curve" for a radial blower if one assumes 100% efficiency ($gH = w$) and invariance of the angle β_2 with flow rate! The equation predicts that, if $\varphi = 0$, the "shutoff head" is $gH_{\text{Shutoff}} = U_2^2$ and, if $\beta_2 < 90°$, the "free-delivery" ($\Psi = 0$) flow is given by $\varphi = \tan^{-1} \beta_2$.

More usefully, Equation 7.16 can be used to investigate the effects of the blade exit angle on performance. Figure 7.30 illustrates three general shapes for the blade exit: *backward leaning* (blade tip slopes away from the direction of rotation [$\beta_2 < 90°$, $\cot(\beta_2) > 0$]); *radial* (blade tip perpendicular to the direction of rotation [$\beta_2 = 90°$, $\cot(\beta_2) = 0$]); and *forward leaning*

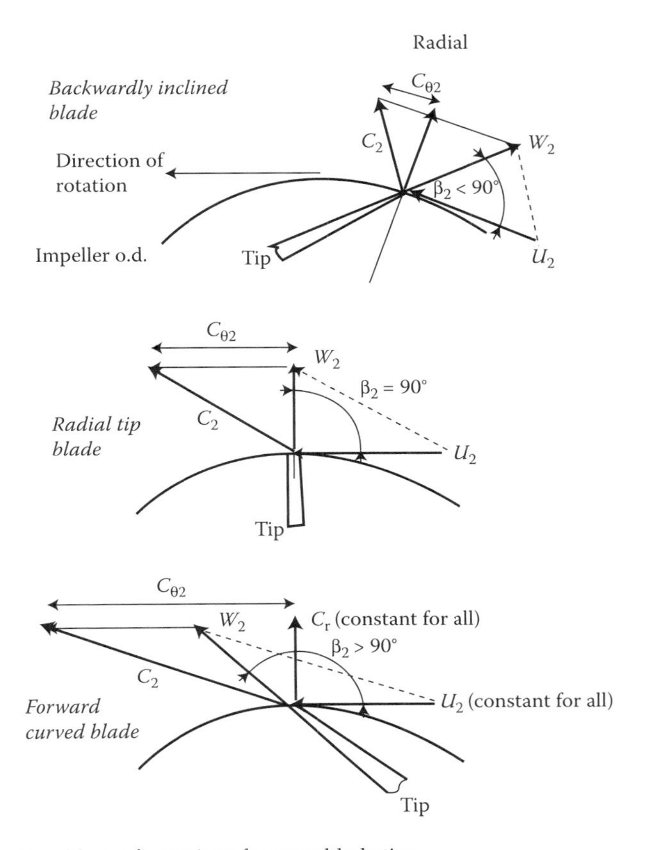

FIGURE 7.30 Possible configurations for rotor blade tips.

(blade tip slopes toward the direction of rotation [$\beta_2 > 90°$, $\cot(\beta_2) < 0$]). Consideration of the corresponding velocity diagrams leads to the following observations:

For the backward leaning blade tip, at a given flow rate (φ) and tip speed (U)

- The work (and head) is relatively small.
- The exit velocity (hence velocity pressure) is the smallest.
- The performance curve has a negative slope.

For the radial blade tip

- The work is constant for any flow rate.
- The exit velocity (hence velocity pressure) is significant.
- The performance curve is flat.

For the forward leaning blade tip, at a given flow rate (φ) and tip speed (U)

- The work (and head) is the largest.
- The exit velocity (hence velocity pressure) is large.
- The performance curve has a positive slope (work/head increase with increasing flow).

Generally, backward leaning blade tips are preferred because the performance curve has a stable negative slope and the efficiency is usually larger because a greater proportion of the added energy appears as static pressure instead of velocity pressure. The fact that the work is relatively smallest is compensated by using a larger rotor diameter. The radial tip design is sometimes used for severe duty applications (because of lower fluid turning and no bending stress on the blades) or when the higher specific work or relative ease of manufacture outweighs the disadvantage of lower efficiency. Forward-curved tip blades are typically used only when a high-velocity stream with low static pressure rise is needed and light weight is important.

Inlet and outlet vector diagrams can be used to examine blade passage diffusion, as was done for axial blade rows. It is necessary to express W_1 and W_2 in terms of geometry, flow, and pressure rise (i.e., specific work) and to consider the de Haller ratio, W_2/W_1, relative to the limiting value of approximately 0.72. W_1 can be calculated by $W_1 = (C_{r1}^2 + U_1^2)^{1/2}$, with $C_{r1} = Q/(2\pi r_1 b_1)$; $U_1 = Nr_1$. W_2 is calculated by $W_2 = (C_{r2}^2 + [U_2 - C_{\theta 2}]^2)^{1/2}$, with $C_{r2} = Q/(2\pi r_2 b_2)$, $U_2 = Nr_2$, and $U_2 - C_{\theta 2} = C_{r2} \cot \beta_2$. After some algebra and trigonometry, one obtains

$$\frac{W_2}{W_1} = \left(\frac{r_1 b_1}{r_2 b_2}\right)\left(\frac{\sin \beta_1}{\sin \beta_2}\right). \tag{7.17}$$

First, consider a constant width impeller with $b_2 = b_1$. If one fixes r_1/r_2 and β_1 (effectively fixing the dimensionless flow Q/ND^3), then $W_2/W_1 = \text{constant}/(\sin \beta_2)$. Thus, increasing β_2 causes a more-or-less linear decrease in W_2/W_1; that is, trying to increase the work by increasing fluid turning leads directly to a decrease in W_2/W_1 and eventual problems in efficiency and flow separation/stall in the blade channels.

Another popular design approach is to specify a constant flow cross-section such that $b_1 r_1 = b_2 r_2$, which tapers the blade row width such that $C_{r1} = C_{r2}$ (i.e., C_r is constant throughout the impeller flow channel). For this case, $W_2/W_1 = (\sin \beta_1)/(\sin \beta_2)$, and again, the de Haller ratio decreases as $(\sin \beta_2)^{-1}$. A typical inlet blade angle for a high specific speed backwardly inclined blade fan might be $\beta_1 = 30°$, giving $W_2/W_1 = 0.5/\sin \beta_2$. Values of W_2/W_1 are shown in Table 7.1; apparently, the maximum value of β_2 to avoid exceeding de Haller's limit is about 44°.

One can make some estimates of the maximum feasible work coefficient (at the design point) to avoid blade passage stall. If $\beta_1 \sim 30°$, then

TABLE 7.1

Diffusion Ratio for Various Exit Blade Angles

β_2	$W_2/W_1 = 0.5/\sin\beta_2$
30	1.000
35	0.872
40	0.778
44	0.720
45	0.707
50	0.653
55	0.610

$C_{r1}/U_1 \sim \tan 30° = 0.58$. For the constant-width ($b_2 = b_1$) impeller, $\varphi = C_{r2}/U_2 = (C_{r2}/C_{r1})(U_1/U_2)(C_{r1}/U_1) = (r_1/r_2)^2(C_{r1}/U_1) = (r_1/r_2)^2 \tan\beta_1$, so assuming $r_1/r_2 \sim 0.65$ gives $\varphi \sim 0.65^2 \times 0.58 = 0.24$ and $\Psi_{Max} = 1 - \varphi \cot \beta_2 \sim 1 - 0.24 \times \cot 44° = 0.75$.

For the tapered, constant C_r, impeller, $\varphi = (r_1/r_2)\tan\beta_1 \sim 0.65 \times 0.58 = 0.375$ and $\Psi_{Max} = 1 - \varphi\cot\beta_2 \sim 1 - 0.375 \times \cot 44° = 0.61$. Clearly, the ratios b_2/b_1 and r_2/r_1 are important parameters for controlling the level of diffusion and hence the efficiency and stall margin in a radial flow impeller.

As was the case for the axial flow layout, it is possible (up to a point) to reduce W_2/W_1 below the level of 0.72 through increases in solidity and careful, detailed design. In particular, for all but the lowest specific speed centrifugals, at low values of W_2/W_1, the flow behavior is dominated by a stable jet-wake pattern as discussed in Chapter 6. In such cases, diffusion is not a dominant factor in design (Johnson and Moore, 1980). However, one can still use de Haller's limit as a conservative level in examining basic blade layouts in centrifugal machines with backwardly inclined blades.

7.9 Radial Flow Layout: A Centrifugal Pump

As a further illustration of the geometry of radial flow machines, consider the layout of the impeller for a centrifugal pump. For a concrete example, suppose that the pump is to deliver 450 gpm of water at 100 ft of head. Choosing a 4-pole direct-drive 1750 rpm ($183\,\text{s}^{-1}$) motor, one obtains $N_s = 183 \times (450/449)^{1/2}/(32.2 \times 100)^{3/4} = 0.429$. Then $D_s = 6.85$ and $\eta_T = 0.81$ from the Cordier correlations. Calculating the diameter from D_s gives $D = 6.85 \times (450/449)^{1/2}/(32.2 \times 100)^{1/4} = 0.91\,\text{ft} = 10.9\,\text{in}$. This size seems a bit large when compared with typical pumps shown in catalogs. Choose a somewhat smaller, less expensive diameter of 10 in. (0.833 ft). Based on the specific speed, the pump will have the general shape shown in Figure 7.31.

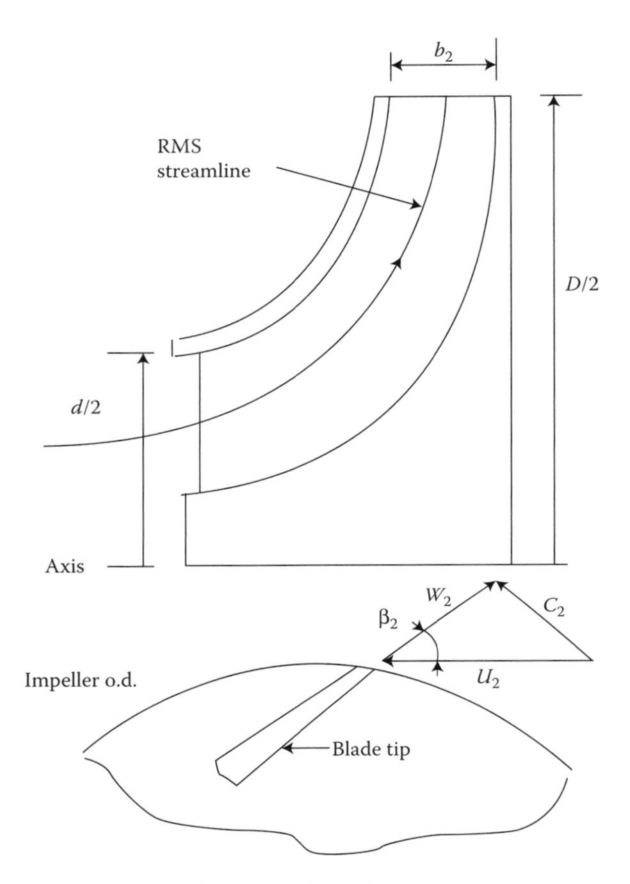

FIGURE 7.31 Pump geometry with an exit velocity diagram.

Note that the flow will enter the blade channel in the axial direction and be turned into the radial direction within the channel.

The key parameters to be determined are the eye diameter, the entrance and exit blade angles (actually, the velocity diagram angles), and the impeller width at exit, b_2. The eye diameter can be determined from Equation 6.33, which was based on minimizing the inlet relative velocity. Then

$$d = D\left(1.53\phi^{1/3}\right) = D\left(1.53\left[\frac{Q}{ND^3}\right]^{1/3}\right) = 1.53 \times \left(\frac{1.002\,\text{ft}^3/\text{s}}{183\,\text{s}^{-1}}\right)^{1/3}$$

$$= 0.270\,\text{ft} = 3.24\,\text{in.}$$

For centrifugal pumps, exit blade angles are kept relatively shallow, typically 20°–35°. This is done to keep fluid turning fairly low and shift a greater portion of the work (head) to the increase in blade velocity, $U_2^2 - U_1^2$ (see the alternate form of Euler's equation, Equation 6.12, and the discussion following). Up to a limit, the result is a better efficiency. For the current pump, use a value toward the lower end, $\beta_2 = 22°$.

Key components of the outlet velocity diagram can now be calculated. $U_2 = Nr_2 = 183 \times (0.83/2) = 75.9\,\text{ft/s}$. From the head and efficiency, $w = U_2C_{\theta 2} = gH/\eta_T = 32.17 \times 100/0.81 = 3972\,\text{ft}^2/\text{s}^2$, so $C_{\theta 2} = w/U_2 = 52.3\,\text{ft/s}$. Then $U_2 - C_{\theta 2} = 75.9 - 52.3 = 23.6\,\text{ft/s}$ and $C_{r2} = (U_2 - C_{\theta 2}) \tan\beta_2 = 23.6 \tan(22°) = 9.54\,\text{ft/s}$. The impeller tip-width is then $b_2 = Q/\pi D_2 C_{r2} = 1.002/(\pi \times 0.83 \times 9.54) = 0.4\,\text{ft} = 0.48\,\text{in}$. Note that a shallower exit angle or a smaller outer diameter would lead to a wider tip-width.

With provisional values for D, b_2, β_2, and d chosen, finish laying out the inlet and outlet vectors (to check diffusion ratio) and the blade angles. As observed, one is using a radial discharge at the outlet and an axial inflow velocity C_{x1} at the inlet. The method for laying out the inlet vector and angle will resemble the technique used for axial flow machines. At the shroud, one can calculate $C_{x1} = Q/[(\pi/4)d^2] = 11.2\,\text{ft/s}$ and $U_1 = Nd/2 = 30.9\,\text{ft/s}$. Then, at the shroud, $W_1 = (C_{x1}^2 + U_1^2)^{1/2} = 32.8\,\text{ft/s}$. From the outlet vector layout, $C_{r2} = 9.54\,\text{ft/s}$ and $U_2 - C_{\theta 2} = 23.6\,\text{ft/s}$, so $W_2 = 25.5\,\text{ft/s}$. The de Haller ratio along the shroud streamline is $W_2/W_1 = 0.78$. To establish a representative "mean" streamline, the root mean square value of the inner diameter (i.e., $r_m = [d^2/2]^{1/2}$) was used because very little activity is associated with the region in which the radius approaches zero. Along this mean streamline, $U_1 = 21.8\,\text{ft/s}$ and $W_1 = 24\,\text{ft/s}$, yielding $W_2/W_1 = 1.06$. The diffusion picture is not very clear-cut, but it is clear that the de Haller ratio is acceptably high. The inlet setting angle for the axial "inducer" section of the blade becomes $\beta_1 = \tan^{-1}(U_1/C_{x1})$, measured relative to the axial direction and is equal to 62.6° at the mean (71%) radial station. Since axial entry geometry coupled with radial exit geometry is being used, there must be a compound twist or curvature of the blade shape to connect the ends in a smooth manner.

7.10 Radial Flow Layout: Turbocharger Components

The layout of a small turbocharger will illustrate some of the additional considerations that arise when compressibility of the fluid must be accounted for. Turbochargers feature a (typically) radial inflow gas turbine in the engine exhaust gas stream coupled to a radial discharge compressor supplying high-pressure air to the induction system of an internal combustion (i.c.) engine. A thorough treatment of these systems is available in i.c. engine textbooks (Obert, 1973; Heywood, 1988).

A radial inflow turbine configuration is chosen on the basis of the exhaust energy available. Then a compressor that is matched in power requirement to the actual shaft power output of the turbine is laid out. Generally, the turbine and the compressor are constrained to run at the same speed on the same shaft and their mass flows are essentially equal, except that the turbine mass flow is increased by the fuel burned in the engine. An exception to this last requirement can exist when a "waste gate" or bypass exhaust flow path allows partial exhaust gas supply to the turbine; but that will be of no concern here.

Begin with the following exhaust gas (turbine inflow) properties, which are typical of a turbocharger for an i.c. engine of about 500 hp:

$$p_{01} = 152\,\text{kPa}; \quad T_{01} = 850\,\text{K}; \quad \rho_{01} = 0.625\,\text{kg/m}^3; \quad \dot{m} = 0.50\,\text{kg/s}.$$

The inlet properties are given as stagnation values. The turbine exhaust has an outlet stagnation pressure $p_{02} = 103.3$ kPa, allowing for a pressure drop in the downstream exhaust system (muffler, tailpipes, or other similar components).

The turbine performance parameters are computed using a compressible fluid formulation:

$$gH = c_p T_{01}\left[1 - \left(\frac{p_{02}}{p_{01}}\right)^{\gamma-1/\gamma}\right] = \frac{\gamma}{\gamma-1}\frac{p_{01}}{\rho_{01}}\left[1 - \left(\frac{p_{02}}{p_{01}}\right)^{\gamma-1/\gamma}\right]. \tag{7.18}$$

For the turbine exhaust gas, use an elevated temperature value, $\gamma = 1.37$, to give

$$gH = \frac{\gamma}{\gamma-1}\frac{p_{01}}{\rho_{01}}\left[1 - \left(\frac{p_{02}}{p_{01}}\right)^{\gamma-1/\gamma}\right]$$

$$= \frac{1.37}{1.37-1}\frac{152{,}000}{0.625}\left[1 - \left(\frac{103.3}{152}\right)^{(1.37-1)/1.37}\right] = 89{,}200\,\text{m}^2/\text{s}^2.$$

The specific speed and specific diameter are based on inlet *stagnation* volume flow, so

$$Q = \frac{\dot{m}}{\rho_{01}} = \frac{0.50}{0.625} = 0.80\,\text{m}^3/\text{s}.$$

Usually, one would begin the layout of the turbine by choosing either specific speed or specific diameter and then using the Cordier correlations to advance the design calculations. In this case, other considerations dictate the design. First, the unit is kept small because of available space in a typical vehicle engine compartment. Small size means high speed. These small, high-speed radial flow turbines are usually designed with radial inlet blade tips (i.e., $\beta_1 = 90°$). This is done for ease of manufacture and, mainly, for eliminating bending stresses on the blades in the severe environment of the hot exhaust gas. Accordingly, assume a radial inlet blade such that $\beta_1 = 90°$ and $C_{\theta 1} = U_1$. Then, assuming that $C_{\theta 2} = 0$, Euler's equation gives

$$w = U_1 C_{\theta 1} = U_1^2 = \left(\frac{ND}{2}\right)^2. \tag{7.19}$$

The work, in turn, is calculated from the head by

$$w = \eta \times gH, \tag{7.20}$$

where η is the turbine total, isentropic efficiency.

Under ordinary circumstances, a rather complex iteration process would now be required. The process would proceed as follows. First, a specific speed N_s (of course, in the "radial machine range") would be selected. Next, the speed would be calculated. Specific diameter and total efficiency would be determined from the Cordier correlations. The Cordier efficiency, interpreted as a polytropic efficiency, would be converted to isentropic efficiency according to Equation 1.11c and the work calculated from Equation 7.20. The diameter would then be calculated from the Cordier specific diameter and the head and volume flow. Finally, a new value of speed would be calculated from Equation 7.19 and the process repeated to convergence. Unfortunately, this iteration process will not converge to any realistic value (the reader is invited to try the calculations to verify this). The reason is simple: a small, high speed, radial tip inlet machine is not the "Cordier machine" for the given values of flow and head—the "Cordier machine" would be larger and slower and has blades sloped *away* from the inlet absolute velocity vector (i.e., "backward-curved" blades). Another path must be taken to arrive at a workable layout.

In order to proceed to a reasonable layout, two assumptions will be made. First, because the pressure ratio is not significantly different from 1, the difference between the polytropic efficiency and isentropic efficiency will be ignored. Second, the efficiency of the turbine is assumed to be 0.80. Note that this value is not completely arbitrary; Figure 2.9 shows that "Cordier efficiency" is about 0.9 over the entire range of radial flow turbines, and a derate of 0.1 seems reasonable for this small, not-quite-optimum (in the Cordier sense) machine.

The work is now calculated as $w = 0.8 \times 89{,}200 = 71{,}360\,\text{m}^2/\text{s}^2$ and, from Equation 7.19, $U_1 = 267.1\,\text{m/s}$ and $ND = 534.3\,\text{m/s}$. Next, either N or D is chosen and the other calculated. A typical speed is $N = 40{,}000\,\text{rpm} = 4189\,\text{s}^{-1}$. This gives $D = 0.128\,\text{m}$ (5.02 in.), which is a reasonable, although somewhat large, size (4 in. would be more typical). For reference, these values of N and D correspond to $N_s = 0.73$ and $D_s = 2.46$; the specific speed is in the proper range for a radial flow machine but the specific diameter is more characteristic of a mixed flow machine.

To do a layout of the turbine itself, one needs to establish values for d, b_1, b_2, and β_2 (recall $\beta_1 = 90°$). The eye diameter ratio is estimated from Equation 6.35, even though this equation was developed for pumping machines. Thus $d/D = 1.53(Q/ND^3)^{1/3}$ so that $d = 1.53(Q/N)^{1/3}$ giving $d = 0.088\,\text{m}$ (3.47 in.). This gives $U_2 = 4189 \times 0.088/2 = 184.6\,\text{m/s}$. Next, make the simplifying, conservative assumption that the blade outlet area is the same as the throat area, that is, $b_2 = d/4$ (at $r = d/2$). This yields $b_2 = 0.022\,\text{m}$. The blade width at the inlet is rather arbitrarily set at the same value, $b_1 = 0.022\,\text{m}$.

The radial velocity at the inlet of the turbine is

$$C_{r1} = \frac{\dot{m}}{\rho_1 \pi D b_1}. \tag{7.21}$$

Because the gas speed at the inlet, C_1, is high, the inlet static density will be somewhat lower than the inlet stagnation density because of compressibility effects. Both ρ_1 and C_{r1} are determined from an iterative calculation, as follows:

1. Guess a value of ρ_1/ρ_{01}, say 0.85.
2. Compute $\rho_1 = (\rho_1/\rho_{01}) \times \rho_{01}$.
3. Compute C_{r1} from Equation 7.21.
4. Compute C_1^2 from $C_1^2 = C_{r1}^2 + C_{\theta 1}^2$.
5. Compute T_1/T_{01} from

$$T_1/T_{01} = 1 - \left(C_1^2/2c_p T_{01}\right) = 1 - [(\gamma - 1)/2\gamma](\rho_{01} C_1^2/p_{01}).$$

6. Compute ρ_1/ρ_{01} from $(\rho_1/\rho_{01}) = (T_1/T_{01})^{1/(\gamma-1)}$.
7. Return to Step 2 and repeat until convergence.

Following this process yields $\rho_1 = 0.556\,\text{kg/m}^3$, $C_{r1} = 101.8\,\text{m/s}$, and $C_1 = 285.9\,\text{m/s}$.

The outlet stagnation density is computed from the inlet stagnation density and the turbine pressure ratio, p_{01}/p_{02}, using the polytropic process equation (Equation 1.6) in the form $\rho_{01}/\rho_{02} = (p_{01}/p_{02})^{1/n}$, where n is the polytropic exponent for an expansion (turbine) process given by Equation 1.11b.

$$\frac{n-1}{n} = \eta_p \frac{\gamma - 1}{\gamma} \approx \eta \frac{\gamma - 1}{\gamma} = 0.8 \frac{1.37 - 1}{1.37}.$$

From this, $n = 1.276$. With these results, one can calculate $\rho_{01}/\rho_{02} = (152/103.3)^{1/1.276} = 1.354$ so that $\rho_{02} = 0.462\,\text{kg/m}^3$. Then the outlet radial velocity is

$$C_{r2} = \frac{\dot{m}}{\rho_2 \pi d b_2}.$$

An iteration similar to that for ρ_1/ρ_{01} gives $\rho_2/\rho_{02} \approx 0.94$, so that $\rho_2 \approx 0.94 \times 0.462 = 0.436\,\text{kg/m}^3$. Then, recalling that $C_{\theta 2} = 0$:

$$C_2 = C_{r2} = \frac{0.5}{0.436 \times \pi \times 0.088 \times 0.022} = 188.8\,\text{m/s}.$$

With these results, the inlet and outlet velocity diagrams and (blade) flow angles can be calculated:

$$W_1 = C_{r1} = 101.8\,\text{m/s},$$

$$W_2 = \left(C_{r2}^2 + U_2^2\right)^{1/2} = (188.8^2 + 184.6^2)^{1/2} = 264.1\,\text{m/s},$$

$$\beta_1 = 90^\circ, \qquad \beta_2 = \tan^{-1}\left(\frac{C_{r2}}{U_2}\right) = \tan^{-1}\left(\frac{188.8}{184.6}\right) = 45.7^\circ.$$

Figure 7.32 shows the geometry developed for the turbine rotor. The actual power available from the turbine shaft (exclusive of mechanical losses) is

$$P = \dot{m}w = \dot{m}\eta gH = 0.5 \times 0.8 \times 89{,}200 = 35.7\,\text{kW}.$$

Now the compressor layout can be done. The compressor and the turbine share three parameters: the input power to the compressor is the output power from the turbine, 35.7 kW (exclusive of mechanical losses, which are ignored here); the compressor and the turbine rotate at the same speed (40,000 rpm) since they are mounted on a common shaft; and the mass flow in the compressor is the mass flow through the turbine minus the mass flow of fuel added downstream of the compressor and burned in the i.c. engine. Assuming that the air-to-fuel mass ratio for the engine is a fairly typical 15 to 1, the compressor mass flow rate will be 0.469 kg/s.

Like the turbine, the compressor is a small, high-speed design that in general, will not fall on the Cordier N_s–D_s or η–N_s correlation lines. Again, two reasonable assumptions will be made in order to proceed with the compressor layout. First, the compressor efficiency will be assumed to be 0.75. Like the turbine efficiency, this is not a completely arbitrary choice. Figure 2.9

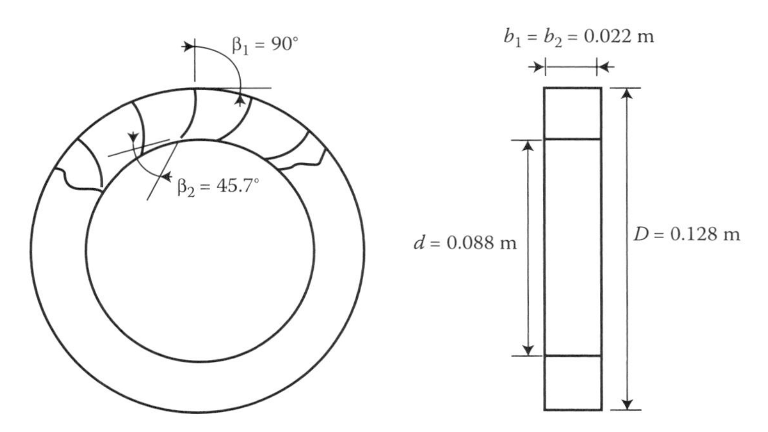

FIGURE 7.32 Small, high-speed radial flow turbine for a turbocharger.

indicates that in the radial flow range of specific speed ($N_s \sim 0.7$), compressor efficiencies are of the order of 0.05 less than turbine efficiencies. The second assumption is that the compressor diameter is the same order of magnitude, but slightly larger than the turbine diameter. Here, a value of $D = 0.145$ m (5.71 in.) will be used.

The specific work in the compressor is

$$w = \frac{P}{\dot{m}} = \frac{35{,}700}{0.469} = 76{,}120\,\mathrm{m^2/s^2},$$

and the head and pressure ratios are

$$gH = \eta w = 0.75 \times 76{,}120 = 57{,}100\,\mathrm{m^2/s^2};$$

$$\frac{p_{02}}{p_{01}} = \left(1 + \frac{gH}{c_p T_{01}}\right)^{\gamma/\gamma-1} = \left(1 + \frac{57{,}100}{1004 \times 288}\right)^{1.4/1.4-1} = 1.879.$$

Standard sea-level values are assumed for the inlet air properties ($p_{01} = 101.3$ kPa, $T_{01} = 288$ K, $\rho_{01} = 1.23\,\mathrm{kg/m^3}$, $c_p = 1004\,\mathrm{m^2/s^2\,K}$, and $\gamma = 1.4$). The (nominal) inlet volume flow rate is $Q = \dot{m}/\rho_{01} = 0.38\,\mathrm{m^3/s}$. From the known speed (40,000 rpm) and the specified diameter, one can calculate $N_s = 0.70$ and $D_s = 3.63$, which is reasonably close to the Cordier line. The blade tip speed is $U_2 = ND/2 = 303.8$ m/s. From Euler's equation, assuming no inlet swirl, $C_{\theta 2} = w/U_2 = 250.5$ m/s. The eye diameter, d, is calculated from Equation 6.33, $d = 1.53(Q/N)^{1/3} = 0.069$ m (2.71 in.). The inlet blade width, b_1, is determined in the usual way, $b_1 = d/4 = 0.017$ m.

Because C_1 is expected to be somewhat high, the inlet static density ρ_1 and the inlet radial velocity C_{r1} ($=C_1$ for $C_{\theta 1} = 0$) are determined from the mass flow rate and inlet dimensions by taking compressibility effects into account. Again an iterative calculation like that for the turbine is required. The calculation is as follows:

1. Guess a value of ρ_1/ρ_{01}, say 0.85.
2. Compute $\rho_1 = (\rho_1/\rho_{01}) \times \rho_{01}$.
3. Compute C_1 from $C_1 = C_{r1} = \dot{m}/\rho_1 \pi d b_1$.
4. Compute T_1/T_{01} from $T_1/T_{01} = 1 - (C_1^2/2c_p T_{01})$.
5. Compute ρ_1/ρ_{01} from $\rho_1/\rho_{01} = (T_1/T_{01})^{1/\gamma-1}$.
6. Return to Step 2 and repeat until convergence.

The converged values are $\rho_1 = 1.16\,\mathrm{kg/m^3}$ and $C_1 = 107.9$ m/s.

The inlet details for the compressor can now be calculated:

$$U_1 = \frac{Nd}{2} = 144.4\,\text{m/s}; \quad W_1 = \left(U_1^2 + C_1^2\right)^{1/2} = 180.2\,\text{m/s};$$

$$\beta_1 = \tan^{-1}\left(\frac{C_{r1}}{U}\right) = 36.8°.$$

The outlet blade width controls the outlet velocity diagram and blade geometry. It can be adjusted to give an appropriate level of diffusion. Here, the value $b_2 = 0.006\,\text{m}$ is selected. The stagnation density at the compressor outlet is calculated from the polytropic process equation, $\rho_{02}/\rho_{01} = (p_{02}/p_{01})^{1/n}$, where n is the polytropic exponent for a compression given by Equation 1.8,

$$\frac{n}{n-1} = \eta_p \frac{\gamma}{\gamma - 1} \approx \eta \frac{\gamma}{\gamma - 1} = 0.75 \frac{1.4}{1.4 - 1},$$

from which $n = 1.615$, so that $\rho_{02}/\rho_{01} = (1.879)^{1/1.615} = 1.478$. Then $\rho_{02} = 1.811\,\text{kg/m}^3$. The outlet radial velocity is calculated from

$$C_{r2} = \frac{\dot{m}}{\rho_2 \pi D b_2}.$$

To account for compressibility, an iteration similar to that for the inlet density is used to give $\rho_2/\rho_{02} = 0.77$ so that $\rho_2 = 1.394\,\text{kg/m}^3$. Then

$$C_{r2} = \frac{0.469}{1.394 \times \pi \times 0.145 \times 0.006} = 123.0\,\text{m/s}.$$

With these results, the exit velocity diagram and (blade) flow angle can be calculated:

$$C_2 = \left(C_{r2}^2 + C_{\theta 2}^2\right)^{1/2} = (123.0^2 + 250.5^2)^{1/2} = 279.1\,\text{m/s},$$

$$W_2 = \left(C_{r2}^2 + [U_2 - C_{\theta 2}]^2\right)^{1/2} = \left(123.0^2 + [303.8 - 250.5]^2\right)^{1/2} = 134.0\,\text{m/s},$$

$$\beta_2 = \tan^{-1}\left(\frac{C_{r2}}{U_2 - C_{\theta 2}}\right) = \tan^{-1}\left(\frac{123.0}{303.8 - 250.5}\right) = 66.6°.$$

The diffusion ratio is $W_2/W_1 = 0.74$, which is a quite acceptable value. This is the result of the "lucky" choice of the exit blade width ($b_2 = 0.006\,\text{m}$); the reader is invited to try other values of exit blade width. In general, increasing b_2 will cause the diffusion ratio to drop to unacceptable values, whereas decreasing b_2 will give an unrealistically narrow blade and an unnecessarily large diffusion ratio.

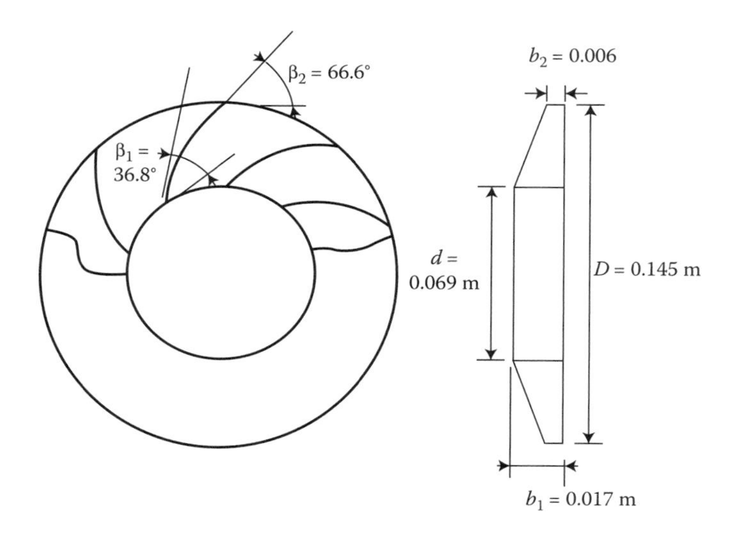

FIGURE 7.33 Small, high-speed radial flow compressor for a turbocharger.

The geometric layout for the impeller is sketched in Figure 7.33. One shortcoming of this compressor layout is that an efficient diffuser will be needed downstream of the impeller to effectively recover the large kinetic energy represented by C_2.

Figure 7.34 shows photographs of the turbine and compressor rotors for an actual turbocharger. The figure also clearly shows that axial inducer and exducer sections are used to maintain careful control of the flow entering or leaving the radial passages of the impellers.

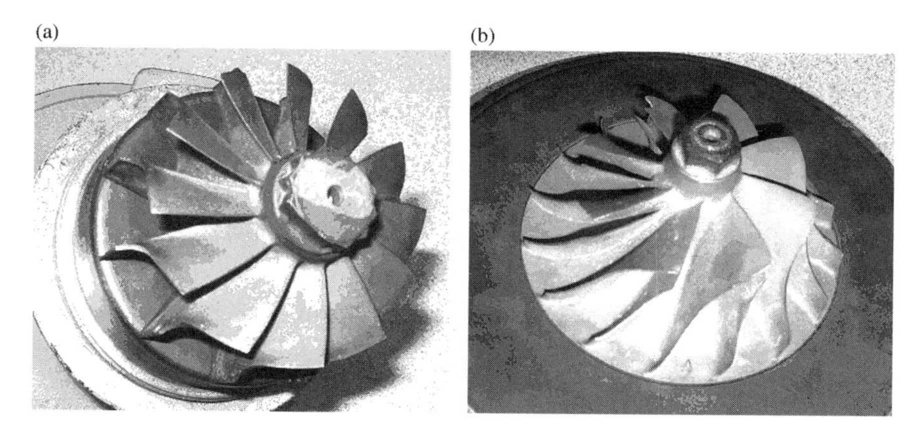

FIGURE 7.34 (a) Turbine and (b) compressor sections of an actual turbocharger. Note the fully developed axial inducer/exducer sections at the impeller eyes and the "half blades" on the compressor.

7.11 Diffusers and Volutes

Layouts of turbomachines examined in Sections 7.3 through 7.10 have mostly considered performance based on the change in total pressure/head across the machine's impeller. Likewise, efficiencies used in the layout process have been exclusively total efficiencies. In some cases, notably the turbocharger turbine and compressor in the immediately preceding section, the fluid exiting the impeller has a rather high velocity, and kinetic energy accounts for a significant fraction of the useful energy rise. In this case, an effective diffuser must be added to the machine so that a significant portion of the kinetic energy is not lost to friction and turbulence.

In other situations, total head is only part of the story. Although one frequently works with the total pressure rise or drop through a machine, the pressure of interest is often the static pressure at the point at which the turbomachine is to be connected to the user's system. This is especially true for fans and hydraulic turbines, in which the device discharges into an arbitrary duct configuration or a large plenum, the atmosphere, or a large body of water. To properly evaluate a machine for such an application, the user must know the *static* pressure at which the required flow rate is delivered and the power required to produce it. A pumping machine must then be designed or specified on the basis of the best static efficiency of a system representing a combination of turbomachine and diffusing device that can match the user's system geometry (Wright, 1984d).

By itself, a turbomachine impeller (perhaps including straightening vanes in an axial flow design) is capable of producing rather high efficiencies through careful design and control of tolerances. However, it is frequently necessary to reduce the absolute discharge velocities to an acceptable level to avoid large frictional or sudden expansion losses of energy and hence efficiency. The actual static or total pressure rise supplied by a pumping machine together with its diffusion device depends on both the diffuser design and the static to total pressure ratio (i.e., the absolute velocity) at the machine discharge into the diffuser. For axial flow machines, one must make a careful examination of the performance capabilities of conical or annular diffusers and their performance limits. For centrifugal machines, one must develop the design and performance characteristics of the scroll diffuser or spiral volute as the equivalent flow discharge device. One must also be concerned with the influence of downstream diffusion on the overall output capabilities of turbines—axial and centrifugal, gas, and hydraulic—as the downstream diffuser establishes the backpressure conditions at the machine exit.

7.12 Axial Flow Diffusers

In axial flow, one is concerned with the ratio of the outlet static pressure to the total pressure delivered at the discharge flange of the machine. This can be expressed as an efficiency ratio η' given by

$$\eta' \equiv \frac{\eta_S}{\eta_T} = \frac{\psi_S}{\psi_T},$$

where ψ is a pressure coefficient defined by $\psi \equiv \Delta p / \rho U^2$. Assuming incompressible flow and using Bernoulli's equation,

$$\psi_S = \psi_T - k'\varphi^2,$$

where φ is the flow coefficient developed earlier in the chapter and k' is a loss coefficient combining a residual velocity pressure at the diffuser discharge and the frictional losses within the diffuser. It is assumed that straightening vanes have removed all of the fluid swirl. Combining these equations yields

$$\eta' = 1 - \frac{k'\varphi^2}{\psi_T}. \tag{7.22}$$

One can develop a functional form for k' in terms of the diffuser geometric parameters. Here, the range of diffuser configurations is restricted to annular diffusers with conical outer walls and cylindrical inner walls* as illustrated in Figure 7.35. For such diffusers, the coefficient k' can be derived by studying the fluid mechanics of the decelerating flow in a rigorous analytical fashion (Weber, 1978; Papailiou, 1975) or by relying on experimental information (McDonald and Fox, 1965; Idel'Chik, 1966; Howard et al., 1967; Sovran and Klomp, 1967; Adenubi, 1975; Smith, 1976). Examination of the literature yields several examples of experimentally derived correlations for predicting diffuser performance and first stall as a function of diffuser geometry.

Figure 7.36 shows a summary of the influence of diffuser geometry on the optimum level of static pressure recovery with minimum diffuser size or length. Also shown in the figure is the limit line for the rate of area increase, above which the flow in the diffuser becomes unstable (the "first stall" condition). Here, L/w is the ratio of diffuser length to inlet annulus height, and AR is the outlet to inlet annular area ratio. The optimal area ratio in terms of L/w as shown can be expressed as

$$\text{AR} \approx 1.03 + 1.85\left(\frac{L}{w}\right) - 0.004\left(\frac{L}{w}\right)^2. \tag{7.23}$$

---

* Although complex wall shapes and wall-bleed schemes can provide significant performance gain or size and weight reductions for some applications, they will generally not provide cost-effective or efficient diffusion for practical situations such as FD or ID, ventilation, or heat exchange. For example, a typical boundary layer control scheme, characterized by a very rapid area increase with wall bleed to prevent stall, will require about 8–10% of the throughflow of the turbomachine to be removed from the walls (Yang, 1975). For most applications of practical interest, a penalty of 8–10 points in static efficiency would be incurred unless the bleed fluid were required for some secondary purpose. A cusp diffuser (Adkins, 1975), which relies on mass bleed to stabilize a standing vortex to turn the flow through a very rapid expansion, again requires an 8–10% bleed rate for annular configurations.

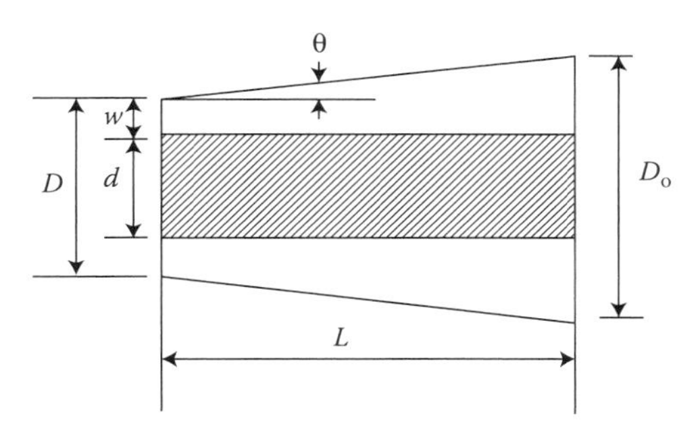

FIGURE 7.35 Geometry and nomenclature for a conical–annular diffuser typically used with axial flow machines.

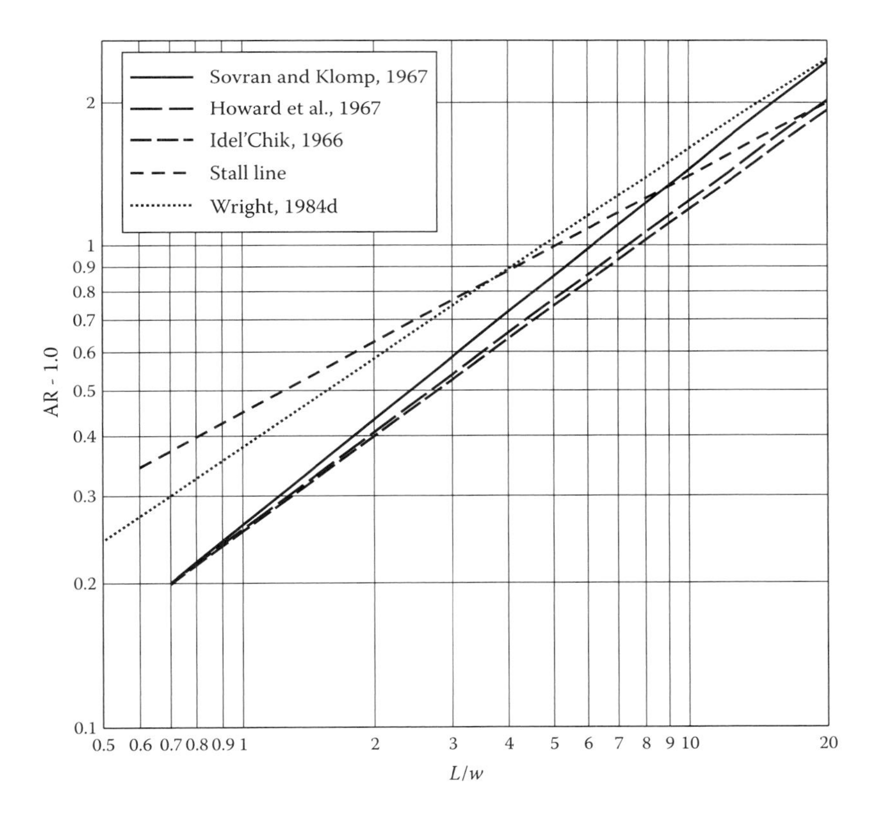

FIGURE 7.36 Summary of the influence of geometry on optimum static pressure recovery with minimum diffuser size.

The corresponding values of k' are shown in Figure 7.37 and can be expressed as

$$k' \approx 0.127 + \frac{1.745}{(L/w) + 2}. \tag{7.24}$$

In a given application, one can use the relationship between k' and L/w together with Equation 7.22 to determine constraints between L/w and the specific diameter of the machine, D_S. This can be done by recognizing that η' can be rearranged to yield

$$D_s = \left(\frac{k'}{1-\eta'}\right)^{1/4}. \tag{7.25}$$

Consider an example wherein an axial fan is capable of a total efficiency of $\eta_T = 0.88$, and an efficiency ratio η' of 0.9 (thus $\eta_s = 0.792$) is required. If constrained to a value of $L/w = 3.0$ (by space considerations), then $k' = 0.476$ and the fan must have a specific speed not less than $D_s = 1.477$. If the length ratio is relaxed to allow $L/w = 6$, then $k' = 0.345$ and D_s must be at least 1.36. By doubling the diffuser length ratio, the fan size is reduced by about 8%; however, the diffuser length increases by 84%. This might represent a favorable trade-off in the cost of rotating equipment versus stationary equipment, provided that there is room to install the longer diffuser.

If, on the other hand, the fan size and specific diameter are constrained by space, noise, or cost considerations, one must settle for either a fixed value of η' if L/w is constrained or an imposed value of L/w if η' is fixed. For example, if one is forced to use a value of $D_s = 1.25$ and η' must be at least 0.92, then the required value of L/w is 23.5, which is a very long diffuser! If one retains the requirement for $D_s = 1.25$ and restricts L/w to be no greater than 10.0, the resulting value of η' becomes $\eta' = 0.89$, only 4% smaller. Clearly,

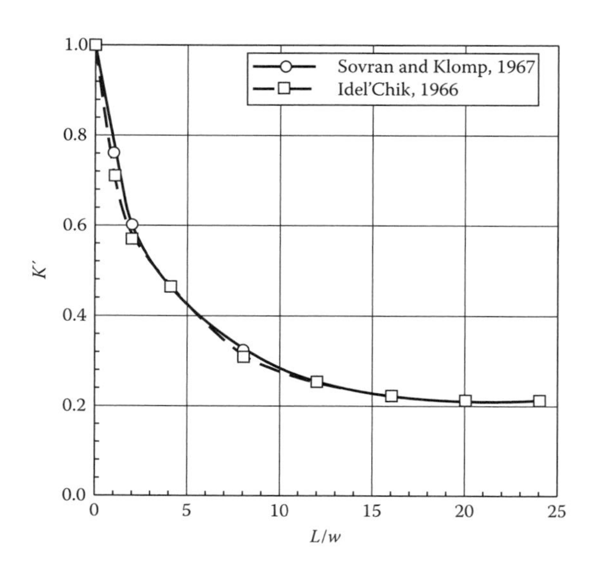

FIGURE 7.37 Diffuser static pressure loss factor (k') as a function of diffuser geometry for conical–annular diffusers.

it pays to examine what one is asking for; in this case, a huge reduction in diffuser length requirement results in a static efficiency reduction of only 0.03. The static efficiency decrease may be the right solution in a space-constrained or initial cost-constrained design.

7.13 Radial Flow: Volute Diffusers

For the radial flow case, one must generate information comparable to that in the previous section for *spiral volutes*. These devices are variously called housings, scrolls, collectors, diffusers, volutes, or casings. The name depends on the type of turbomachine being discussed (pump, compressor, etc.) and, to a degree, on the point of emphasis being made. Although all of these machines collect and direct the flow to a discharge station, not all of them diffuse the flow in the process. In a general way, one can consider the layout and performance of all diffusing collectors and refer to them as volutes.

A prominent difference that exists from machine to machine is the existence or the absence of straightening vanes in the diffuser. These vanes are primarily used to recover the swirl energy in the discharge flow of highly loaded or high-speed machines and convert it to a useable static pressure rise; in this sense, they perform the same function as straightening vanes in axial flow machines. For more lightly loaded machines, the task of swirl recovery is included in the fluid collection process. The total energy in the lightly loaded machines is more equally distributed among static pressure rise through the impeller, velocity head associated with the through flow, and the velocity head associated with the fluid swirl. Figure 7.38 defines the dominant features of the geometry of a vaneless volute diffuser. The cross-section of the device can be circular, square, or more or less rectangular.

This geometry is more or less common to radial flow machines and will be modified to include diffusion vanes in a later discussion. Here, the impeller radius is shown as r_2; a radius characteristic of the volute is shown as r_V. The volute radius is a function of angular position in the volute and usually varies linearly between the volute "entrance" and the volute "exit." The entrance/exit points as seen in Figure 7.38 relate to the so-called "cutoff" of the volute. Various other terms are used for the cutoff, including "tongue," "cutwater," and "splitter." There is a clearance gap, ε_C, between the stationary volute and the rotating impeller at the cutoff. In general, for high-pressure and high-efficiency applications, this clearance might be specified as nearly the smallest value that can be achieved subject to fit-up and manufacturing tolerances required to avoid a "rub." At times, using the smallest possible clearances can result in serious acoustic pulsations in fans, blowers, and compressors and potentially destructive pressure pulsation in liquid pumps and high-pressure gas machines. A more frequent, conservative design choice is to set the clearance ratio, ε_C/D, in the range of 10–20%. For example, most

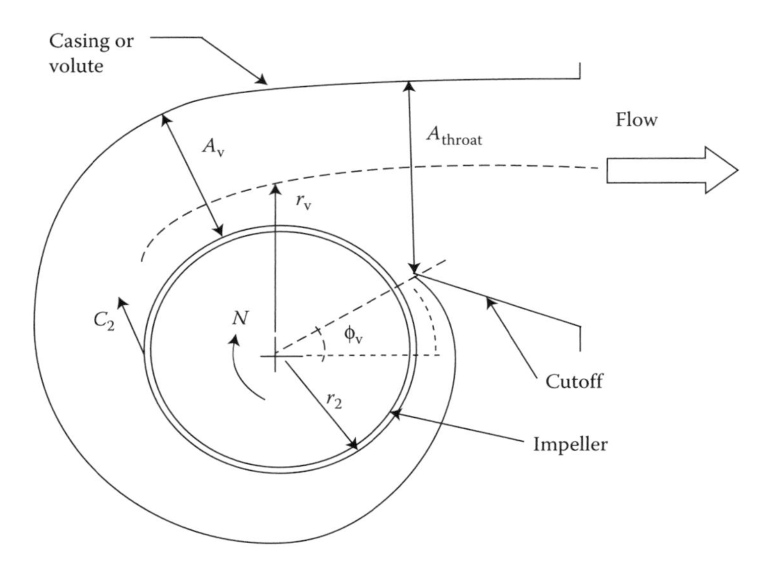

FIGURE 7.38 Geometry and nomenclature for a vaneless volute diffuser.

centrifugal fans of moderate pressure rise will employ a cutoff clearance ratio between 9% and 15%. High-pressure blowers and compressors may use slightly smaller values, perhaps 6–10%, and liquid pumps use a "reasonably safe distance" (Karassik et al., 2008) of 10–20% of impeller diameter.

Common design practice for shaping the volute relies on maintaining simple conservation of angular momentum along the mean streamline of the volute (at $r = r_V$), so that

$$\frac{C_2 r_2}{C_V r_V} = \text{constant}, \tag{7.26}$$

where the constant is 1.0 for simple conservation but may be less than 1.0 if diffusion is accomplished within the volute. C_2 is the impeller (absolute) discharge velocity. Its components, $C_{\theta 2}$ and C_{r2}, are determined by total pressure rise and volume flow rate, respectively. For linear volutes (the most common type), the volute area is related to the throat area by $A_V = A_{\text{throat}}(\phi_V^o/360)$. The throat area is related to C_V and Q according to $C_{V,\phi=0} = Q/A_{\text{throat}}$. The flow in the volute is at best a nondiffusing frictional flow; in reality, it includes total pressure losses from friction and diffusion. As modeled by Balje (1981), the magnitude of the loss depends on the parameters closely related to the specific speed of the machine, such as the impeller width ratio (b_2/D), the absolute or blade-relative discharge angles, and the degree of diffusion incorporated into the volute design. Typical values (Shepherd, 1956; Balje, 1981; Wright, 1984c) can be defined in terms of a loss coefficient defined by either

$$\zeta_V \equiv \frac{h_L}{C_2^2/2g} \tag{7.27a}$$

or

$$\omega_V \equiv \frac{h_L}{U_2^2/2g}. \tag{7.27b}$$

For low specific diameter centrifugal machines such as fans and blowers, $0.2 \leq \zeta_V \leq 0.4$; for large specific diameter machines such as high-pressure pumps and gas compressors, $0.6 \leq \zeta_V \leq 0.8$. As pointed out by most investigators or designers in the field, the details of vaneless diffuser or volute scroll design are largely based on experience and craft.

Low specific speed centrifugal machines, both high-pressure gas compressors and high-pressure liquid pumps, often include a vaned diffuser section in the volute. For a pump, this section consists of a number of vanes set around the impeller discharge as shown in Figure 7.39. Following the procedure outlined by Karassik et al. (2008), the flow can be considered to have several throats rather than just one as in the vaneless volute. The volute outlet radius, r_4, is now smaller, and the volute velocity, C_V, would be commensurately larger except that a "diffusing constant" of about 0.8 is generally used in pump design. The diffusing vane channel may be straight or curved but, either way, it generally results in a gradual rate of increase in the cross-sectional area A_V. Although the sidewalls of these passages are generally designed to be parallel, or even divergent, a slight converging taper in passage width with increasing radius would provide a more conservative design. Karassik et al. suggest a number of vanes equal to the number of impeller blades plus one; this is to avoid a harmonic pulsation due to blade-passing frequency.

In a high-speed compressor volute, the vane section will appear somewhat like the vane row shown in Figure 7.40, with a sharper "point" at the leading edge. The throat area must be carefully configured to match the relatively high-speed approach flow. The channels are usually straight and configured to yield outlet flow angles around 3° to 5° greater than the inlet flow angles.

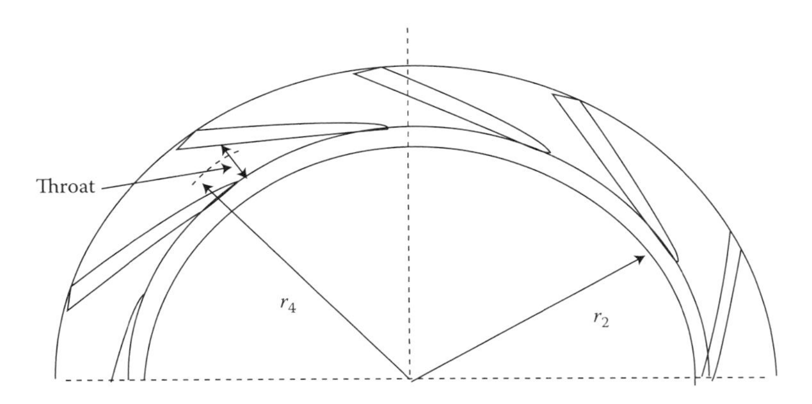

FIGURE 7.39 Vaned diffuser section in a pump volute.

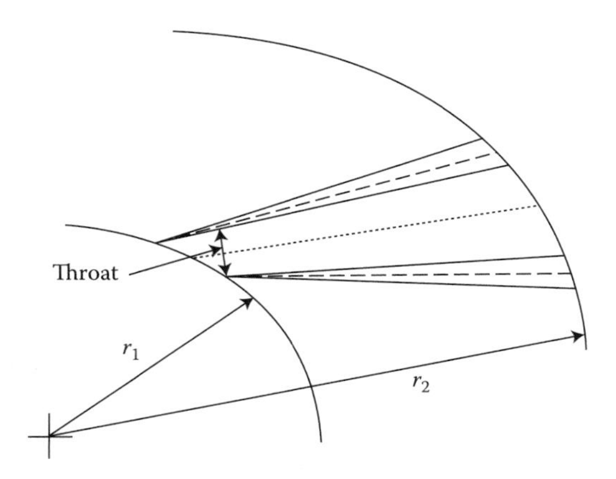

FIGURE 7.40 Diffuser vanes typical of a high-speed centrifugal compressor.

The static pressure achieved in these passages must generally be calculated using compressible isentropic or polytropic flow models. Diffuser pressure recovery values are, as usual, dependent on the inflow angle and impeller width ratio (thus, intrinsically on the specific speed). Recovery values may lie in the range of 70% of $C_2^2/2$ for narrow impellers ($w_2/d = 0.02$) up to about 80–85% for wider impellers ($w_2/D = 0.06$) (Balje, 1981).

In the case of both pumps and compressors, overall diffuser performance must be modeled as a combination of all of the diffusing elements: an initial vaneless space, followed by a vaned section, followed by a final collector/diffuser section to the volute discharge.

7.14 Summary

This chapter introduced the formal process of initial design layout of turbomachinery based on velocity vector diagrams. Both dimensional and dimensionless forms of the diagrams were used. Two simplifying assumptions were used throughout the chapter; first, machine blade shapes can be inferred by assuming that the blade-relative flow is tangent to the blade camber line (especially at the blade entrance and exit) and second, Cordier correlations between specific speed and specific diameter and between specific speed and total efficiency were applied in nearly all cases.

First, axial flow pumping machines and turbines were discussed to develop a familiarity with the terms and concepts of the interacting components, including IGVs, OGVs, and nozzles. The degree of reaction definition was developed and used to differentiate between different approaches to fans, pumps, and turbine layouts. Several examples of design procedure were provided, including a parametric layout of a hydraulic turbine. Initially, consideration was limited to conditions along a mean radius streamline but was ultimately extended to "three dimensions" by the idea of "stacking" blade sections.

The concepts, definitions, and assumptions were then extended to mixed flow and radial flow machines. Although it was necessary to modify the velocity diagram procedure slightly, the same conceptual form was applied and used to develop example layouts. An example of a mixed flow fan was used to illustrate techniques for diffusion levels in nearly axial flow configurations. Extensive examples for a centrifugal fan at rather high specific speed and a centrifugal pump were used to provide insight into the layout variables and procedures for radial discharge flow. Lastly, an example was presented for the fully compressible flow process in a radial turbocharger design problem. This last example provided an opportunity to examine the matching requirements between components of a combined system.

Finally, stationary diffusing devices were considered as the necessary adjunct equipment used to control the amount of discharge energy leaving the machine. For axial flow machines, the geometry and performance of annular diffusers were discussed in concept and by example. For radial flow machines, the volute scroll or collector was examined. Layouts and design techniques were examined and discussed with examples to illustrate the difference between vaneless and vaned diffuser layouts.

EXERCISE PROBLEMS

7.1. Flow enters a centrifugal blade row (Figure P7.1), with $\rho = 1.2\,\text{kg/m}^3$ and a mass flow rate of 1200 kg/s. If $C_1 = C_2 = 100\,\text{m/s}$, calculate the power and pressure rise. Note that the vectors shown are absolute and $N = 1000$ rpm.

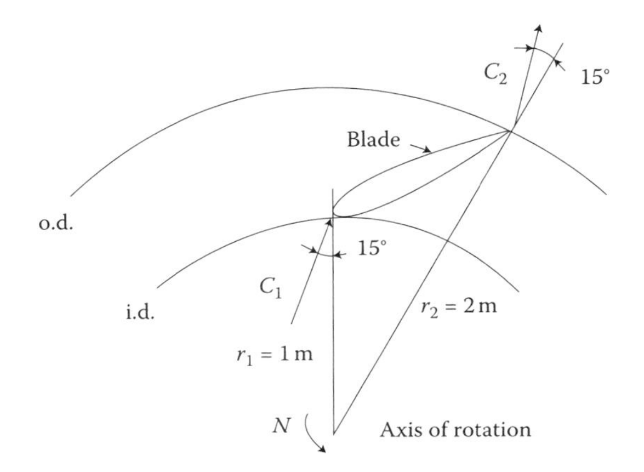

FIGURE P7.1 Centrifugal fan blade layout.

7.2. Repeat Problem 7.1 for inlet absolute flow angles of 0° and +15° (relative to the radii).

7.3. Determine the absolute outlet flow angles for Problem 7.2 such that the total pressure rise is zero.

7.4. A centrifugal blower has a blade exit angle β_2, and $C_{r1} = C_{r2}$. Use Euler's equation to show that

$$\psi_{\text{ideal}} = \frac{\Delta p_T}{(1/2)\rho U^2} = 2(1 - \varphi_c \cot \beta_2),$$

where $\varphi_c = C_{r2}/U_2$.

7.5. Repeat the pump layout example (Figure 7.31) using $Q = 600$ gpm and $H = 75$ ft.

7.6. Define the frictional effects in a fan impeller in terms of a total pressure loss, δp_T. Using the concept of a force coefficient, or "loss coefficient," one can specify δp_T as $\delta p_T = (1/2)\rho W_1^2 C_{wb} N_B$, where C_{wb} is a blade (or wake) loss coefficient and N_B is the number of blades. Use the formulation for ψ in Problem 7.4 to show that the losses can be written as

$$\delta\psi_{\text{loss}} = \frac{\delta p_T}{(1/2)\rho U_2^2} = \left(\left(\frac{d}{D}\right)^2 + \phi_c^2\right) N_B C_{wb},$$

where $\delta\psi_{\text{loss}}$ is the viscous correction to ψ_{ideal}.

7.7. Devise a means of estimating C_{wb} as used in Problem 7.6. On what parameters should C_{wb} depend? (Hint: Is the Cordier diagram of any help?)

7.8. In working with swirl recovery vanes (OGVs) behind heavily loaded rotors, the limit in performance for the stage may become the de Haller ratio in the vane row cascade.

a. For example, if $C_{\theta 1} = C_{x1}$ entering the vane row and $C_{x2} = C_{x1}$, show that the inflow angle to the vane is 45° and the de Haller ratio becomes $C_2/C_1 = 0.707$.

b. For $C_{\theta 1} > C_{x1}$ (very heavy loading), the approach angle to the vanes, α, becomes less than 45° and $C_2/C_1 < 0.707$ until, if α is further decreased, C_2/C_1 becomes unacceptably small. Show that, for $C_2/C_1 = 0.64$ as a lower limit, α_1 must be greater than 40°.

7.9. One way to alleviate the vane overloading described in Problem 7.8 is to allow residual swirl to exit the vane row. That is, $C_{\theta 2}$ may be nonzero and $\alpha_2 = \tan^{-1}(C_{x2}/C_{\theta 2}) < 90°$.

a. Show that to maintain a given level of (C_2/C_1), one must constrain α_2 to

$$\alpha_2 = \sin^{-1}\left\{\frac{\sin \alpha_1}{C_2/C_1}\right\}.$$

b. If α_2 is allowed to be less than 90° as specified in part (a), what penalty in performance will result (qualitatively)?

c. For the example above with $C_{x2} = C_{x1}$ and C_2/C_1 constrained to 0.707, develop a table and curve of α_2 versus α_1 for $20° < \alpha_1 < 45°$.

7.10. Derive a quantitative expression for the reduction of static pressure rise and static efficiency associated with the vanes of Problem 7.9, which allow residual swirl in the flow discharge. [Hint: Account for the velocity pressure of a residual swirl ($\rho C_{\theta 2}^2/2$) as a decrement to total pressure along with the axial velocity pressure ($\rho C_{x2}^2/2$).]

7.11. For the fan of Problem 6.16, define the velocity triangles in both the absolute and relative reference frames. Do so for both blades and vanes at hub, mean, and tip radial stations. Calculate the degree of reaction for all stations.

7.12. A hydraulic turbine generator set is supplied with water at $H = 67$ ft at a rate of $Q = 25{,}000\ \text{ft}^3/\text{s}$. The turbine drives a 72-pole generator at $N = 100$ rpm. Calculate the nondimensional performance parameters N_s, D_s, ϕ, and ψ (note: the latter two parameters are the "Chapter 2" definitions), and estimate the efficiency. Develop a velocity diagram for the mean radial station of the turbine.

7.13. A fan specification calls for $7\ \text{m}^3/\text{s}$ of air at a pressure rise of 1240 Pa with a density of $1.21\ \text{kg/m}^3$. Using a 4-pole motor as a cost constraint, lay out the fan blade-relative velocity vectors at the hub, mean, and tip radial stations.

7.14. A small fan is required to supply $0.1\ \text{m}^3/\text{s}$ of air with a total pressure rise of 2.0 kPa in air with $\rho = 1.10\ \text{kg/m}^3$. Lay out a centrifugal impeller, with a single-inlet, single-width configuration. Construct the outer shape of a volute scroll to work with the impeller layout. (Hint: Neglect compressibility.)

7.15. Select the layout parameters for an annular diffuser for the fan of Problem 7.13. Choose a length/diameter ratio that will provide optimal pressure recovery, and estimate the diffusion losses (total pressure) and the resulting static efficiency.

7.16. For the small centrifugal blower analyzed in Problem 7.14, select a reasonable cutoff location and shape, and estimate the total pressure losses in the volute. Compare these total pressure losses with the total pressure rise through the impeller.

7.17. In Problem 7.14, the performance requirements for a centrifugal fan were given as $0.1\ \text{m}^3/\text{s}$ of air at a pressure rise of 2.0 kPa with $\rho = 1.10\ \text{kg/m}^3$. These requirements can be imposed on smaller high-speed axial fans if one includes a "mixed flow" character by allowing the hub and tip radius to increase between the inlet and outlet stations of the fan. Begin an analysis with $N_s = 3.0$ and $D_s = 1.5$, using $d/D = 0.55$, and calculate $(W_2/W_1)_{hub}$. Then, hold N and d/D constant and allow D_s at the exit plane of the fan to increase to values greater than 1.5, recalculating the hub de Haller ratio. Sketch several such flow paths and compare them with the original and the one that allows the de Haller ratio to approach 0.7 at the hub station.

7.18. Following the concept of Problem 7.17, let the flow path be conical as sketched in Figure 7.24. At the entrance point of the blade row, let $D = 0.1$, $d/D = 0.5$ m, and let the length, L, along the axis be 0.1 m. With $Q = 0.2\,\text{m}^3/\text{s}$, $N = 18{,}000$ rpm, and with C_m held constant, calculate the maximum ideal pressure rise possible, using a flow path slope of 45°. Constrain the pressure so that $W_2/W_1 = 0.72$ on the mean streamline.

7.19. Examine the influence of the narrowing of the impeller of Problem 6.19 on the shape and efficiency of a volute scroll designed to accompany the design variations. Sketch the volute shapes and discuss their differences.

7.20. The results for Problem 7.4 give an ideal pressure rise coefficient as $\psi_{\text{ideal}}\ [\equiv \Delta p_T/(\rho U_2^2/2)] = 2(1 - \varphi_c \cot \beta_2)$. $\varphi_c \equiv C_{r2}/U_2$ and β_2 is the blade exit angle for the centrifugal blower. Replace ψ_{ideal} with the more realistic value of $\psi_c = \eta_T \psi_{\text{ideal}} = 2\eta_T(1 - \varphi_c \cot \beta_2)$. Show that the slope of the fan characteristic curve of ψ_c versus φ_c can be estimated by $d\psi_c/d\varphi_c = -2\eta_T \cot \beta_2$. Also show that the blade angle is $\beta_2 = \tan^{-1}[\varphi_c/(1 - \psi_c/(2\eta_T))]$. Note that this result can be used to plot an approximation to the characteristic curve in the region of the BEP to examine off-design performance.

7.21. Use the conventional definitions of $\psi_T = \Delta p_T/(\rho N^2 D^2)$ and $\varphi = Q/(ND^3)$ to show that $\psi_T = \psi_c/8$ and $\phi = \pi\varphi_c/8$ using the ψ_c and φ_c definitions in Problem 7.20.

7.22. The dimensionless curves of Figure 5.11 are based on experimental performance data for a centrifugal fan. Choose a point near the BEP and estimate the slope $d\psi_T/d\phi$. Compare this value to an estimate, based on the Euler equation, that $d\psi_T/d\phi = -2\eta_T \cot \beta_2/\pi$.

7.23. Generalize the result of Problem 7.4 to include axial flow pumping machines by using the 70% radial station value for $U = U_{0.7}$ in definitions for $\varphi_c = C_x/U_{0.7}$ and $\psi_c = \Delta p_T/(\rho U_{0.7}^2/2)$.

7.24. Based on the stability concepts considered in Chapter 10, a fan with a "steeper" characteristic curve is inherently more stable when the flow is perturbed from an equilibrium condition. Comment on the influence of ψ_c and φ_c at the BEP on the stability of a fan. Sketch the expected qualitative influence of ψ_c (with fixed φ_c) on the stability and the influence of φ_c (with fixed ψ_c).

7.25. A small, high-speed centrifugal blower has $Q = 0.1\,\text{m}^3/\text{s}$, $\rho = 1.10\,\text{kg/m}^3$, $\Delta p_T = 2.0$ kPa at the design point. The diameter of the fan is $D = 0.150$ m and the speed is $N = 7900$ rpm. This fan is connected to a system of ductwork with $d = 0.1$ m, $L = 75$ m, and a friction factor $f = 0.030$.

a. Use the results of Problem 7.20 to form an approximation to this fan's $\Delta p_T \sim Q$ curve for a range about the BEP from $0.8Q_{\text{BEP}} < Q < 1.2Q_{\text{BEP}}$.

b. Calculate the point of operation (flow and pressure rise) of the fan and system.

c. What will the performance become if the system resistance degrades to $f = 0.06$?

7.26. In many axial flow fan designs, a family of designs is extracted from a basic fan size by fixing the hub size and varying the blade length to achieve larger fans. If the basic design is a uniformly loaded free-vortex blade shape (Chapter 8), then addition of more blade length does not affect the pressure rise. That is, the pressure rise is the value at the hub station. However, the annulus area of the fan increases strongly as the outer diameter increases. Subject to similarity of the velocity vectors at the hub and assuming a uniform axial component, the flow rate will be proportional to the annulus area. Quantitatively show that the performance can be scaled according to $Q_2 = Q_1(N_2d_2^3/N_1d_1^3)\{[(D/d)_2^2 - 1]/[(D/d)_1^2 - 1]\}$ and $\Delta p_{T2} = \Delta p_{T1}(\rho_2/\rho_1)(N_2d_2)^2/(N_1d_1)^2$.

7.27. Use the results of Problem 7.26 to compare the performance curves given in Figure 5.18. Note that the performance of this fan can be scaled from the lower pitch angle curves to those at higher pitch angles. Select performance near the BEP values for 30° or 35° pitch and try to predict results for higher pitch angles (or lower). The loading of the fan may deviate significantly from free vortex and the scaling assumptions of Problem 7.26 begin to fail.

7.28. Further examine the results of Problem 7.26 by choosing representative performance numbers from Figure 5.17. Choose for each fan with $d = 630$ mm at $N = 1470$ rpm a value of flow rate at $\Delta p_T = 1250$ Pa. To try to represent the 35° pitch angle value use $Q = 9.8\ \text{m}^3/\text{s}$ for the 1000 mm fan. This point lies one-third of the 1000 mm o.d. box width from the left edge in Figure 5.17. That is, choose a point one-third of the way across each box as the "representative" value. Normalize the flow by the value chosen for the largest fan and compare this variation with the proposed dependence on annulus area ratio.

7.29. A fan, similar to those represented in Figure 5.17, must provide $2.5\ \text{m}^3/\text{s}$ of air ($\rho = 1.21\ \text{kg/m}^3$) at 1250 Pa total pressure rise, running at 1470 rpm. From the figure, the fan hub diameter must be at least 630 mm to generate the pressure rise. However, the smallest choice, with a tip diameter of 891 mm, gives a minimum flow rate of about $5\ \text{m}^3/\text{s}$. To limit the flow rate, we can reduce the diameter to less than 891 mm using the results of Problem 7.26. Determine the required fan diameter to match the performance requirements.

7.30. The fan of Problem 7.29 should have had a diameter of 772 mm. Use this value with the performance parameters of Problem 7.29 to analyze this fan in terms of specific speed and diameter, and the hub–tip ratio. Comment on the match to Figures 5.1 and 6.10. Is the layout conservative?

7.31. A method to predict blade stall for lightly loaded axial fans was developed (Ralston and Wright, 1987) for use in a computer code for preliminary design. The method, based on correlation of the National Advisory Committee for Aeronautics

(NACA) blade cascade data (Emery et al., 1958), can be reduced to a simple, approximate algorithm: $\omega_{1\text{stall}} \approx \omega_{1\text{BEP}} + [4 + 0.2(1 - \omega_{1\text{BEP}}/45)\theta_{\text{fl BEP}}] f(t/c)$, with $\omega = 90° - \beta, \theta_{\text{fl}} = \beta_1 - \beta_2$, and $f(t/c) = 0.6 + 4(t/c)$. Here, if φ and ψ (based on the definitions of Problem 7.20) are known at the BEP, $\omega_1 = \tan^{-1}(1/\varphi)$ and $\omega_2 = \tan^{-1}[(1 - \psi)/\varphi]$. The thickness ratio, t/c, must also be known or approximated. The method can be used with the linear approximation to the $\psi - \phi$ curve developed in Problems 7.20 and 7.21 (the slope passing through the design point is $d\psi/d\phi = -2\eta_T \cot \beta_2$). Develop a $\psi - \phi$ curve with $\psi = 0.3$, $\phi = 0.3$, and $\eta_T = 0.85$ at the design point. Use $t/c = 0.05$ and find the slope and the stall point from $\phi_{\text{stall}} = \cot \beta_{1\text{stall}}$. Plot the characteristic curve.

7.32. A small-scale model of a controllable-pitch axial fan (CPAF) was tested with $D = 0.508$ m, $d/D = 0.631$, and $N = 3600$ rpm. The fan had 16 blades and 23 vanes, with a blade thickness of $t/c = 0.07$. The performance curves are shown in Figure P7.32. Use the methods developed in Problems 7.20, 7.21, and 7.31 to simulate the three $\Delta p_T - Q$ curves for the model CPAF and compare the results with the test data.

7.33. A steam turbine is being designed with normal stages. Blade speed is 700 ft/s and the axial component of steam velocity is 400 ft/s. For three different degrees of reaction—0.0, 0.4, and 0.7—compute the stage specific work and lay out the velocity diagrams and nozzle and rotor blade shapes.

7.34. One stage of the turbine section of a gas turbine engine has the following characteristics: hub diameter: 1 ft; blade length: 4 in.; and rotational speed 18,000 rpm. The stage inlet pressure is 32.5 psia and the outlet pressure is 15.0 psia. The inlet temperature is 900°F. The isentropic efficiency of the stage is estimated to be 89.0%. Stage design will be done assuming a normal stage with a constant axial velocity $C_x = 850$ ft/s and constant specific work along the blade length. The gas has $R = 53.3$ ft lb/lbm°R, $\gamma = 1.37$.

   a. Calculate the specific work done in this turbine stage.
   b. Using the average density, calculate the mass flow rate through the stage.
   c. Calculate the loading coefficient and the degree of reaction required at the hub, mean radius, and blade tip.
   d. Lay out velocity diagrams for hub, mean radius, and tip sections.
   e. Sketch nozzle and blade shapes at hub, mean radius, and tip.

7.35. For the following situations, sketch velocity diagrams and blade shapes. Point out important features of the diagrams and blade shapes.

   a. Axial flow turbine (normal stage): flow coefficient $\varphi = Cx/U = 0.4$ and degree of reaction = 0.3.
   b. Centrifugal pump: no preswirl, backward leaning vanes ($\beta_2 = 25°$), no slip.

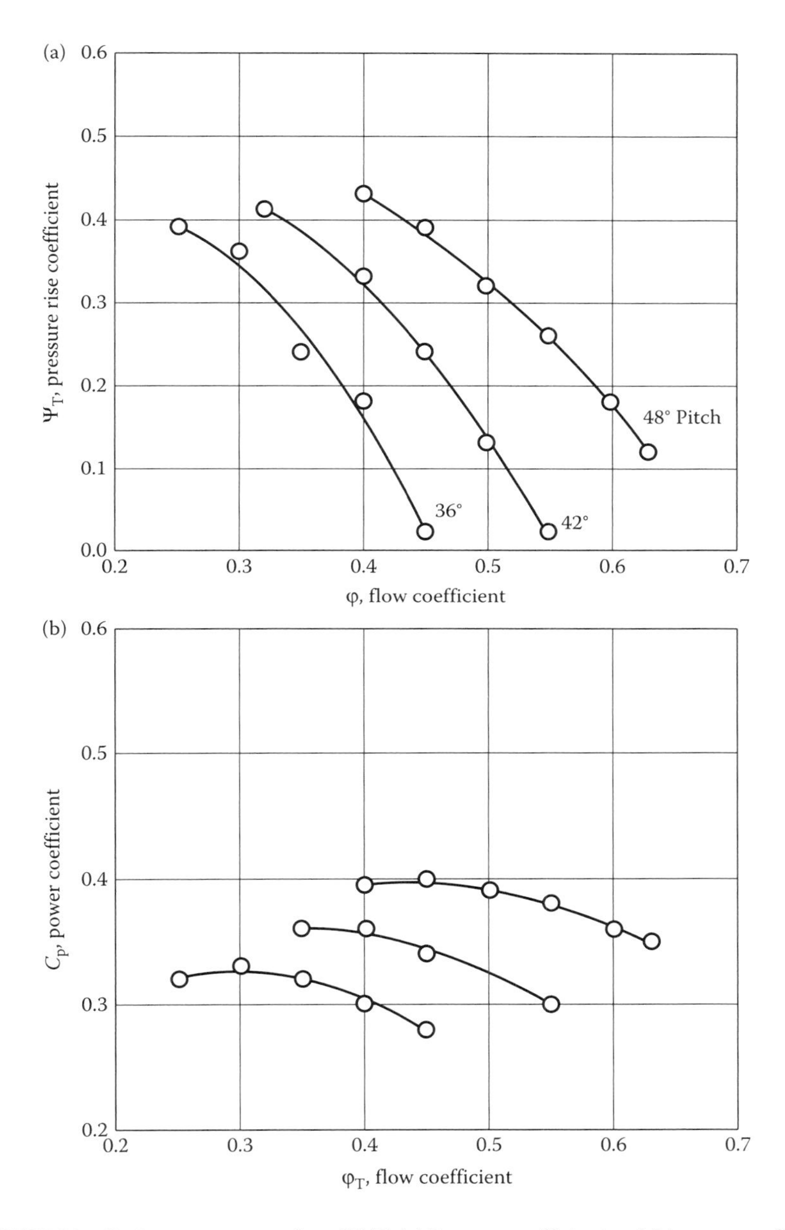

FIGURE P7.32 Performance curves for a CPAF. (a) Pressure coefficient and (b) power coefficient.

c. Axial flow compressor: normal stage, 50% reaction, zero incidence, and 5° deviation.

7.36. A *Parsons turbine* is a steam turbine in which the rotor blades and stator (nozzle) blades have identical shape, but with the angles measured in the opposite direction. Stated another way, the absolute flow angle leaving the nozzles (α_2) is equal to the relative flow angle leaving the rotor (β_3). A single stage of a Parsons turbine has a nozzle exit angle (measured relative to the moving blade direction) of 20°, the absolute velocity of the steam leaving the nozzles is 525 ft/s, and the blade speed is 500 ft/s.

a. Draw stage velocity diagrams and blade shapes. Is this a normal stage?
b. What is the degree of reaction of a Parsons turbine?
c. What is the loading coefficient ($\Psi = w/U^2$)?
d. A particular turbine stage group has 10 of these Parsons stages. The group receives steam at 220 psia and 570°F. The total-to-total isentropic efficiency of the stage group is 80%. Modeling steam as an ideal gas with $R = 85.74$ ft-lb/lbm°R and $\gamma = 1.3$, estimate the stagnation pressure and temperature of the steam as it exits the stage group. What is the polytropic efficiency of the Parsons stage?

8

Two-Dimensional Cascades

8.1 One-, Two-, and Three-dimensional Flow Models

The real flow field in a turbomachine is three dimensional and (necessarily) unsteady, at least due to the periodic passing of the blades. In order to make a tractable analysis, it is traditional to model the flow as two coupled quasi-steady flows, one in a reference frame fixed to the casing (the "absolute" coordinates) and the other in a reference frame fixed to the rotating impeller (the "relative" coordinates). These two systems are connected by the velocity of the impeller, U.

The geometric complexity is handled by a variety of models. In Chapters 6 and 7, an essentially one-dimensional model was used; all flow streamlines were assumed to be parallel to the camber lines of the blade passages in which the fluid is flowing. It should be noted that some would consider this as a two-dimensional model because the streamlines and velocity vectors are represented in a two-dimensional plane. The important point is that the flow direction was presumed known at every point and that there was no variation in the velocity in the direction perpendicular to the streamlines.

One of the most popular models of the true 3D flow field is the so-called quasi-three-dimensional (Q3D) model. In this model, the flow is treated as the superposition of two two-dimensional flows. One of these flows is the blade-to-blade flow whose streamlines lie on a meriodional surface between adjacent blades (often modeled as a cylinder coaxial with the machine shaft). The second flow occurs in the hub-to-tip plane (axial flow) or the "backplate-to-shroud" plane (radial flow). In the second plane, the axial and radial velocity components appear explicitly while the swirl velocity appears as the average from the blade-to-blade analysis.

A somewhat simpler model for the axial flow hub-to-tip plane was suggested in Section 7.5, namely neglecting variations of axial velocity, assuming zero radial velocity, and simply "stacking" velocity diagrams in the radial direction by using the value of blade velocity (U) appropriate to the actual radius. The simplest model of all was employed in the remainder of the chapter—the use of a single representative streamline at an "average" radius, or, for radial flow, a point midway across the impeller depth.

The subject of this chapter is the blade-to-blade flow. Attention will be focused on a *blade cascade*. An axial flow cascade is a planar, repeating row of identical blades. A radial cascade is a radial array of identical blades, arranged in a complete circle. In a simple sense, one obtains an axial flow machine by "wrapping" a finite-height planar cascade around a cylinder. The typical radial flow machine is a radial cascade when viewed looking down the shaft. Simple physical models, analytical models, and, especially, models based on empirical data will be used to relate actual fluid velocity vectors to cascade geometry (blade shape, curvature, solidity, and setting angle).

Additionally, data and models for flow energy losses in two-dimensional cascades will be considered. The simple "average radius/width" or, at most, "radial stacking" models for the hub–tip or backplate–shroud planes will be retained until the following chapter. As in the previous chapter, consideration will be limited to flows very near the BEP; that is, the focus will be on "design" performance, nearly to the exclusion of "off-design" performance.

8.2 Axial Flow Cascades: Basic Geometry and Simple Flow Models

In prior analyses of the flow through blade rows, it was implicitly assumed that the velocity vectors along a blade or vane were constrained to lie along the mean camber line of the blade or vane. Even if this is true in the average sense, exactly at the blade surface, the flow generally follows the actual surface instead of the camber line. More generally, if the blade channel is not fairly narrow compared to its length, the flow nearer the center of the channel will not be constrained completely by the blade shape through the flow channel. Figure 8.1 illustrates the situation while Figure 8.2 shows the typical layout and

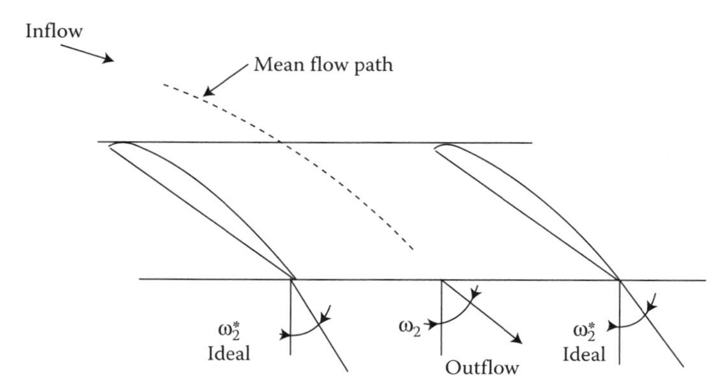

FIGURE 8.1 Flow in a two-dimensional axial flow cascade, illustrating the concept of average velocity vectors.

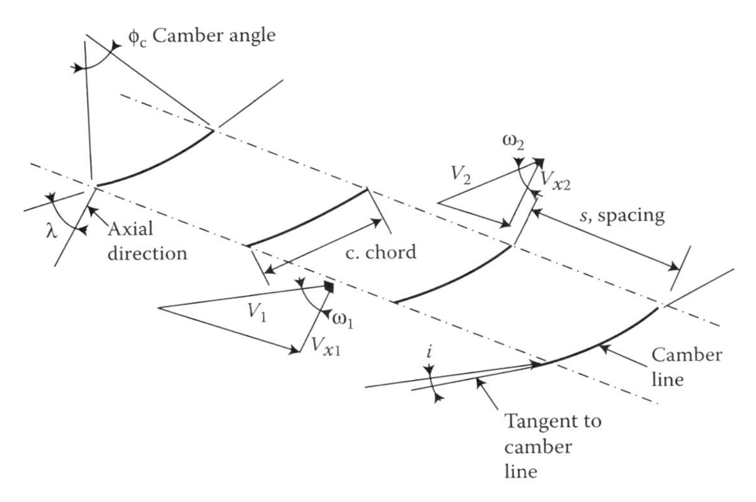

FIGURE 8.2 Geometric layout and nomenclature for an axial flow cascade. Finite thickness blades are collapsed to their camber line.

nomenclature for an axial flow cascade (see also Figure 6.2). Two issues arise. First, the velocity vectors will vary in magnitude and direction across the blade channel—velocities right at the blade surface will be zero, those "near" the top surface of a blade will be aligned with that surface, and those near a bottom surface will be aligned with that surface. Typically, this issue is handled by averaging the velocity magnitude and direction across the channel width. If the channel width is of the same scale as the channel length, then the (average) flow will be fairly closely guided by the (mean) camber line shape. As the length increases relative to the width, the average flow becomes even more closely aligned with the ideal exit angle (i.e., the tangent to the camber line).

The second issue is that, for large channel widths, the flow is not well constrained, and the average outflow angle will deviate significantly from the mean camber angle as shown in Figure 8.2. Define ω_2 as the average exit angle or flow angle, measured from the axial direction.* Similarly, ω_2^* denotes the angle associated with the blade shape (the so-called ideal exit angle or blade angle). In general, $\omega_2 > \omega_2^*$ such that the true fluid turning angle, $\theta_{fl}(\omega_1 - \omega_2)$, is less than the ideal fluid turning angle, $\theta_{fl}^*(\omega_1^* - \omega_2^*)$. As a result, for a specific cascade geometry, $C_{\theta 2}$ is smaller than previously assumed. That is, a specific blade does not do as much work on the fluid as we would expect from the simple calculation we have been using thus far. The difference between the ideal exit angle and the actual (averaged) exit angle is called the *deviation*, δ:

$$\delta \equiv \omega_2 - \omega_2^*. \tag{8.1}$$

* In the literature, the symbol β is most frequently used for this angle. In this book, β is usually used to represent the angle between the blade velocity (U) and the relative velocity (W), so the relatively little-used symbol ω is employed for the axial-direction-referred flow angle.

The deviation, as defined, is always positive, that is, the flow is turned less than the blade curves.

One may wonder about the relation between the flow angle and the blade angle at the entrance to the cascade. The difference between the entering flow velocity vector and the tangent to the camber line at the blade leading edge is called the incidence angle, usually shortened to simply the *incidence*, *i*. Incidence is positive as shown, with the vector rotated toward the pressure surface of the blade. The direction of the approach flow is physically determined upstream of the cascade, before the fluid is influenced by the blades, so for a given blade shape and blade setting angle, the incidence is determined before the flow enters the cascade. In the analyses of Chapters 6 and 7, the incidence was assumed to be zero, with the blade-relative flow tangent to the camber line at the leading edge. This is not necessarily the optimum flow condition; in fact, data show that, often, cascade performance is optimum when the incidence is somewhat positive. Including all factors, the actual fluid turning in a cascade is given by

$$\theta_{\mathrm{fl}} = i + \phi_{\mathrm{C}} - \delta, \tag{8.2}$$

where ϕ_{C} is the blade (camber line) turning angle

$$\phi_{\mathrm{C}} = \omega_1^* - \omega_2^*. \tag{8.3}$$

To increase the accuracy of the performance predictions and design calculations for axial flow machines, one needs to formulate quantitative models for cascade flow. Early models (see, e.g., Howell, 1945) began with two assumptions: (1) the optimum incidence is zero (inlet flow is tangent to the camber line at the leading edge) and (2) camber lines are simple shapes such as straight lines or circular arcs. Effort was then concentrated on finding a useful model for the deviation. Early potential flow models suggested that the deviation is proportional to the blade turning angle and inversely proportional to a fractional power of the solidity σ ($= c/s$). The following simple semiempirical formula was suggested by Constant (Constant, 1939)

$$\delta \approx \frac{0.26\phi_{\mathrm{C}}}{\sigma^{1/2}}. \tag{8.4}$$

Equation 8.4 is applicable to circular arc diffusing (as opposed to accelerating) blade rows and is called, appropriately enough, "Constant's rule."

It was soon recognized that Constant's value (0.26) was not fixed but is a function of the orientation of the blades in the cascade itself, and the model was generalized to

$$\delta = \frac{m\phi_{\mathrm{C}}}{\sigma^{1/2}}. \tag{8.5}$$

Howell (1945) correlated the parameter m with the trailing edge angle of the blade, ω_2^*, and the position of maximum camber (at mid-chord for circular arc camber lines):

$$m \approx 0.23 + 0.1\left(\frac{\omega_2^*}{50}\right) \quad \text{(Howell, circular arc).} \tag{8.6}$$

Carter and Hughes (1946) related m to the blade setting angle, λ; their graphical relationship can be approximated (for a circular arc camber line) by

$$m \approx 0.216 + 0.046\left(\frac{\lambda}{50}\right) + 0.056\left(\frac{\lambda}{50}\right)^2 \quad \text{(Carter, circular arc).} \tag{8.7}$$

Equation 8.5 is referred to as the "Simple Howell rule" or "Carter's rule," depending on the formulation used for m, and is limited to circular arc-cambered diffusing blades.

For a typical cascade camber angle of, say, 20° and a solidity of $\sigma = 1.0$, Constant's rule gives $\delta = 5.2°$. The Simple Howell rule gives the same value when the trailing edge angle is 15° as does Carter's rule when the blade setting angle is about 30°. Clearly, 5.2° is not negligible when compared with the "theoretical" flow turning of 20°. The early deviation "rules" thus provided a needed measure of conservatism in cascade layout for axial compressors and fans. The rules in any one of the three forms clearly show the expectation of large deviation angles at low solidity and the ability to reduce deviation through increased solidity. The secondary influence of blade orientation,* as well as that of the solidity σ, on δ can be seen in Figure 8.3 in which both solidity and blade orientation are varied (keeping the blade camber angle at 20°). The range of deviation is seen to be substantial compared with the simplistic assumption of total guidance of the flow. The deviation angles are all directly proportional to the camber angle and range from about 20% to 40% of ϕ_c.

8.3 Systematic Investigation of Axial Cascade Flow

The development of practical axial compressors for gas turbine and jet engines in the mid-twentieth century gave great impetus to the search for more accurate models of cascade flows. Much of the research and development was done at national laboratories such as the U.S. NACA Langley Aeronautical Laboratory and Lewis Laboratory† and the British National Gas Turbine Establishment (NGTE). Developments included both theoretical models and

* For a circular arc camber line, the blade orientation parameters are related by $\omega_2^* = \lambda - (\phi_c/2)$.

† Now the National Aeronautics and Space Administration (NASA) Langley Research Center and Glenn Research Center, respectively.

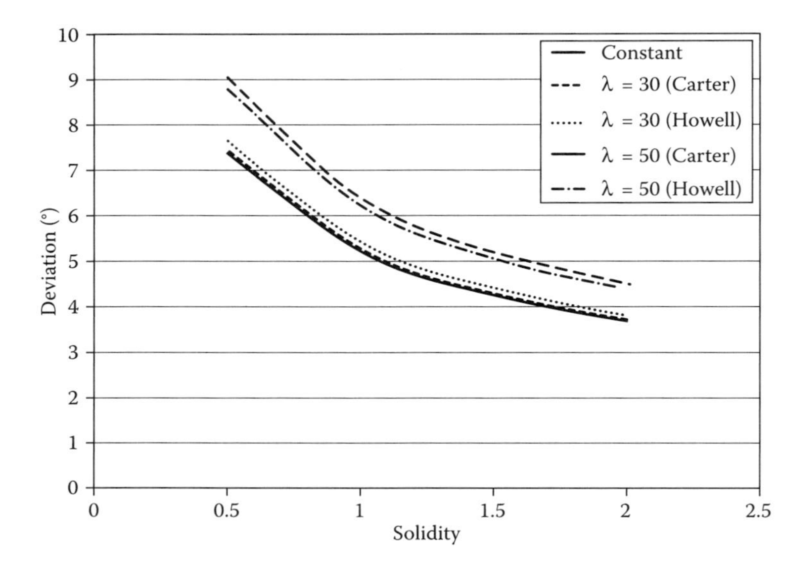

FIGURE 8.3 Deviation angles for 20° camber angle using various deviation rules.

experimental investigations. An outstanding summary of U.S. work is NASA SP 36 *Aerodynamic Design of Axial-Flow Compressors* (Johnsen and Bullock, 1965), which is a declassified version of a series of reports compiled in 1956. Here, only empirical models will be discussed.

The most useful design data were that based on systematic tests in a cascade wind tunnel. Figure 8.4 is a schematic of such a wind tunnel. The primary data from the wind tunnel tests are (average) flow turning angle and stagnation pressure loss as functions of cascade parameters. The testing program was conducted as follows. First, a basic blade geometry (a *blade series*) was selected (more detail shortly). Then, for that blade series, the following parameters were systematically varied:

- Solidity, σ.
- Blade (camber line) curvature, ϕ_c.
- Blade thickness (magnitude and shape details).
- Blade setting (stagger) angle, λ.
- Approach flow angle, ω_1.

Note that both the blade setting angle (λ) and the approach flow angle (ω_1) are referred to the axial (throughflow) direction. Using these angles, together with knowledge of the blade camber line, the flow orientation can alternately be expressed as angle of attack (angle between approach velocity vector and chord line) or incidence (angle between approach velocity vector and the tangent to the camber line at the leading edge) (Figure 6.1).

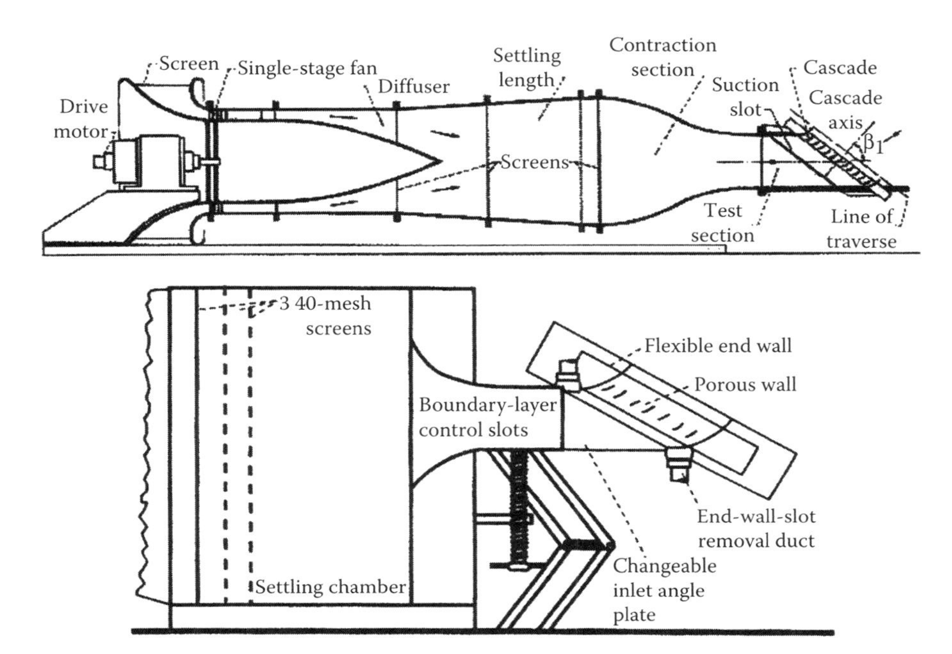

FIGURE 8.4 Schematic of a cascade wind tunnel. (From Johnsen, I. A. and Bullock, R. O. (1965), "Aerodynamic design of axial flow compressors," NASA SP-36. With permission.)

The basic test procedure was to set up a specific cascade with certain values of blade shape (including camber), solidity, and approach flow angle. Then the blade setting angle was varied, and (averaged) fluid turning angle and blade trailing edge boundary layer characteristics were measured. The resulting data were converted to deviation angle (using $\delta = i + \phi_c - \theta_{fl}$) and total pressure loss coefficient ζ ($\zeta \equiv \Delta p_{T,Loss}/(1/2)\rho V_1^2$), which can be plotted versus incidence angle (i). A typical data plot is shown in Figure 8.5.

For design purposes, the most important points on such a plot are the values that occur when the loss is minimum (according to Johnsen and Bullock, this point is generally selected as the center point of the relatively flat low-loss region). The values corresponding to this minimum loss point are then designated the "design values" (i.e., optimum performance values) for the particular cascade. These design values are expressed as the following:

- (Minimum-loss) incidence angle, i
- Deviation
- Loss coefficient
- Maximum permissible diffusion ratio.

These design parameters are correlated in terms of the various cascade parameters—as a set of simple equations supported by graphs in Johnsen and Bullock and with the graphs replaced by curve fits by later workers.

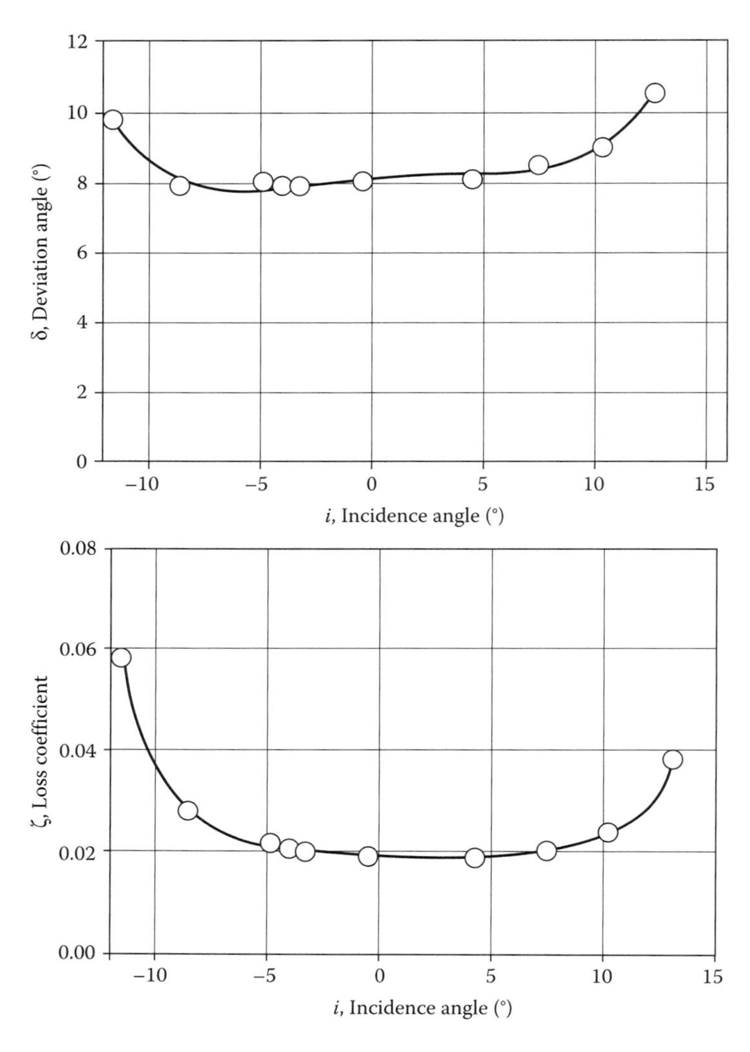

FIGURE 8.5 Typical cascade test results plot. (From Johnsen, I. A. and Bullock, R. O. (1965), "Aerodynamic Design of Axial Flow Compressors," NASA SP-36. With permission.)

Clearly, a very large amount of experimental work was required to generate these data. In the NACA research, a large array of experimental configurations was examined with camber angles (ϕ_c) varying from 0° to about 70°, solidity (σ) varying from 0.5 to 1.75, and the inlet flow angle (ω_1) varying from 30° to 70°. All of this is in addition to variations in the basic blade shape, which must now be considered.

Blade profiles in a cascade are essentially airfoil shapes (Figure 6.1). The aeronautical research establishments in various countries (NACA in the United States) developed and tested a wide variety of airfoil shapes in the decades of the 1920s, 1930s, and 1940s. A particular family of airfoil shapes is called an *airfoil series*. Typically, an airfoil series is defined by a certain family of curves that specify the camber line shape and a second family of curves that specify the thickness distribution. Shape variations, and hence airfoil performance characteristics, are obtained by selecting, essentially, the magnitude of maximum camber, the location of maximum camber, and the maximum thickness, all as a percentage of the airfoil chord. Airfoil shapes are typically identified by a "code designation," in which the parameters that define the airfoil shape and/or performance are listed.

The NACA cascade tests were run using a particular family of airfoils called the NACA 65-Series Compressor Blade Section, based on the original NACA 65-Series airfoil (Emery et al., 1958). The 65-Series airfoil is a high-performance airfoil designed to maintain laminar flow over a sizable portion of its surface. This type of airfoil was used on a high-performance fighter aircraft built during World War II, perhaps most notably on the P-51 "Mustang." The key to the code designation for 65-Series airfoils is 65-(x)(yy), where "65" refers to the series (i.e., a specific functional form for the camber line and thickness distribution), x is 10 times the "design lift coefficient" (at zero angle of attack), and yy is the blade thickness in percent of chord, (t/c)%. For example, a 65-010 has a design lift coefficient of 0 (zero) and the maximum thickness is 10% of the chord length. A 65-810 has a design lift coefficient of 0.8 and 10% thickness. All cascades tested in the initial work were assembled from 10% thick airfoils. Figure 8.6 shows the shapes of the airfoil sections used in the

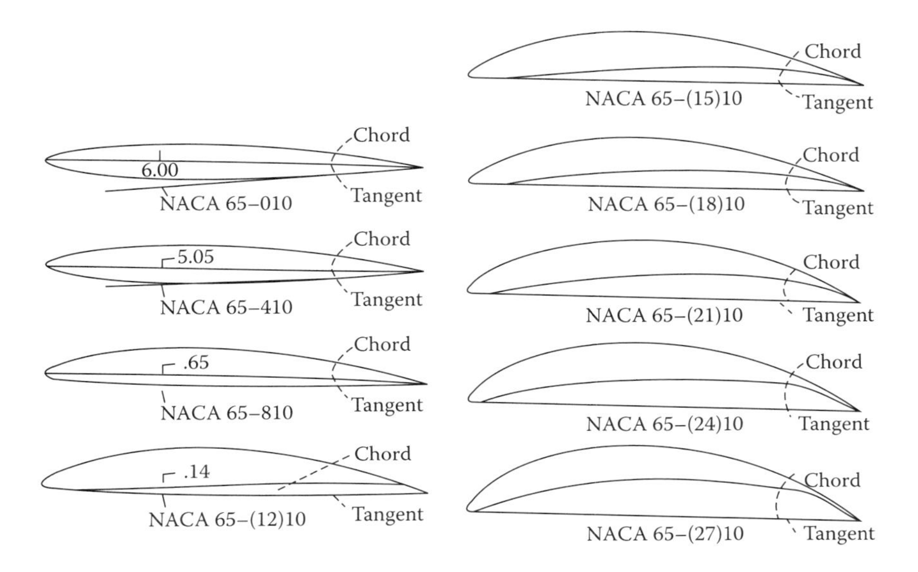

FIGURE 8.6 Blade shapes used in cascade tests by NACA. (From Emery, J. C., et al., (1958), *Systematic Two-Dimensional Cascade Tests of NACA 65-Series Compressor Blades at Low Speed*, NACA-TR-1368. With permission.)

blade cascades. Information for laying out the shapes can be found in several references (Emery et al., 1958; Abbott and von Denhoff, 1959; Ladson et al., 1996).

Johnsen and Bullock point out that the 65-Series airfoils tested actually do not have a circular arc camber line and that, in fact, the camber line slope is theoretically infinite at the leading and trailing edges. This is handled in the data by using an equivalent circular arc drawn through the leading edge, the trailing edge, and the point of maximum camber, which is conveniently at the half-chord point. Incidence, deviation, and blade turning angle are then determined from this equivalent arc. The blade turning angle (camber angle) is related to the design lift coefficient by the equation

$$\phi_c^\circ = 25\,C_{L0} \tag{8.8}$$

where C_{L0} is the design lift coefficient ("x"/10 from the airfoil code designation).

8.4 Correlations for Cascade Performance

To facilitate convenient use by fan and compressor designers, the massive quantity of cascade data was reduced to a set of correlations (Johnsen and Bullock, 1965). In this section, correlations for the "angle" parameters, incidence and deviation, are presented. Correlations for loss- and diffusion-related parameters are considered in a later section. It is essential to recall that *the correlations refer to performance parameters for the minimum loss condition only*. One must understand the basic correlation strategy, which is as follows.

- Correlations are based on simple functional forms, similar to the "Carter/Howell rule" for deviation.
- The parameters in the formulae (e.g., "m" and the exponent on σ) are themselves functions of the cascade and flow variables.
- The correlations are based on the performance characteristics of NACA 65-A10 blades ("A" is arbitrary) with "correction factors" for shapes other than the NACA 65 series and maximum thickness other than 10%.

First consider the (minimum loss) incidence angle, i. Incidence is presented as a linear function of camber in a form suggested by Carter's rule (i.e., proportional to the blade camber angle):

$$i = i_0 + n\phi_c, \tag{8.9}$$

where i_0 is the incidence angle required for a zero-camber (flat) blade profile and n is a "slope factor" relating incidence to camber. Both i_0 and n are complicated functions of the cascade variables $\omega_1, t/c$ (thickness–chord ratio), and σ. i_0 is expressed in terms of the zero-camber incidence for the 65-A10 profiles by

$$i_0 = (K_i)_{sh}(K_i)_t(i_0)_{10}. \tag{8.10}$$

This is a messy function intended to be very general.* The "t" subscript refers to variation with relative thickness of the blade (other than 10%) and the "sh" subscript refers to behavior using shapes other than the 65-Series airfoils. In the limiting case, for very thin blades (t/c approaching zero), the value of $(K_i)_t$ goes to zero so that i_0 goes to zero. For 10% thick 65-Series blades, the function reduces to

$$i_0 = (i_0)_{10}, \tag{8.11}$$

so that the problem becomes determining $(i_0)_{10}$ and n as functions of ω_1 and σ. The original correlations were presented as graphs of $(i_0)_{10}$ and n versus ω_1 with σ as a parameter. More recent efforts have reduced these extensive graphs to a set of curve fits (Wilson, 1984; Wright, 1987). For example, the use of $i = i_0 + n\phi_c$ with $n = n_0/\sigma^c$ yields a form very similar to the original correlations of Howell and Carter:

$$i = i_0 + \left(\frac{n_0}{\sigma^c}\right)\phi_c. \tag{8.12}$$

For small values of t/c, the formula for i reduces to

$$i = \left(\frac{n_0}{\sigma^c}\right)\phi_c, \qquad \frac{t}{c} \to 0, \tag{8.13}$$

which is the same form as the simple Howell/Carter equation, but with the square root of σ replaced by a variable exponent c. To fit the data, n_0 and c must be expressed as functions of ω_1 and σ itself.

For the 10% thick 65-Series blades, a fairly simple set of curve fits results in the approximate expressions:

$$i_0 = \sigma\left[8.0\left(\frac{\omega_1}{100}\right) - 1.10\left(\frac{\omega_1}{100}\right)^2\right] \tag{8.14}$$

and

$$n_0 = -\left[0.0201 + 0.3477\left(\frac{\omega_1}{100}\right) - 0.5875\left(\frac{\omega_1}{100}\right)^2 + 1.0625\left(\frac{\omega_1}{100}\right)^3\right] \tag{8.15}$$

* Note that no "K" functions are provided in this book; thus consideration here is limited to 10% thick 65-Series blades.

and

$$c = 1.875\sigma\left[1.0 - \left(\frac{\omega_1}{100}\right)\right], \quad \text{for } \sigma \leq 1.0 \tag{8.16}$$

or

$$c = 1.875\left[1.0 - \left(\frac{\omega_1}{100}\right)\right], \quad \text{for } \sigma > 1.0. \tag{8.17}$$

This group of equations becomes the algorithm for a more formal and accurate treatment for establishing the optimal leading edge incidence angles for blade layout. They replace the rather simplistic tangency condition ($i = 0$) used earlier, since that condition is reasonably accurate for only high-solidity cascades of very thin blades.

To complete the more accurate treatment of the blade cascade, one should replace the Howell/Carter rule for flow deviation with the more general forms given in NASA SP-36. As in the simpler rules, the deviation is assumed to be a linear function of camber angle:

$$\delta = \delta_0 + m\phi_c. \tag{8.18}$$

The parameters δ_0 and m are functions of the cascade parameters ω_1 and σ. The expression for δ_0 is based on the value for 10% thick 65-Series airfoils with corrections for thickness and shape, as was i_0, so that

$$\delta_0 = (K_\delta)_{sh}(K_\delta)_t(\delta_0)_{10}.$$

As before, $(K_\delta)_{sh}$ goes to 1 for 65-Series blade shapes and $(K_\delta)_t$ goes to 1 for 10% thick blades. $(K_\delta)_t$ becomes zero for very thin blades. Using the Howell/Carter form for the slope factor m as $m = m_0/\sigma^b$, the deviation expression becomes

$$\delta = \delta_0 + \left(\frac{m_0}{\sigma^b}\right)\phi_c, \tag{8.19}$$

or, for very thin blades,

$$\delta = \left(\frac{m_0}{\sigma^b}\right)\phi_c, \quad \frac{t}{c} \to 0. \tag{8.20}$$

For the 10% thick 65-Series airfoils, the values of δ_0, m_0, and b can be approximated by curve fits to the graphs presented in NASA SP-36 as:

$$\delta_0 = 5.0\sigma^{0.8}\left(\frac{\omega_1}{100}\right)^2 \tag{8.21}$$

and

$$m_0 = 0.170 - 0.0514\left(\frac{\omega_1}{100}\right) + 0.3592\left(\frac{\omega_1}{100}\right)^2 \tag{8.22}$$

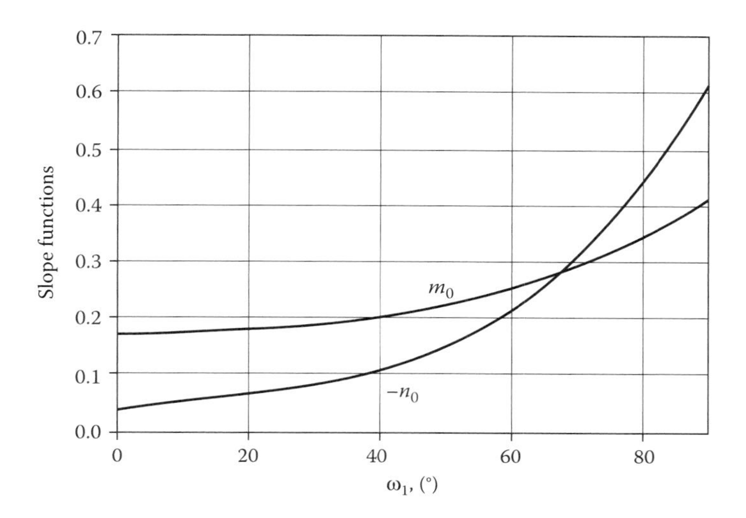

FIGURE 8.7 Slope functions for incidence and deviation correlations.

and

$$b = 0.965 - 0.0200\left(\frac{\omega_1}{100}\right) + 0.1249\left(\frac{\omega_1}{100}\right)^2 - 0.9720\left(\frac{\omega_1}{100}\right)^3, \quad \text{for } \sigma > 1.0 \tag{8.23}$$

or*

$$b = \sigma\left[0.965 - 0.0200\left(\frac{\omega_1}{100}\right) + 0.1249\left(\frac{\omega_1}{100}\right)^2 - 0.9720\left(\frac{\omega_1}{100}\right)^3\right], \quad \text{for } \sigma < 1. \tag{8.24}$$

This set of equations provides an accurate estimate for the deviation angle in preparing blade layouts.

The polynomial curve fit equations for the incidence/deviation slope functions (n_0 and m_0), and the solidity exponents (b and c), Equations 8.15 through 8.17 and 8.21 through 8.24, are quite useful for computer or programmable calculator use. If one is making hand calculations, the graphical representation provided in Figures 8.7 and 8.8 might be more useful.

Finally, one needs to assemble all of this accurate detail for cascade performance into a rational scheme for turbomachine blade layout. The basic objective of the layout process is, of course, to pick a geometry that will achieve the required specific energy rise or fluid turning, θ_{fl}, at a given flow rate. From Equation 8.2, the fluid turning is calculated from the blade camber together with the incidence and deviation:

$$\theta_{fl} = \phi_c + i - \delta.$$

* Equation 8.24 is not a misprint; it is Equation 8.23 multiplied by σ.

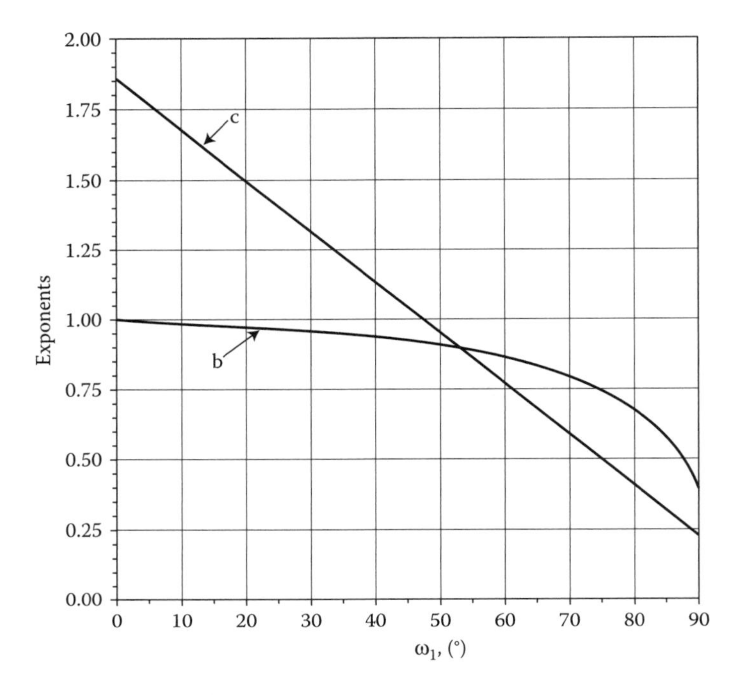

FIGURE 8.8 Solidity exponents for incidence and deviation correlations.

Using Equations 8.12 and 8.19 gives

$$\theta_{fl} = \phi_c + i_0 + \left(\frac{n_0}{\sigma^c}\right)\phi_c - \left[\delta_0 + \left(\frac{m_0}{\sigma^b}\right)\phi_c\right]. \tag{8.25}$$

The amount of camber required to achieve a particular value of θ_{fl} (typically derived from laying out velocity diagrams) is

$$\phi_c = \frac{\theta_{fl} - i_0 + \delta_0}{1 + \left[(n_0/\sigma^c) - (m_0/\sigma^b)\right]}. \tag{8.26}$$

Consider a situation in which a fluid turning θ_{fl} of 20° is required. If one assumes zero incidence and perfect guidance of the fluid (as was done in Chapters 6 and 7), then the required blade camber would be 20°. If the rather simple Constant's rule is adopted for deviation and the assumption of zero incidence is retained, then, from Equation 8.26, the required camber would be a function of solidity only

$$\phi_c = \frac{\theta_{fl}}{1.0 - \left(0.26/\sigma^{1/2}\right)} = \frac{20°}{1.0 - \left(0.26/\sigma^{1/2}\right)}.$$

For $\sigma = 1.0$, $\phi_c = 26.3°$; for $\sigma = 0.5$, $\phi_c = 31.6°$; and so on.

Next, one can use the more complex and more accurate correlations for i and δ to calculate the camber and compare the results with those using Constant's rule. Using 10% thick 65-Series airfoils and assuming an inlet flow angle $\omega_1 = 45°$, one calculates, for $\sigma = 1.0$:

$$i_0 = 3.37; \quad \delta_0 = 1.01; \quad b = 0.894$$

$$c = 1.031; \quad m_0 = 0.2196; \quad n_0 = -0.1344$$

$$\phi_c = \frac{\theta_{fl} - i_0 + \delta_0}{1 + \left[(n_0/\sigma^c) - (m_0/\sigma^b)\right]} = \frac{20 - 3.37 + 1.01}{1 - \left[(0.1344/1^{1.031}) - (0.219/1^{0.894})\right]} = 27.3°.$$

Thus, the camber must be 27.3°, that is, extremely close to the Constant value, differing by only 1°. If $\sigma = 0.5$, the detailed method gives $\phi_c = 34.6°$; this time, there is a somewhat larger difference from the Constant's rule-based result, a difference of just over 3°. Of course, both calculations give a camber considerably different from the "ideal" 20°.

Example: Axial Fan Layout

Determine the blade and vane shapes for an axial fan mean radius blade station. The fan performance requirements are $Q = 10{,}000$ cfm and $\Delta p_T = 1.8$ in. wg at density $\rho g = 0.0748\ \text{lb/ft}^3$.

Calculation of N_s and D_s for various choices suggests a vane axial fan with $N = 1425$ rpm, $D = 2.5$ ft, and $\eta_T = 0.89$. Figure 6.10 suggests $d/D = 0.6$ giving $d = 1.5$ ft and thus $r_m = (d + D)/4 = 1$ ft. Then calculate $U_m = 149$ ft/s and $C_x = 4Q/\pi(D^2 - d^2) = 53$ ft/s. Now, assuming no preswirl and using the Euler equation, $C_{\theta 2}$ is calculated as

$$C_{\theta 2} = \frac{\Delta p_T/\rho}{U_m \eta_T} = 30\ \text{ft/s}.$$

Laying out the velocity diagrams gives

$$\beta_1 = 19.6°,\ \omega_1 = 70.4°; \quad \beta_2 = 24.0°,\ \omega_2 = 66.0°; \quad \theta_{fl} = 4.4°.$$

These parameters provide nearly all the information necessary to lay out the mean blade station. However, one does have to select a value of σ. The de Haller ratio can be evaluated as $W_2/W_1 = 130.3/158.1 = 0.82$. This is rather conservative, so choose a moderate value of solidity of σ, say 1.0. Now proceed to calculate the required values for the camber angle, ϕ_c, and the blade setting angle, λ. For simplicity (or, perhaps for low cost), use stamped sheet metal blades whose t/c value is very small, perhaps about 0.005. This allows one to set i_0 and δ_0 to zero and work with the simplified equations

$$i = \left(\frac{n_0}{\sigma^c}\right)\phi_c \quad \text{and} \quad \delta = \left(\frac{m_0}{\sigma^b}\right)\phi_c.$$

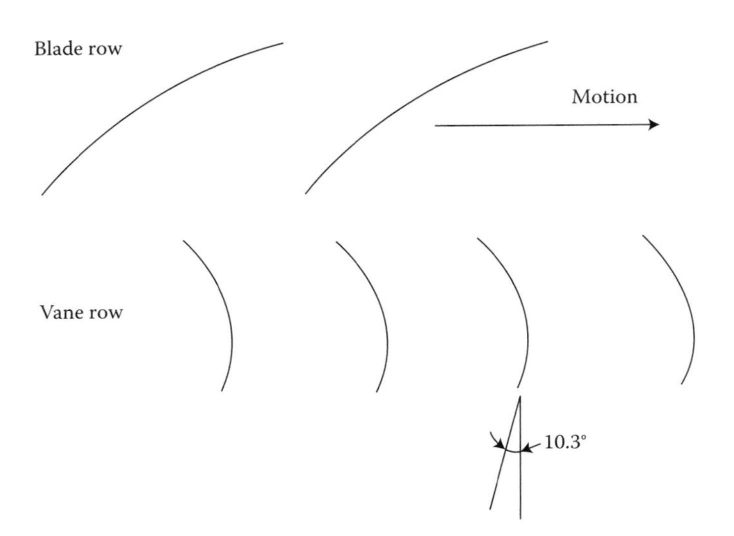

FIGURE 8.9 Fan blade and vane shapes at mean radius.

Then calculate from the curve fits or read from Figures 8.7 and 8.8

$$m_0 = 0.312; \quad b = 0.674; \quad n_0 = -0.340; \quad c = 0.555,$$

so

$$\phi_c = \frac{\theta_{fl} - i_0 + \delta_0}{1 + \left[(n_0/\sigma^c) - (m_0/\sigma^b)\right]} = \frac{4.4 - 0 + 0}{1 - \left[(0.340/1^{0.555}) - (0.312/1^{0.674})\right]} = 12.6°.$$

Even with a moderately high level of solidity, one observes significant effects due to deviation and optimal incidence setting (−4.28° and 3.93°, respectively) with 12.6° of circular arc camber required to achieve only 4.4° of fluid turning. The blade pitch or setting angle is calculated as

$$\lambda = \omega_1 - i - \left(\frac{\phi_c}{2}\right) = 68.4°$$

The layout for this mean blade station is shown in Figure 8.9 (the vanes, to be developed below, are also shown).

To complete the rotor design, one must select some number of blades to finalize the geometry. Let the number of blades be $N_B = 8$ and calculate the blade chord, c_m, and spacing from the solidity. At the mean radius, the solidity is $\sigma_m = (N_B)c_m/(2\pi r_m)$ so that the blade chord $c_m = (2\pi r_m \sigma_m)/N_B = 0.785$ ft. The blade length, L_B, from the hub to the tip is $(D - d)/2 = 1$ ft so that the blade is nearly square in appearance, with an aspect ratio of $L_B/c_m = 1.274$. This is a somewhat "stubby" blade, as aspect ratios of 2 to 3 are not unusual for a vane axial fan.

The methods developed here can also be used to lay out a vane section for the fan. Again, working at the mean radial station of the fan, determine the inflow properties for the vane row directly from the outflow from the blade row. The inlet velocity for the vane row, C_3, is the absolute velocity at the rotor discharge, C_2. The discharge velocity requirement for the vane row is to simply provide a purely axial flow at the outlet (complete swirl recovery). The inlet angle for the vane, ω_3, is set by the value of C_x and $C_{\theta 2}$ from the blade row. These were $C_x = 53$ ft/s and $C_{\theta 2} = 30$ ft/s, so that $C_3 = 60.9$ ft/s. At the discharge from the vane row, the value of C_4 is just C_x, so $C_4 = 53$ ft/s. The de Haller ratio for the vanes is

$$\frac{C_4}{C_4} = \frac{53}{60.9} = 0.87.$$

This is a very conservative value; thus, one can assign a lower solidity value to set up the vane mean station cascade, say, $\sigma = 0.75$. The vane inflow angle is

$$\omega_3 = \tan^{-1}\left(\frac{C_{\theta 2}}{C_x}\right) = 29.5°.$$

The fluid turning of the vane must drive the outflow angle to zero, so $\theta_{fl} = \omega_3 = 29.5°$. Using a thin airfoil section, calculate

$$m_0 = 0.186; \quad b = 0.709; \quad n_0 = -0.0908;$$
$$c = 0.991; \quad i_0 = 0; \quad \delta_0 = 0;$$

to yield

$$\phi_c = 45.3°.$$

The number of vanes, N_V, is chosen primarily to avoid a number that is evenly divisible by a multiple of the number of blades so as to avoid the natural resonances of the acoustic frequencies of the two rows. Since $N_B = 8$ for this design, consider, say, $N_V = 17$. Secondarily, try to keep the aspect ratio of the vane (L_V/c_{mV}) between 1 and 6. With 17 vanes, the mean chord (based on a vane row solidity of 0.75) is 0.227 ft, with a height of 1 ft (to match the blade height). This gives an aspect ratio of about 4.4. Seven vanes would require a chord of 0.551 ft with an aspect ratio of 1.8; 23 vanes would give a chord of 0.168 ft and an aspect ratio of nearly 6.0. Either the 7-vane or 17-vane configuration looks all right, so stay with $N_V = 17$. Values of i and δ for the thin-vane mean station are calculated as

$$i = \left(\frac{n_0}{\sigma^c}\right)\phi_c = -5.47° \quad \text{and} \quad \delta = \left(\frac{m_0}{\sigma^b}\right)\phi_c = 10.33°.$$

Finally,

$$\lambda = \omega_1 - i - \frac{\phi_c}{2} = 12.4°.$$

The mean radius blade and vane shapes and orientations for this preliminary design layout are shown in Figure 8.9. Note that the vane would appear to turn the flow well past the axial flow direction. This appearance led to the term "over-turning" commonly seen in the early cascade literature. This is clearly a misnomer since the extra 10.33° of blade camber simply accounts for the deviation and causes the average velocity across the channel to be purely in the axial direction, rather than "overturned."

8.5 Blade Number and Low-Solidity Cascades

Suppose that one decides to use the same hub and blade/vane designs developed in the previous example but wants to modify the fan performance by selecting fewer or greater numbers of blades of the same size and shape. For example, one might market a line of fans with $N_B = 4, 6, 8, 10, 12, 14$, and 16 blades to amortize engineering, development, and tooling costs over a broader range of products. The corresponding values of mean radius solidity are $\sigma_m = 0.5, 0.75, 1.0, 1.25, 1.50, 1.75$, and 2.0. (Note that the question of variable solidity will also arise when considering blade sections at different radii along a single blade.) Intuitively, one should expect that removing blades from the fan (e.g., going from eight to four blades) will reduce the amount of work done on the fluid at any particular flow rate and hence the pressure rise will be reduced. On the other hand, if one adds blades (e.g., from 8 to 16), the work and pressure rise should be increased. We specify that, in addition to hub and tip diameter and blade and vane shape, the blade setting angle and the design flow rate remain the same, so that λ, ϕ_c, ω_1, and i are unchanged.* Only the value of δ changes, due to the change in σ. Of course, the flow turning θ_{fl} changes as well because

$$\theta_{fl} = \phi_c + i - \delta.$$

Using the (unchanged) values of m_0, b, i, and ϕ_c from the example yields

$$\theta_{fl} = 8.32° - \frac{3.93°}{\sigma^{0.674}}$$

This equation for turning angle is used to revise the pressure rise as follows

$$\beta_2 = \beta_1 + \theta_{fl} = 19.6° + \theta_{fl},$$
$$C_{\theta 2} = U_m - C_x \cot\beta_2 = 149.0 - 53.0 \cot\beta_2,$$
$$\Delta p_T = \eta_T \rho U_m C_{\theta 2} = 0.89 \times 0.00233 \times 149.0 \times C_{\theta 2}.$$

* We also assume that the total efficiency is unchanged. This is a bit more subtle because the specific speed will change when the pressure rise changes at constant flow rate.

TABLE 8.1

Effect of Blade Number on Pressure Rise

N_B	σ	δ^o	θ°_{fl}	β°_2	$C_{\theta 2}$ (ft/s)	Δp_T (in. wg)
2	0.25	10.0	−1.7	17.9	−15	−0.89
4	0.50	6.27	2.05	21.7	15.5	0.92
6	0.75	4.77	3.55	23.2	24.8	1.49
8	1.00	3.93	4.40	24.0	30.0	1.78
10	1.25	3.38	4.94	24.5	32.9	1.95
12	1.50	3.00	5.33	24.9	35.3	2.09
14	1.75	2.69	5.62	25.2	36.5	2.17
16	2.00	2.46	5.85	25.5	37.7	2.24

Table 8.1 summarizes the calculations and Figure 8.10 displays the results graphically.*

Several observations can be made. First, the pressure rise does indeed increase with N_B (or σ), but the relationship is decidedly nonlinear. Using half of the eight blades from the initial design layout yields 52% of the initial pressure rise (i.e., half the solidity gives about half the pressure rise); however, using 16 blades instead of 8 gives only 126% of the initial pressure rise. Doubling the solidity does not come close to doubling the pressure rise. The overall behavior appears to be asymptotic as solidity becomes large; this is a clear reflection of the $1/\sigma^c$ behavior of δ. The deviation is slowly approaching zero, so that the pressure rise is gradually reaching an upper limit, namely, "ideal" zero-deviation perfect guidance of the fluid. When δ approaches zero, θ_{fl} nears the limit of $\phi_c + i$ (=8.82°), giving $\Delta p_T = 2.90$ in. wg. Of course, this value requires that all of the inherent assumptions prevail at infinite solidity. It has already been noted that a practical upper limit for solidity is about $\sigma = 2.0$ or less because of flow channel blockage due to blade thickness and viscous boundary layer thickness. In addition, at very high solidity, losses due to surface fluid friction would be very high and the efficiency would fall well below the Cordier value. The 16-blade case is therefore a probable upper limit of performance for this design.

Referring again to Table 8.1, one sees that the values for swirl velocity, fluid turning, and pressure rise all become negative for a 2-blade configuration. This seems to imply that at very low solidity, the fan is operating as a turbine and is capturing energy from the airflow. This is decidedly not true. What is happening is that the algorithms for incidence and deviation are breaking down because of the paucity of experimental performance data for really low values of solidity. Rather than extrapolate these correlations toward zero solidity, one must recognize the singular nature of the limit, both physically and mathematically. In the curve-fit correlations, the use of the form $1/\sigma^a$

* The continuous curve must be interpreted as a plot of pressure rise versus solidity; a blade count of, say, $N_B = 6.7$ is not possible.

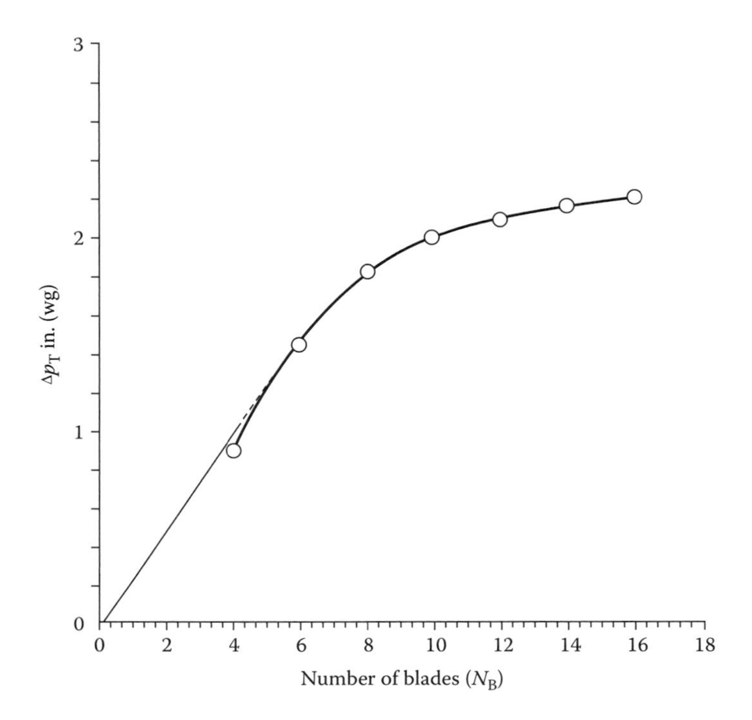

FIGURE 8.10 Variation of total pressure rise with the number of blades. All other parameters are held constant. All calculations use standard cascade correlations.

yields nonfinite results as σ becomes vanishingly small. Physically, for very low solidity, if performance requirements are held at a fixed level, one is forced to use very high values of camber angle or blade incidence to achieve any fluid turning at all as blade channels become very short (i.e., the blade chord is very small in proportion to the blade spacing).

For the blade "design" problem, the solution to this difficulty lies in establishing an analytical limit based on the airfoil theory for "isolated" (not in cascade) airfoils for very low-solidity cascades, say $\sigma \ll 0.5$. Working in the area of low-solidity cascades, Myers and Wright (1993) provide some guidance and a means of circumventing the difficulty. The approach relies on the classical, isolated airfoil theory as developed in many references (e.g., Mellor, 1959). Mellor defined the limiting ability of a low-solidity cascade to turn the flow in terms of lift coefficient, which is a function of the cascade geometry and flow variables. This value C_{Lm}, from theoretical considerations, is

$$C_{Lm} = \left(\frac{2}{\sigma}\right) \cos \omega_m (\tan \omega_1 - \tan \omega_2), \tag{8.27}$$

where ω_m is a mean value of ω defined by

$$\omega_m = \tan^{-1}\left(\frac{\tan\omega_1 + \tan\omega_2}{2}\right). \tag{8.28}$$

The camber value for isolated or very low-solidity blades is established on the basis of the lift coefficient at an arbitrarily low solidity (i.e., the so-called blade element theory) such that

$$\phi_c \approx \phi_{c,\text{isolated}} = \frac{2}{\pi}C_{Lm}(\text{rad}) = \frac{360}{\pi^2}C_{Lm}(^\circ) \tag{8.29}$$

with the incidence angle given by

$$i = -\frac{\phi_c}{2}. \tag{8.30}$$

These equations should be used to make calculations for low σ values ($\sigma < 0.5$) because the "standard" correlating equations give erroneous values, such as those shown for $\sigma = 0.25$ in Table 8.1. Some "smoothing" between the values so calculated and those for higher solidity ($\sigma > 0.5$) may be required; Meyers and Wright (1993) recommend a linear interpolation.

For a "performance" analysis carried out for an existing blade shape and hub, as was done to arrive at Table 8.1, one uses a slightly different kind of limit to interpolate for the deviation at very low solidity. Here, one can use the true physical limit, recognizing that the amount of flow turning for a zero-solidity cascade is actually zero. In terms of the deviation angle δ, with $\theta_{fl} = 0$:

$$\delta_{\sigma=0} = (\phi_c + i) = \frac{\phi_c}{2} + \beta_1 - \lambda. \tag{8.31}$$

For the second limit for interpolation, calculate $\delta_{0.6}$ using the "standard" correlation with $\sigma = 0.6$. The interpolation for very low values of σ is then:

$$\delta = \left(\frac{\sigma}{0.6}\right)(\delta_{0.6} - \delta_{\sigma=0}) + \delta_{\sigma=0}. \tag{8.32}$$

Now, returning to the product line of fans with different blade numbers, re-examine the low-solidity performance of the two-blade fan. The use of $\sigma = 0.25$ and the "standard" cascade algorithms predicted a deviation angle of 10°, resulting in a negative turning angle and a negative pressure rise. Actually, one must use low-solidity interpolation method since σ is <0.5. First, calculate $\delta_{\sigma=0} = (\phi_c + i) = 8.23^\circ$. Then, the deviation at $\sigma = 0.6$ is given by $\delta_{0.6} = 3.93^\circ/\sigma^{0.674} = 5.55^\circ$. Finally, interpolate for $\sigma = 0.25$ by $\delta_{\sigma=0.25} = (0.25/0.6)(5.55^\circ - 8.23^\circ) + 8.23^\circ = 7.4^\circ$.

Completing the remaining calculations, the turning angle is a positive 1.12°, β_2 is 21.6°, $C_{\theta 2}$ is 8.88 ft/s, and Δp_T is 0.527 in. wg, a realistic, consistent, positive value. Carrying out the calculations for a four-blade configuration ($\sigma = 0.5$) using the very low-solidity interpolation yields a pressure rise of 0.97 in. wg, only slightly higher than that obtained with the "standard" correlations. The new values are shown in Figure 8.10 to form the dashed line to the limit of zero pressure rise for $N_B = 0$ (no blades, zero solidity).

8.6 Diffusion Limitations and Selection of Solidity

In the models for deviation in axial cascades (Equations 8.4, 8.5, and 8.19), deviation is a function of ω_1, σ, and ϕ_c. These rules seem to imply that a low value of σ can simply be compensated by increasing the amount of blade camber, ϕ_c, to meet the requirements on flow turning, θ_{fl}. Unfortunately, it is really not that simple because a more detailed consideration reveals that the allowable diffusion depends, in part, on the solidity of the cascade. In Chapter 6, the de Haller ratio and the diffusion analogy established the rule that maintaining $V_2/V_1 > 0.72$ would yield an acceptable blade loading level,* at least for a solidity value of the order of one. Although it was indicated that increasing solidity could allow some adjustment of the minimum V_2/V_1 value, the idea was not developed much further than a simple correlation based on the "diffuser chart."

It is possible to develop a more realistic diffusion limit by making a better estimate of velocity, V, at the blade surface (Johnsen and Bullock, 1965). Figure 8.11 shows the surface velocity distribution for a typical airfoil. It can be seen that diffusion (actually, fluid deceleration) occurs most sharply on the low pressure "suction" surface. Denoting the maximum value of V on the suction surface of a blade (or vane) as V_p (p for "peak"), an alternate diffusion parameter can be defined as

$$D_p \equiv \frac{(V_p - V_2)}{V_p} = 1 - \frac{V_2}{V_p}. \tag{8.33}$$

D_p is called the *NACA local diffusion parameter*.

The theoretical potential flow velocity distributions on the airfoil surface can be used to estimate the boundary layer relative momentum thickness Θ_c/c as a function of D_p, employing any standard two-dimensional boundary layer calculation procedure (White, 2005). Θ_c/c is the ratio of the momentum thickness of the boundary layer at the trailing edge of the blade suction surface to the blade chord length; it is a measure of blade wake thickness and momentum loss (and hence, drag) due to viscous effects on the blade flow.

* Recall that $V = W$ for moving blades and $V = C$ for fixed vanes.

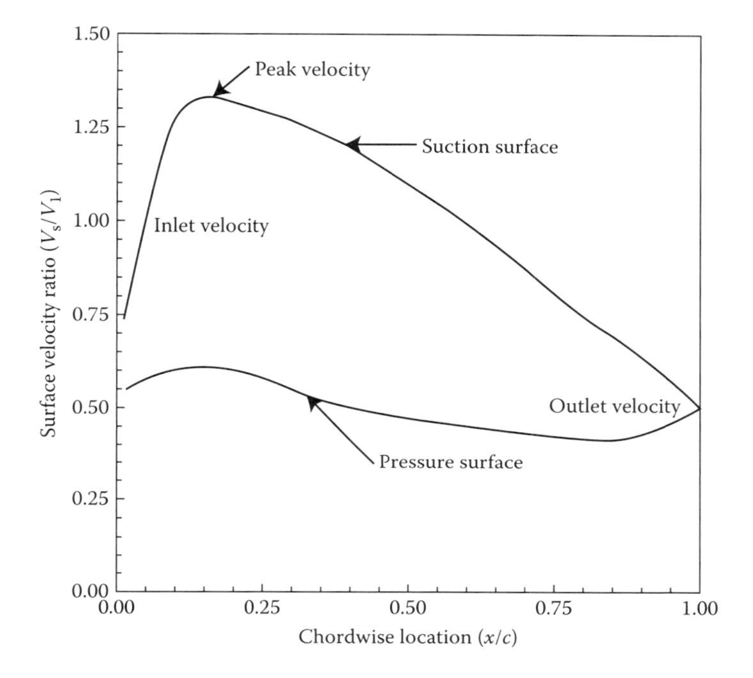

FIGURE 8.11 Blade surface velocities showing peak suction surface velocity for local diffusion model.

Calculated data showing Θ_c/c as a function of D_p for the NACA 65-Series airfoils are summarized in Figure 8.12. As seen in the figure, Θ_c/c increases monotonically as D_p increases. When D_p reaches a value of slightly above 0.5 (i.e., when V_2/V_p becomes <0.5), the magnitude of Θ_c/c begins to increase rapidly. This sudden change indicates flow separation from the blade suction surface. The blade cascade experiences stall and drag, and total pressure loss increases rapidly as well. Clearly, operation of the cascade with $D_p < 0.5$ would not be advisable.

The D_p-based concept is a clear and useful illustration of the viscous flow behavior, and of considerable utility in establishing diffusion limits, but calculation of D_p from potential flow analysis and Θ from boundary layer analysis is a nontrivial requirement. Seeking a simpler approach, Lieblein (1956) developed an approximate form for V_p such that

$$V_p \approx V_1 + \frac{\Delta V_\theta}{2\sigma}, \tag{8.34}$$

where $\Delta V_\theta/2$ is the mean circulation velocity, additive on the suction surface and subtractive on the pressure side. The magnitude of ΔV_θ can be extracted from a velocity diagram for the blades as shown in Figure 8.13. This velocity

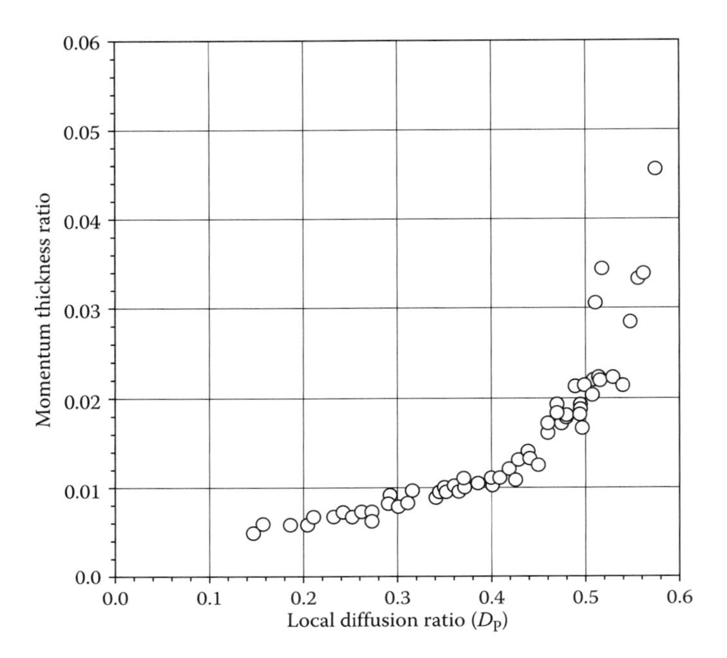

FIGURE 8.12 Momentum thickness variation with local diffusion parameter (using calculated values of velocity and momentum thickness).

difference is, qualitatively speaking, the source of the difference in pressure between the two surfaces illustrated in Figure 8.11.

Lieblein defined a new form for the diffusion factor as

$$D_L = 1 - \frac{V_2}{V_1} + \frac{\Delta V_\theta}{2V_1\sigma}. \tag{8.35}$$

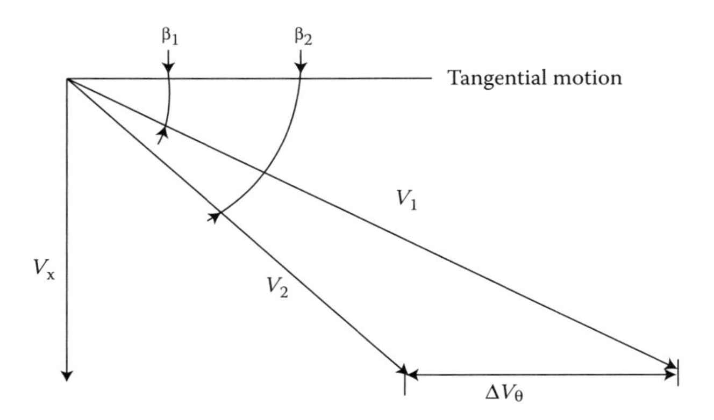

FIGURE 8.13 Velocities used in Lieblein's diffusion model.

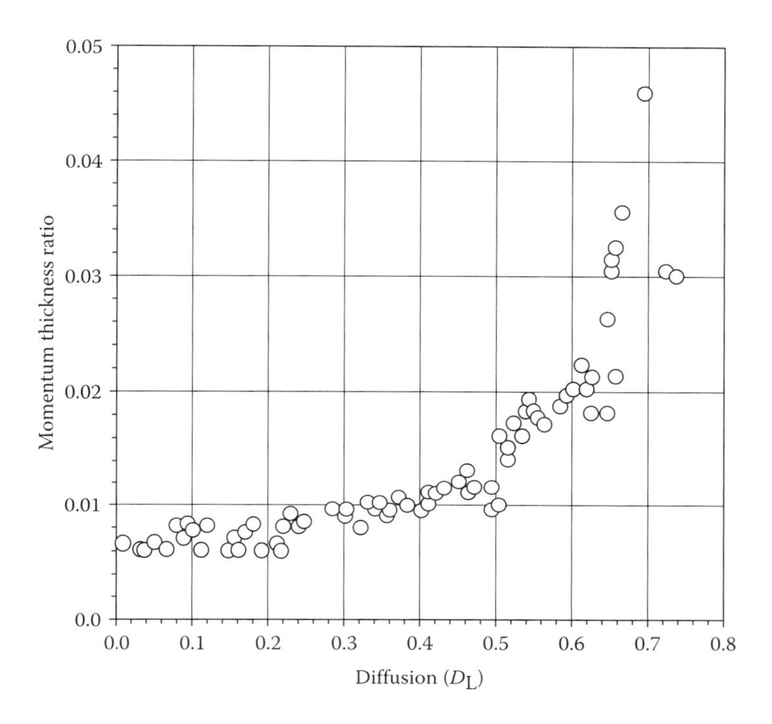

FIGURE 8.14 Momentum thickness variation with the Lieblein diffusion factor (correlation of experimental data).

He then correlated existing, extensive cascade data for Θ/c with D_L. Results are presented in Figure 8.14, which shows excellent collapse of the data. Significantly, one observes a gradual increase of Θ/c as D_L increases up to about $D_L = 0.6$, where Θ/c has increased by about fourfold. Beyond $D_L \approx 0.6$, Θ/c increases very rapidly and shows greater scatter, indicating the onset of blade stall. Lieblein's results can be used as a solid guideline as to what constitutes an acceptable level of blade loading; namely, D_L should clearly be <0.6, and for reasonably conservative design practice, the value of D_L should not exceed the range 0.45–0.55. For relatively high levels of blade solidity, the last term in Equation 8.35 becomes small and D_L reduces to one minus the (or $1 - 0.72 = 0.28$) de Haller ratio, the previously employed criterion for maximum blade loading.

If a maximum level of D_L that can be allowed as a design limit, $D_{L,max}$, is chosen, then the required minimum value of solidity can be determined as

$$\sigma_{min} = \frac{\Delta V_\theta / 2V_1}{(V_2/V_1) - (1 - D_{L,max})}. \tag{8.36}$$

Determination of minimum solidity can be illustrated by considering an axial flow fan blade section (recall $V = W$ for a moving blade) with $C_{\theta 1} = 0$,

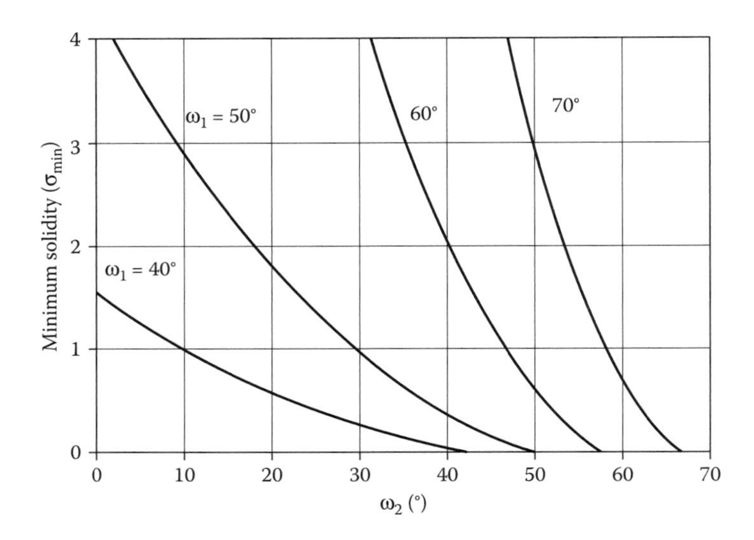

FIGURE 8.15 Variation of minimum solidity with ω_1 and ω_2.

$C_{\theta 2} = 12\,\text{m/s}$, $U = 35\,\text{m/s}$, and $C_x = 10\,\text{m/s}$. For these values, $W_1 = 36.4\,\text{m/s}$ and $W_2 = 25.1\,\text{m/s}$, and $W_2/W_1 = 0.690$. Then, specifying $D_{L,max} = 0.5$ yields

$$\sigma_{min} = \frac{12/2 \times 36.4}{0.69 - (1 - 0.5)} = 0.868.$$

This is a moderate value of solidity, which is consistent with a fairly conservative value of the de Haller ratio.* Choosing $D_{L,max}$ as 0.45 ($\sigma_{min} = 1.18$) would provide a more conservative level of design by allowing a more substantial margin away from stall conditions. The penalty for this conservatism would be a heavier, more expensive machine (larger blade count or longer blades).

Using the geometry of the velocity diagrams, an alternate equation for the minimum solidity in terms of the flow angles and minimum diffusion parameter can be derived from Equation 8.36:

$$\sigma_{min} = \frac{\cos\omega_1}{2}\frac{\tan\omega_1 - \tan\omega_2}{(\cos\omega_1/\cos\omega_2) - (1 - D_{L,max})}. \tag{8.37}$$

Figure 8.15 shows typical results for minimum solidity as a function of the outlet flow angle with the inlet flow angle as parameter, calculated using Equation 8.37. The value of $D_{L,max}$ is 0.5, a "moderately conservative" value.

These results show that, for large ω_1, the value of ω_2 does not vary much from ω_1; that is, the flow turning θ_{fl} ($=\omega_1 - \omega_2$) is restricted to small values if reasonable diffusion levels are to be maintained, even if very high values of

* The earlier correlation, Equation 6.17, would imply $\sigma = 1.16$.

solidity are used. On the other hand, at smaller ω_1, the diffusion level is not nearly as significant a constraint, even for very large values of θ_{fl} (small ω_2). At smaller ω_1, the required minimum solidity takes on much smaller values. To put this behavior in perspective, note that large values of ω_1 correspond to small values of C_x/U ($\omega_1 = \tan^{-1} C_x/U$),that is, low flow rate and low specific speed (for an axial machine). Smaller ω_1 corresponds to larger C_x/U, that is, higher flow rate and specific speed.

If ω_1 is fairly large and, in addition, the desired θ_{fl} is high, one is attempting to obtain a high pressure rise at a low flow rate and can expect to have difficulty laying out an acceptable axial cascade to do the job—a radial flow machine would be more suitable. This difficulty is consistent with previous experience using the Cordier guidelines and arriving at unacceptable de Haller ratios (Section 6.6). If an axial flow machine is desired, one will be forced into using unacceptably large solidity to try to keep the diffusion level under control. The problem can be alleviated somewhat by selecting a very large hub ($d/D \approx 1$) to force C_x higher and ω_1 lower. This would result in increased viscous losses and a high level of velocity pressure through the blade row, resulting in low static efficiency.

Finally, note that, in the process of design, the solidity can be chosen directly, perhaps because of geometric constraints such as a linearly tapered blade. It is, however, very important to calculate the diffusion level that results from the chosen solidity to make sure that the design will be below the stall limitation.

8.7 Losses in Diffusing Cascades

Thus far, the only method that has been used to relate the work done on or by the fluid and the useful energy change (head or pressure change) has been to multiply or divide by the efficiency. Information on efficiency has not been related to any specific characteristics of the actual machine layout; the efficiency was given, assumed, or estimated from the Cordier correlations. While this practice resulted in a degree of realism in calculations, it is not rigorous nor does it relate directly to the parameters of the actual machine being designed or analyzed. A more detailed analysis must be based on the determination of the various losses that occur in the flow through the machine. These losses, which are typically expressed as a loss in stagnation pressure, must be related to the actual machine geometry and flow. Accurate prediction of these losses is highly complicated, requiring modeling of complex, 3D, viscous (usually turbulent) flow. Here, losses will be considered in a fairly general fashion, and only approximate, semiempirical models will be presented.

Koch and Smith (1976) conducted a generalized study of losses in axial flow compressors; their work is also applicable to axial flow fans. They identified four significant types of loss for compressors:

- Blade and vane profile losses due to surface friction and diffusion
- Losses associated with the end-wall boundary layers and blade end clearances
- Losses due to shocks within the blade passages
- Drag losses due to part-span shrouds and supports.

In general, all of these losses must be considered for compressors, but only the first two are significant in the preliminary design of low-speed axial fans or pumps, which will be the focus here. Strictly speaking, only profile and shock losses are associated with 2D cascades, while end-wall and shroud/support losses arise from 3D effects and interaction between a rotating impeller and the stationary casing. All are listed here for completeness although only cascade profile losses will be considered in detail.

Profile losses that occur in a blade cascade are a result of boundary layer development along the blade surfaces. Profile losses for a specific blade and vane elements can be estimated on the basis of the blade loading work of Lieblein (1957), as expanded by Koch and Smith (1976). The technique applies boundary layer theory to conventional blade cascades in order to relate the profile losses with the conventional boundary layer parameters of blade trailing edge momentum thickness and form factor.* The designer estimates Lieblein's equivalent diffusion parameter, D_{eq}, as modified by Koch and Smith:

$$D_{eq} = \left(\frac{\cos\omega_1}{\cos\omega_2}\right) 1.12 + 0.0117\, i^{1.43} + 0.61 \left(\frac{\cos^2\omega_1}{\sigma}\right)(\tan\omega_1 - \tan\omega_2). \quad (8.38)$$

This parameter is then used to estimate the blade-outlet momentum thickness from Equation 8.39 and the blade trailing edge form factor from Figure 8.16

$$\frac{\Theta}{c} \approx 0.00210 + 0.00533\, D_{eq} - 0.00245\, D_{eq}^2 + 0.00158\, D_{eq}^3. \quad (8.39)$$

Both these relations are based on a nominal blade chord Reynolds number of 1.0×10^6 and an inlet Mach number of 0.05. The blade-outlet momentum-thickness and trailing edge form factor values must be corrected to account for conditions other than these. Corrections are obtained from Figures 8.17 and 8.18. Figure 8.17 presents the variation of blade-outlet momentum-thickness ratio (lower half of the figure) and trailing edge form factor (upper half of the figure) as functions of inlet Mach number and the equivalent diffusion factor. Figure 8.18 presents the variation of blade-outlet momentum thickness as a function of blade chord Reynolds number. After the designer has determined

* The definitions of these parameters are as follows: Momentum thickness $\Theta \equiv \int_0^\delta u/V[1-(u/V)]\,dy$; displacement thickness $\delta^* \equiv \int_0^\delta (1-(u/V)\,dy$; form factor $H \equiv \delta^*/\Theta$; where δ is the boundary layer thickness.

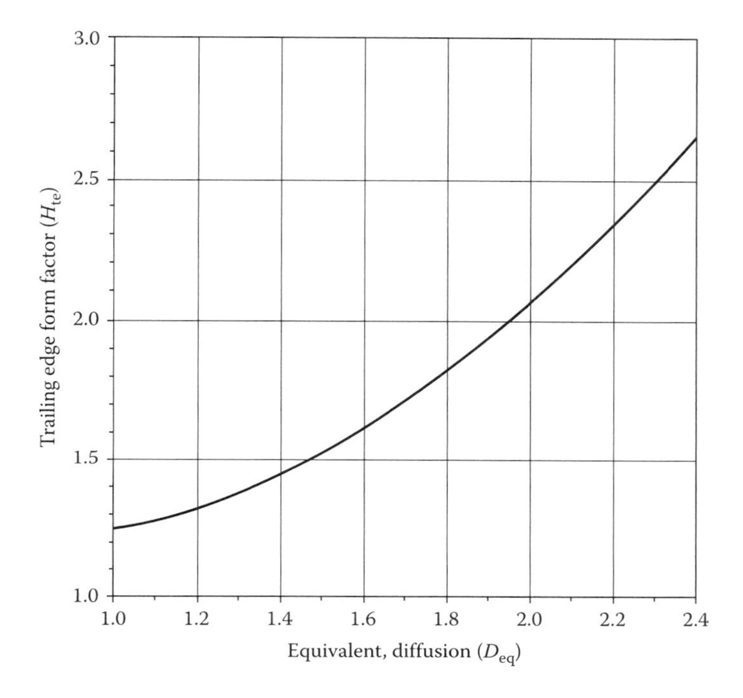

FIGURE 8.16 Trailing edge form factor as a function of equivalent diffusion parameter. (From Koch, C. C. and Smith, L. H., Jr. (1976), "Loss sources and magnitudes in axial-flow compressors," *ASME Journal of Engineering for Power*, 98(3), 411–424. With permission.)

the boundary layer parameters, the loss in total pressure for the blade or vane section is calculated using (Lieblein and Roudebush, 1956; Koch and Smith, 1976)

$$\zeta_{\text{blade}} = \frac{\Delta p_T}{\rho V_1^2/2} = \frac{2(\Theta/c)(\sigma/\cos\omega_2)(\cos\omega_1/\cos\omega_2)^2(2/[3-(1/H)])}{1-(\Theta/c)(\sigma H/\cos\omega_2)^3}. \quad (8.40)$$

The net pressure loss across the blade row is determined by performing a mass-averaged integration of the individual blade element pressure loss distribution across the entire blade row.

If outlet guide vanes are used to recover the rotor exit swirl velocity, the profile losses for the vane row are calculated in the same manner as the blade profile losses. In designs where outlet guide vanes are not employed, the discharge swirl velocity comprises an additional loss in total pressure. This loss is determined using a mass-averaged integration of the tangential kinetic energy distribution. The swirl kinetic energy can be the most significant loss in a fairly heavily loaded design without outlet guide vanes.

The loss induced by the hub and casing boundary layers and tip-clearance effects is not considered in detail here. Typically, such losses contribute significantly to the overall reduction in axial flow turbomachine performance

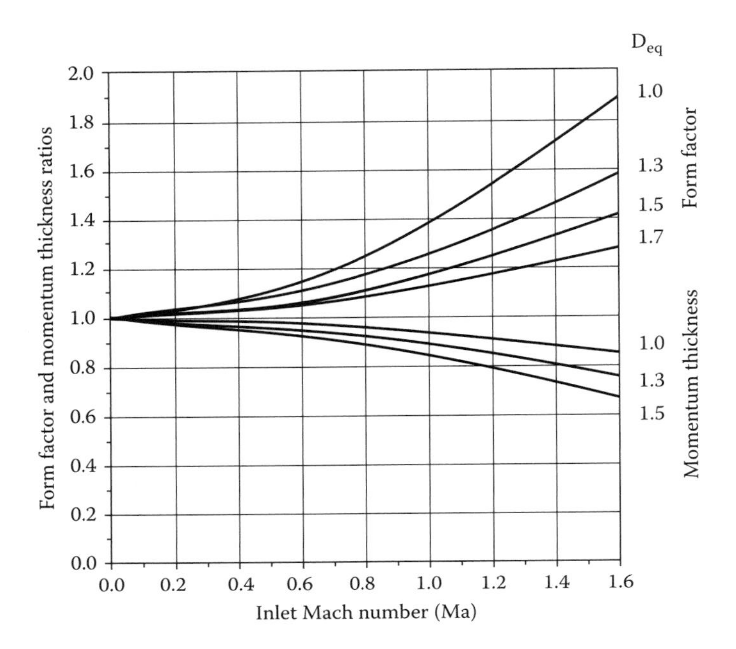

FIGURE 8.17 Influence of Mach number on the form factor and momentum thickness. (From Koch, C. C. and Smith, L. H., Jr. (1976), "Loss sources and magnitudes in axial-flow compressors," *ASME Journal of Engineering for Power*, 98(3), 411–424. With permission.)

(Hirsch, 1974; Horlock, 1958; Mellor and Wood, 1971). Extensive studies indicate that thick, irregular end-wall boundary layers can lead to the inception of stall (McDougall et al., 1989). Several techniques have been developed to estimate these losses (Lakshminarayana, 1979; Comte et al., 1982; Hunter and Cumpsty, 1982; Wright, 1984d). The semiempirical method of Koch and Smith (1976) is based on classical boundary layer theory and is relatively simple to use. The reader is referred to their paper for details.

Having estimated the losses, efficiency is calculated as the ratio of the reduced total pressure rise (ideal minus losses) to the ideal value. One must, of course, make sure to refer all stagnation pressure losses to the same dynamic pressure. Here, the dynamic pressure corresponding to the blade speed is used. The (dimensionless) reduced total pressure rise is

$$\frac{\Delta p_T}{\rho U^2} = \frac{w}{U^2} - \zeta_{blade}\frac{C_1^2}{2U^2} - \zeta_{vane}\frac{C_2^2}{2U^2} - \zeta_{end}\frac{C_1^2}{2U^2}, \tag{8.41}$$

where w is the (mass-averaged) specific work (calculated from the Euler equation) and ζ_{end} is a loss coefficient representing end-wall boundary layer and tip-clearance effects. The total efficiency is then

$$\eta_T = 1 - \zeta_{blade}\frac{C_1^2}{2w} - \zeta_{vane}\frac{C_2^2}{2w} - \zeta_{end}\frac{C_1^2}{2w}. \tag{8.42}$$

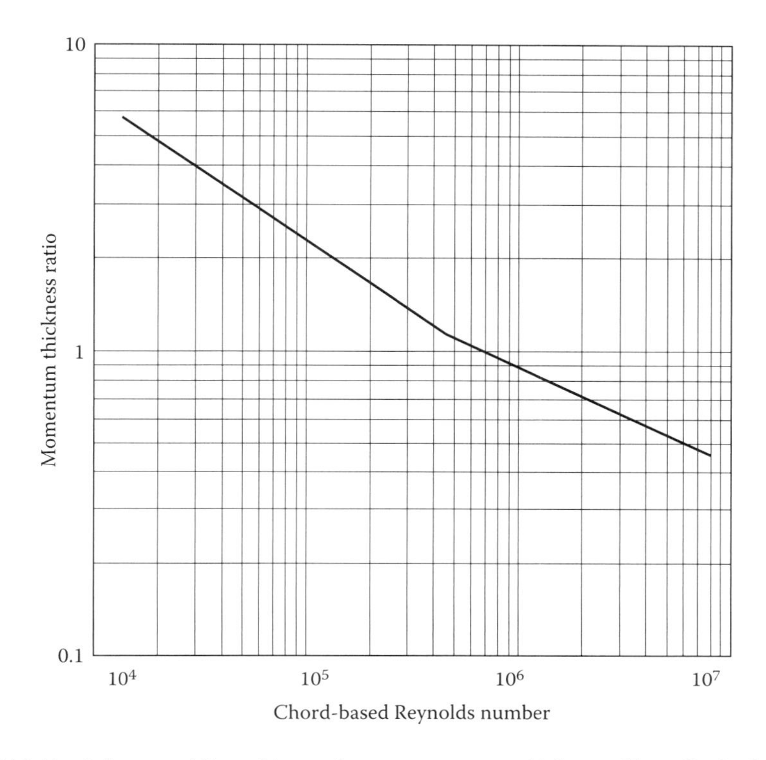

FIGURE 8.18 Influence of Reynolds number on momentum thickness. (From Koch, C. C. and Smith, L. H., Jr. (1976), "Loss sources and magnitudes in axial-flow compressors," *ASME Journal of Engineering for Power*, 98(3), 411–424. With permission.)

Static efficiency is estimated by

$$\eta_s = \eta_T - \frac{C_3^2}{2w} = \eta_T - \frac{C_{3\theta}^2 + C_{3x}^2}{2w}, \tag{8.43}$$

where $C_{\theta 3}$ is the unrecovered swirl velocity averaged across the annulus. For full outlet guide vane swirl recovery, the value of $C_{\theta 3}$ is, of course, zero.

8.8 Axial Flow Turbine Cascades

All of the material discussed thus far in this chapter applies only to diffusing (compressors, fans, and pumps) cascades. Similar consideration, although not nearly as extensive, is needed for accelerating (turbine) cascades.

In turbine applications, one does not have to deal with diffusion limitations, but the phenomena of incidence, deviation, and cascade profile losses must be addressed to maintain design accuracy. Also, simple rules for selecting solidity are needed. The first and simplest consideration is incidence. In

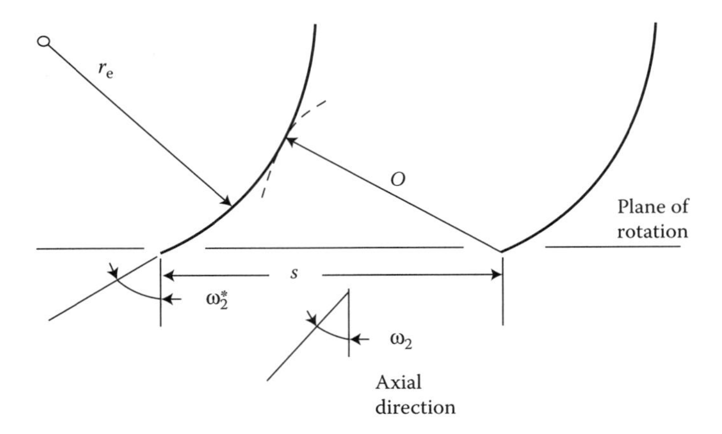

FIGURE 8.19 Turbine cascade parameters for the Ainley–Mathieson deviation model.

turbine cascades, the optimum incidence angle is usually taken as zero ($i = 0$); in other words, the blade is usually designed and set so that the inlet velocity vector is tangent to the camber line at the leading edge.

Next, consider the deviation. A fairly simple method by Ainley and Mathieson (1951) [as modified by D. G. Wilson (1984)] can be used to estimate blade trailing edge angle requirements in both blade and nozzle rows for axial flow turbines. It appears to be useable for hydraulic, gas, and steam turbines for Mach numbers up to 1, and it is based on geometric arguments, tempered by comparisons with experimental data from actual turbine testing. For purposes of illustration, the "rule" can be simplified to a mean camber line analysis. As shown in Figure 8.19, the calculation of the actual fluid outlet angle is expressed in terms of an opening ratio, o/s, and a curvature ratio, r_e/s. The effective outlet angle or fluid angle,* ω_2, is given by

$$\omega_2 = \left[\left(\frac{7}{6}\right)\left(\cos^{-1}\left(\frac{o}{s}\right) - 10^\circ\right) + 4^\circ\left(\frac{s}{r_e}\right)\right], \quad \text{for } Ma_2 \leq 0.5, \tag{8.44}$$

$$\omega_2 = \left[\cos^{-1}\left(\frac{o}{s}\right) - \left(\frac{s}{r_e}\right)^{1.787+4.128(s/r_e)}\right]\sin^{-1}\left(\frac{o}{s}\right), \quad \text{for } Ma_2 = 1.0. \tag{8.45}$$

Angles for Mach numbers between 0.5 and 1.0 are found by interpolation. Denoting the value for $Ma_2 = 0.5$ as $(\omega_2)_{0.5}$ and the sonic value as $(\omega_2)_{1.0}$, estimates for Mach numbers between 0.5 and 1.0 are made by

$$\omega_2 = (\omega_2)_{0.5} - (2Ma_2 - 1)\left[(\omega_2)_{0.5} - (\omega_2)_{1.0}\right], \quad \text{for } 0.5 \leq Ma_2 \leq 1.0. \tag{8.46}$$

* Recall that the angle ω is referred to the axial direction; it is the complement of the angle β (for the relative velocity vector, W) or α (for the absolute velocity vector, C) that is referred to the blade (tangential) direction that was used to lay out velocity diagrams in Chapters 6 and 7.

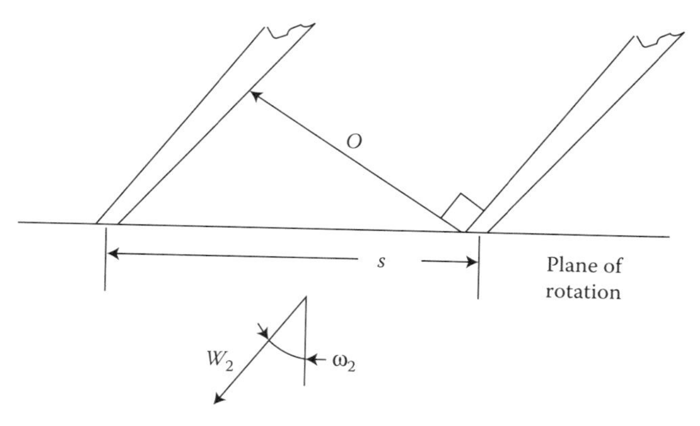

FIGURE 8.20 Turbine nozzle geometry: flat blades in the trailing edge region.

The absolute value of the calculation must be used.

These correlations are often simplified by assuming that the blades are flat near the trailing edge (Figure 8.20). This drives r_e to very large values, reducing the flow angle equation (for $Ma_2 \le 0.5$) to

$$\omega_2 = \left(\frac{7}{6}\right)\left[\cos^{-1}\left(\frac{o}{s}\right) - 10^\circ\right]. \tag{8.47}$$

From Figure 8.20, $\cos^{-1}(o/s) = \omega_2^*$, the blade metal angle. The equation can be reformulated in terms of deviation as

$$\delta = 10^\circ - \frac{\omega_2^*}{6}. \tag{8.48}$$

The relationship between blade and fluid angles is shown in Figure 8.21.

As an unexpected bonus, the Ainely–Mathieson-Wilson method can be used to lay out IGVs for pumping machines when preswirl of the incoming fluid is desired (i.e., $C_{\theta 1} \neq 0$).

The remaining major blade layout parameter, the cascade solidity, is not strictly prescribed, as there is no diffusion limitation. The choice of solidity can be adequately constrained by holding the blade lift coefficient to some upper limit (Zweifel, 1945). From momentum considerations (for a constant axial velocity), the blade lift coefficient is related to the flow angles by

$$C_L = \left(\frac{2\cos\lambda}{\sigma}\right)\cos^2\omega_2(\tan\omega_2 - \tan\omega_1)$$

giving

$$\sigma = \left(\frac{2\cos\lambda}{C_L}\right)\cos^2\omega_2(\tan\omega_2 - \tan\omega_1). \tag{8.49}$$

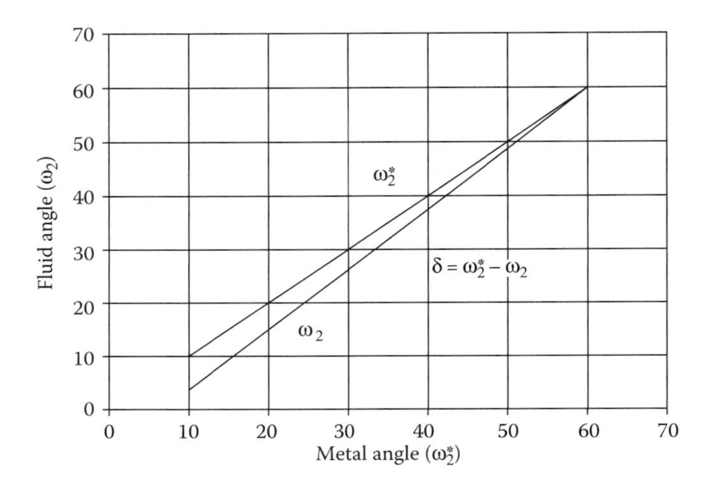

FIGURE 8.21 Trailing edge angles and deviation in a turbine cascade.

Fixing C_L to an upper limit yields an initial choice of solidity and hence the blade or nozzle chord when spacing is given (recall $\sigma = c/s$). Zweifel suggests $C_{L,max}$ approximately equal to 0.8, while Wilson recommends $C_{L,max} = 1.0$ as a less conservative value. Typical numbers for an inlet nozzle might be $\omega_1 = 0°$ and $\omega_2 = 65°$. With $C_{L,max} = 1.0$, one obtains $\sigma = \cos(\omega_2/2)$ so that $\sigma = 0.84$. For other realistic values of ω_2, σ would range from near 1.0 to perhaps 0.75. For low reaction blading, the required solidity increases substantially. With, say $\omega_1 = 30°$, $\omega_2 = -60°$ (90° turning), and $C_{L,max} = 1.0$, one requires $\sigma = 1.5$. Clearly, these much harder working blades need more solidity for reasonable loading.

A variety of methods exist for estimating the profile losses for accelerating cascades. The method of Soderberg (Soderberg, 1949) is reasonably simple and accurate for initial design. The method employs loss coefficients based on the *exit velocity* from a blade row. Also, because turbine blade rows often operate with a significant energy change in a compressible fluid, losses are evaluated in terms of enthalpy, that is, "loss" $= h_2 - h_{2s}$, where h_2 is the actual enthalpy at the blade row exit and h_{2s} is the enthalpy that would result if the flow through the blade row were isentropic. If the flow is nearly incompressible, then $h_2 - h_{2s} \approx \Delta p_{T,Loss}/\rho$.

The loss across a blade or vane row is given by

$$h_2 - h_{2s} = \zeta \times \frac{V_2^2}{2}, \tag{8.50}$$

where V_2 is either W_2 or C_2, depending on whether a blade or vane (nozzle) row is considered. The loss coefficient is calculated as follows. First, calculate

ζ^* from

$$\zeta^* = 0.4 + 0.6\left(\frac{\theta^\circ_{\text{fl}}}{100}\right)^2. \tag{8.51}$$

ζ^* is the loss coefficient for a blade row of aspect ratio, L_B/b, of 3 and Reynolds number, based on hydraulic diameter $Re = 10^5$. L_B is the blade height, hub to tip, b is the cascade width in the axial direction, $b = c\cos\lambda$, and $Re = \rho_2 V_2 D_H/\mu_2$, where $D_H = 2sH\ \cos\omega_2^*/(s\ \cos\omega_2^* + H)$. Then correct the loss coefficient using

$$\zeta = \left(\frac{10^5}{Re}\right)^{1/4}\left[(1+\zeta^*)f\left(\frac{b}{L_B}\right) - 1\right], \tag{8.52}$$

where

$$f\left(\frac{b}{L_B}\right) = 0.0993 + 0.021\frac{b}{L_B}, \quad \text{for nozzles} \tag{8.53a}$$

and

$$f\left(\frac{b}{L_B}\right) = 0.0975 + 0.075\frac{b}{L_B}, \quad \text{for rotors.} \tag{8.53b}$$

Excluding end-wall and tip-clearance losses, the total efficiency for a turbine stage can be calculated by

$$\eta_{\text{T,No endloss}} = \frac{w}{w + \text{Losses}} = \left(1 + \zeta_{\text{Nozzle}}\frac{C_2^2}{2w} + \zeta_{\text{Rotor}}\frac{W_3^2}{2w}\right)^{-1}.$$

According to Dixon (Dixon, 1998), end-wall and tip-clearance effects can be included approximately by the simple expedient of multiplying this efficiency value by the ratio of blade area to total area (i.e., blade area + nozzle area + casing area + hub area), so the complete stage total efficiency is

$$\eta_{\text{T,Stage}} \approx \left(\frac{A_{\text{Blades}} + A_{\text{Nozzles}}}{A_{\text{Blades}} + A_{\text{Nozzles}} + A_{\text{Casing}} + A_{\text{Hub}}}\right) \times \left(1 + \zeta_{\text{Nozzle}}\frac{C_2^2}{2w} + \zeta_{\text{Rotor}}\frac{W_3^2}{2w}\right)^{-1}. \tag{8.54}$$

8.9 Radial Flow Cascades

Having considered axial flow cascades rather extensively, turn now to radial flow. For the layout of a radial flow turbomachine, one must consider essentially the same issues as for axial flow; namely, "incidence" (the alignment of the inlet flow with the leading edge of a blade), "deviation" (the difference between the flow direction and the blade direction at the trailing edge of a blade), "solidity" (the ratio of blade spacing and blade length), and various issues affecting losses such as diffusion ratio and blade boundary layer development. It is obvious that a radial cascade is very different from an axial cascade because of the radial geometry as opposed to the planar geometry, but there are other differences as well. In most cases, radial cascades are not formed from airfoils. Although there have been several experimental studies of radial-blade systems, there is nothing like the systematic wind tunnel investigations of Emery et al. As a result, most design calculations are based on models derived from theory or numerical studies. Rarely does one consider a static radial cascade; studies and models almost always address the rotating cascade (i.e., the impeller of a pump, fan, compressor, or turbine).

Following the pattern established for axial cascades, the first and simplest consideration is incidence. No simple model to predict optimum incidence angles is known to the authors. It is recommended that the leading edge of the blade be matched to the blade-relative velocity vector at the blade row entrance, based on uniform radial velocity into the cascade; that is, the incidence angle is taken as zero. Although uniformity of radial velocity is a relatively simplistic approach, refinement beyond this assumption requires a detailed knowledge of the three dimensionality of the flow passing into the impeller inlet and approaching the blade leading edge. A rudimentary preliminary design analysis can be carried out with the simple assumption stated here. In the next chapter, this restriction is alleviated somewhat by the introduction of approximate Q3D analysis for estimating the distribution of the flow across the blade inlet (in the axial direction). To obtain a truly optimum design, the geometry developed for a centrifugal impeller using such preliminary design techniques should be refined by a computational fluid dynamics analysis of the impeller flow field or through an experimental development program, or both.

As found for the flow through axial cascades, incomplete guidance of the fluid leads to a reduction in the impeller outlet swirl, $C_{\theta 2}$, compared to an ideal value based on the assumption that the outlet relative velocity is tangent to the blade camber line. The actual value of $C_{\theta 2}$ is always less than the ideal value, denoted by $C^*_{\theta 2}$, as shown in Figure 8.22. In a radial flow cascade/impeller, this phenomenon is called *slip*. Slip is quantified by a *slip coefficient*, μ_E, defined as

$$\mu_E \equiv \frac{C_{\theta 2}}{C^*_{\theta 2}}. \tag{8.55}$$

There are many models, mostly based on numerical flow solutions and/or geometric arguments, to estimate μ_E. The more commonly used methods used to calculate slip include the method of Busemann (Busemann, 1928), which is used in the initial design and layout for blades with both large and

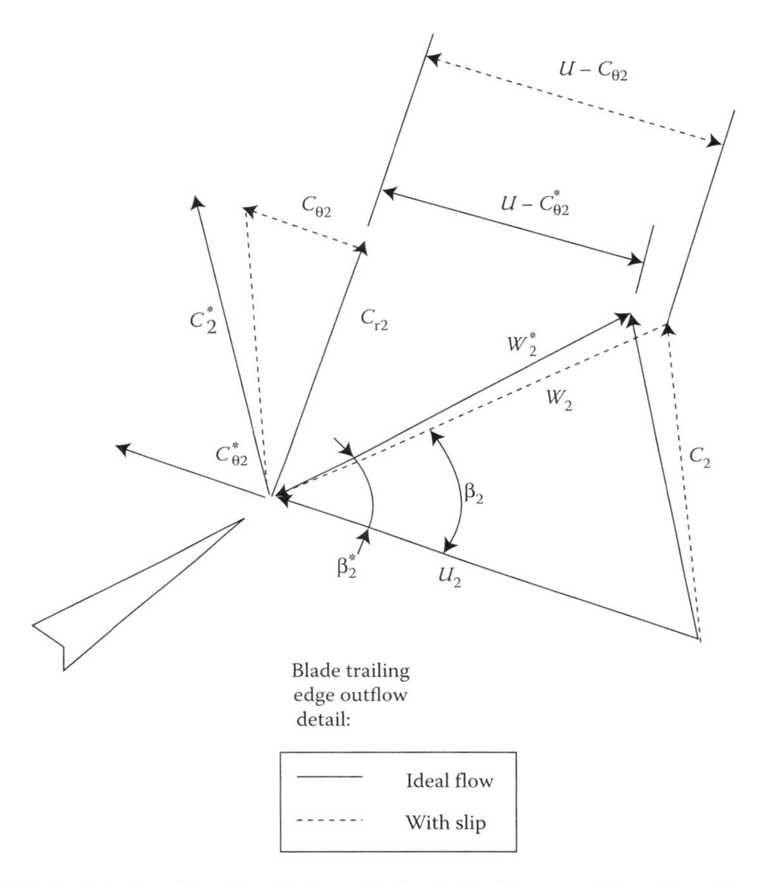

FIGURE 8.22 Actual and ideal (perfectly guided) velocity diagrams at impeller exit.

moderate discharge angles, in the range of $45^\circ \leq \beta_2^* \leq 90^\circ$. Another method used extensively in the compressor layout is the method by Stanitz (Stanitz, 1951), who used relaxation methods to solve the inviscid potential flow in radial cascades ($\beta_2^* = 90^\circ$). Stanitz' results, which are more accurate for higher values of β_2^*, are approximated by the equation

$$\mu_E \approx 1 - \frac{0.63\pi}{N_B} \quad \text{(Stanitz)}. \tag{8.56}$$

Balje (Balje, 1981) suggests a similar model for radial blades:

$$\mu_E \approx 1 - \frac{0.75\pi}{N_B} \quad \text{(Balje)}. \tag{8.57}$$

The method developed by Stodola (Stodola, 1927) is perhaps the oldest in common use. Consideration of the details of Stodola's model gives considerable insight into the mechanism of slip. The model is based on the concept of the "relative eddy," defined as the cylindrical fluid element with diameter ℓ that just fits between adjacent blades at the impeller exit (Figure 8.23). If the number of blades is very large and the blades are very thin, then all of the fluid in the impeller would be brought into a solid-body-like rotation with the impeller and every fluid particle would be rotating with the same angular velocity as the impeller, N. For a finite number of vanes, the fluid particles do not rotate at the impeller angular velocity; Stodola assumed that they do not rotate at all and therefore have angular velocity relative to the impeller of $-N$. The fluid particle at the outer edge of the relative eddy, just at the impeller exit, thus has a relative velocity opposite to the impeller velocity of $\Delta W_\theta = -N(\ell/2)$. From the diagram, $\Delta C_\theta = \Delta W_\theta$ and $\ell \approx (\pi D_2/N_B)\sin\beta_2^*$. Since $U_2 = ND/2$, the slip velocity is $\Delta C_\theta = -N(\pi D/2N_B)\sin\beta_2^* = \pi U_2 \sin\beta_2^*/N_B$. From the "ideal" diagram,

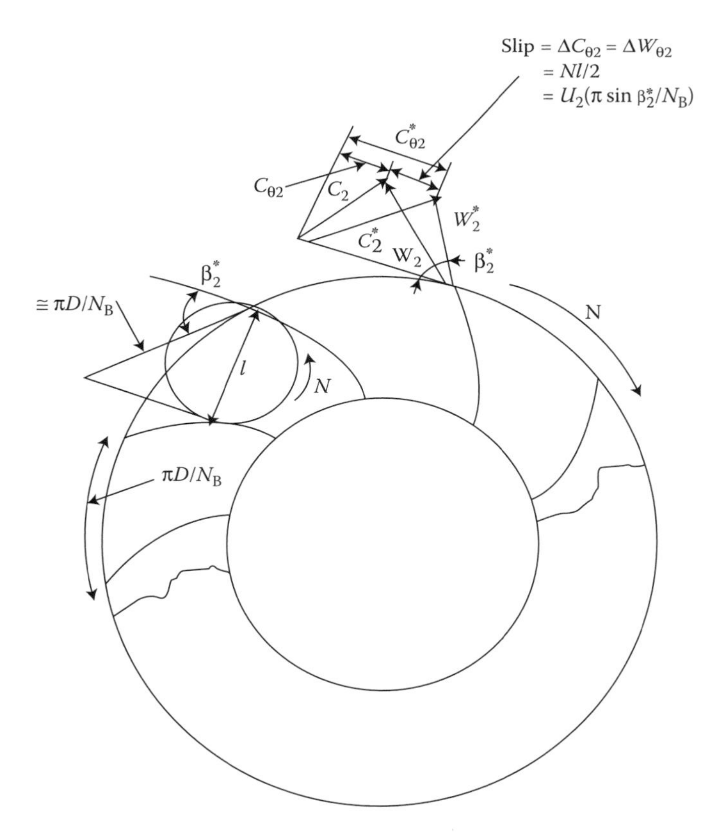

FIGURE 8.23 Details of Stodola's model for slip.

$C^*_{2\theta} = U_2 - W_{2\theta} = U_2 - C_{r2} \cot \beta^*_2$ so Stodola's slip coefficient is

$$\mu_E = \frac{C_{\theta 2}}{C^*_{\theta 2}} = 1 - \frac{(\pi \sin \beta^*_2)/N_B}{1 - ((C_{r2} \cot \beta^*_2)/U_2)} \quad \text{(Stodola)}. \tag{8.58a}$$

Stodola's formula is somewhat conservative (it tends to overestimate slip and hence μ_E is somewhat low), but it has the advantage of simplicity and it applies across the entire range of values of β^*_2; therefore, it is the one recommended here as a preferred choice in preliminary design blade layout for radial flow machines.

For moderate values of C_{r2}/U_2, Stodola's formula is often simplified to

$$\mu_E = 1 - \left(\frac{\pi \sin \beta^*_2}{N_B}\right) \quad \text{(Simplified Stodola)}. \tag{8.58b}$$

This form is most relevant for blades that are very flat in the trailing edge region. Finally, Wiesner (Wiesner, 1967) suggested a purely empirical formula that he showed matched Busemann's method fairly closely. Wiesner's formula is:

$$\mu_E = 1 - \frac{(\sin \beta^*_2)^{1/2}}{N_B^{0.7}}. \tag{8.59}$$

Figure 8.24a provides a comparison of the different methods for calculating μ_E for radial-bladed or radial-tipped compressors or blowers ($\beta^*_2 = 90°$). The methods of Stodola, Busemann, and Balje are in substantial agreement, while the Stanitz method provides the highest estimates for μ_E by about 5%. The Balje method is moderately conservative at blade numbers >10. Figures 8.24b and 8.24c compare the slip coefficient for lower blade angles 20° and 40°, respectively. The Balje and Stanitz methods are not shown as they are limited to larger blade angles. The methods of Busemann and Stodola are in excellent agreement for the lower range of blade angles when the blade number is greater than about eight. The Stodola formula, with its relatively simple form, is easy to use, and its accuracy is adequate for use in preliminary blade layout for centrifugal machines. Again, refinement of a final choice of layout, by laboratory development or detailed computational fluid dynamics analysis, would be required in an advanced phase of design.

Example: Effects of Slip on a Radial Pump Design

A water pump is to develop 30 m of head at a flow rate of 0.015 m^3/s. The pump is driven at 3000 rpm by a small gasoline engine. Investigate the effects of slip and number of blades on the trailing edge angle of the blades.

The given performance data yield a specific speed of $N_s = 0.54$. Cordier analysis gives $D_s = 5.3$, $D = 0.155$ m, and $\eta_T = 0.84$. d/D should be about 0.35 based on Equation 6.33. From this, $d = 0.055$ m, $U_1 = 8.62$ m/s, $U_2 = 24.3$ m/s, and

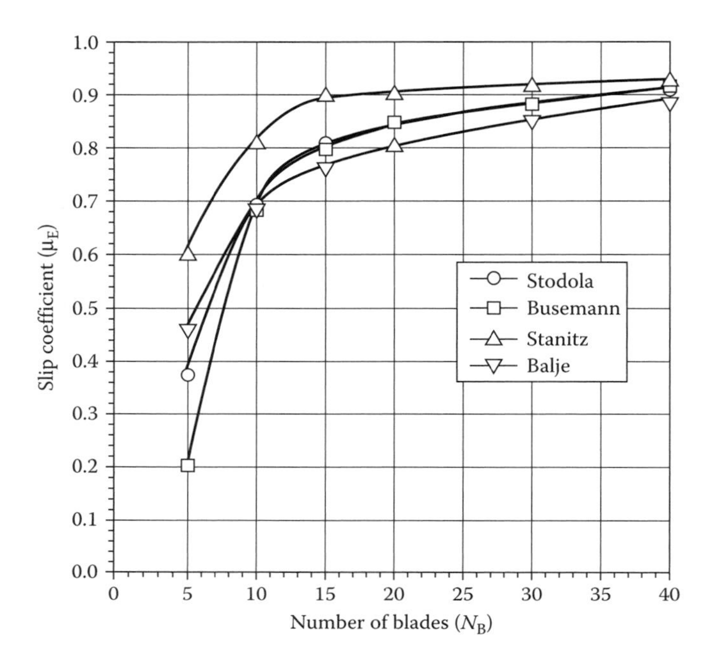

FIGURE 8.24a Slip coefficient comparison for radial blades ($\beta_2^* = 90°$).

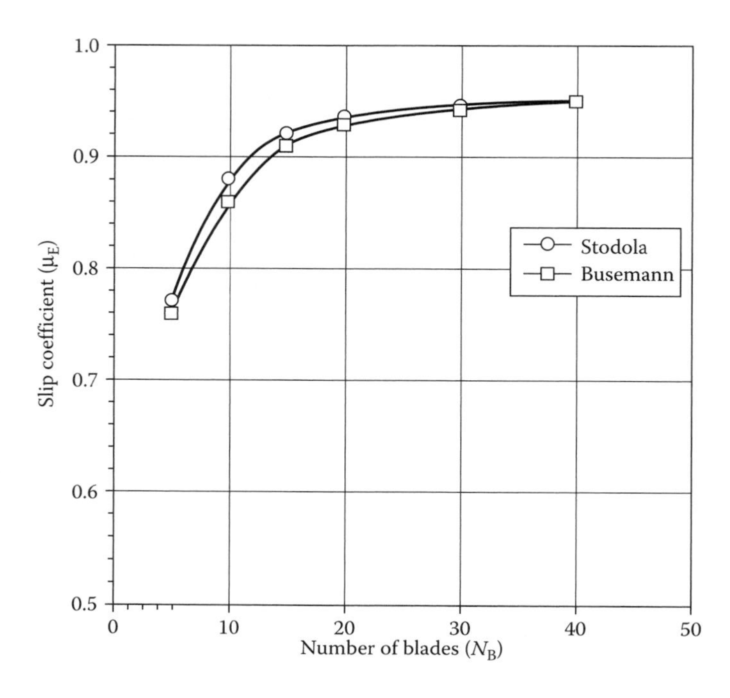

FIGURE 8.24b Slip coefficient comparison for $\beta_2^* = 20°$.

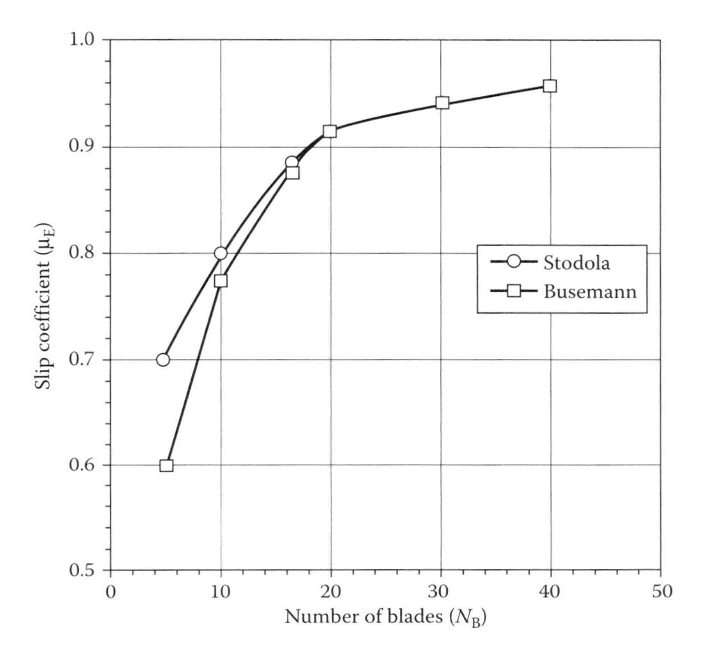

FIGURE 8.24c Slip coefficient comparison for $\beta_2^* = 40°$.

$C_{\theta 2} = gH/(\eta_T U_2) = 14.4$ m/s. If C_{r1} $(=C_1)$ is set to match the throat velocity, then $C_{r1} = Q/(\pi d^2/4) = 6.35$ m/s. Next, choose a reasonable value for $C_{r2} = 4.0$ m/s and calculate a conservative de Haller ratio of $W_2/W_1 \approx 1$. The values of C_{r2} and C_{r1} will size the inlet and outlet widths of the impeller. Summarizing the results

$$U_1 = 8.62 \text{ m/s}; \quad U_2 = 24.3 \text{ m/s};$$

$$C_{r1} = 6.35 \text{ m/s}; \quad C_{r2} = 4.0 \text{ m/s}; \quad C_{\theta 2} = 14.4 \text{ m/s}.$$

$C_{\theta 1}$ has been set to zero.

The relative velocity angle (angle between W_2 and U_2) is

$$\beta_2 = \tan^{-1}\left(\frac{C_{r2}}{U_2 - C_{\theta 2}}\right) = 22.0°.$$

If there were no slip, this would be the blade angle as well; however, it is necessary to correct the blade angle by using the slip coefficient. The blade angle required to produce the calculated exit velocities is

$$\beta_2^* = \tan^{-1}\left[\frac{C_{r2}}{U_2 - (C_{\theta 2}/\mu_E)}\right] = \tan^{-1}\left[\frac{4.0}{24.3 - (14.4/\mu_E)}\right].$$

Anticipating a rather small blade angle, Stodola's model (Equation 8.58a) is used to estimate the slip coefficient

$$\mu_E = 1 - \frac{\pi \sin \beta_2^*/N_B}{1 - [(C_{r2} \cot \beta_2^*)/U_2]} = 1 - \frac{\pi \sin \beta_2^*/N_B}{1 - [(4.0 \cot \beta_2^*)/24.3]}.$$

Because μ_E depends rather strongly on β_2^*, the calculation is iterative. Because the slip coefficient depends on the number of blades, the blade angle will depend on the number of blades one chooses. If 10 blades are chosen, one calculates $\mu_E = 0.758$ and $\beta_2^* = 37.0°$. That is a substantial change—15°—from 22.0° and is clearly non-negligible. Higher blade numbers will increase the slip coefficient and reduce the required angle for the blade. Figure 8.25 shows the effect of blade number on the slip coefficient and the blade trailing edge angle. The discharge fluid angle, β_2, is constant to meet the required head rise and is included in the figure for emphasis. It is important to note that the iteration for β_2^* does not converge for blade numbers less than eight; this means that, within the accuracy of the model, a pump with the selected speed, outer and eye diameters, and blade widths cannot produce the desired head and flow with less than eight blades.

The behavior of slip can be illustrated further by examining the head developed by a centrifugal machine as blades are added or removed. Once the blade angle, β_2^*, is fixed to some value, one can recalculate $C_{\theta 2}$ based on a different number of blades. Using the simplified Stodola rule (Equation 8.58b), the ratio of the new value, $C_{\theta 2,\text{new}}$, to the original value, $C_{\theta 2,\text{orig}}$, will be

$$\frac{C_{\theta 2,\text{new}}}{C_{\theta 2,\text{orig}}} = \frac{\mu_{E,\text{new}}}{\mu_{E,\text{orig}}} = \frac{1 - [(\pi \sin \beta_2^*)/N_{B,\text{new}}]}{1 - [(\pi \sin \beta_2^*)N_{B,\text{orig}}]}.$$

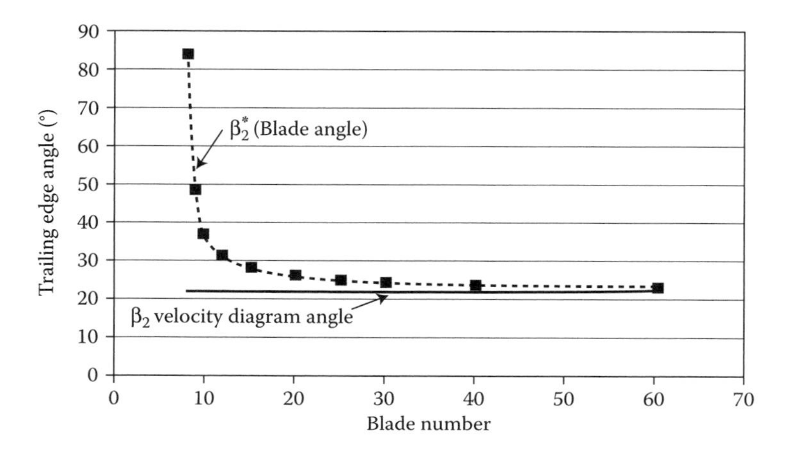

FIGURE 8.25a Blade angle and velocity angle, showing effects of slip.

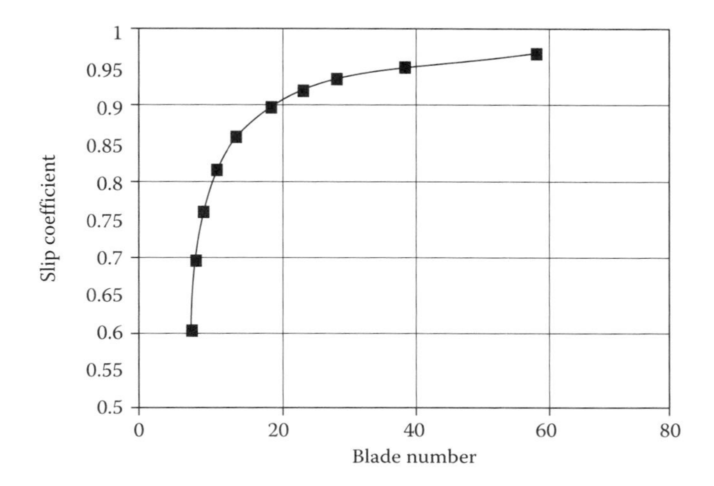

FIGURE 8.25b Effect of blade number on slip coefficient.

Following the previous example and using $N_{B,orig} = 40$ with $\beta_2^* = 23.6°$, and noting that efficiency and blade tip speed are to be held fixed, one obtains

$$\frac{H_{new}}{H_{orig}} = \frac{\eta_T U_2 C_{\theta 2,new}}{\eta_T U_2 C_{\theta 2,orig}} = \frac{\eta_T U_2 \mu_{E,new} C_{\theta 2}^*}{\eta_T U_2 \mu_{E,orig} C_{\theta 2}^*} = \frac{\mu_{E,new}}{\mu_{E,orig}} = 1.032 - \frac{1.299}{N_{B,new}}.$$

Table 8.2 gives the head ratios and the head in meters for blade numbers beginning at 8 and increasing to 40.

The results shown in Figure 8.25 and Table 8.2 indicate that the pump design should probably not be executed with more than 10–15 blades. The gains above this level are achieved by doubling or tripling the numbers of blades and do not justify the extra cost and complexity of the impeller layout for

TABLE 8.2

Variation of Head with Blade Number

N_B	H_{new}/H_{orig}	H (m)
8	0.87	26.09
9	0.89	26.63
10	0.90	27.06
12	0.92	27.71
15	0.95	28.36
20	0.97	29.01
25	0.98	29.40
30	0.99	29.66
40	1.00	30.00

the higher blade numbers. Additionally, a higher blade number would lead to increased losses from surface friction due to the larger surface area. If a head of 30 m were absolutely necessary, impeller diameters or speed should be adjusted.

8.10 Solidity of Centrifugal Cascades

The term "solidity" is not commonly used in centrifugal cascades nor is solidity a primary design variable used in blade layout or impeller design. As shown by the various formulas for slip coefficient, the number of blades becomes a very important parameter in determining both outlet angle and pressure rise. Because a finite number of blades must be arranged in a circular impeller, something like solidity must be considered. The solidity of a radial flow cascade can be defined, if one so chooses, in more than one way. The most common method is to define the solidity as the ratio of the angular segment subtended by the leading and trailing edges of a blade, ϑ_{Blade}, to the angular spacing between adjacent leading edges, ϑ_{LE} $(= 360°/N_B)$, as shown in Figure 8.26. A general rule in centrifugal cascade layout is to keep this angular solidity at about 1 or slightly greater.

As seen in the previous example, the use of a small number of blades leads to high slip and a small slip coefficient. The design pressure rise is then achieved by raising the blade-outlet angle to rather large values to overcome the inherently low turning. Unfortunately, substituting high blade turning angles for large blade densities can lead to a very heavily loaded blade configuration and, hence, to unnecessarily high total pressure losses from thick blade boundary layers or even separation. Using a greater number of blades

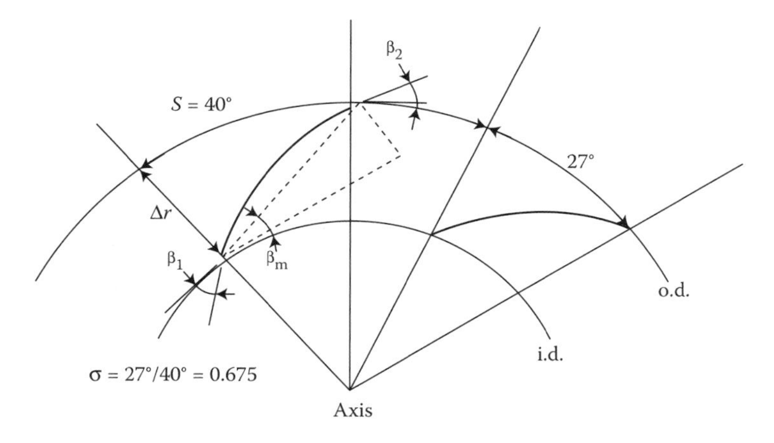

FIGURE 8.26 Geometry for defining approximate solidity in a centrifugal cascade.

reduces the requirement on blade turning, distributes the work to a larger number of blades, and leads to a higher angular solidity for the cascade. This might increase friction losses because of the larger surface area. Examination of typical centrifugal blade layouts will show that the slip coefficient is usually set above 0.85 so that a reasonably large number of blades are required and a fairly high angular solidity is achieved.*

Using the simplified Stodola formula (Equation 8.58a) and rearranging yields

$$N_B = \mathrm{INT}\left(\frac{\pi \sin \beta_2^*}{1-\mu_E}\right) + 1, \tag{8.60}$$

where INT stands for the "greatest integer." Clearly, for larger outlet blade angles, the blade number increases as the sine of the angle; and for larger slip coefficients, the required number of blades increases rapidly. Both trends drive the cascade to high angular solidity in order to control the blade loading. To that degree, the use of solidity in radial impellers is analogous to that in axial cascades.

Centrifugal cascades for impellers at the high range of specific speeds ($N_s > 1$) can be analyzed in much the same manner as an axial configuration using solidity and diffusion coefficient as parameters. One can approximate the solidity of a centrifugal cascade for fairly a high blade number as

$$\sigma_c \approx \frac{[1-(d/D)]N_B}{[2\pi \sin \beta_m]}, \tag{8.61}$$

where

$$\beta_m = \sin^{-1}\left(\frac{\sin\beta_1 + \sin\beta_2}{2}\right). \tag{8.62}$$

The formulation is similar to the axial solidity where the chord of the blade is divided by the arc length between adjacent trailing edges (Figure 8.26). If one chooses to select the blade solidity on the basis of diffusion, then again employ the Lieblein formulation in a form relevant to the centrifugal cascade:

$$D_{Lc} = 1 - \frac{W_2}{W_1} + \frac{\Delta W_\theta}{2W_1\sigma_c}. \tag{8.63}$$

If we assume that the boundary layer flow on the blade suction surface behaves the same as in an axial cascade (one must neglect curvature and Coriolis effects), then the value of D_{Lc} should be kept <0.5–0.6 to avoid flow separation. As it turns out, this is a rather conservative constraint as illustrated in Table 8.3, which presents calculations for a fan with

* A notable exception is low specific speed centrifugal pumps, which usually have only 4–8 blades.

TABLE 8.3

Selection of Solidity for a Centrifugal Fan

N_B	μ_E	ΔW_θ	W_2/W_1	β_2^*	c/D	σ_c	D_L
10	0.892	135.1	0.620	21.4	0.467	1.48	0.638
12	0.915	131.9	0.636	20.8	0.467	1.78	0.575
14	0.925	130.4	0.645	20.6	0.467	2.08	0.550
16	0.933	129.6	0.65	20.4	0.467	2.38	0.554

$Q = 150\,\text{ft}^3/\text{s}$, $\Delta p_T = 60\,\text{lbf}/\text{ft}^2$, $\rho = 0.00233\,\text{slug}/\text{ft}^3$, $D = 3\,\text{ft}$, $d/D = 0.727$, and $N = 158\,\text{rad/s}$ (1506 rpm). Here, $N_s = 0.95$ and $D_s = 3.1$. $C_{\theta 1}$ is assumed to be zero so that ΔW_θ is simply $C_{\theta 2}$. The blade widths are $b_1 = d/4$ and $b_2 = (d/D)b_1$. The calculations imply that the diffusion parameter becomes >0.6 for fairly moderate values of blade number, implying that the fan must have $N_B > 10$ and $\sigma_c > 1.48$. In truth, most centrifugal fans in this range of specific speeds will have a solidity of perhaps 1.1–1.25, and very high total efficiencies as well, indicating very clean boundary layer flows. Twelve blades would be typical so that the Lieblein diffusion constraint requires blades about 40% longer than are commonly used. If one considers a lower specific speed machine, the requirements for solidity can become totally unacceptable. For example, one might quadruple the pressure rise requirement for the fan and use $N_s = 0.40$ and $D_s = 7.3$. The solidity requirement with 30 blades becomes about 8, which is really not workable. For this case, the de Haller ratio would be 0.53—an unacceptably low value according to previous standards. Why is this acceptable here?

What happens is that the suction side of a blade in a centrifugal impeller often operates with separated flow. In axial flow machines, flow separation is unacceptable because of flow stability problems. However, the flow in the blade-to-blade channel of a centrifugal impeller with boundary layer separation is relatively benign, with a well-ordered jet-wake flow pattern that remains stable through the channel and out the exit (Johnson and Moore, 1980). Permitting separated flow in the impeller channel greatly relaxes the restriction on diffusion and allows much higher blade loading with acceptable performance. Typically, a solidity near 1 with adequate blade number to keep the slip coefficient near 0.9 is satisfactory.

8.11 Summary

The study of the detailed flow in blade cascades was conducted to provide a greater depth of understanding of the flow behavior. Employing the relationships for prediction of cascade flows also leads to significant improvement in the accuracy with which one can lay out blade and vane rows for a specified performance. Although the final design and analysis is being done now with computational fluid mechanics analysis techniques, the established cascade methods can still provide good preliminary design and trade-off information prior to refinement and final optimization.

The information on performance of cascades in axial flow is extensive. Most of the information is based on the studies at the Langley and Lewis (now Glenn) research centers. The information reviewed in this chapter was initially centered on determining the flow incidence and deviation from the prescribed geometry as a function of the cascade variables: solidity, camber, and pitch. The study provided a set of algorithms by which one could size the blades and vanes of a machine with reasonable accuracy. Diffusion limitations and their effects on cascade solidity were discussed.

Next, the generation of viscous boundary layer losses was examined for cascades. The classical prediction techniques were presented with simple methods for estimating the efficiency of a given cascade layout. These loss predictions were based on the cascade geometry, the improved estimation of diffusion presented here, and the blade-relative Mach numbers on the machine.

Attention then shifted to flow in turbine blade cascades. Incidence, deviation, and solidity were considered. Relatively simple methods for flow prediction and design were presented for axial flow cascades that depend on the blade-outlet geometry and the flow Mach number. A simple method for estimating losses in turbine cascades was discussed.

The final consideration was for cascades of radial or centrifugal blades. There, the base of empirical information was found to be less extensive and came to rely on a semitheoretical treatment of flow deviation (called *slip*). Several methods were presented and compared with good agreement as to prediction of slip. Slip coefficient was found to depend strongly on the number of blades and the outflow angle requirements for the impeller. The concept of solidity in centrifugal cascades was found to be less important in design than was the case for axial flow machines.

EXERCISE PROBLEMS

8.1. Repeat Problem 6.4, selecting appropriate solidity, camber, and pitch, using the diffusion factor limitation and Howell's rule for estimating deviation.

8.2. Air flows into an axial fan cascade through an IGV row, as sketched in Figure P8.2, with an absolute fluid outlet angle of $-60°$. The blade row is designed to generate a pure axial discharge flow.

a. Select solidity and camber values for the IGVs.

b. Select solidity and camber values for the blades.

c. Estimate the ideal power.

Show all work and state all assumptions. (Hint: The Howell–Carter rule is for diffusing cascades only.)

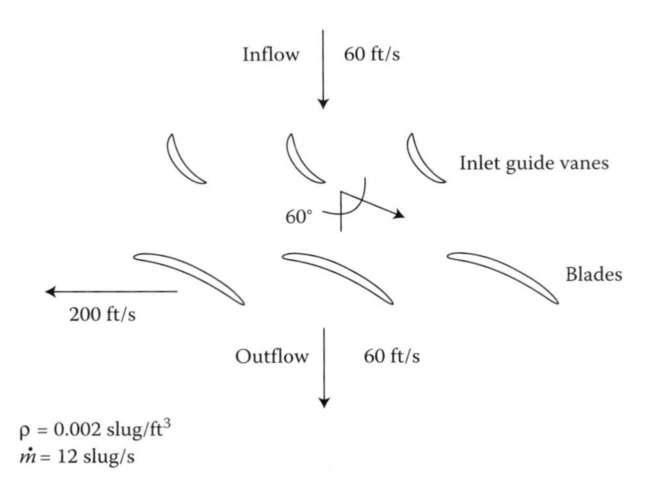

FIGURE P8.2 An axial fan cascade with IGVs.

8.3. An SWSI centrifugal fan has the following geometry: $b_1 = 5$ in., $b_2 = 3$ in., $d = 12$ in., $D = 20$ in., $\beta_1 = 25°$, and $\beta_2 = 35°$ (actual fluid angles). Other parameters are $N = 1800$ rpm and $\rho = 0.0233\,\text{slug/ft}^3$.

a. Estimate the ideal performance as Q and Δp_{Ti}.

b. Estimate the total efficiency for the fan, modify Δp_{Ti} to Δp_T, and estimate the required power.

c. Select a blade-outlet angle and compatible blade number (β_2^* and N_B) that will yield Δp_T and Q. Use both the full Stodola and the simplified Stodola forms for μ_E. (Hint: Use enough blades to keep $\mu_E > 0.85$.)

8.4. For the fan of Problem 6.5, select a reasonable mean station solidity for the blade row and determine the camber angle, ϕ_c, required to generate Δp_T, including the effects of efficiency and deviation. Sketch the vector diagram and the blade shape and layout.

8.5. One can correct ψ_{ideal} (Problem 7.6) for nonviscous effects by including the Stodola formula for slip. Show that ψ_T, corrected for slip and viscous loss, becomes

$$\psi_T = 2\left[1 - \frac{(\pi \sin\beta_2)/N_B}{1 - \varphi_c \cot\beta_2}\right](1 - \varphi_c \cot\beta_2) - \left[\left(\frac{d}{D}\right)^2 + \varphi_c^2\right] N_B C_{wb}.$$

8.6. Use the result given in Problem 8.5 to show that an optimum number of blades for a given value of d/D, β_2, and ϕ_c is given by the following equation:

$$N_{B,opt} = \left[\frac{2\pi \cos\beta_2}{\left((d/D)^2 + \varphi_c^2\right) C_{wb}}\right]^{1/2}.$$

8.7. Using the equation from Problem 8.6, explain how you might estimate C_{wb}.

8.8. For a blower with $\beta_2 = 60°$, $d/D = 0.7$, $\phi_c = 0.3$, and $C_{wb} = 0.025$, what is the "best" number of blades to use?

8.9. An axial water pump has a total pressure rise of 20 psi and a volume flow rate of 6800 gpm. Tip and hub diameters are 16 and 12 in., respectively, and the rotational speed is 1100 rpm.

a. Estimate the total efficiency and calculate the required power.

b. Construct a mean station velocity diagram to achieve the pressure rise at the desired flow and the efficiency from part (a).

c. Lay out the blade shape at the mean station using Constant's rule for deviation.

8.10. Redo the blade layout of Problem 8.9, part (c), using the more complex and more accurate Howell's rule. Compare the result with the "Constant rule" result. What would the error in the pressure rise be using the simplified rule? Discuss the impact on pump design.

8.11. Use the Ainley–Mathieson method to detail the blade and vane shapes to achieve the performance of Problem 7.12 (at the mean radius only).

8.12. Rework the fan mean station of Section 8.4, using 12,000 cfm and 2.0 in. wg. Also lay out hub and tip stations using $d = 0.6D$.

8.13. Consider the rather well-accepted Stodola formula for predicting slip in centrifugal blade cascades: $\mu_E = 1 - (\pi/N_B)\sin\beta_2^*/(1 - \phi\cot\beta_2^*)$ and $\phi = C_{r2}/U_2$. As β_2^* approaches 90°, the formula reduces to $\mu_E = 1 - (\pi/N_B)\sin\beta_2^*$, or less flexibly, $\mu_E = 1 - (\pi/N_B)$.

a. Compare this result to the other common expressions for radial-bladed impellers ($\beta_2^* = 90°$) given by

$$\text{Stanitz, 1951:} \quad \mu_E = 1 - 0.63\left(\frac{\pi}{N_B}\right),$$

$$\text{Balje, 1962:} \quad \mu_E = \frac{1}{1 + [(6.2(d/D)^{2/3})/N_B]},$$

$$\text{Buseman, 1928:} \quad \mu_E = \frac{N_B - 2}{N_B}.$$

b. Identify the more conservative methods and discuss the design implications of using the methods producing higher values of μ_E. When does the Balje method seem to offer an acceptable level of conservatism?

8.14. A centrifugal fan operating in air with $\rho = 0.00235\ \text{slug/ft}^3$ is running at 1785 rpm with a total efficiency of 85%. Fan diameter is $D =$ 30 in.; inlet diameter is $d = 21$ in. The blades have an outlet width $b_2 = 7.5$ in. and an inlet width $b_1 = 9.0$ in. If $\beta_1^* = 32°$, $\beta_2^* = 35°$, and $N_B = 12$ (number of blades),

a. Estimate the volume flow rate of the fan.

b. Estimate the ideal head rise or pressure rise of the fan.

c. Estimate the slip coefficient and the net total pressure rise (include effects of both slip and efficiency).

8.15. A blower is to be designed to supply $5\,\text{m}^3/\text{s}$ of air at a pressure rise of 3000 Pa with an air density of $1.22\,\text{kg/m}^3$. Design a centrifugal fan impeller to achieve this specified performance:

a. Select D, d, b_1, b_2, and N.

b. Determine the efficiency using the modified Cordier value, including Reynolds number and clearance gap ($C/D = 0.002$) effects.

c. Lay out the blade shapes (β_1^*, β_2^*, and N_B) accounting for slip.

d. Provide a table of all final dimensions with a dimensioned sketch or drawing.

e. Make an accurate representation of a blade mean camber line and channel to show blade angles and solidity.

8.16. A double-width centrifugal blower has the following geometry: $D = 0.65\,\text{m}$; $b_2 = 0.15\,\text{m}$; $b_1 = 0.20\,\text{m}$; $d = 0.45\,\text{m}$; and $\beta_2^* = 32°$. The fan has 15 blades and runs at 1185 rpm.

a. If the total flow rate is $8\,\text{m}^3/\text{s}$, determine the correct value of β_1.

b. With an inlet fluid density of $\rho_{01} = 1.19\,\text{kg/m}^3$, estimate the pressure rise of the fan neglecting viscous losses. (Hint: Assume that the flow is incompressible for an initial solution. Then use p/ρ^γ = constant to correct the pressure rise associated with an increased outlet density.)

c. Estimate the efficiency and use this value to correct the net pressure rise of the blower.

8.17. A vane axial fan is characterized by $D = 0.80\,\text{m}$, $d = 0.49\,\text{m}$, and $N = 1400\,\text{rpm}$. The performance of the fan is $Q = 5.75\,\text{m}^3/\text{s}$, $\Delta p_T = 750\,\text{Pa}$, with $\rho = 1.215\,\text{kg/m}^3$ and $\eta_T = 0.85$. Select cascade variables to define the geometry for the blades and vanes. Allow the local diffusion factors, D_L, at hub, mean, and tip station to be 0.60, 0.45, and 0.30, respectively, for the blades and vanes.

a. Choose blade and vane solidity distributions that can satisfy these D_L constraints.

b. Assume that the blades and vanes are very thin. With the solidities and flow angles determined, calculate deviation, camber, and incidence and stagger angles for the blade at hub, mean, and tip stations.

c. Repeat (b) for the vane row.

8.18. The performance requirements for a blower are stated as flow rate, $Q = 9000\,\text{cfm}$; total pressure rise, $\Delta p_T = 8.0\,\text{in. wg}$; ambient air density, $\rho = 0.0022\,\text{slug/ft}^3$.

a. Determine a suitable speed and diameter for the blower impeller and estimate the total efficiency. (Work with a wide centrifugal design, but justify this choice by your analysis.)

b. Correct the total efficiency for the effects of Reynolds number and a generous running clearance of 2.5× ideal (i.e., $C/D = 0.0025$) and the influence of a belt drive system (assume a drive efficiency of 95%).

8.19. Use the dimensionless performance curves of Figure 5.11 to select three candidates for the requirements of Problem 8.18 that closely match the results of your preliminary design decision in that problem.

8.20. Choose a suitable throat diameter ratio, d/D, and width ratios, b_1/D and b_2/D, for the fan of Problem 8.18. Define the velocity. Do so for both blade row inlet and outlet. Calculate the de Haller ratio for the blade.

8.21. For the centrifugal fan analyzed in Problem 8.20, select cascade variables to define the geometry for the blades. Select the number of blades needed using Stodola's slip factor. (Hint: Choose μ_E between 0.85 and 0.9 to maintain good control of the discharge vector.) Assume thin blades and use the flow angles already determined for the blade row inlet. On the basis of the chosen μ_E, define the outlet angle for the blade metal, β_2^*, that will increase the previously calculated $C_{\theta 2}$ to $C_{\theta 2}/\mu_E$.

8.22. Develop a set of drawings that describe two views of the impeller of Problem 8.21. In the "front" view, show at least three blades to define camber shape and the blade-to-blade channel. For the "side" view, show the shape of the sideplate and shroud (if any) relative to blades, illustrating b_1, b_2, and d/D.

8.23. The performance requirements for an air mover are stated as flow rate, $Q = 1\,\mathrm{m^3/s}$; total pressure rise, $\Delta p_T = 500\,\mathrm{Pa}$; and ambient air density, $\rho = 1.175\,\mathrm{kg/m^3}$. The machine is constrained not to exceed 82 dB sound power level.

a. Determine a suitable speed and diameter for the fan and estimate the total efficiency. (Work with a vane axial design, but justify this choice with your analysis.)

b. Correct the total efficiency for the effects of Reynolds number, a running clearance of 2.5× ideal (i.e., $C/D = 0.0025$), and the influence of a belt-drive system (assume a drive efficiency of 95%).

8.24. A small blower is required to supply 250 cfm with a static pressure rise of 8 in. wg and an air weight density of $\rho g = 0.053\,\mathrm{lbf/ft^3}$. Develop a centrifugal flow path for an SWSI configuration. Use the optimal d/D ratio for minimum W_1, estimate the slip using the Stodola formula ($\mu_E \approx 0.9$), assume polytropic density change across the blade row (using Cordier efficiency), and size the impeller outlet width by constraining $25° < \beta_2^* < 35°$.

8.25. For the fan of Problem 8.23, let 500 Pa be the static pressure requirement, and:

a. Estimate the change in power, total and static efficiency, and the sound power level, L_W.

b. Choose a suitable hub–tip diameter ratio for the fan and repeat the calculations of part (a).

8.26. For the fan of Problem 8.23, with $\Delta p_T = 500$ Pa, define the velocity triangles. Do so for both blade and vane rows at hub, mean, and tip stations. Calculate the degree of reaction for all stations.

8.27. For the vane axial fan analyzed in Problem 8.26, select cascade variables to define the geometry for the blades and vanes. Allow the local diffusion factors, D_L, at hub, mean, and tip stations to be 0.60, 0.45, and 0.30, respectively, for the blades and vanes.

a. Choose blade and vane solidity distributions that can satisfy these D_L constraints.

b. Assume that the blades and vanes are very thin. With the solidities and flow angles already determined, calculate deviation, camber, and incidence and stagger angles for the blade at hub, mean, and tip stations.

c. Repeat (b) for the vane row.

8.28. Develop a set of curves for the blade geometric variables, versus radius, of the fan of Problem 8.27, using the calculated blade angles and the blade chord based on 17 blades. Develop curves for the vanes too, using nine vanes to determine the chord lengths.

8.29. For the turbine generator set of Problem 7.12, do a complete layout of the nozzles and blades at the hub, mean, and tip radial stations. Account for the flow deviation using the Ainley–Mathieson method. Develop a plot of the nozzle and blade parameters as a function of radius.

8.30. For the centrifugal fan impeller of Problem 7.14, develop an exact shape for the blades.

a. Use the Stodola formula with $\mu_E = 0.9$ to select the blade number and the outlet metal angle for the blades.

b. Use the Stanitz formula with 11 blades to lay out the blade exit angles and compare to the impeller shape in (a).

8.31. For the fan in Problem 8.26, use $\sigma = 0.4$ at the blade tip to estimate the required camber. Use the NACA correlations and compare with the result using the low-solidity corrections.

8.32. A heat exchanger requires 2500 cfm of air with the static pressure at 2.2 in. wg above ambient pressure. The ambient density is $\rho = 0.00225$ slug/ft^3. The installation for the fan requires that the sound power level be no greater than 85 dB.

a. Determine a suitable speed and diameter for a vane axial fan and estimate the total efficiency.

b. Modify the total efficiency for the influence of Reynolds number and a running clearance of $C/D = 0.002$.

c. Select a thin circular arc blade for the mean station of the blade and vane using Constant's rule, with a d/D ratio chosen from Figure 6.10. (Choose solidity on the basis of an equivalent diffusion of 0.5.)

d. Use the Howell correlation to select these sections and compare the result with the estimation of part (c).

e. For a 10% thick NACA 65-Series cascade, define the blade and vane mean station properties using the full NACA correlation for blade angles and compare with the previous calculations.

8.33. A vane axial fan has $Q = 12\,\text{m}^3/\text{s}$, $N = 1050\,\text{rpm}$, $\rho = 1.18\,\text{kg/m}^3$, $D = 1.0\,\text{m}$, and $d = 0.7\,\text{m}$. Constrain the fan blade hub station to not exceed $D_L = 0.6$ with the hub solidity, σ_h, set to 1.50. Calculate the maximum total pressure rise the fan can develop with these constraints. Assume that the blade is uniformly loaded and correct the results using efficiency based on specific speed.

8.34. Further develop the calculations of Problem 8.33 by recalculating the maximum total pressure rise using $\sigma_h = 1.0$ and 2.0. Develop a curve of maximum total pressure rise as a function of hub solidity. Discuss the limitations of the trend shown by this study.

8.35. Repeat the calculations of Problem 8.33 with diffusion limitations of $D_L = 0.55$ and 0.50. These values represent increasingly conservative levels of design. Compare the results for $D_L = 0.55$ and 0.50 with the result from Problem 8.33, and compare the $D_s \sim N_s$ values of each set with the Cordier curve.

8.36. For a vane axial fan with $Q = 12\,\text{m}^3/\text{s}, N = 1050\,\text{rpm}, \rho = 1.18\,\text{kg/m}^3$, $D = 1.0\,\text{m}$, and $d = 0.7\,\text{m}$, constrain the vane hub station to not exceed $D_L = 0.58$. Determine the maximum total pressure rise achievable with the vane hub solidity constrained to $\sigma_{v-h} = 1.25$.

8.37. The first stage of a low-pressure steam turbine is to be designed for a volume flow of 2000 ft^3/s and a specific work of 45 Btu/lb. The stage outer diameter is 5 ft and the speed is 3600 rpm. The stage is to be normal, axial flow. Complete the design by choosing/determining the following:

a. Blade length, hence hub diameter.

b. Number of blades.

c. Blade and nozzle profile at hub, tip, and mean radius.

8.38. A normal, axial flow stage with air as the working fluid has the following specifications:

Axial velocity C_x:	425 ft/s
Degree of reaction:	0.6
Blade speed:	650 ft/s
Inlet stagnation properties:	24 psia, 650°R
Stage (total–total) polytropic efficiency:	88.7%
Solidity:	1.1

Draw a complete set of velocity diagrams, lay out reasonable blade/vane shapes (correcting for deviation as appropriate), and calculate the stage exit stagnation pressure if the stage is:

a. A turbine.

b. A compressor.

8.39. Design a water pump to deliver 500 gpm against a head of 50 ft. The pump will be called on to deliver a wide range of flow rates (say +15%/−25% of design flow) with little change in head, so a fairly "flat" performance curve is desired. For economic reasons, the pump will be driven by a synchronous electric motor. Your design should include speed, outer and eye diameters, blade widths and blade shapes and include the effects of slip and efficiency. The design is expected to be "typical," not "radical."

8.40. Design a DWDI "airfoil blade" centrifugal fan for Forced Draft in a power plant. The specified total flow is 5000 ft^3/s and the pressure rise is 15 in. wg. Fan speed is 500 rpm and compressibility can be ignored. Determine

a. Impeller layout, including inner and outer diameters, blade width (use constant blade width for manufacturing economy), number of blades, and blade shape (account for slip).

b. Fan efficiency and power.

c. Approximate shape of casing.

8.41. Start with the specifications given in Problem 7.13. Design an axial flow fan with flow straightener/diffuser vanes. Specify fan hub and tip diameters, number of blades, and rotor/stator blade shapes at hub, tip, and mean radius. Your design must have reasonable values of de Haller ratio and must account for fan efficiency and blade/flow deviation. [Hints: Use Cordier information to estimate (total–total) efficiency and to determine a starting value for rotor diameter. Use Constant's rule for estimating deviation.]

8.42. A centrifugal water pump is desired to pump 500 gpm against a head of 50 ft. The impeller will be a "simple" design with no inlet preswirl vanes. It is anticipated that the pump will be called on to deliver a wide range of flow rates (say ±25 of design flow) with little change in head so a fairly "flat" performance curve is desired. Which of the following specific speeds ("pump form" (n_s): $N(\text{rpm})\sqrt{Q(\text{gpm})}/H(\text{ft})^{3/4}$) would be a better design value: 1500, 3000, or 6000?

a. Using the appropriate specific speed, determine the pump rotational speed, (estimated) hydraulic efficiency, and (outer) diameter.

b. Calculate pump work per unit mass and the tangential component of absolute velocity ($C_{\theta 2}$) at the impeller exit.

c. Determine a reasonable ("optimum") value for the eye diameter.

d. Assume that $C_1 = C_{r1} = 0.7U_1$ and lay out the inlet velocity diagram and compute W_1 and the inlet width of the impeller, b_1.

e. Let the impeller width at discharge be half the impeller width at the inlet ($b_2 = b_1/2$). Find C_{r2}. Next, use U_2, C_{r2}, and $C_{\theta 2}$ to lay out the exit velocity diagram.

f. Calculate W_2 and check the de Haller ratio.

g. Choose a reasonable number of blades.

h. Lay out the impeller blade shape.

8.43. Rework Problem 7.14. Specify a drive motor with two, four, or six poles. Account for fan efficiency and slip. The layout must include rotor tip and eye diameters, rotor tip and eye widths, and blade number and blade shape.

8.44. A large axial flow pump must supply 0.45 m^3/s water flow rate with a pressure rise of 140 kPa. The hub and tip diameters for the impeller are $D = 0.4$ m and $d = 0.3$ m, respectively, and the running speed is 1200 rpm. Assume that at the hub station the blades are 15 cm long (chord) and the solidity is 1.0.

a. Use the Cordier diagram to estimate the pump's total efficiency, correcting the Cordier number for the influence of Reynolds number.

b. Lay out an axial flow velocity diagram for the mean radial station of the blade for this pump to achieve the required pressure rise, allowing for the influence of efficiency on the vectors.

c. Approximate the shape of the mean station of this blade based on the vector diagram of part (b).

d. Is the de Haller ratio for this station acceptable?

e. Lay out blade shape and straightening vane (for the mean radius only) using the Howell rule to estimate deviation.

8.45. Develop vector diagrams, blade shapes, and de Haller ratios for the hub and tip stations of the pump examined in Problem 8.44.

8.46. A radial flow pump is to be connected to a system. The system's head-flow characteristic is $H = 5 + 950 \times Q^2$ with H in m and Q in m^3/s. The pump has the following characteristics:

Impeller diameter:	0.25 m
Rotational speed:	1450 rpm
Number of blades:	8
Blade exit angle:	30° (backward leaning)
Inlet preswirl:	None
Impeller tip width:	2.0 cm
D/d:	3.0
b_2/b_1:	0.33333
Hydraulic efficiency:	0.82
Mechanical efficiency:	0.90

a. Develop an equation for the pump's head-flow characteristic. Use Stodola's slip factor.

b. Find the flow rate through the pump/system.

c. Calculate the impeller blade inlet angle.

d. Sketch the pump impeller.

e. Calculate the shaft power input.

8.47. A centrifugal water pump is to be designed to the following specifications:

Head: 120 ft   Flow: 900 gpm   Speed: 1750 rpm

There is no inlet preswirl and use seven vanes. The inlet will be optimum/standard with $A_1 = A_0$. The blade width will be tapered to maintain constant radial velocity.

a. Estimate the hydraulic efficiency and the specific work.
b. Investigate the relationship between blade tip angle (β_2^*) and impeller diameter by calculating impeller diameter for angles of 25° and 60°. Use the Weisner model for slip.

8.48. A small portable gas turbine power plant is being designed. For simplicity and ruggedness, the unit will have a single-stage centrifugal compressor and a single-stage axial flow turbine. The unit is designed for a net shaft power output of 75 kW (100 hp) with standard sea-level inlet conditions (59°F, 14.696 psia, and 0.0764 lbm/ft^3). This problem will work through a preliminary design for the unit.

Part I: *Thermodynamic Calculations*

The following parameters are chosen for the preliminary design:

Compressor (total) pressure ratio:	3 : 1
Total pressure drop in combustor (between compressor and turbine):	2 psi
Combustor discharge (turbine inlet) total temperature:	1500°F
Fuel/air ratio:	2%
Turbine outlet total pressure:	14.8 psia

The working fluid will be assumed to be air, with $R = 53.35$ ft × lb/lbm° R. In the compressor, $\gamma = 1.4$. In the turbine, $\gamma = 1.35$. Where appropriate, design parameters will be selected such that the compressor has maximum efficiency; however, Cordier values should be "derated" by 8 points (0.08) owing to the small size of the components.

a. Calculate the specific work (Btu/lb) required in the compressor.
b. Calculate the turbine inlet total pressure and the specific work (Btu/lb) produced by the turbine wheel. Assume that the turbine efficiency is 3% points greater than the compressor efficiency.
c. Noting that the turbine mass flow rate is 2% higher than the compressor mass flow rate (the fuel flow is added between the compressor and turbine), calculate the air (compressor intake) mass flow required to produce 100 hp net.
d. Determine the compressor rotational speed necessary to achieve the maximum design efficiency.

Part II: *Compressor Design*

The following choices have been made: The compressor will have radial tip blades ($\beta_2^* = 90°$). The compressor inlet will have axial inducer vanes and there will be no preswirl ($\alpha_1 = 90°$, $C_1 = C_{1x}$). The compressor has 11 full blades plus 11 "tandem half blades" (i.e., 22 blades at discharge and 11 blades at inlet).

a. Calculate the slip coefficient and determine the blade tip speed (U_2) needed to do the required compressor work.

b. Calculate the necessary compressor diameter.

c. Determine a reasonable value for the height of the inlet inducer blade (i.e., the eye diameter).

d. Draw the inlet velocity diagram and determine the inducer blade angle (β_1^*) at the inducer mean radius.

e. Assuming that the discharge radial velocity is the same as the inlet axial velocity ($C_{r2} = C_{x1}$), estimate the impeller tip width. Use $\rho = 0.10$ lbm/ft^3.

Part III: *Turbine Design*

Assume that the turbine wheel outer diameter is the same as the compressor outer diameter. The turbine and compressor share a common shaft, so both rotate at the same speed. The turbine blades are 0.5 in. high. Turbine design will be done for the mean blade radius only. Recall that the turbine flow rate is 102% of the compressor flow rate.

a. Calculate the blade speed, the work coefficient, and the required degree of reaction for the turbine.

b. Using a mean density of 0.035 lbm/ft^3, calculate the axial velocity for the turbine (C_x).

c. Lay out the turbine velocity diagrams.

d. Sketch nozzle and rotor blades. Account for imperfect guidance.

8.49. Design a power station condensate pump. Work out a preliminary design for a centrifugal pump to produce a water flow of 5.0 ft^3/s (0.14 m^3/s) against a head of 650 ft (200 m). Speed is to be 3550 rpm. Desired information is:

a. Pump efficiency.

b. Power requirement.

c. Choose number of blades.

d. Impeller layout including eye and outer diameters, blade width, and blade shape (account for slip).

e. Sketch of casing.

8.50. Design a power station circulating water pump. Work out a preliminary design for an axial flow pump to produce a water flow of 170.0 ft^3/s (4.8 m^3/s) against a head of 25 ft (7.6 m). Speed is to be 370 rpm. Determine:

a. Pump efficiency.

b. Power requirement.

c. Number of blades.

d. Impeller layout including hub and outer diameters (mean radius), blade shape, and straightening vane shape (account for deviation and diffusion limits).

e. Sketch of casing.

8.51. An axial flow turbine stage is a normal stage with constant axial velocity. The flow coefficient, $\varphi = C_x/U$, has a value of 0.6 and the nozzle exit angle is $\alpha^* = 21.8°$

a. Sketch the velocity diagrams.

b. Sketch the nozzle and rotor blade shapes.

c. Calculate the stage loading coefficient, $\Psi = C_\theta/U$, and the degree of reaction.

d. Using Solderberg's method, estimate the turbine's total–total efficiency.

8.52. Observation of a wind tunnel fan (including the "nameplate") gives the following information: Fan type=Full stage tubeaxial; speed = 3500 rpm; flow = 9600 cfm; total pressure rise = 4.121 in. wg; fan static pressure rise = 3.0 in. wg; total efficiency = 56%; static efficiency = 41%. Using as much of this information as you need, design a fan for these specifications.

9

Quasi-Three-Dimensional Flow

9.1 Quasi-Three-Dimensional Flow Model

As discussed at the beginning of Chapter 8, the actual flow in a turbomachine is three-dimensional. Because the machine is generally symmetric about its axis of rotation, it is most convenient to describe the flow in cylindrical coordinates, with three space coordinates x (axial), r (radial), and θ (circumferential). In general, there are three corresponding velocity components, C_x, C_r, and C_θ in the absolute frame of reference and W_x, W_r, and W_θ in the relative (rotating) frame of reference. Of course, the two sets of velocity components are not independent, being related by $\boldsymbol{W} = \boldsymbol{C} - U\boldsymbol{e_\theta}$ with U the rotor linear speed ($U = rN$) and $\boldsymbol{e_\theta}$ a unit vector in the rotational (circumferential) direction.

The complexity of the actual flow requires that various simplified models be used for analysis and design. For example, in the analyses and calculations performed thus far in this book, it was generally assumed that simple, uniform inlet and outlet conditions exist for a blade or vane row and we generally ignored flow details within the row itself. Specifically, in axial flow machines, it was assumed that meridional flow surfaces through the machine are a simple set of concentric cylinders with a common axis—the axis of rotation. It was also assumed that the mass or volume flow rate is uniformly distributed in the annulus between the inner hub and the outer casing. That is, one assumes that C_x is a constant value across the inlet, across the outlet, and throughout the machine; functionally, $C_x \neq f(x, r, \theta)$. In addition, it was also assumed that there is no radial velocity anywhere, $C_r = 0$ (everywhere). For radial flow machines, the corresponding assumptions are uniformity of flow across the rotor depth and that at least one velocity component is zero (typically C_r in the inlet eye and C_x at the rotor exit). In the previous three chapters, this flow model allowed us to concentrate on the interaction between the blades and the fluid in the so-called *blade-to-blade* plane.

In many machine flow fields, these assumptions are well justified, although one naturally expects some distortion of the flow near the end walls (the hub and casing surfaces) due to the viscous/turbulent boundary layers there. However, in a more general treatment, one must examine more carefully the flow along the curved surfaces associated with the axial motion and the rotating motion due to swirl (axial velocity and tangential velocity in the

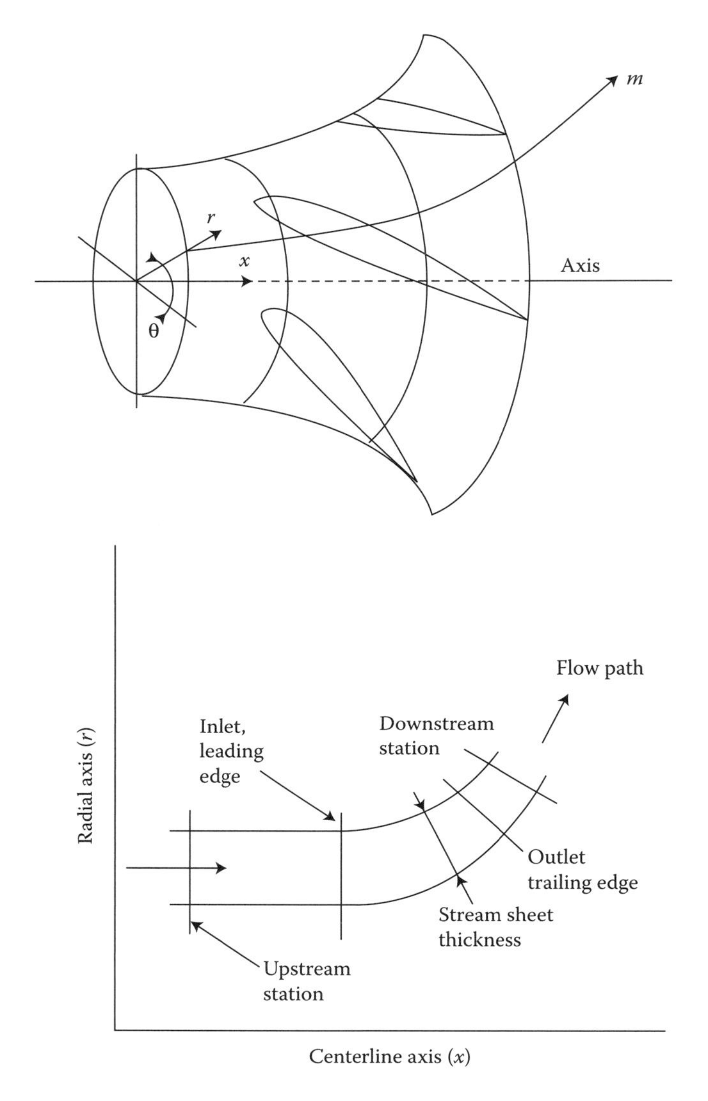

FIGURE 9.1 Meridional (S1) and planar (S2) surfaces for Q3D flow model.

absolute frame of reference, C_x and C_θ). One widely used solution to this dilemma is the *quasi-three-dimensional (Q3D) flow model* (popularly designated the Q3D model) (Wu, 1952; Johnsen and Bullock, 1965; Katsanis, 1969; Wright et al., 1982).

The Q3D method seeks to determine an approximate 3D solution of the flow by solving two fully compatible, two-dimensional flow problems on mutually perpendicular surfaces. The Q3D model is illustrated in Figure 9.1. One of the surfaces is the meridional surface of revolution on which one can examine the region between cross-sections of adjacent blades cut by this meridional surface shown in Figure 9.1. This is called the *S1 surface*; the solution on this surface is the blade-to-blade solution. The other surface, called the *S2 surface* or "throughflow surface," is planar and passes through the axis of rotation. It is on this surface that one analyzes and determines the shapes of the meridional stream surfaces by solving for a circumferentially averaged velocity field. Typically, the flow in the S2 surface is modeled as inviscid. There are three velocities associated with the S2 surface, roughly axial, radial, and tangential. These velocities are functions of only two space coordinates; the tangential variation is interpreted as an average of the blade-to-blade flow. Losses can be included in the S1 (blade-to-blade) solution, usually by employing semiempirical models like those discussed in Chapter 8. The S2 solution drives the flow solution on the S1 surface by determining the shape and location where the blade-to-blade flow is analyzed. For best results, the two solutions are iterated with the meridional (S1) solution supplying tangential flow information (C_θ and loss) to the circumferentially averaged flow on the (S2) surface until one solution no longer changes the other. Such a solution, although not rigorously accurate nor fully 3D, can provide an excellent approximation to the fully 3D flow (Wright et al., 1984b). The method provides substantial reduction in effort over a fully 3D analysis and is most useful when the geometry is to be changed again and again. The model is still rather complex, typically requiring the solution of partial differential equations of fluid motion in the S2 plane, and numerical computer-based calculation is almost always required. This will be discussed in Chapter 10.

9.2 Simple Radial Equilibrium for Axial Flow Machines

For preliminary design work, for establishing limiting parameters, and to seek some fundamental insights into flow behavior, much can be learned by describing the throughflow in an axial flow machine using the concept of simple radial equilibrium (SRE). Here, the complex task of establishing a detailed description of the two-dimensional flow (including both the axial and radial motion) through a blade or vane row (i.e., on the complete S2 surface) is abandoned in favor of establishing patterns of equilibrium flow at locations upstream and downstream of the blades. At these stations, we assume that radial velocity components are negligibly small ($C_r = 0$) and require that axial and tangential components be compatibly matched through the simple balancing of the "centrifugal force" and a radial pressure gradient. This SRE concept is illustrated in Figure 9.2. The "centrifugal force" (per unit volume) acting on the fluid particle shown is $\rho(C_\theta^2/r)$. The net force toward the axis arises from the pressure gradient and is $\mathrm{d}p/\mathrm{d}r$ (per unit volume). If we ignore any viscous forces and assume that there is no streamline curvature in

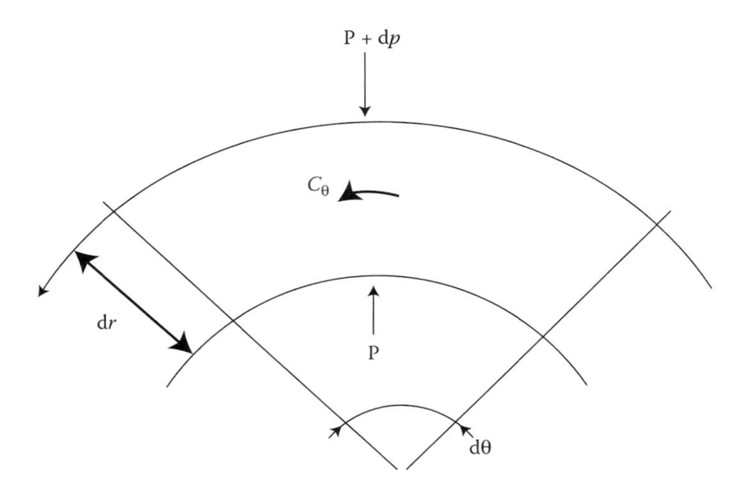

FIGURE 9.2 Fluid particle in SRE.

the S2 plane (Figure 9.3), then radial equilibrium of the fluid particle requires

$$\frac{dp}{dr} = \rho \frac{C_\theta^2}{r}. \tag{9.1}$$

Even though some detail is lost in the SRE model, according to Johnsen and Bullock, the flow in an annular cascade can be analyzed with reasonable accuracy based on applying it to only the inlet and outlet conditions (Johnsen and Bullock, 1965). Figure 9.3 further illustrates the geometry. In Figure 9.3, the "radial flow region" represents passage through the blades. At the "inlet station" and the "outlet station," SRE (Equation 9.1) is assumed.

It must be noted that there is no physical requirement that a machine flow field satisfies the SRE requirement. If the condition stated by Equation 9.1 is not true, then there will be additional radial accelerations [of the form

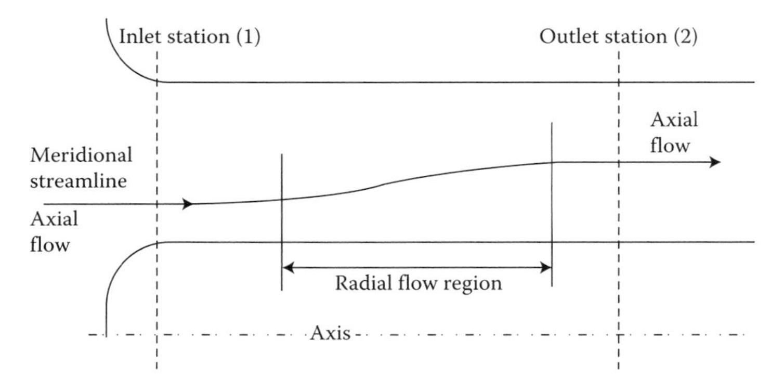

FIGURE 9.3 The SRE concept and flow geometry.

$C_r(\partial C_r/\partial r)$ and $C_x(\partial C_r/\partial x)]$, C_r may not be zero, and the streamlines will be curved. The utility of the SRE model is that it leads to an equation that allows the designer to specify distributions of C_x and C_θ that assure radial equilibrium and straight streamlines in the spaces between blades. Although the appropriate equation may be developed for compressible flows (Dixon, 1998; Japikse and Baines, 1994), consideration here will be limited to incompressible flows, in keeping with the primary emphasis of this book.

For steady, incompressible flow, the total pressure rise can be evaluated from the Euler equation (Chapter 6), together with the total efficiency:

$$\Delta p_T = \eta_T \rho N r (C_{\theta 2} - C_{\theta 1}). \tag{9.2}$$

At planes "1" and "2," the radial velocity is zero and

$$\Delta p_T = \left[p + \frac{1}{2}\rho C_x^2 + \frac{1}{2}\rho C_\theta^2\right]_2 - \left[p + \frac{1}{2}\rho C_x^2 + \frac{1}{2}\rho C_\theta^2\right]_1. \tag{9.3}$$

Now substitute Equation 9.3 into Equation 9.2, differentiate with respect to r, and rearrange the terms to yield

$$\frac{dp_2}{dr} - \frac{dp_1}{dr} + \rho C_{x2}\frac{dC_{x2}}{dr} - \rho C_{x1}\frac{dC_{x1}}{dr} + \rho C_{\theta 2}\frac{dC_{\theta 2}}{dr} - \rho C_{\theta 1}\frac{dC_{12}}{dr}$$
$$= \rho N\left(\frac{d(\eta_T r C_{\theta 2})}{dr} - \frac{d(\eta_T r C_{\theta 1})}{dr}\right).$$

Next, impose the radial equilibrium condition, Equation 9.1, at locations "1" and "2" and rearrange

$$\frac{d(C_{x2}^2/2)}{dr} = N\left(\frac{d(\eta_T r C_{\theta 2})}{dr} - \frac{d(\eta_T r C_{\theta 1})}{dr}\right) - \frac{d(C_{x2}^2/2)}{dr} + \frac{d(C_{x1}^2/2)}{dr}$$
$$- \frac{d(C_{\theta 2}^2/2)}{dr} + \frac{d(C_{\theta 1}^2/2)}{dr} + \frac{C_{\theta 1}^2 - C_{\theta 2}^2}{r}. \tag{9.4}$$

In general, C_{x1} and $C_{\theta 1}$ are presumed known from upstream calculations. To achieve a more manageable form for Equation 9.4, assume that C_{x1} is a constant (uniform inflow) and that $C_{\theta 1}$ is zero (simple inflow). Also assume that η_T is independent of r to obtain*

$$\frac{d(C_{x2}^2/2)}{dr} = N\eta_T\frac{d(rC_{\theta 2})}{dr} - \frac{d(C_{\theta 2}^2/2)}{dr} - \frac{C_{\theta 2}^2}{r}. \tag{9.5}$$

* It is not assumed that there are no losses, which would give $\eta_T = 1$, just that the efficiency for the particular blade row does not vary with radius.

It is sometimes helpful to nondimensionalize this equation. One way to do so is to divide the velocities by the blade tip speed, $ND/2$, and the local radius by the rotor radius $D/2$. Then, defining Ψ and φ as in Chapter 7

$$\varphi_2 = \frac{C_{x2}}{ND/2} \quad \text{and} \quad \Psi = \frac{C_{\theta 2}}{ND/2}, \tag{9.6}$$

and defining a dimensionless radius as

$$z = \frac{r}{D/2}, \tag{9.7}$$

the dimensionless SRE velocity equation becomes

$$\frac{\mathrm{d}(\varphi_2^2/2)}{\mathrm{d}z} = \eta_T \Psi + \eta_T z \frac{\mathrm{d}\Psi}{\mathrm{d}z} - \Psi \frac{\mathrm{d}\Psi}{\mathrm{d}z} - \left(\frac{\Psi^2}{z}\right). \tag{9.8}$$

The velocity equation in either dimensional form (Equation 9.5) or dimensionless form (Equation 9.8) appears difficult because it is a nonlinear ordinary differential equation. Worse than that, it has two independent variables, either φ_2 and Ψ or C_{x2} and $C_{\theta 2}$, and of course one cannot determine two variables from a single equation! Recall now that the purpose of the SRE model is as a design tool; the way it works is that the designer specifies the variation of one parameter ($\varphi_2 \sim C_{x2}$) or ($\Psi \sim C_{\theta 2}$), and the solution to the appropriate velocity equation prescribes the radial variation of the other parameter necessary to achieve radial equilibrium. Typically, the designer specifies a variation of the swirl and uses the equation to determine the necessary distribution of the axial velocity.

A fairly trivial solution can be found by specifying $\Psi = 0$. This is the case for no inlet swirl and no outlet swirl, and φ_2 = constant satisfies the differential equation. Then, by conservation of mass, $\varphi_2 = \varphi_1$. This would, of course, correspond to flow in an open pipe with no blades to turn the flow.

Another simple but extremely important solution exists when one specifies $\Psi = a/z$, where a is a constant. Substituting $\Psi = a/z$ into the SRE velocity equation, one obtains

$$\begin{aligned} \frac{\mathrm{d}\varphi_2^2}{\mathrm{d}z} &= 2\left(\eta_T\left(\frac{a}{z}\right) + \eta_T \frac{\mathrm{d}}{\mathrm{d}z}\left(\frac{a}{z}\right) - \left(\frac{a}{z}\right)\frac{\mathrm{d}}{\mathrm{d}z}\left(\frac{a}{z}\right) - \left[\frac{a^2/z^2}{z}\right]\right), \\ &= 2\left(\eta_T\left(\frac{a}{z}\right) - \eta_T\left(\frac{a}{z}\right) + \left(\frac{a^2}{z^3}\right) - \left(\frac{a^2}{z^3}\right)\right) = 0. \end{aligned} \tag{9.9}$$

Thus, φ_2 = constant satisfies the equation. Reverting to dimensional variables, C_x = constant when $C_{\theta 2}$ = constant/r. Recall from the study of fluid mechanics that the flow field of a simple, potential vortex is described by

$C_\theta = \Gamma/2\pi r$, where Γ is the (constant) strength of the vortex. Such a flow is usually referred to as a free-vortex velocity distribution. In turbomachinery parlance, this is shortened to "free-vortex flow." From the Euler equation, the distribution of total pressure rise is

$$\Delta p_T = \eta_T \rho N r (C_{\theta 2} - C_{\theta 1}).$$

Using simple inflow ($C_{\theta 1} = 0$) together with the free-vortex distribution, $C_{\theta 2} = a/r$, yields

$$\Delta p_T = \eta_T \rho N r \left(\frac{a}{r}\right) = a \eta_T \rho N = \text{constant}. \tag{9.10}$$

We may summarize the results as follows:
For a free-vortex outlet swirl distribution,

- The axial velocity distribution is uniform.
- Each streamline experiences the same work and the same change in stagnation pressure.

The bulleted results were used as assumptions for analyzing axial flow fans, pumps, compressors, and turbines in the earlier chapters, especially in the "blade stacking" model of 3D flow in Section 7.5. The free-vortex distribution makes the flow analysis very easy to deal with, and many machines, especially steam turbines and early multistage axial flow compressors, have been designed as free-vortex machines. The free-vortex distribution does have its drawbacks. Since $C_{\theta 2}$ must increase as $1/r$ with decreasing radial position, a machine with a relatively small hub will require very large values of C_θ near the hub to achieve the free-vortex conditions. For pumping machines, when C_θ is large and U is small, the de Haller ratio for the blade becomes unacceptably small and the flow turning (and, hence, deviation and blade twist) can become excessive.

A second, reasonably obvious, design choice for swirl is the *forced vortex distribution*: $C_{\theta 2} = ar$, or, in dimensionless form, $\Psi = az$. This is the velocity distribution that would result if the fluid were rotating like a solid body. For forced vortex flow, the SRE velocity equation becomes

$$\frac{\mathrm{d}(\varphi_2^2/2)}{\mathrm{d}z} = \eta_T a z + \eta_T z a - aza - \left(\frac{a^2 z^2}{z}\right) = 2(\eta_T a z - a^2 z) = 2az(\eta_T - a),$$

which integrates to

$$\varphi_2 = \left[2a(\eta_T - a)z^2 + K\right]^{1/2}.$$

K is a constant of integration, which must be determined from the upstream flow by a mass balance

$$\pi \varphi_1 \left(1 - z_h^2\right) = 2\pi \int_{z_h}^{1} \varphi_2 \, z \, dz = \int_{z_h}^{1} \left[2a(\eta_T - a)z^2 + K\right]^{1/2} z \, dz.$$

Determination of K will be delayed until the next section—for now note that the axial velocity varies in a nonlinear fashion.

The total pressure rise for a forced vortex distribution is

$$\Delta p_T = \eta_T \rho N r (C_{\theta 2} - C_{\theta 1}) = \eta_T \rho N r (ar - C_{\theta 1}) = \eta_T \rho N (ar^2 - r C_{\theta 1}), \qquad (9.11)$$

a parabolic distribution. This means that quite a bit more work is done at the blade tip than at the hub. In effect, the hub region is not contributing very much to the overall pressure rise, which makes the forced-vortex distribution relatively ineffective. Nevertheless, some early axial flow compressors were designed for a forced-vortex distribution.

One might explore free-vortex and forced vortex distributions more extensively, but for now it appears that neither of them is a perfect choice for axial turbomachine design. To investigate the matter further requires that one deal with the nonlinearity of the SRE velocity equation and its solutions.

9.3 Approximate Solutions for SRE

As we have seen, the typical application of the SRE model is for the designer to specify a radial distribution for the swirl velocity and then to determine the appropriate axial velocity distribution to assure that simple equilibrium is maintained. For two specific distributions, zero swirl (trivial) and the free-vortex distribution, the mathematics involved is fairly simple. For any other case, the SRE velocity differential equation and its algebraic solutions are prohibitively difficult because they are nonlinear. One option for solving them is to use numerical methods, supported by a digital computer or, perhaps a programmable calculator. Although useful for specific problems, this method lacks generality and it is difficult to obtain general insight and guidance for design. A useful alternate is to investigate closed-form, approximate solutions—the topic of this section.

We begin by postulating a general form for Ψ using a polynomial algebraic description, retaining the a/z term (free vortex) as a baseline and adding additional terms to permit a systematic deviation from the simplest case:

$$\Psi = \frac{a}{z} + b + cz + dz^2 + \cdots . \qquad (9.12)$$

The coefficients a, b, c, d, and so on are all constants that can be chosen to manipulate the swirl velocity (effectively, the loading) along the blade. For simplicity, as well as reasonable generality in the swirl distribution, only three terms will be retained; d and all coefficients of higher order will be set to zero. Thus, the swirl distribution will be a combination of the free and forced vortex distributions, with a component of constant swirl ($\Psi = a/z + b + cz$). Specifying this distribution, the equilibrium equation becomes

$$\frac{d(\varphi_2^2/2)}{dz} = \eta_T b + 2c\eta_T z - \frac{ab}{z^2} - \left(\frac{2ac}{z}\right). \quad (9.13)$$

The equation has been simplified by dropping the higher-order terms in b and c (i.e., b^2, bc, and c^2). In doing so, we assume (actually, specify) that b and c are small compared to a, b^2 is small compared to b, and so on (formally, $b \ll 1, c \ll 1, b \ll a$, and $c \ll a$).

This device for simplification of the ensuing mathematical manipulations is a common technique referred to as "order analysis" or "small perturbation theory." One "perturbs" the free-vortex theory for equilibrium by including small deviations from $C_\theta = a/r$ by using small values for b and c. This implies an inherent restriction on the model but will yield a simple algebraic result that can then be examined in depth to gain some insight into the general flow behavior.

Equation 9.13 can be integrated directly to give

$$\frac{\varphi_2^2}{2} = \eta_T bz + \eta_T cz^2 - 2ac\ln(z) + \frac{ab}{z} + K, \quad (9.14)$$

where K is the constant of integration. Thus,

$$\varphi_2 = \left[2\left(\eta_T bz + \eta_T cz^2 - 2ac\ln(z) + \frac{ab}{z} + K\right)\right]^{1/2}. \quad (9.15)$$

One must use K to enforce conservation of mass, so the constant is not at all arbitrary and we require

$$\int_{z_h}^{1} 2\pi\varphi_1 z\,dz = \int_{z_h}^{1} 2\pi\varphi_2 z\,dz,$$ which gives, upon substituting,

$$\pi\varphi_1\left(1 - z_h^2\right) = 2\pi\int_{z_h}^{1}\left[2\left(\eta_T bz + \eta_T cz^2 - 2ac\ln(z) + \frac{ab}{z} + K\right)\right]^{1/2} z\,dz. \quad (9.16)$$

Here, z_h is the hub–tip ratio, d/D, and it is assumed that φ_1 is not a function of z.

Unfortunately, the integral cannot be evaluated in the closed form. In order to proceed toward the goal of obtaining a generalized analysis, a further approximation is required.

What must be done is to find an acceptable way to linearize the messy integrand in Equation 9.16 so that the calculus can be handled more readily. One can do so by introducing a perturbation variable form for φ_2 according to

$$\varphi_2 = [1 + \varepsilon(z)]\varphi_1. \tag{9.17}$$

$\varepsilon(z)$ is a function of z that describes the deviation of the outlet axial velocity component from the uniform inlet flow ($\varphi_1 = \text{constant}$). That is, $\varepsilon(z) = 0$ for all z describes a uniform outflow, while $\varepsilon(z) = mz$ would describe a linear redistribution of flow at the outlet.

Reverting back to Equation 9.14 and substituting Equation 9.17

$$\begin{aligned} \varphi_2^2 &= [1 + \varepsilon(z)]^2\varphi_1^2 = \left[1 + 2\varepsilon(z) + \varepsilon^2(z)\right]\varphi_1^2, \\ &= 2\left(\eta_T bz + \eta_T cz^2 - 2ac\ln(z) + \frac{ab}{z} + K\right). \end{aligned} \tag{9.18}$$

Now we require that $\varepsilon(z) \ll 1$ (i.e., the departure from a uniform flow is small) so that $\varepsilon^2(z) \ll 2\varepsilon(z)$, then

$$[1 + 2\varepsilon(z)]\,\varphi_1^2 \approx 2\left(\eta_T bz + \eta_T cz^2 - 2ac\ln(z) + \frac{ab}{z} + K\right). \tag{9.19}$$

Solving for $\varepsilon(z)$

$$\varepsilon(z) = \left(\frac{1}{\varphi_1^2}\right)\left[\eta_T bz + \eta_T cz^2 - 2ac\ln(z) + \frac{ab}{z} + K - \frac{1}{2}\right]. \tag{9.20}$$

Now, substituting Equation 9.17 into the conservation of mass and canceling 2π

$$\int_{z_h}^{1} \varphi_1 z\,dz = \int_{z_h}^{1} \varphi_2 z\,dz = \int_{z_h}^{1} [1 + \varepsilon(z)]\varphi_1 z\,dz = \int_{z_h}^{1} \varphi_1 z\,dz + \int_{z_h}^{1} \varepsilon(z)\varphi_1 z\,dz. \tag{9.21}$$

Upon canceling the terms, conservation of mass requirement reduces to

$$\int_{z_h}^{1} \varepsilon(z) z\,dz = 0. \tag{9.22}$$

Note that Equation 9.22 does not itself require that $\varepsilon(z)$ be small. When Equation 9.20 (which does require $\varepsilon(z) \ll 1$) is substituted into Equation 9.22, the integration can be performed and then K can be determined. Following this, the final expression for $\varepsilon(z)$ can be determined. Leaving the details to the reader as an exercise, the result is

$$\varepsilon(z) = \left(\frac{\eta_T}{\varphi_1^2}\right)\left[\begin{array}{l}(z - f_1[z_h])\,b + \left(z^2 - f_2^2[z_h]\right)c + \left(\dfrac{ab}{\eta_T}\right)\left(\dfrac{1}{z} - \dfrac{1}{f_3[z_h]}\right) \\ -\left(\dfrac{2ac}{\eta_T}\right)(\ln(z) + f_4[z_h])\end{array}\right], \tag{9.23}$$

where f_1, f_2, f_3, and f_4 are functions of z_h (=d/D) given by

$$\begin{aligned}
f_1(z_h) &= \left(\frac{2}{3}\right)\left(\frac{1 - z_h^3}{1 - z_h^2}\right),\\
f_2(z_h) &= \left(\frac{1 - z_h^2}{2}\right)^{1/2},\\
f_3(z_h) &= \frac{1 + z_h^2}{2},\\
f_4(z_h) &= \frac{1 + z_h^2\left[\ln(z_h^2) - 1\right]}{2\left(1 - z_h^2\right)}.
\end{aligned} \tag{9.24}$$

Using Equations 9.17, 9.23, and 9.24 allows us to calculate C_{x2} (from $C_{x2} = U\varphi_2 = U\varphi_1[1 + \varepsilon(z)]$) as a function of r for various blade loading distributions, as determined by choices of a, b, and c subject to $(b, c) \ll a$ and $(b, c) \ll 1$. Note that when $b = c = 0$, C_{x2} reverts to C_{x1} (=constant), so one has a limit-case check.

Examination of the errors associated with the approximations of these results (Wright, 1988) shows that the errors in estimating φ_2 begin to exceed about 5% when the magnitude of ε exceeds a maximum value of 0.25, in comparison with the result of "exact" numerical integration of the nonlinear equation (Kahane, 1948; Wright and Ralston, 1987). Figure 9.4 shows the influence of a series of loading distributions $[\Psi(z)]$ on the nonuniformity of φ_2. The design study is constrained to yield the same total pressure rise for every case for the same flow rate. The hub–tip ratio is taken as $z_h = 0.5$, the dimensionless flow rate is $\varphi_1 = 0.5$, and the total efficiency is set at $\eta_T = 0.9$. Each case corresponds to a different choice of (a, b, and c). Case 2 is the free-vortex baseline case ($a = 1$ and $b = c = 0$), and cases 1, 3, 4, and 5 represent increasing departure from the free-vortex loading. Figure 9.4 also shows the corresponding variations of φ_2 with z. The free-vortex distribution produces a uniform outflow at $\varphi_2 = 0.5$, with the succeeding distributions showing greater and greater deviation from uniform outflow. Generally, the flow shifts to the outside of the annulus while reducing the flow near the hub. Shifting the swirl generation load to the outer region of the blade (as in a forced

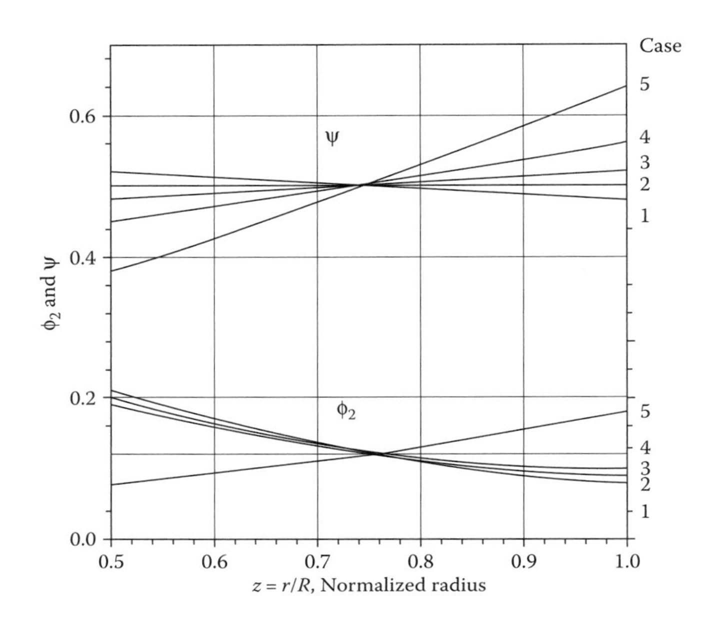

FIGURE 9.4 Comparison of swirl and axial velocity distributions for the linearized SRE model; $\varphi_1 = 0.5, z_h = 0.5$, and $\eta_T = 0.9$. (From Wright, T., 1988. "A closed-form algebraic approximation for quasi-three-dimensional flow in axial fans," ASME Paper No. 88-GT-15. With permission.)

vortex distribution) can lead to very low axial velocities near the hub and even to negative values, or even flow reversal—a clearly unacceptable design condition.

Figure 9.5 shows a comparison for the predicted velocities associated with a blade design with constant swirl generation across the blade. Here, $\Psi = b = 0.2$ and $a = c = 0$ with $\varphi_1 = 0.4, z_h = 0.5$, and $\eta_T = 0.9$. The linearized solution for φ_2 is shown along with an "exact" calculation by numerical integration (Wright and Ralston, 1987). A certain amount of error is seen to be creeping into the linearized solution as a result of the assumptions. The largest error appears to be about 4% in φ_2 at the blade tip, $z = 1.0$. The corresponding value of ε is about 0.12.

Figures 9.6 and 9.7 show similar comparisons for linear swirl distributions (i.e., a forced vortex distribution) with $c = 0.23$ and $c = 0.5$ ($\Psi = 0.23z$ and $\Psi = 0.50z$ with $a = b = 0$), respectively. The first case, Figure 9.6, shows a worst-case error of about 6%, while the second case, Figure 9.7, shows a worst-case 11% error, suggesting that the limits of the allowable perturbation have been exceeded when b is as high as 0.5. Note that several other parameters have been changed in Figure 9.7.

The approximate solution is compared with experimental data as well using the test results of a study at NACA (Kahane, 1948) in which several fans with highly 3D, tip-loaded (constant swirl) flows were designed and tested (Wright, 1988). The results showed that maximum errors generated are

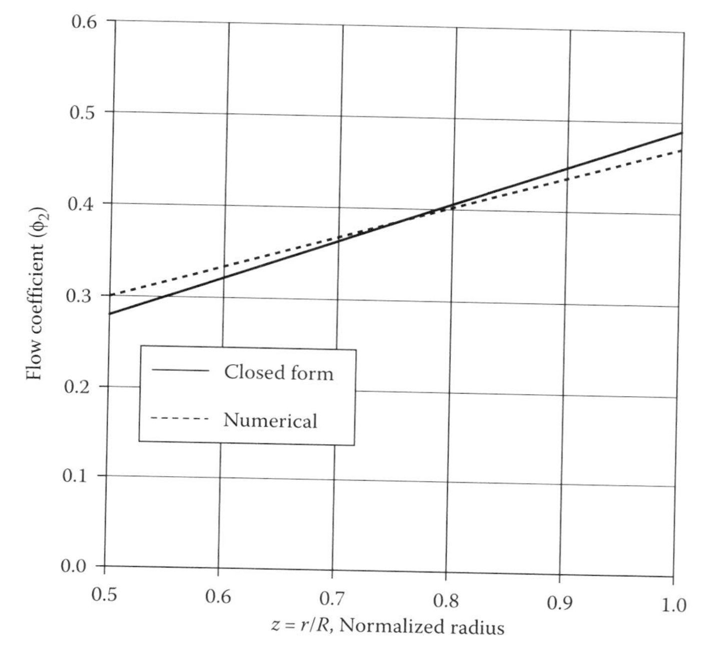

FIGURE 9.5 Comparison of linearized and exact numerical results for axial velocity profile for the uniform swirl distribution: $\varphi_1 = 0.4$, $z_h = 0.5$, $\eta_T = 0.9$, $b = 0.2$, and $a = c = 0$.

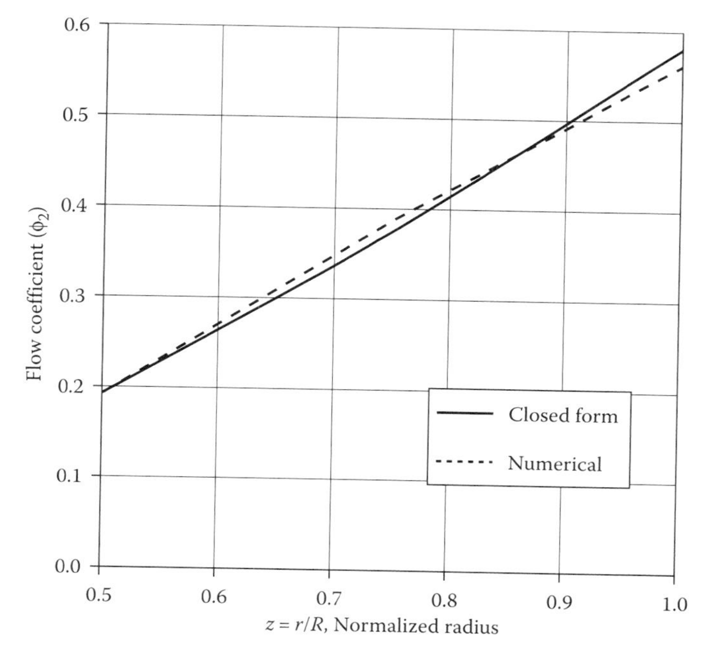

FIGURE 9.6 Comparison of linearized and exact numerical results for axial velocity profile for the forced vortex swirl distribution: $\varphi_1 = 0.4$, $z_h = 0.5$, $\eta_T = 0.9$, $c = 0.23$, and $a = b = 0$.

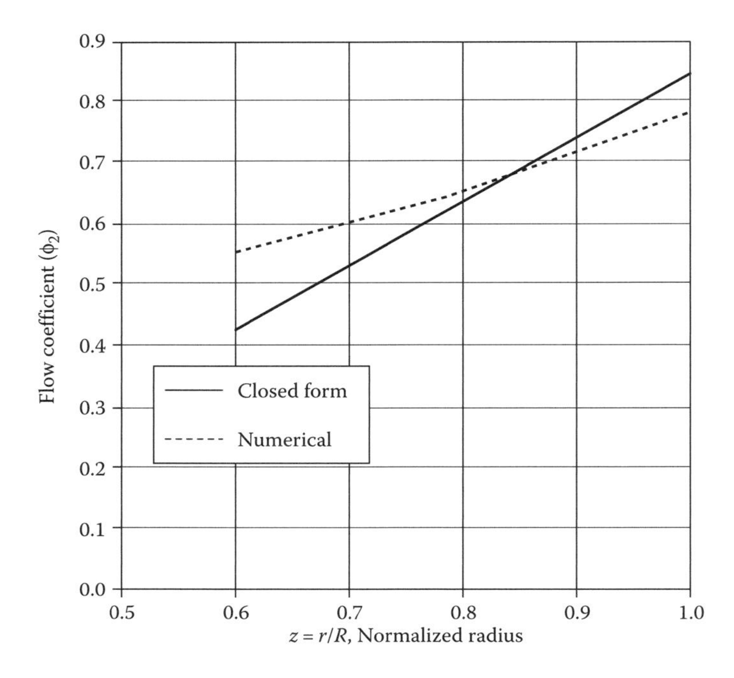

FIGURE 9.7 Comparison of linearized and exact numerical results for axial velocity profile for the forced vortex swirl distribution: $\varphi_1 = 0.667$, $z_h = 0.6$, $\eta_T = 0.9$, $c = 0.5$, and $a = b = 0$.

between 5% and 14% (near stall) when the linearized SRE calculations are used for the S2 solutions. To put these results into better perspective, overall performance prediction using the linearized and numerical SRE solutions is compared with the experimental data from NACA in Figures 9.8a and b. Of course, there are many components to the complete performance predictions, not only the SRE/S2 surface model, but also the S1 (blade-to-blade) model, the loss/efficiency prediction, and the overall validity of the Q3D assumption. Nevertheless, agreement with experiment is good for both methods of calculating the S2 solution, suggesting that local differences in the φ_2 distributions are submerged in the integrated performance results.

9.4 Extension to Nonuniform Inflow

The SRE model developed in the previous section was for uniform inflow to the blade row. Such a situation might well characterize the inlet to a single-stage machine or to the first stage of a multistage machine; however, the inflow to OGVs or later stages of a multistage machine is most likely nonuniform. As one might suspect, this makes the determination of the outlet velocity

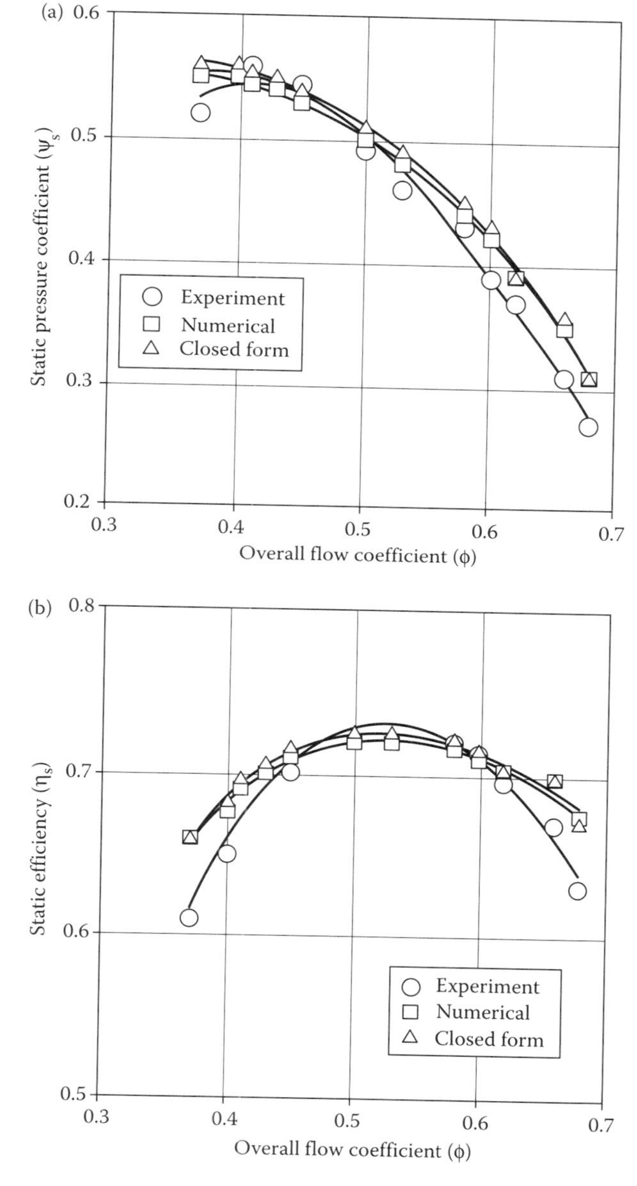

FIGURE 9.8 Comparison of the measured static pressure (a) and efficiency (b) of an NACA constant swirl axial fan to predictions using linearized and numerical throughflow models. (From Wright, T., 1988. "A closed-form algebraic approximation for quasi-three-dimensional flow in axial fans," ASME Paper No. 88-GT-15. With permission.)

distribution more complicated and often a numerical solution is the only option. One exception is the flow in a stator vane row. Jackson and Wright (1991) extended the linearized model of radial equilibrium flow to include the variation of axial velocity into the vane row. The fundamental assumptions, beyond the existence of radial equilibrium, were as follows:

- The total pressure through the vane row is constant (no energy addition and negligible losses) and density is constant so that the Bernoulli equation applies.
- The vane inlet flow distribution is identical to the rotor outlet distribution.
- The vane row removes all of the swirl.

Assigning the subscript 3 to the vane inlet station and 4 to the vane outlet station, one obtains

$$\Delta p_T = p_{T4} - p_{T3} = \left(p_4 + \frac{\rho C_{x4}^2}{2} + \frac{\rho C_{\theta 4}^2}{2}\right) - \left(p_3 + \frac{\rho C_{x3}^2}{2} + \frac{\rho C_{\theta 3}^2}{2}\right) = 0. \tag{9.25}$$

Differentiation with respect to r, using the radial equilibrium specification, Equation 9.1, and putting $C_{\theta 4}$ as identically zero yields, on some rearrangement,

$$\frac{1}{2}\frac{dC_{x4}^2}{dr} - \frac{C_{\theta 3}^2}{r} - \frac{1}{2}\frac{dC_{x3}^2}{dr} - \frac{1}{2}\frac{dC_{\theta 3}^2}{dr} = 0. \tag{9.26}$$

If one normalizes the velocities and radius according to $\varphi = C_x/U$, $\Upsilon = C_\theta/U$, and $z = r/(D/2)$, then Equation 9.26 can be rewritten in the differential form as

$$\frac{d\varphi_4^2}{2} = \left(\frac{\Upsilon_3^2}{z}\right) dz + \varphi_3\, d\varphi_3 + \Upsilon_3\, d\Upsilon_3. \tag{9.27}$$

It is assumed that $\Upsilon_3 = \Upsilon_2 = \Psi = a/z + b + cz$ so that

$$\varphi_4\, d\varphi_4 = \eta_T b\, dz + 2\eta_T cz\, dz, \tag{9.28}$$

where the higher-order terms have again been neglected. Integration yields the now familiar form

$$\frac{\varphi_4^2}{2} = \eta_T bz + \eta_T cz^2 + K, \tag{9.29}$$

where K is again the constant of integration to be determined from conservation of mass.

Using the perturbation form for φ_4,

$$\varphi_4 = [1 + \varepsilon_4(z)]\varphi_3. \tag{9.30}$$

A subscript "4" has been included on ε_4 as a reminder that this ε is different from the one that describes the blade outlet flow (which might be written as ε_2). Applying the principle of conservation of mass and solving for K and ε_4 allow the vane outlet velocity distribution to be expressed as

$$\varphi_4 = \varphi_3 + \left(\frac{\eta_T}{\varphi_3}\right)\left[b\left(z - f_1[z_h]\right) + c\left(z^2 - f_2[z_h]\right)\right], \tag{9.31}$$

with

$$f_1[z_h] = \left(\frac{2}{3}\right)\frac{1 - z_h^3}{1 - z_h^2} \quad \text{and} \quad f_2[z_h] = \frac{1 + z_h^2}{2}. \tag{9.32}$$

Recall that φ_3 is not a constant but is itself a function of z and z_h.

9.5 Q3D Model for Centrifugal Machines

As is the case for axial flow machines, the flow in a mixed flow or radial flow machine is complicated and 3D. Simpler approaches to analysis of centrifugal flow fields include the sort of one-dimensional calculations used in earlier chapters, based on mean flow considerations. However, considering the rather large variations of the flow properties along the blade span seen in most studies, even at the design point, it seems that an accurate analysis must deal in some manner with the inherent three dimensionality of the flow in centrifugal impellers.

A number of computational fluid dynamics techniques have been applied to these flows, including finite-difference, finite-volume, and finite-element methods. Both inviscid and viscous fluid models have been used. A very brief review of these methods will be included in the following chapter.

As in axial flow machine analysis and design, the Q3D flow model provides a somewhat simpler, although less rigorous, approach to the solution of these flows. Unlike axial flow, however, there are essentially no closed-form solutions for the Q3D model in centrifugal machines—all are also numerical and computer based. This does not mean that one might just as well resort to the fully 3D methods for design as the Q3D models are often more easily applied to design studies.

A comparison of fully 3D and Q3D methods is available (e.g., Wright, 1982; Wright et al., 1984b) along with extensive comparisons to experimental data. In Wright et al. 1984b, inviscid, incompressible calculations of the velocities and surface pressures on the blade and shroud surfaces of the impeller of

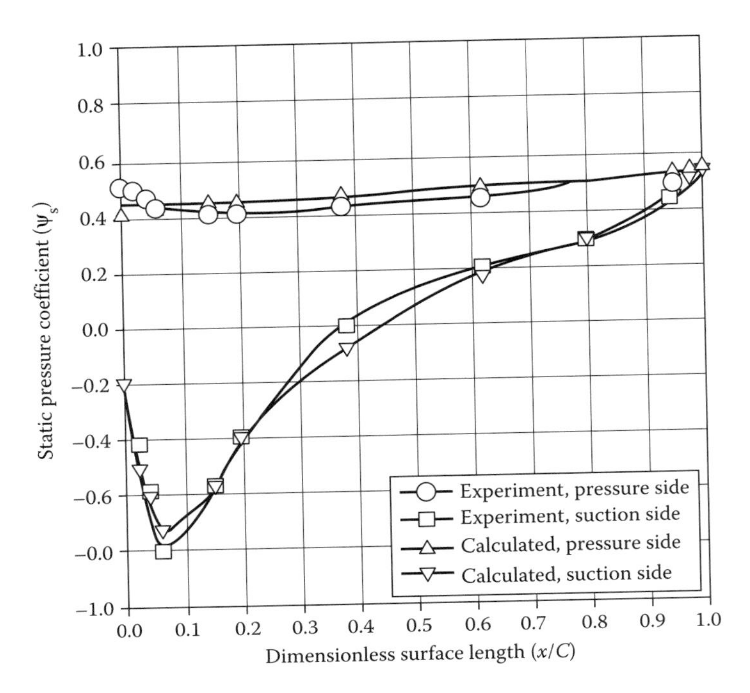

FIGURE 9.9 Comparison of the predicted blade surface pressures and experimental results (design point flow at midspan). (From Wright, T., Tzou, K. T. S., and Madhavan, S., 1984b, *ASME Journal for Gas Turbines and Power*, 106(4). With permission.)

a large centrifugal fan were computed using a 3D finite-element potential flow analysis and a finite-difference Q3D analysis of the same flow based on the "streamline curvature" method (Katsanis, 1977). An instrumented fan impeller fitted with 191 internal pressure taps provided experimental evaluation of the computational results. The fan was a wide-bladed centrifugal fan with specific speed $N_s = 1.48$ and $D_s = 2.14$. Experimental data and analytical prediction comparisons are shown in Figures 9.9 and 9.10. Figure 9.9 shows the results for the blade surface at the midspan at design flow rate (i.e., "S1" data). Figure 9.10 shows the results for the shroud and inlet bell mouth on a path midway between the camber lines of adjacent blades at design flow (an "S2" solution). The blade and shroud static pressures are given in the coefficient form, where

$$\psi_s = \frac{p_s}{\rho U_2^2/2}.$$

In general, the predictions of the more rigorous, fully 3D finite-element method yielded predictions that are in no better agreement with the experimental data than those of the Q3D method. Stated more positively, both methods provide good predictions of the surface velocities and pressures measured in the fan. As was true in the axial flow studies, both of these

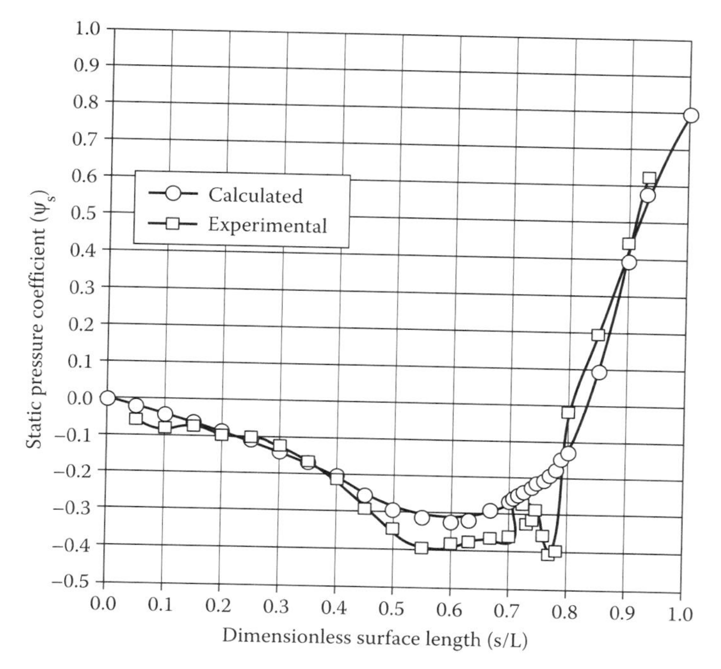

FIGURE 9.10 Comparison of the predicted and experimental surface pressures on the impeller shroud and the inlet bell mouth (design flow). (From Wright, T., Tzou, K. T. S., and Madhavan, S., 1984b, *ASME Journal for Gas Turbines and Power*, 106(4). With permission.)

numerical methods required extensive input data preparation and refinements. For the purposes of preliminary design or systematic parametric analysis of design features, simpler and perhaps less rigorous Q3D-based methods of analysis seem desirable.

9.6 Simpler Solutions

The simple approach we used for axial flow analysis—SRE—cannot be readily adapted for mixed or radial flow analysis. Difficulty arises from the fact that a meridional streamline generally has a very significant amount of curvature. The pressure gradient term must then involve the C_m value and its related radius of curvature, r_{mc}, and the C_θ value and its local radius of curvature, the radial coordinate r. There have been relatively successful attempts to model equilibrium flows using a full streamline curvature model that accounts for both terms (Novak, 1967; Katsanis, 1977). These models are collectively called "streamline curvature" analyses, in which the pressure gradient must be balanced by two "centrifugal force" terms $\rho C_m^2/r_{mc}$ and $\rho C_\theta^2/r$. The differential equation for velocity is quite complicated, having the form

(Japikse and Baines, 1994)

$$C_m \frac{dC_m}{dq} = \text{Radial equilibrium terms} + \text{Streamline curvature terms} + \text{Blade force terms}, \tag{9.33}$$

where q is a coordinate (nearly) perpendicular to a streamline and lying in the S2 plane. The right-side terms involve derivatives of velocities along the streamline trajectory, and angles between the streamline and the coordinates and radial velocities must be accounted for. The differential equation cannot be solved by analytical means, so a numerical (computer) solution is necessary. To complicate matters, the meridional streamlines interior to the machine must adjust their locations in terms of the governing equations, so that the streamline shape is known only at the shroud and hub surfaces prior to the solution of the flow field. Therefore, an iterative approach with fairly complicated relations between flow and geometry is required to generate a converged solution. Convergence is of course strongly linked to the designer's choice of the distribution of the generation of swirl velocity, C_θ, along the stream surface as well.

As always, iterated numerical procedures can provide good information and design guidance, but they may become difficult or troublesome to work with because of their input requirements and the ever-present worries of convergence and accuracy of the solutions. To illustrate general principles, to carry out parametric studies, or to investigate preliminary design layouts, a simple, reasonably accurate analysis would be very useful. There have been many attempts to develop such an analysis, and it is fair to say that most of these methods have been only partially successful. For example, Davis and Dussourd (1970) developed a calculation procedure based on a modified mean flow analysis for radial flow machines by modeling the viscous flow in the impeller channels to provide "mean" velocities at an array of positions along the flow path. Simplifying assumptions concerning the rate of work input along the flow path allowed the calculation of distributed diffusion and estimation of loss through the impeller. Meridional curvature was modeled by interpolating streamline shapes between the shroud and hub surfaces. Because of this approximation, the results were valid primarily for very narrow impeller channels. In spite of the approximations, Davis and Dussourd were able to establish reasonable predictions for the performance of several different compressor impellers. An example is shown in Figure 9.11, in which predicted and experimental values are in reasonable agreement. The work, which was carried out in a proprietary environment, showed good promise but, unfortunately, many details were missing from the paper.

Some success has been achieved in analytical models of important portions of a centrifugal flow field, as opposed to a full impeller flow model. These are particularly successful when coupled with the results of a more extensive numerical model. As a simple but important example, consider the development of a simple correlation for the distribution of approach velocity

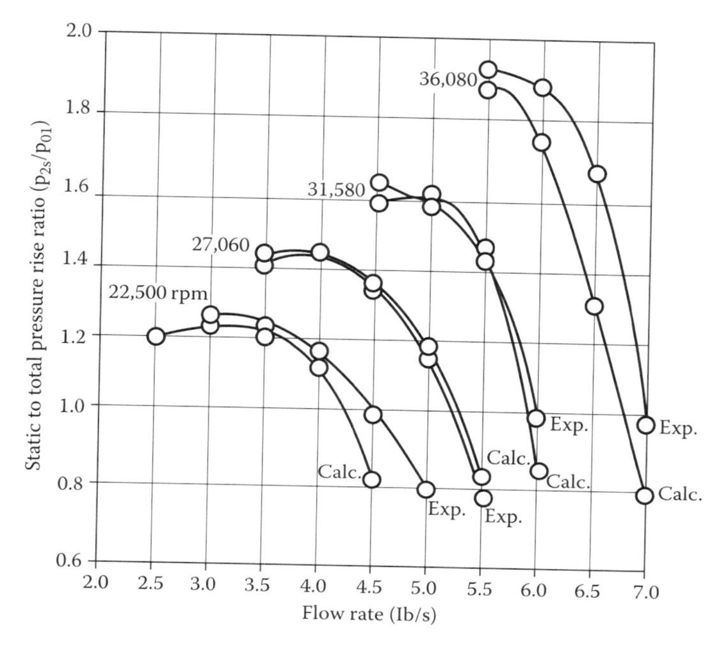

FIGURE 9.11 Comparison of the Davis and Dussourd model to measured compressor performance. (From Davis, R. C. and Dussourd, J. L., 1970, "A unified procedure for calculation of off-design performance of radial turbomachinery," ASME Paper No. 70-GT-64. With permission.)

to blade leading edge in a mixed flow or centrifugal flow machine. Wright et al. (1984b) began by using a streamline curvature method to analyze a systematically designed set of shroud configurations. The geometry of the inlet region was modeled as a hyperbolic shroud shape characterized by its minimum radius of curvature, as shown in Figure 9.12. Surface velocities were calculated along the shroud contour to estimate the ratio of the maximum value of C_m compared to the mean value $\bar{C}_m$. The mean value at any station along the shroud is related to the inlet velocity $C_x = Q/(\pi d^2/4)$ and the cross-sectional area at that location on the shroud. It was assumed that the maximum value of $C_m/\bar{C}_m$ would occur at the minimum radius of curvature, as indicated in Figure 9.12. Obviously, the ratio would be greater than unity. Samples of the surface velocity calculations are shown in Figure 9.13.

Considering the typical inlet velocity diagram and using the impeller tip speed ($U_2 = ND/2$) to normalize the velocity, the maximum relative velocity (at the shroud), W_{max}, can be expressed as

$$\frac{W_{max}}{U_2} = \left[\left(\frac{d}{D}\right)^2 + F^2\left(\frac{\bar{C}_m}{U_2}\right)^2\right]^{1/2}, \quad (9.34)$$

where F is a "curvature function," $F > 1$.

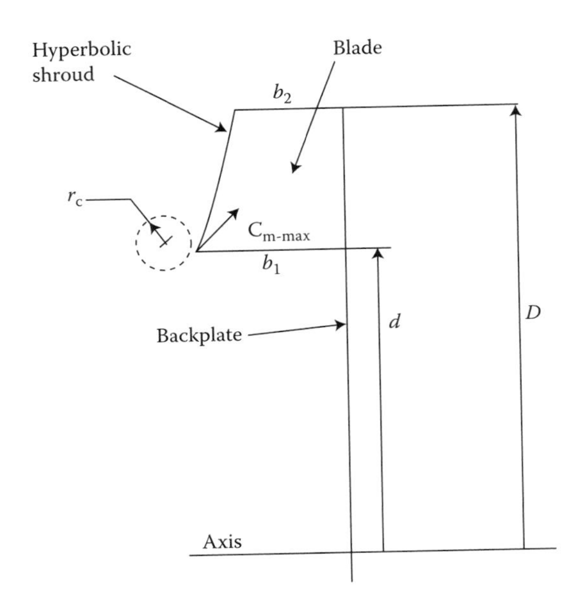

FIGURE 9.12 Geometry used in the hyperbolic shroud study.

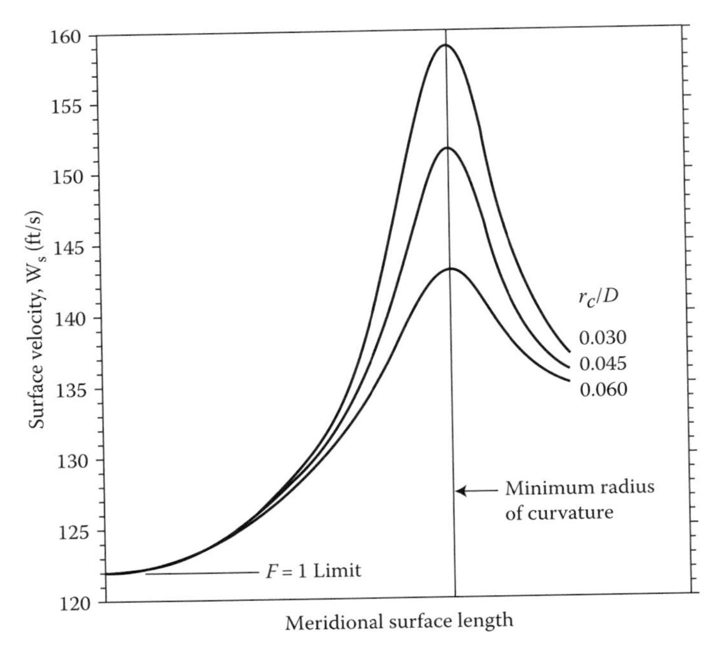

FIGURE 9.13 Surface velocity results for hyperbolic shrouds of different minimum curvature.

Based on simple ring vortex modeling of the computed results, F was assumed to take the form

$$F = 1.0 + \frac{\text{Constant}}{(r_c/D)^n}. \tag{9.35}$$

F would be expected to approach 1.0 for very large values of r_c/D (e.g., in a cylindrical duct). The functional model for F proposed by Equation 9.35 is compared with values from "exact" numerical calculations in Figure 9.14.

With these concepts in place, one is now in a position to begin constructing a heuristic model for the cross-channel distribution of $C_m(y)$ in the region of a blade leading edge (y is a coordinate measured from the shroud toward the hub and perpendicular to the shroud at the blade inlet—see Figure 9.15). Now assume that the distribution of C_m from shroud to hub ($0 < y < b_1$) varies according to $C_m/\bar{C}_m = 1 + ay^n$. Fitting a curve to the curvature function shown in Figure 9.14 gives $\left(C_m/\bar{C}_m\right)_{max} = 1 + [0.3/(r_c/D)^{1/3}]$ so that the velocity distribution is

$$\frac{C_m}{\bar{C}_m} = \frac{1 + \left[0.3/(r_c/D)^{1/3}\right]}{1 + ay^n}. \tag{9.36}$$

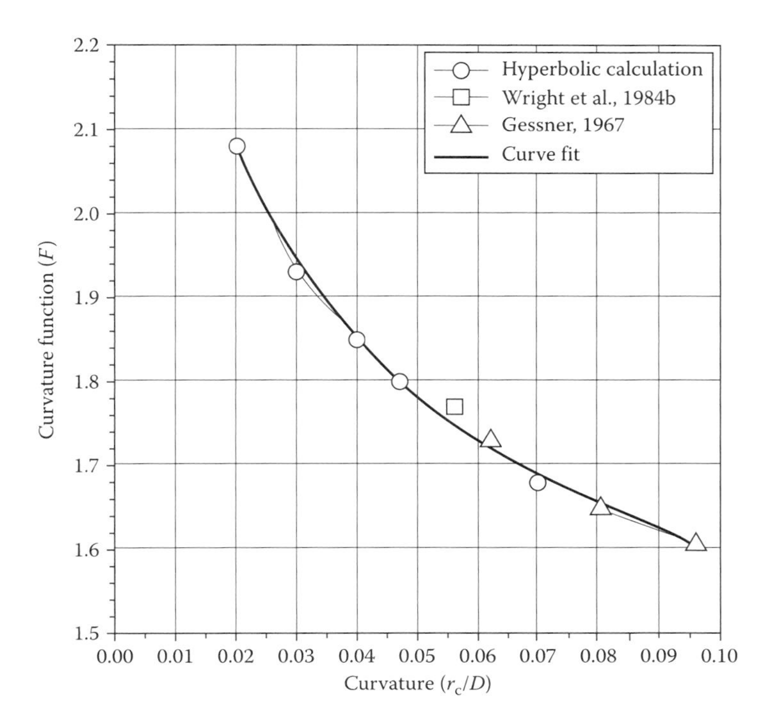

FIGURE 9.14 The curvature function, F, based on the hyperbolic shroud model.

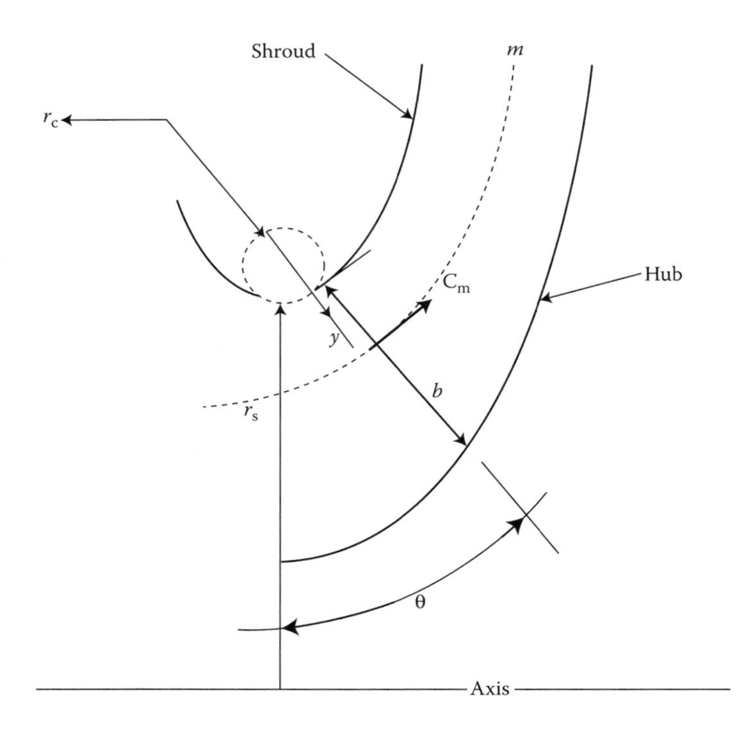

FIGURE 9.15 Geometry for the blade inlet velocity distribution model.

Mass conservation is satisfied by requiring

$$\frac{\int C_{\mathrm{m}}\,\mathrm{d}A}{\int \bar{C}_{\mathrm{m}}\,\mathrm{d}A} = \frac{\int C_{\mathrm{m}}\,\mathrm{d}A}{\bar{C}_{\mathrm{m}}A} = 1.0. \tag{9.37}$$

For the geometry defined by Figure 9.15, we can express $\bar{C}_{\mathrm{m}}$ in terms of volume flow rate as

$$Q = \bar{C}_{\mathrm{m}}\left[2\pi r_{\mathrm{s}}^2(1-(b'/2)\cos\Phi)b'\right]; \quad b' = b_1/r_{\mathrm{s}}.$$

Q is also calculated by

$$Q = 2\pi r_{\mathrm{s}}\int_0^{b_1} C_{\mathrm{m}}\,\mathrm{d}y - 2\pi\cos\Phi\int_0^{b_1} C_{\mathrm{m}}y\,\mathrm{d}y. \tag{9.38}$$

Defining dimensionless variable $y' = y/r_{\mathrm{s}}$ yields

$$Q = C_{\mathrm{mu}}2\pi r_{\mathrm{s}}^2\left(\int\left(\frac{C_{\mathrm{m}}}{C_{\mathrm{mu}}}\right)\mathrm{d}y' - 2\pi\cos\Phi\int\left(\frac{C_{\mathrm{m}}}{C_{\mathrm{mu}}}\right)y'\,\mathrm{d}y'\right). \tag{9.39}$$

Now write $C_m/\bar{C}_m = (C_{m0}/\bar{C}_m)(1 + ar_s y')$ (i.e., $n = 1$), where $C_{mo} = Q/A_o$ and A_0 is the area of the "eye," where $\Phi = 0$. In terms of r_c and r_s, noting that $r_s = d/2$ and the machine represented in Figures 9.13 and 9.14 had $d/D = 0.7$, one obtains

$$\left(\frac{C_m}{\bar{C}_m}\right)_{\max,0} = 1.0 + \frac{0.426}{(r_c/r_s)^{1/3}}. \tag{9.40}$$

Finally, defining for convenience $g = ar_s$, the constraint on mass conservation reduces to

$$\frac{b'[1 + ((b' \cos\Phi)/2)]}{1 + [0.426/(r_c/r_s)^{1/3}]} = \int \left[\frac{1}{1 + gy'}\right] dy' - \cos\Phi \int \left(\frac{y'}{1 + gy'}\right) dy', \tag{9.41}$$

with integration limits from 0 to b'. This relationship provides a solution for g. The integrals evaluate readily to yield

$$\frac{b'[1 + ((b' \cos\Phi)/2)]}{1 + [0.426/(r_c/r_s)^{1/3}]} = \frac{1}{g}\left[\ln(1 + gb')\left(1 + \frac{\cos\Phi}{g}\right) - b' \cos\Phi\right]. \tag{9.42}$$

For a specific choice of geometry, the left side of this equation is simply a constant. The right side is badly transcendental and cannot be solved directly for g. Fortunately, it can be solved iteratively for a given set of geometric choices, yielding a parametric function of the three variables b', Φ, and r_c/r_s. A computer code was used to solve the transcendental equation for a wide range of the variables ($0 < b' < 1.25$, $0° < \Phi < 90°$, $0.05 < (r_c/r_s) < 1.2$). A sample of results for g, with $\Phi = 90°$, is shown in Figure 9.16.

The results are rather nonlinear in both r_c/r_s and b'. Computer-assisted linear regression produced the following functional form for g:

$$g \approx \left[0.264(\cos\Phi)^{1.18} + \frac{0.955}{b'}\right]\left(\frac{r_c}{r_s}\right)^{-0.38}, \tag{9.43}$$

with a correlation coefficients of 0.999.

Thus, the heuristic model for the blade inlet velocity distribution is

$$\frac{C_m}{\bar{C}_m} = \frac{C_{m0}}{1 + gy'}, \tag{9.44}$$

with

$$\bar{C}_m = \frac{Q}{2\pi r_s^2 b'[1 - (b' \cos\Phi/2)]}, \tag{9.45a}$$

$$\frac{C_{m,\max,0}}{\bar{C}_{m,0}} = 1.0 + \frac{0.426}{(r_c/r_s)^{1/3}}, \tag{9.45b}$$

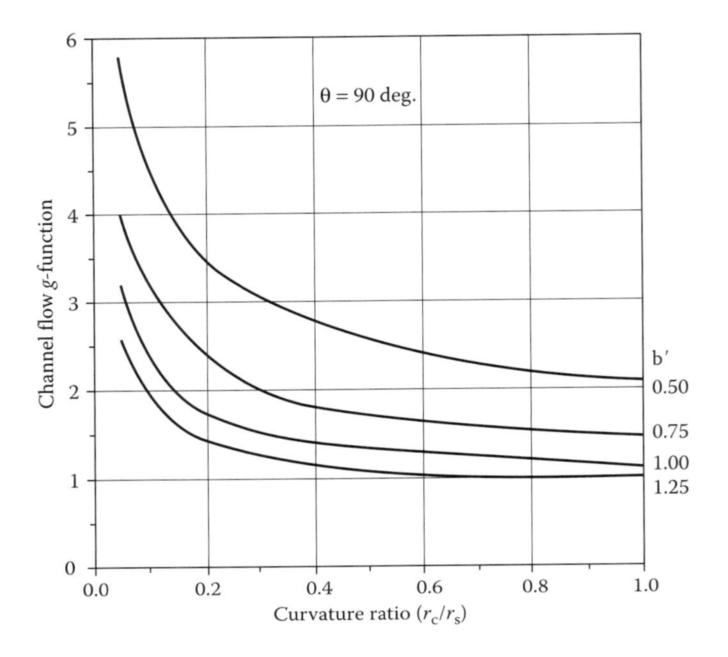

FIGURE 9.16 Sample results for the g-function.

$$y' = \frac{y}{r_s} \qquad (9.45c)$$

$$b' = \frac{b}{r_s}. \qquad (9.45d)$$

with g calculated from Equation 9.43.

Comparison with experimental data (Wright et al., 1984b) seems to validate the heuristic model, although it is somewhat conservative. The approximate equation for $C_m/\bar{C}_m$ overpredicts the value at $y = 0$ in every case, although not greatly (about 7–8%). The flow model is useful for illustration and preliminary design layout in radial and mixed flow geometries, at least for predicting the inflow velocity distribution near the blade leading edge. It is suspected that the increasing importance of viscous influences further downstream in the blade channel will significantly reduce the relevance of the simple model as the flow approaches the channel exit.

Example: Investigation of a Series of Impeller Inlet Configurations

Consider a centrifugal pump impeller in which $\bar{C}_m$ is constant along the inlet channel, implying a constant flow area along lines emanating from the center of the shroud radius of curvature for values of Φ between 0° and 90°; that is,

$\bar{C}_m = \bar{C}_m(\Phi = 0) = \bar{C}_0$. If one defines b_0 as b at $\Phi = 0$, this requires that

$$\bar{C}_m = \bar{C}_o = \frac{Q}{\pi[r_s^2 - (r_s - b_0)^2]} = \frac{Q/\pi r_s^2}{1 - (1 - b_0')^2}, \tag{9.46}$$

so that

$$b' = \frac{1 - (1 - B_1 \cos\Phi)^{1/2}}{\cos\Phi}, \tag{9.47}$$

where

$$B_1 = 1 - (1 - b_0')^2, \tag{9.48}$$

to ensure the constant average meridional velocity.

For this example, select $b_0' = 0.8$ so that as a function of Φ, b' behaves as shown in Figure 9.17. Note that at $\Phi = 90°$, the equation is indeterminate and must be evaluated in the limit. One can describe the flow path in detail if values for r_s and D are chosen. Choosing $D = 0.5$ m, $r_s = 0.333$ m, and $Q = 4\,\text{m}^3/\text{s}$, one obtains $\bar{C}_m = 12.9$ m/s. The impeller's cross-section shape then depends on r_c, as sketched in Figure 9.17 for $r_c = 0.05$, 0.10, and 0.20 m.

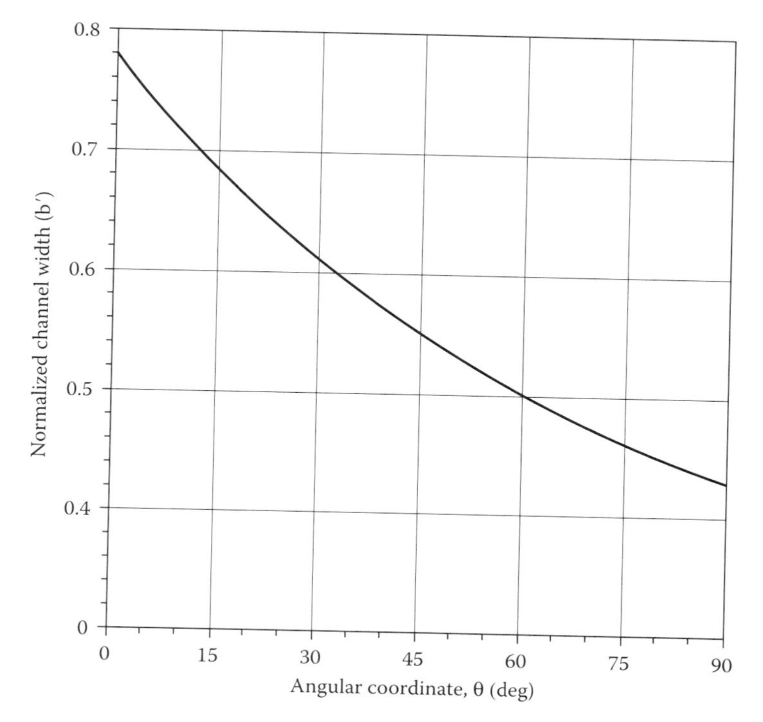

FIGURE 9.17 Normalized channel width distribution in the entrance region of the impeller.

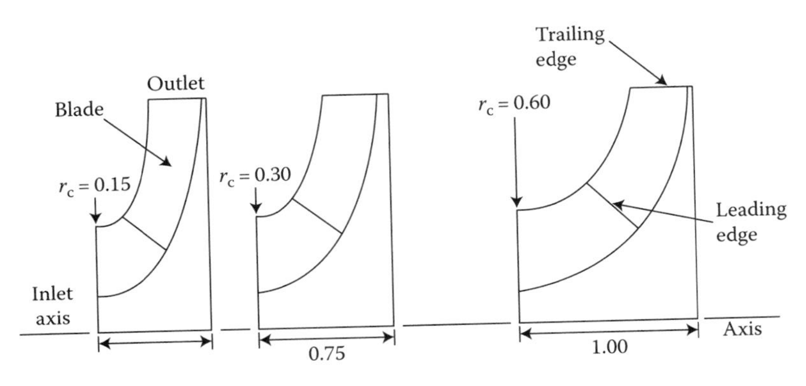

FIGURE 9.18 Three geometric channel layouts for the example study (dimensions in meters).

From the viewpoint of mechanical design and cost, it is desirable to keep the axial length of the impeller as small as possible; however, a constraint on axial length can sometimes lead to negative effects on the velocity distribution in the blade channel. Choose the leading edge of the blade to lie along the $\Phi = 45°$ line as shown in Figure 9.18 and use the previously developed equations to analyze the variation of the blade approach velocity, C_m, across the blade (from hub to shroud). Figure 9.19 shows the calculated results as C_m versus y/b (fraction of channel width). For the configuration with $r_c = 0.05$ m (the impeller with the smallest axial length), C_m ranges from 27 m/s at the shroud to 8 m/s at the hub. For the largest radius of curvature considered, $r_c = 0.40$ m, C_m ranges from 21 to 11 m/s. The

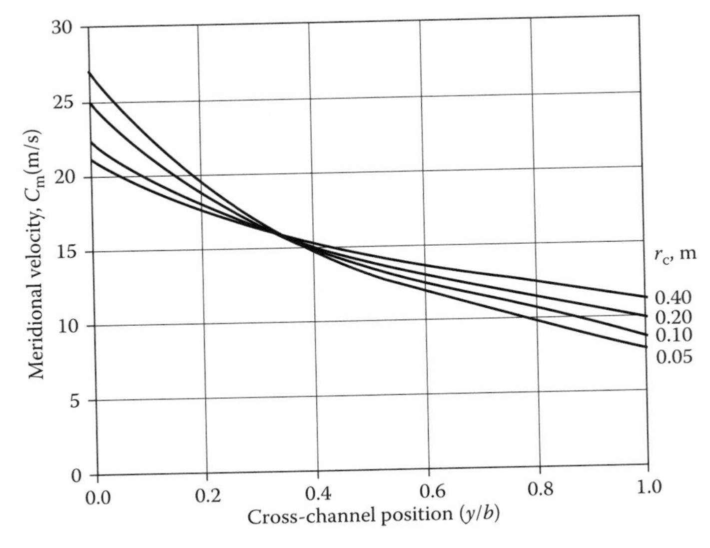

FIGURE 9.19 Blade approach velocity distribution across the channel for the blade channels shown in Figure 9.18.

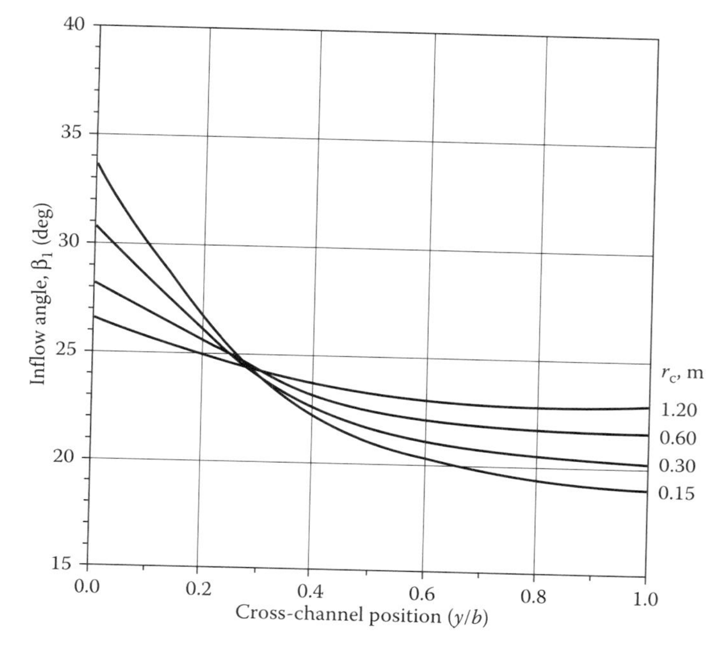

FIGURE 9.20 Inflow angles across the channel for the blade channels shown in Figure 9.18.

ratio of minimum to maximum meridional velocity has been reduced from 3.4 to 1.9 by changing the radius of curvature.

This relatively large change in C_m is reflected in the variation of the relative inflow angle β_1 $(=\tan^{-1}(C_{m1}/U_1))$. Assuming that $N = 1200$ rpm, the variation of inflow angle β_1 across the channel is shown in Figure 9.20. The smallest curvature, $r_c = 0.05$ m, yields the largest range, $18° < \beta_1 < 34°$ (16° of blade twist), while the largest curvature, $r_c = 0.40$ m, gives the smallest range, $23° < \beta_1 < 27°$ (4° of twist). This small amount of twist will lead to a much simpler blade, and, as will soon be demonstrated, lower levels of diffusion in the impeller. However, the impeller will be longer in the axial direction, leading to higher costs and perhaps to excessive overhang of the rotating mass or large bearing spans.

Assuming that the impeller is designed to generate 60 m of head and that the efficiency is 0.9, one can calculate a relative outlet velocity of $W_2 = 18.8$ m/s (recall that $D = 0.5$ m and C_m is constant). Using the mean value $\bar{C}_m$ at the blade leading edge gives a mean value of $W_1 = 22.4$ m/s, which leads to a very mild (mean) de Haller ratio of 0.84. The de Haller ratio for streamlines at the hub and the tip is different; Table 9.1 shows values for the different shroud curvatures. Values along the shroud are marginal at best. Thus, in addition to inducing a highly twisted blade, a small radius of curvature can lead to high levels of diffusion, unstable flow along the shroud, and perhaps reduced efficiencies as well.

As implied earlier, the larger radius of curvature allows use of an untwisted blade without the penalties associated with large leading edge incidence or increased de Haller ratios.

TABLE 9.1

De Haller Ratio Dependence on Shroud Radius of Curvature

r_c (m)	W_2/W_1 Shroud	W_2/W_1 Hub
0.05	0.63	0.93
0.10	0.65	0.97
0.20	0.66	1.00
0.40	0.67	1.04

9.7 Summary

This chapter introduced the concepts of fully 3D flow in the blade and vane passages of a turbomachine. Because of the inherent great complexity of these flows, the idea of using an approximate solution involving the superposition of two physically compatible flow fields in mutually perpendicular planes was developed. This Q3D approach was pursued throughout the chapter to provide a tractable calculation scheme to aid in understanding the flows and in generating reasonable design geometry.

The classical results developed and applied here allow the relaxation of the rather simple but convenient assumption of uniform throughflow velocities, either purely axial or purely radial. Rather, one is able to estimate radial distributions of axial velocities in axial flow impellers and streamline-normal distributions of meridional velocities in centrifugal machines. In either case, the flow is constrained to maintain equilibrium between stream-normal pressure gradients and centrifugal force on the fluid along streamlines.

For axial flow machines, this concept is further simplified to SRE and a nonlinear flow modeling equation is developed. This rather intractable form is linearized using the concept of a perturbation solution around the known uniform solution. The result is an approximate, somewhat restricted, closed-form algebraic solution for nonuniform axial flow in a turbomachine. Comparison of the approximate result with numerical solutions and experimental results illustrates the utility of the method and its limitations.

Q3D flow solutions are also available for centrifugal flow fields; however, they generally require numerical computer-based solutions. Prior analytical and experimental results were used to illustrate some of the approximate but still rather intractable approaches to the problem. Simpler solutions are sought for use in broad studies of geometric influences on flow and for approximate layout of a machine in preliminary design. Such an approximate method was outlined and developed. A heuristic analytic model for the flow approaching the inlet of a centrifugal impeller was developed and shown to yield valuable design information.

The approximate nature of these simpler prediction tools to estimate 3D flow effects was emphasized throughout the chapter. Although they are capable of creating useful information for preliminary design layout and yield insight into the complex flow fields, they will in practice be followed by rigorous, fully 3D, fully viscous, and fully compressible computational flow analyses for detailed design calculations.

EXERCISE PROBLEMS

9.1. As outlined in the chapter, the work coefficient can be written as

$$\Psi = \frac{a}{z} + b + cz,$$

where a, b, and c are constants. Let $\Psi = b$, with $a = c = 0$. Impose the streamline equilibrium condition on $d\phi_2^2$ with the conservation of mass constraint and show that

$$\varphi_2 = \varphi_1 \left[1 - z_1 \left(\frac{\eta_T \Psi}{\varphi_1^2}\right) + \left(\frac{\eta_T \Psi}{\varphi_1^2}\right) z\right],$$

where

$$z_1 = \frac{(2/3)(1 - z_h^3)}{1 - z_h^2}.$$

9.2. With $\eta_T = 0.85$, $\Psi = b = 0.35$, and $\varphi_1 = 0.625$, use the results of Problem 9.1 to explore the influence of nonuniform loading and hub size (z_h) on the outlet velocity distribution. Calculate and graph the distribution of φ_2 for these parameters for $z_h = 0.25, 0.375, 0.5, 0.625, 0.75$, and 0.875.

9.3. Reproduce the results shown in Figure 9.5 for the constant swirl solutions by the small perturbation method and by numerical integration. Use the results of Problem 9.1 to evaluate the other methods shown in Figure 9.5.

9.4. As was done in Problem 9.1, use the radial equilibrium method to describe a special case outlet flow and swirl distribution. Here, use the single term $\Psi = cx$ with c as a constant (i.e., a forced vortex distribution). Show that

$$\frac{\varphi_2}{\varphi_1} = 1 - \left(\frac{\eta_T c}{\varphi_1^2}\right) z_2^2 + \left(\frac{\eta_T c}{\varphi_1^2}\right) z^2,$$

where

$$z_2^2 = \frac{1 + z_h^2}{2}.$$

9.5. Consider the example shown in Figure 9.6. With $z_h = 0.5$, $c = 0.23$, $\eta_T = 0.9$, and $\phi_1 = 0.4$, use the results of Problem 9.4 to evaluate the numerical method and perturbation method solutions shown in Figure 9.6.

9.6. When one allows the outlet swirl velocity distribution to deviate from the free-vortex (a/z) form, one loses the simplicity of the two-dimensional cascade variables across the blade or vane span. Although the flow must now follow a path that has a varying radial position through the blade row, one can still simplify the analysis by considering the inlet and outlet stations only. A common approximation for a fixed radial location on the blade is to consider the flow to be roughly two dimensional by using the averaged axial velocity entering and leaving, along with U $(= U_1 = U_2)$ and $C_{\theta 2}$ to form the velocity triangles. That is, the mean axial velocity

$$C_{xm} = \frac{C_{x1} + C_{x2}}{2},$$

can be used with U to calculate

$$\omega_1 = \tan^{-1}\left(\frac{U}{C_{xm}}\right).$$

Similarly

$$\omega_2 = \tan^{-1}\left(\frac{U - C_{\theta 2}}{C_{xm}}\right).$$

One retains the character of the simple-stage assumptions by a minor adjustment in definition. Use these definitions and the results of Problem 9.2 to lay out the blade sections at hub, mean, and tip radial stations. Use a hub–tip ratio of 0.5.

9.7. Compare the camber and pitch distribution of the blade developed in Problem 9.6 with an equivalent blade for a fan meeting the same performance specifications with a free-vortex swirl distribution. Lay out the blade with $\Psi = a/z$ to give the same performance.

9.8. Determine the minimum hub size achievable for both the free vortex and the constant swirl fans of Problem 9.7. Use a de Haller ratio at the fan hub equal to 0.6 as the criterion for acceptability.

9.9. Repeat the study of Problem 9.8 using $D_L \leq 0.6$ with $\sigma \leq 1.5$.

9.10. Use the approximation for the mean axial velocity (Problem 9.6) with the fan specified in Figure 9.6 to lay out a blade to meet the specified performance. Use the properties given in Figure 9.6, including $z_h = 0.50$.

9.11. Explore the blade design of Problem 9.10 to find the smallest allowable hub size for the fan. Use the de Haller ratio criterion with $W_2/W_1 \geq 0.6$.

9.12. Design a vane row for the fan of Problem 9.5. Use the mean axial velocity across the blade at the entrance of the vane row. Assume that there is no further skewing of axial velocity component within the vane row.

9.13. Design a vane row for the fan of Problem 9.6. Use the same assumptions as were used in Problem 9.12.

9.14. Repeat the design of the blade row of Problem 9.12 using the nonuniform inflow from the rotor to calculate the nonuniform vane row outflow according to

$$\varphi_4 = \varphi_3 + \left(\frac{\eta_T}{\varphi_3}\right)\left[b(z - z_1) + c\left(z^2 - z_2^2\right)\right],$$

with

$$z_1 = \frac{(2/3)\left(1 - z_h^3\right)}{1 - z_h^2} \quad \text{and} \quad z_2^2 = \frac{1 + z_h^2}{2}.$$

9.15. Repeat the design of the vane row from Problem 9.13 using the specifications of Problem 9.14.

9.16. Initiate the design of a vane axial fan to provide a flow rate of $0.5\,\text{m}^3/\text{s}$ with a total pressure rise of 600 Pa. Select the size, speed, and hub–tip ratio and do a preliminary free-vortex swirl layout for the blades and vanes.

9.17. Extend the work of Problem 9.16 by preparing a swirl velocity distribution that is distributed according to $C_{\theta 2} = kr^{1/2}$. (Hint: Although this is clearly not one of the terms in the series adopted for swirl distribution, one can approximate the distribution by fitting three values, one each at the hub, mean radius, and tip, of this function to $\Psi = a/z + b + cz$.)

9.18. With the swirl distribution of Problem 9.17, solve the outlet flow distributions for both blade and vane row and determine the velocity triangles for blade and vane at the hub, mean, and tip stations. Set the hub size by requiring $W_2/W_1 \geq 0.6$.

9.19. Lay out the cascade properties for both blades and vanes for the fan design of Problem 9.17. Use the Q3D flow for the "square-root" swirl distribution and compare the physical shape of these blades and vanes with those of the constant swirl fan of Problem 9.5.

9.20. The mass-averaged pressure rise for an arbitrary swirl distribution can be calculated from the Euler equation with a weighted integration of Ψ across the blade span. Using $\Psi = a/z + b + tcz$, show that

$$\Psi_{\text{avg}} = \frac{\int \Psi \varphi_1 z\,\text{d}z}{\int \varphi_1 z\,\text{d}z},$$

and

$$\Psi_{\text{avg}} = \frac{2a}{1 + z_h} + b + \frac{(2/3)c\left(1 - z_h^3\right)}{1 - z_h^2}.$$

9.21. The usefulness of the work coefficient or swirl velocity form, $\Psi = a/z + b + cz$, and its approximate throughflow solution, can be greatly extended by fitting the constants a, b, and c to an "arbitrary" set of the three values: Ψ_H (at $z = z_h$), Ψ_M (at $z = (1 + z_h)/2$), and

Ψ_T (at $z = 1$) at hub, mean, and tip stations of the blade (as was suggested for a specific case in Problem 9.17). Show that

$$a = (\Psi_H + \Psi_T - 2\Psi_M)\,\frac{z_h(1+z_h)}{(1-z_h)^2},$$

$$b = \Psi_H - \frac{a}{z_h} - cz_h,$$

$$c = \frac{\Psi_T - \Psi_H}{1 - z_h} + \frac{a}{z_h}.$$

9.22. Using the notation of Section 9.5, consider a fan that has $\phi_1 = 1.0$, $\eta_T = 0.90$, $\Psi_H = 0.4$, $\Psi_M = 0.4$, $\Psi_T = 0.3$, and $z_h = 0.5$. This kind of flattened load distribution is sometimes used in heavily loaded fans to relieve severe work requirements near the hub station. A similar free-vortex fan would have $\Psi_H = 0.6$ for the same ψ_M and ψ_T, for a 50% heavier load at the hub. Use the result of Problem 9.21 to determine $\Psi(z)$ (i.e., find a, b, and c to model the "flattened" load distribution).

9.23. For the fan of Problem 9.22, the loading distribution is $\Psi = -0.30/z + 1.40 - 0.80z$.

   a. Use Ψ with the results of Problem 9.20 to find Ψ_{avg}.

   b. Use Equations 9.23 and 9.24 to calculate the axial outflow $\phi_2 = [1 + \varepsilon(z)]\phi_1$ for the fan.

   c. Verify the conservation of mass for the ϕ_2 distribution of part (b).

   d. Note that the Ψ distribution does not have $b, c \ll a$. What effect does this have on the results?

9.24. For the fan of Problem 9.23, the average value, Ψ_{avg}, should have been 0.378 and the outflow distribution should be $\phi_2 = 1.26z - 0.72z^2 - 0.42/z - 0.48 \ln z + 0.901$.

   a. With $\Psi(z) = -0.30/z + 1.40 - 0.80z$ (from Problem 9.22) and using the local value of the axial velocity, calculate the distributions of W_2/W_1 for the blade. (b) Compare these results to the equivalent free-vortex fan with $\phi_1 = \phi_2 = 1.0$ and $\Psi(z) = 0.289/z$. Also calculate and compare the fluid turning angles.

9.25. For the two fans of Problem 9.23, use $D_L = 0.55$ at the hub station, $D_L = 0.45$ at the mean stations, and $D_L = 0.35$ at the tip station. From the results for W_2/W_1 and $\Psi = C_{\theta 2}/U$ for these fans, calculate the required distributions of solidity and compare the results for the two designs.

9.26. From the geometry of Figure 9.17, derive Equations 45a–d. (Hint: Recall that $\bar{C}_m$ is defined as the integrated mean value of C_m across the channel. Assume that $r_c \ll r_s$.)

9.27. Beginning with Equations 45a–d, verify Equations 9.46 and 9.47.

9.28. Show that, in the limit case for θ approaching 90° (Figure 9.15), Equation 9.47 reduces to $b' = B_1/2$.

9.29. In the example given in Section 9.6, the channel width, b, was developed from the eye ($\theta = 0°$) to the horizontal level ($\theta = 90°$). If we assume that the channel area remains constant ($\bar{C}_m$ = constant) beyond the 90° level, determine the outlet widths (at $r = D/2$) for all three impellers shown in Figure 9.18.

9.30. Repeat the outlet width calculations of Problem 9.29 requiring the outlet area to match the eye of the impeller (the inlet area at $\theta = 0°$). Can you explain the apparent discrepancy in the two results?

10

Advanced Topics in Performance and Design

10.1 Introduction

Previous chapters treated the flow inside a turbomachine relatively simply. Even when we included the first-order effects of three dimensionality, we restricted our models and analyses to inviscid flow. Although agreement was often rather good, at other times there were some significant differences between the predicted and measured flow behavior. To attempt to achieve a completely realistic description of turbomachinery flows or to seek significant improvement in flow modeling and design, one must address some of the following complexities: the flow may be unsteady, three dimensional, highly viscous/turbulent, and frequently separated; the machines are run at significant off-design conditions, which might lead to stall and loss of stability.

This chapter will survey several of these advanced topics. It will be rather different from the previous chapters in that very few usable analytical or design tools will be presented and, accordingly, there are no exercise problems at the end of the chapter. The focus is primarily on identifying and describing the physical phenomena involved and the advanced computational tools available. The chapter is fairly brief and does not attempt to completely describe the "state of the art" at the time of publication. It is assumed that the reader is reasonably familiar with graduate-level topics in fluid mechanics—especially viscous flow and turbulence.

10.2 Freestream Turbulence Intensity

One of the more vexing characteristics of turbomachinery flows is the existence of high levels of turbulence intensity imbedded in the core flow. This turbulence results from blade, vane, or end-wall boundary layers, wake shedding, and the influence of less than ideal inflows with obstructions, bends,

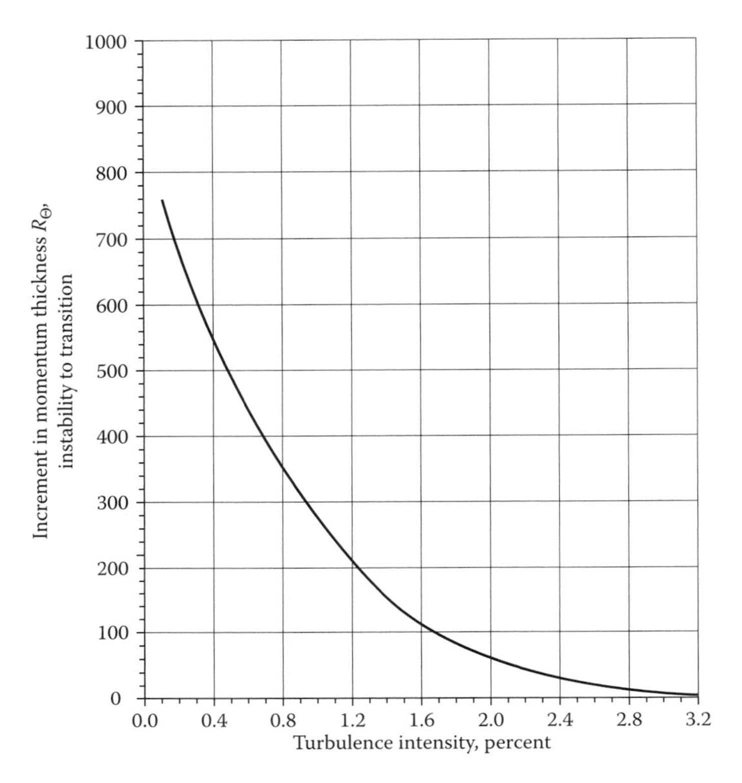

FIGURE 10.1 Influence of freestream turbulence level on transition. (From Schlichting, H., 1979, *Boundary Layer Theory*, 7th ed., McGraw-Hill, New York. With permission.)

or other interferences. This high turbulence level in the core flow influences the transition to turbulence on blades and other surfaces. In terms of classical boundary layer considerations (Schlichting, 1979; White, 2005), the so-called instability development length, that is, the distance between the first boundary layer instability and the point of completed transition to turbulent flow, decreases remarkably in the presence of high turbulence levels (Granville, 1953). This behavior is illustrated here in Figure 10.1, adapted from Schlichting. The instability point is characterized by the magnitude of the Reynolds number based on the momentum deficit thickness, Re_Θ. The fully turbulent condition occurs downstream from instability, at some larger values of Re_Θ. As shown in Figure 10.1, the increment between instability and a fully turbulent layer can be reduced by an order of magnitude with freestream turbulence levels as low as $T_u = 2\%$, where

$$T_u = 100(u'^2/\bar{u}^2)^{1/2}.$$

u' is the random, unsteady fluctuating velocity component and $\bar{u}$ is the time-averaged velocity.

The influence of high turbulence level on convective heat transfer and skin friction is also very important (Kestin et al., 1961 in Schlichting, 1979), and it is capable of causing multiple orders of magnitude changes in heat transfer at values of freestream turbulence intensity from about 2.5% upward. To place these numbers in the proper context, turbulence intensity in a compressor typically ranges from about 2.5% at the inlet station to levels near 20% at stations downstream of blade and vane row elements (Wisler et al., 1987). This implies that classical considerations of laminar flow instability and gradual transition may not provide an accurate assessment of such phenomena inside the typical turbomachine.

An extensive review and a critical study of laminar to turbulent transition in turbomachines were carried out by Robert Mayle (Mayle, 1991). He ably traces some of the history of the earlier efforts in this field and summarizes the manner in which transition can be expected to occur. Importantly, "natural transition," outlined above in terms of stability and boundary layer development, is largely irrelevant to the flow in a gas turbine engine or other types of turbomachines. Typically, transition in turbomachine boundary layers occurs at values of Re_Θ of an order of magnitude smaller than in the classical smooth-flow results. With high levels of freestream turbulence, the sequential process of instability and vortex formation and growth is replaced with the rapid formation of turbulent spots, triggered or created directly by the freestream turbulence. For cases of practical interest, this "bypass" of the normal development process narrows the range of significant phenomena. Further, since blade and vane flows are frequently characterized by strong pressure gradients and low Reynolds numbers, the occurrence of laminar separation near the leading edge can also lead to a short cutting of the transition process. Often, laminar-turbulent transition may take place in a free shear layer above a separation bubble, with subsequent reattachment to the blade surface as a turbulent boundary layer. To complicate matters, the flow in a turbomachine is usually unsteady owing to the motion of imbedded wakes shed by the blade, vane, and inlet elements in the upstream flow, so boundary layer processes are clearly far from simple. Further work in the area of the effects of unsteadiness and "freestream" turbulence can be found in Mayle and Dullenkopf (1990), Mayle (1991), Addison and Hanson (1992), Wittig et al. (1988), and Dullenkopf and Mayle (1994).

10.3 Secondary and 3D Flow Effects

As was discovered fairly early in the development of turbomachinery technology, the flow in and around the rotating elements of a machine is characterized by complex secondary flows. These effects are associated with blade-to-blade and radial pressure gradients, centrifugal force effects, spanwise flows, and flows through the seals and clearance gaps of the machines. Johnsen and

Bullock (1965) present an early review of these messy effects and provide a clear discussion of the flows along with contemporary work by Smith (1955) and Hansen and Herzig (1953).

These early investigations confirmed the presence of "passage vortex" flows driven by the roll-up motion forced on the low momentum end-wall boundary layer fluid by the basic flow turning process and blade-to-blade pressure gradient. The presence of spanwise radial flow of vane boundary layers (inward flow due to pressure gradients) and blade boundary layer fluid (outward flow due to centrifugation) showed even higher levels of complexity. End-wall effects in regions where a clearance gap exists between vane or blade and an end wall (casing or hub) can lead to the roll-up of additional vortices in

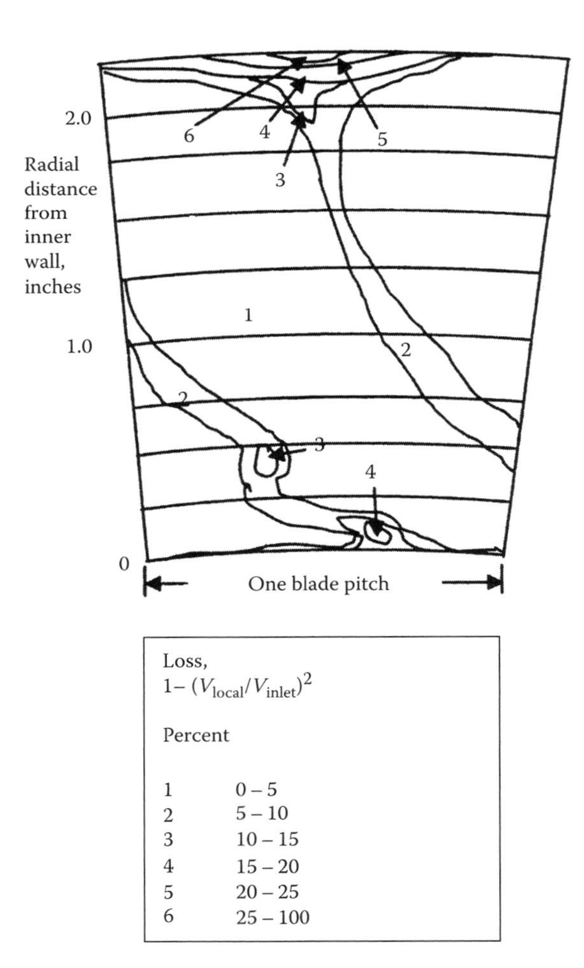

FIGURE 10.2a Discharge velocity measurements for an axial machine. (Kinetic energy deficit contours, from Johnsen, I. A. and Bullock, R. O., 1965, "Aerodynamic design of axial flow compressors," NASA SP-36. With permission.)

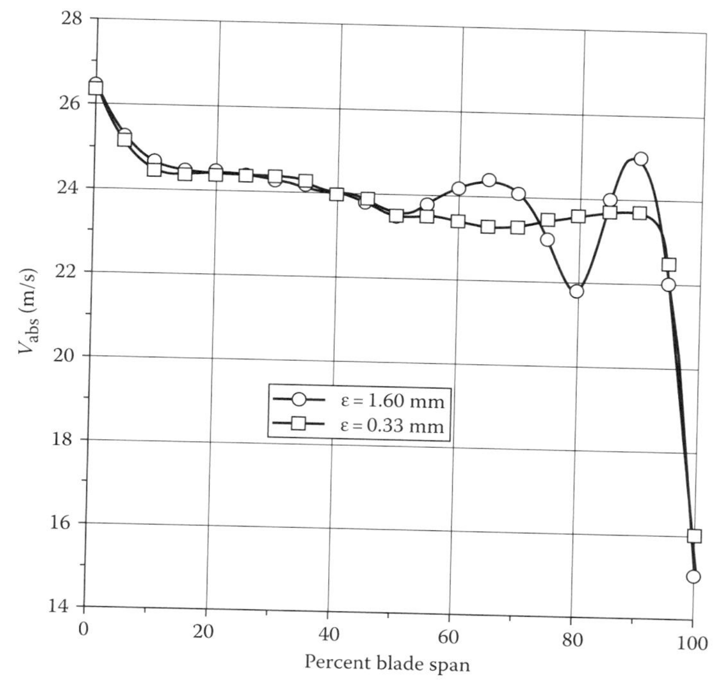

FIGURE 10.2b Absolute discharge velocity from a centrifugal impeller with large and small inlet clearance. (From Wright, T., 1984c, *ASME Journal for Gas Turbines and Power*, 106(4). With permission.)

the flow passages. Figure 10.2a shows kinetic energy loss contours in a blade flow passage of an axial compressor downstream of the rotor. Buildup of centrifuged boundary layer fluid and tip flow effects as well as passage vortex formation near the inner end wall can be seen clearly. Figure 10.2b shows a somewhat similar effect in a centrifugal blower passage (Wright et al., 1984c), which includes the influence of impeller inlet gap. In both cases, the basic flow is significantly perturbed by the secondary effects.

In the decades since the work reported by Johnsen and Bullock, intensive investigation of these secondary flow phenomena has continued. Improved measurement and flow tracing techniques (Denton and Usui, 1981; Moore and Smith, 1984; Gallimore and Cumpsty, 1986a,b) contributed to the basic understanding of these flows. The concurrent modeling and computational work of these years led to two major concepts of how to handle the problem.

Adkins and Smith (1982) and Wisler et al. (1987) concentrated on convective transport models of secondary flow to provide explanation and modeling. On the other hand, Gallimore and Cumpsty (1986a,b) pursued equally successfully a model based on the turbulent diffusion process as the dominant mechanism of spanwise mixing. The two camps appeared to be rather different and nor was willing to accept the notions of the other. In a later paper, Wisler et al. (Wisler et al., 1987) published something of a landmark paper, with commentaries by Gallimore and Cumpsty, L. H. Smith, Jr., G. J. Walker, B. Lakshminarayana, and others, which reconciled the two views as components of the same problem and its solution; apparently, either convective secondary flow or diffusive turbulent mixing can dominate the mixing processes, depending on the region of the passage. Therefore, an adequate model must include both flow mechanisms.

Subsequently, Leylek and Wisler (1991) re-examined the physical measurements and flow tracing data in terms of the results of extensive computational fluid dynamics (CFD) work based on a detailed 3D Navier–Stokes solution using high-order turbulence modeling. The results (and those of Li and Cumpsty, 1991a,b) are extensive and support the view that both convective transport and turbulent mixing are important to a clear understanding of fully viscous, 3D internal flows.

10.4 Low Reynolds Number Effects in Axial Flow Cascades

An area that has historically received little attention in the open literature is the effect of very low Reynolds number on the performance of moderately loaded axial flow equipment. The term "low Reynolds number" describes the conditions of the flow in which the (chord-based) Reynolds number ($Re = \rho Vc/\mu$) lies within the range of 10^3–10^5. The effect of low Reynolds number on the flow appears as the formation of a separation bubble, typically near the leading edge on the blade suction surface. A separation bubble is a detachment of the boundary layer from the blade surface, with the formation of a region of recirculating flow. In many cases, the separating boundary layer is laminar, transition occurs in the separated flow, and the flow reattaches as a turbulent boundary layer. This effect is highly undesirable in axial flow equipment because the separation bubble may cause performance to severely degrade or may lead to stall. Even if there is no stall, the flow turning and losses are affected. Figure 10.3, adapted from the results of Roberts (1975b), illustrates the global effect of low Reynolds number on the flow in an axial cascade. As the Reynolds number decreases, fluid turning, θ_{fl}, decreases rapidly and the total pressure loss coefficient, ζ_1, increases rapidly, as Re_c falls below about 10^5.

These phenomena have been dealt with by Mayle (1991), who also provides an excellent review bibliography. Investigations have been made into this unique phenomenon by Roberts (1975a,b), Citavy and Jilek (1990), Cebeci (1983), Gostelow (1995), Pfenninger and Vemura (1990), Mayle (1990), O'Meara and Mueller (1987), and Schmidt and Mueller (1989). Although there have been methods presented for predicting the flow under these conditions, substantial disagreement on particular aspects of the flow remains. Also,

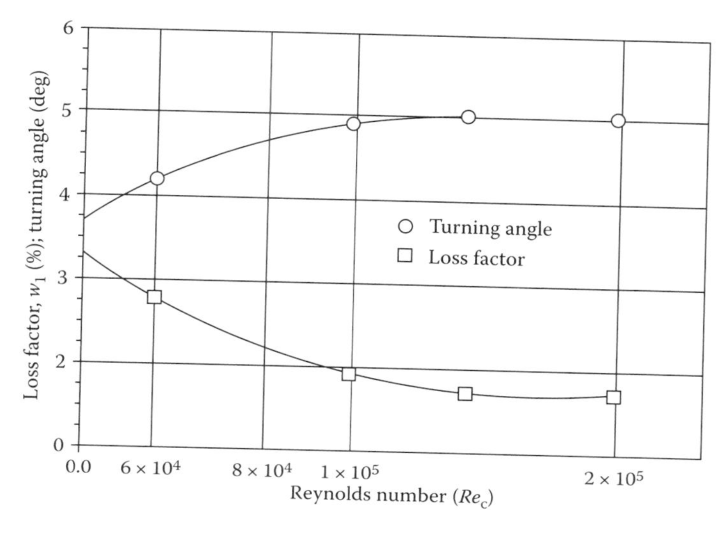

FIGURE 10.3 Effects of low Reynolds number on flow turning and loss. (From Roberts, W. B., 1975a, *ASME Journal of Engineering for Power*, 97, 261–274; Roberts, W. B., 1975b, *ASME Journal of Engineering for Power*, (July). With permission.)

many of these treatments require rather complex methods of integration and interpolation and are not very straightforward in their use. Although other methods are more straightforward in their use, they sacrifice accuracy in the end results. What is missing is a method, short of full 3D Navier–Stokes CFD methods, that will yield a fairly accurate description of these low Reynolds number flows for design purposes.

Wang (1993) used the results of Roberts, and Citavy and Jilek to create a procedure for estimating the changes to losses and deviation angles caused by low Reynolds number separation bubbles that can be used for preliminary design. Wang's results, which were checked against the results of available test data, can be incorporated into existing axial fan design procedures.

Ordinarily, prediction of losses involves many parameters including camber angle, inlet flow angle, outlet flow angle, solidity, Reynolds number, deviation angle, angle of incidence, angle of attack, and turbulence intensity. In order to simplify the process by which an approximation of the losses can be achieved, several of the values are held constant or disregarded in Wang's method of prediction. The angle of incidence is set near zero in most preliminary design routines and should not strongly affect the results. Turbulence intensity was considered negligible, since in most of the test data used it was $< 5\%$. Reynolds number, inlet flow angle, outlet flow angle, camber angle, solidity, and deviation angle are either known values or determined in the design process.

Approximating the losses incurred in an axial flow cascade operating in a low Reynolds number regime begins with the methods of Lieblein (1957) and

Roberts. As discussed in Chapter 8, Lieblein's method was not developed for low Reynolds number loss prediction; in Wang's model, it is used to determine the loss just prior to laminar separation. This can be done if it is assumed that the losses up to separation are not heavily dependent on the Reynolds number. From Figure 10.3, the losses depend weakly on the Reynolds number until separation occurs, after which they become dependent on the Reynolds number.

The loss values up to the point of separation can be approximated as constant; therefore, the loss at separation can be determined from Lieblein's method. As discussed in Chapter 8, this loss is a function of the local diffusion factor and is calculated from a curve-fit equation:

$$\zeta_{1,s} = \zeta_1 = \left(\frac{2\sigma}{\cos\omega_2}\right)\left(\frac{\cos\omega_1}{\cos\omega_2}\right)^2 \times [0.005 + 0.0049(D_L) + 0.2491(D_L)^5], \tag{10.1}$$

where

$$D_L = 1.0 - \left(\frac{\cos\omega_1}{\cos\omega_2}\right) + \left(\frac{\cos^2\omega_1}{2\sigma}\right)(\tan\omega_1 - \tan\omega_2). \tag{10.2}$$

Wang's model is based on the work by Roberts (Roberts, 1975a). Roberts' method can be summarized in the following equations. He proposed calculating the losses beyond separation by adding an increment to the value of loss just prior to separation:

$$\zeta_{1sb} = \zeta_{1b} + \Delta\zeta_1 = \zeta_{1b} + K_1\left(\frac{\phi_c}{a}\right)\left(\frac{t/c}{b}\right)\left(\frac{1}{\sigma}\right)(k\Delta F_D), \tag{10.3}$$

where ϕ_c is the airfoil camber ($\phi_c \geq 5°$), t/c is the thickness ratio, and σ is the solidity. K_1 is a constant and a, b, and k are reference constants determined by Roberts. ΔF_D is a change in the "deviation factor." The deviation factor is related to the deviation angle, camber angle, and solidity by $F_D \equiv \sigma\delta/\phi_c$, thus

$$\Delta F_D = F_{D,sb} - F_{D,s}. \tag{10.4}$$

From the available data, the deviation δ becomes increasingly dependent on Re for values below about 10^5 (Figure 10.3). To express this effect in a simple function, ΔF_D was rewritten as

$$\Delta F_D = \frac{\sigma\delta_{sb}}{\phi_c} - \frac{\sigma\delta_s}{\phi_c} = \left(\frac{\delta_{sb}}{\delta_s - 1}\right)\left(\frac{\sigma\delta_s}{\phi_c}\right). \tag{10.5}$$

From Equation 8.18, $\delta_s = \delta_0 + m\phi_c$, so $(\sigma\delta_s/\phi_c) = m/\sigma^{b-1} + \sigma\delta_0/\phi_c$. For moderate camber and solidity, $\sigma^{b-1} \approx 1$ and $\sigma\delta_0/\phi_c \approx 0$. If a moderate inlet angle (around $\omega_1 = 50°$) is chosen, then $m \approx 1/4$. Collecting all of this, ΔF_D may be approximated as

$$\Delta F_{\mathrm{D}} \approx \left(\frac{1}{4}\right)\left(\frac{\delta_{\mathrm{sb}}}{\delta_{\mathrm{s}}} - 1\right), \tag{10.6}$$

which is accurate within the prevailing uncertainty of the complete model.

Returning attention to $\Delta\zeta_1$, one can replace ϕ_c in Equation 10.3 by $\phi_c = \text{constant} \times C_{L0}$ and then absorb all of the constants into a single value to get

$$\Delta\zeta_1 = K_f C_{L0}\left(\frac{t}{c}\right)\frac{(\delta_{\mathrm{sb}}/\delta_{\mathrm{s}}) - 1}{\sigma}. \tag{10.7}$$

This is as far as "theory" can go. The constant K_f must be provided empirically, and the "deviation ratio" δ_{sb}/δ_s must also be curve fit to data. Using the data

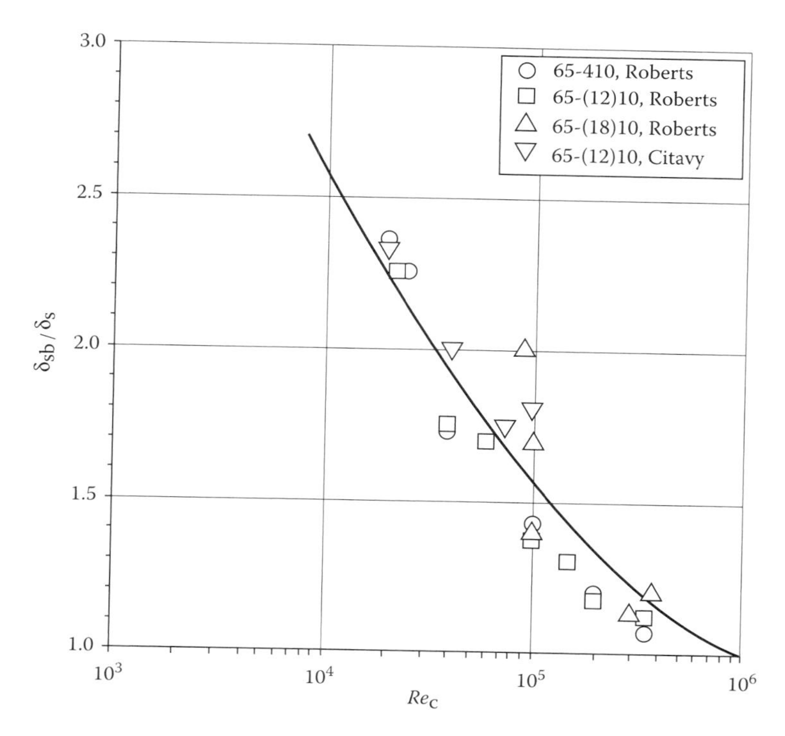

FIGURE 10.4 Comparison and correlation of deviation angles. (From Roberts, W. B., 1975b, *ASME Journal of Engineering for Power*, (July); Citavy, J. and Jilek, J., 1990, ASME Paper No. 90-GT-221. With permission.)

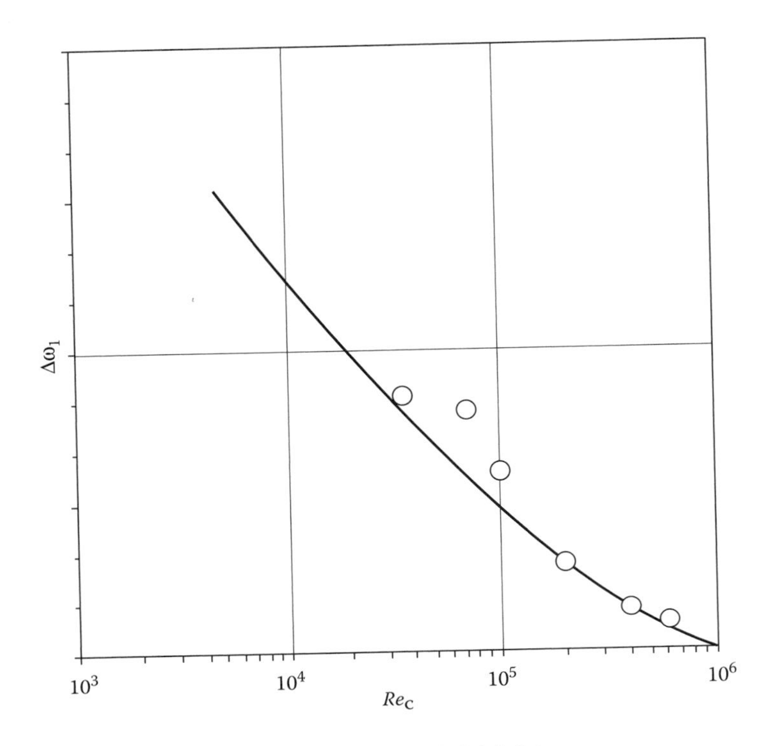

FIGURE 10.5 Comparison to Roberts NACA 65–410 airfoil data.

of Citavy and Jilek (1990), with $C_{L0} = 1.2$, δ_{sb}/δ_s can be fit in a simple form as

$$\frac{\delta_{sb}}{\delta_s} \approx \left(\frac{10^6}{Re_c}\right)^{0.2}. \tag{10.8}$$

This curve, along with the data of Citavy and Jilek (1990) and of Roberts (1975b), is shown in Figure 10.4. The curve fit was constrained to conservatively estimate the Citavy and Jilek data, with a forced fit at $Re = 10^6$.

Further data from Roberts, representing a low camber airfoil, $C_{L0} = 0.4$, are shown in Figure 10.5. The value of K_f is selected to provide a good match to these data; the resulting value is $K_f \approx 2$, so that the final equation for the loss coefficient increment is

$$\Delta\zeta_1 \approx 2C_{L0}\left(\frac{t}{c}\right)\frac{(10^6/Re_c)^{0.2} - 1}{\sigma}. \tag{10.9}$$

Equations 10.8 and 10.9 are the model for estimating both the loss increment and the deviation angle drop that occur in axial flow cascades at low Reynolds numbers ($Re_c \leq 10^6$). The predictions can be tested against other information

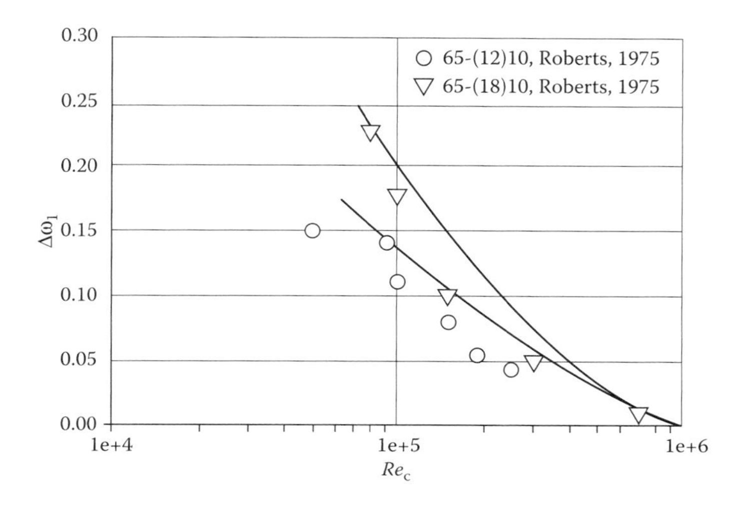

FIGURE 10.6 Comparison to Roberts' NACA 65-(12)10 and NACA 65-(18)10 airfoil data.

published by Roberts, Citavy, and Jilek, and Johnsen and Bullock (NASA SP-36) to establish the consistency and accuracy of the results.

Figure 10.6 shows comparisons of predictions with the data of Roberts for $C_{L0} = 1.2$ and 1.8. One would expect reasonable agreement since Roberts $C_{L0} = 0.4$ data were used to determine K_f and that is the case. There is, however, a general overestimation by the correlation.

Figure 10.7 is a comparison of the correlation with two additional data sets, that of Citavy and Jilek at $C_{L0} = 1.2$ and the British C4 data, quoted

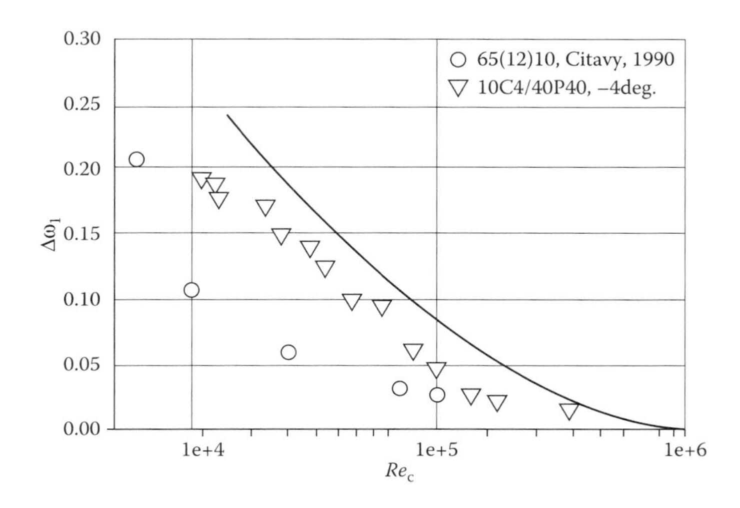

FIGURE 10.7 Comparison to Citavy's data for the NACA 65-(12)10 airfoil and British C4 airfoil data.

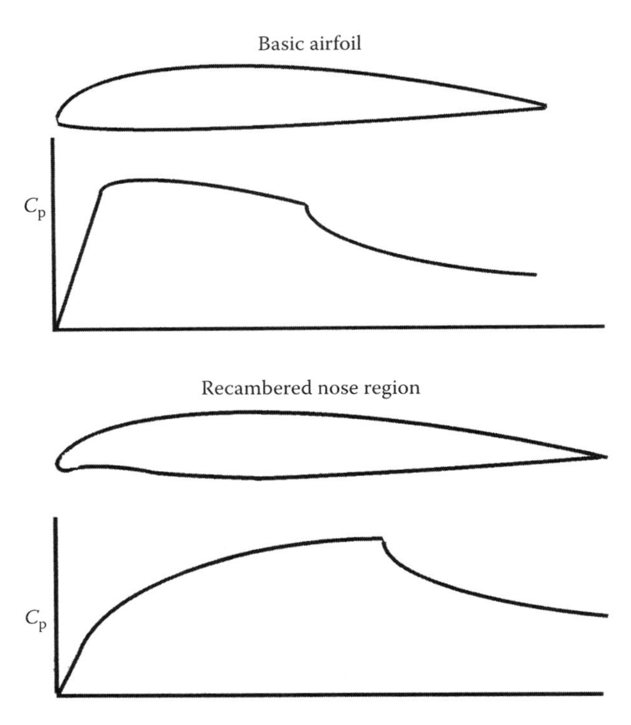

FIGURE 10.8 Airfoil shapes and pressure distributions for good performance at low Reynolds number.

from NASA SP-36 (Johnsen and Bullock, 1965). Agreement is only fair with a significant overestimation of loss in both cases.

Clearly, the results of this model must be taken only as a rough approximation at the lowest values of Re_c ($\leq 10^4$). Although the loss estimation becomes doubtful at very low Reynolds numbers, the prediction of deviation angle is reasonably good, particularly for moderate blade loading.

Airfoils can be designed to handle flows that are within the low Reynolds number regime. Studies performed by Pfenninger and Vemura (1990) address the problems of optimizing the design of low Reynolds number airfoils. These investigations demonstrate that an airfoil can be designed to operate under low Reynolds number conditions and still provide acceptable performance with optimum control of transition. The successful airfoils had a redesigned airfoil whose leading edge region was modified to significantly change the pressure distribution on the suction surface. Figure 10.8 qualitatively illustrates both the airfoil shape and the change in pressure distribution. The fundamental result is that the camber line in the nose region is modified to create a locally negative leading edge incidence of a few degrees. This adequately reduces the adverse pressure gradient on the suction surface responsible for the formation and subsequent bursting of the laminar separation bubble.

These results might be incorporated into design procedures using conventional cascade algorithms by simply adding a few degrees of camber to an existing airfoil selection and reducing the blade pitch angle slightly to avoid excess fluid turning. The resulting configuration should achieve a negative leading edge incidence sufficient to suppress bubble formation or bursting. Unfortunately, this concept has not been verified experimentally.

10.5 Stall, Surge, and Loss of Stability

A fundamental principle for the operation of a turbomachine is that the steady-state operating point of the machine is determined by the intersection of the machine characteristic curve with the system characteristic curve. Only at this "match point" is the flow through the machine equal to the flow through the system, and the specific energy supplied or used by the machine is equal to the specific energy used or supplied by the system. This principle was established as early as Chapter 1 in this book. While this is true for both pumping machinery and turbines, discussion in this section is limited to pumping machinery only (pumps, fans, and compressors) in which the turbomachine supplies energy (or head or pressure) and the system consumes it.

At the match point, the machine and system are in equilibrium and, thus far, the equilibrium has been assumed to be stable. When one considers the limitations of actual pumping systems, then one must investigate the conditions under which the equilibrium is actually stable and the consequences, if any, of instability.

One begins by considering the influence of a small disturbance imposed on the machine. If the disturbance is a small change in flow rate and the machine and system return to the original match point (flow and energy), the system and machine are in stable equilibrium. If the flow rate, when disturbed, continues to change, the system is said to be *statically unstable*. This is illustrated in Figure 10.9 (adapted from Greitzer, 1981). Note that the system curves are typical monotonically increasing "parabolas," but the machine curve in this particular case has a region of negative slope, a region of positive slope, and a region with essentially zero slope.

First, suppose that the point of operation prior to the disturbance is at point A. If the flow were to increase slightly, the pressure supplied by the machine will drop. At the same time, the flow increase will cause the pressure drop in the system to be greater than it was. Because the machine cannot supply the increased pressure required by the system, the flow in the system will decrease, moving back to point A. One can also establish that a momentary decrease in flow will drive the operating point back to A. This return to the unperturbed condition defines the classic requirement for (static) stability.

Now suppose that the machine and system were operating at point B (there is now a higher resistance system). If there were a small increase in flow,

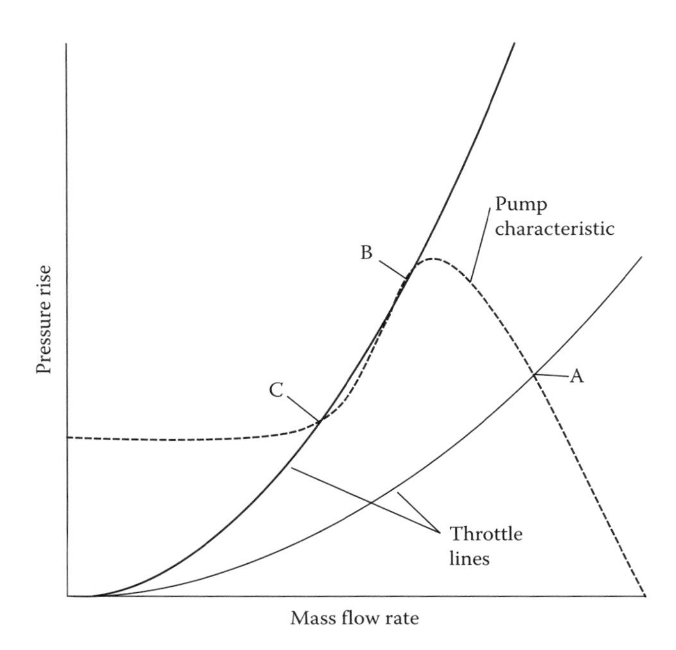

FIGURE 10.9 Illustration of static and dynamic instability conditions for a pumping machine. (From Greitzer, E. M., 1981, *ASME Journal of Fluids Engineering*, 103, 193–242. With permission.)

the machine would generate an increase in pressure rise greater than the corresponding increase in pressure drop due to system resistance. The "extra" pressure would lead to a further increase in flow.* The machine and system do not tend to return to their unperturbed equilibrium and, by definition, are said to be unstable.

This may be summarized as follows:

At point B, the slope of the machine characteristic curve is steeper, or more positive, than the slope of the system resistance curve resulting in unstable operation.

At point A, the slope of the pump curve is much less (in fact, it is negative) than that of the system resistance and the operation is stable.

These slope "rules" are the simplest criteria static stability of a pumping system.

Static stability is not the only concern; one must also be very wary of conditions that can lead to *dynamic instability*, which is the occurrence and growth of oscillations of the flow rate around the initial setting. Referring to Figure 10.9, dynamic instability can occur when the pump is trying to operate very nearly at the peak pressure rise of the pump.

---

* Again, one can work out a corresponding situation for a momentary decrease in flow.

A simple pumping system involving a closed volume of fluid $\forall_f$, a flow length L, and a flow cross-sectional area A can be analyzed to determine criteria for both static and dynamic stability. Greitzer (1996) derived the following criteria

$$\left(\frac{dp_{machine}}{d\dot{m}}\right) < \left(\frac{dp_{system}}{d\dot{m}}\right) \quad \text{(for static stability)} \tag{10.10}$$

and

$$\left(\frac{dp_{machine}}{d\dot{m}}\right) < \frac{K_s}{dp_{system}/d\dot{m}} \quad \text{(for dynamic stability).} \tag{10.11}$$

K_s is a system property defined as

$$K_s \equiv \left(\frac{\gamma p_1 L}{\rho \forall_f A}\right), \tag{10.12}$$

where γ is the ratio of specific heats. K_s is positive definite and may be quite small for systems of large volume or short flow path length. This implies that dynamic instability may occur when $dp_{machine}/d\dot{m}$ is slightly positive. Dynamic instability can and often does occur near the zero slope condition, shown as point C in Figure 10.9. The condition expressed by Equation 10.11 is clearly less restrictive than the criterion for static instability (Equation 10.10) and more readily encountered in a system with increasing resistance.

Greitzer (1996) has provided a simple and clear explanation of the underlying process leading to dynamic instability in terms of energy required to sustain oscillation. He examines the product of the increments to performance, $\delta\dot{m}$ and $\delta\Delta p$, which are the perturbations in mass flow and pressure rise. There are, as stated, two possible conditions of interest: the case of positive machine curve slope and negative machine curve slope. Assuming quasi-steady behavior in the machine, the first case of interest is when the change in $\dot{m}$ is positive and the change in Δp is positive so that the product is always positive. In this case, energy is being supplied by the pump to drive the oscillation. For a negative machine slope, the change in $\dot{m}$ is always of opposite sign to the change in Δp, the product is always negative, and the energy is being removed from the flow by a dissipation process, damping out the oscillation.

Now consider the flow conditions within the machine that correspond to the development of unstable performance. The required positive, or at least zero, slope condition on the pump characteristic is normally accompanied by a stalled (separated) flow condition in the passages of the machine, as discussed in terms of diffusion limits in Section 6.8. Frequently, the stall or flow separation is of the rotating type, where one or more flow passages or groups of flow passages experience separation. The flow leaves the blade surfaces and/or the passage end walls, and these stall "cells" move (rotate) relative to the impeller itself.

Rotating stalls have been documented in the turbomachinery technical literature (Greitzer, 1981; O'Brien et al., 1980; Laguier, 1980; Wormley et al.,1982; Goldschmeid et al., 1982; Madhavan and Wright, 1985) in both axial and centrifugal pumping machines. If the initial flow separation occurs on the suction surface of a blade, as is the case for flow rates below the design flow, then the stalled cell has a natural tendency to move to the next suction surface in a direction opposite to the direction of rotation of the impeller. Figure 10.10 illustrates how this happens. At reduced flow rates, the flow approaches a blade at increased incidence. This creates a very steep pressure gradient on the suction surface, leading to flow separation. As the main flow is diverted to the passage behind the first-stalled passage, the leading edge incidence is further increased on the blade ahead and decreased on the blade behind, resulting in separation on the blade ahead and mitigation of separation on the blade behind. This forward motion of the stall cell or cells is the form taken by classical rotating stall at reduced flow rate. Stall cells can also propagate in the opposite direction when flow is excessively above the design flow rate, or when the flow has been deliberately preswirled in the direction of motion of the impeller (as in the use of prespin control vanes to reduce blower performance) as shown by Madhavan and Wright (1985).

The speed of movement (i.e., the cell rotating frequency) observed in the stationary frame of reference differs from the rotating speed or frequency by the relative motion of the cells. When cells of rotating stall are fully established, the pressure pulsation observed in the stationary frame occurs at two-thirds of running speed for the classical suction surface stall and at four-thirds of running speed for the pressure surface stall at high flow or preswirled inflow.

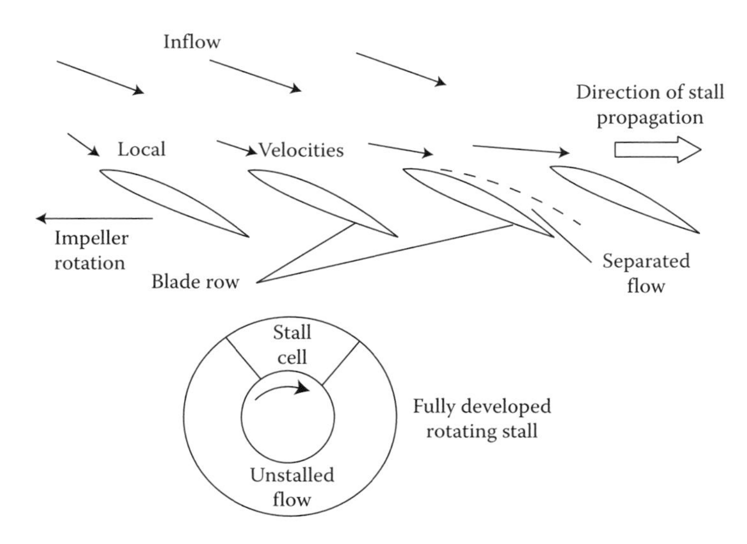

FIGURE 10.10 Sketch illustrating stall cell formation and propagation.

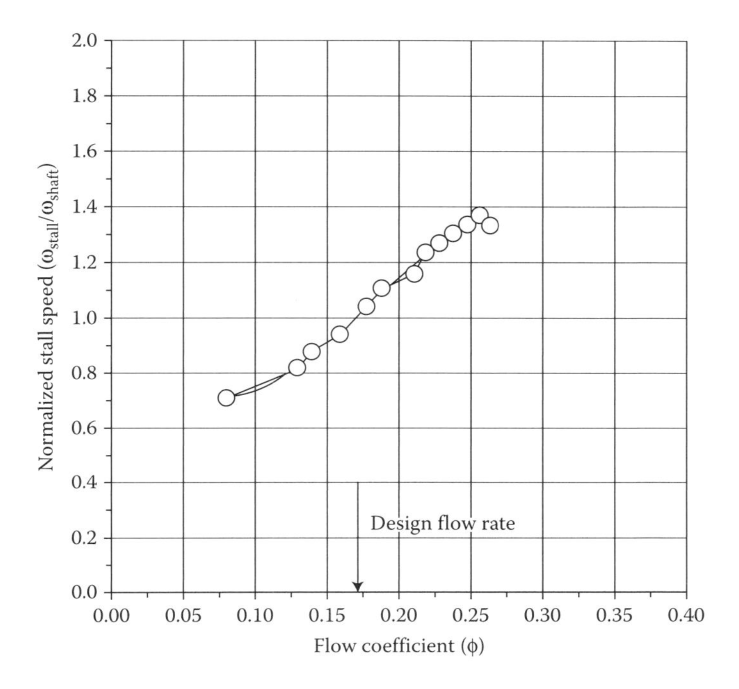

FIGURE 10.11 Stall cell speed and pressure pulsation frequency in a centrifugal fan. (From Madhavan, S. and Wright, T., 1985, *ASME Journal of Engineering for Gas Turbines and Power*. With permission.)

Figure 10.11, from centrifugal fan experiments, shows the variation of pressure pulsation frequency as a function of flow rate variation about the mean flow.

Conditions leading to stall may be investigated by using the analogy between blade passages and diffusers. Based on the diffuser data of Sovran and Klomp (1967), as discussed in Chapter 7, one can estimate the maximum static pressure increase in a cascade as a function of geometry. One interprets the parameter $\psi_{s\text{-}max}$ as the stalling value of the cascade and L/w_1 as the geometric parameter (Figure 7.37), where

$$\psi_{s\text{-}max} = \frac{\delta p_s}{\rho U_2^2}. \tag{10.13}$$

This concept is reinforced in an earlier correlation by Koch (1981), where he estimates $\psi_{s\text{-}max}$ as a function of the length ratio L/g_2, where L is the blade chord and g_2 is the "staggered pitch," $g_2 = s\cos\omega_2^*$ (in Section 6.8, we used simply s as the diffuser width). Koch's correlation is shown in Figure 10.12 (without the supporting experimental data) and shows that one can expect to achieve, at most, a pressure rise coefficient of perhaps 0.6 or less before stall.

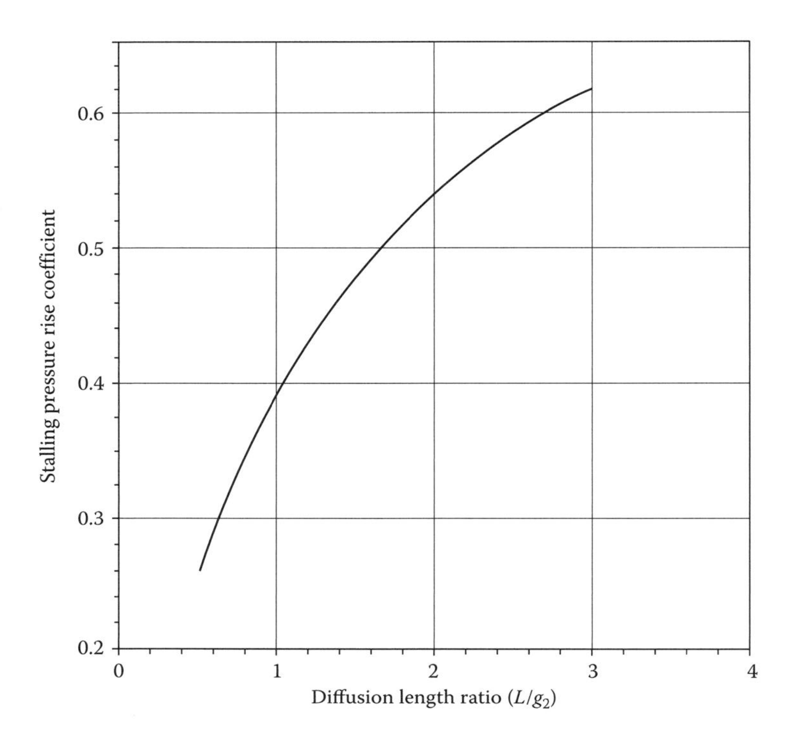

FIGURE 10.12 Koch's correlation for the stalling pressure rise coefficient.

As pointed out by Greitzer (1981), Wright et al. (1984a), and Longley and Greitzer (1992), the onset of stall in the impellers of both axial and centrifugal machines can be significantly influenced by the existence of flow distortion or nonuniformity in the impeller inlet. Most distortions encountered in practice include radially varying or circumferentially varying steady-state inflows and unsteadiness in the fixed coordinates as well. An example, a centrifugal impeller tested by Wright et al. (1984a) shows the level of reduction in stall margin that can result from poor installation or inattention to inlet conditions. As seen in Figure 10.13, the stall pressure rise can be reduced by 5–10% in the presence of significant distortion of the fan inlet flow velocity profile. The "Distortion Parameter," V_{rms}, is essentially the normalized standard deviation of the measured inlet velocity profile:

$$V_{\mathrm{rms}} = \sum_{i=1}^{n} \frac{\left[(V_i - V_{\mathrm{m}})^2/n\right]^{1/2}}{V_{\mathrm{m}}}, \tag{10.14}$$

where V_{m} is the mean (average) velocity, V_i is the velocity at measurement point i, and n is the number of points sampled in the inlet cross-section. The stall pressure rise is normalized by the stall pressure rise with a uniform inlet flow.

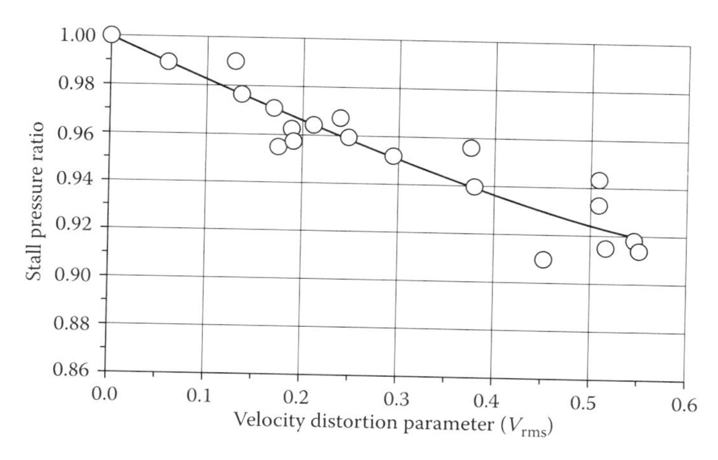

FIGURE 10.13 Influence of inlet flow distortion on stall margin.

The existence of rotating stall cells, by themselves, is not necessarily a great hazard to operation of the machine. In a fan, the accompanying pressure pulsations are more nearly a low-frequency acoustical problem than a structural danger to the machine and its system, with pressure amplitudes of the pulsation ranging from about 20% to 40% of static pressure rise. More seriously, if the pulsation frequency coincides with a natural acoustic frequency of the system, a resonance condition may arise.

The greatest risk to performance and noise, or even the structural integrity of the turbomachine or its system, is the possibility that operation with stall can trigger the major excursions in pressure rise and flow rate known as *surge*. The range of pressure pulsations generated in surge can be as high as 50% of design values, and flow excursions may include significant levels of total flow reversal through the machine. This can lead to very high levels of stress in both machine and system, which can cause severe damage or destruction of the equipment. If the turbomachine in surge is the compressor of a gas turbine or jet engine, surge can lead to a "flameout," with potentially disastrous consequences.

In an analysis of resonator compression system models, Greitzer (1996) (employing modeling techniques pioneered by Emmons et al., 1955) defined a parameter to assess the proclivity of the entry into surge conditions. Greitzer's parameter is

$$B \equiv \left(\frac{U}{2a}\right)\left(\frac{\forall_C}{A_C L_C}\right)^{1/2}, \tag{10.15}$$

where $\forall_C$, A_C, and L_C describe the compression system geometry in terms of an external plenum volume, flow cross-sectional area, and length. U is the impeller tip speed, and a is the speed of sound. For a given system layout, there is a critical value, B_{crit}, above which a compression system will exhibit large, oscillating excursions in flow rate and pressure rise—that is, surge. Below the critical value, the system will experience a transient change to a steady rotating stall operating condition with poor performance and high noise levels, but not the wild behavior of surge.

Determination of B_{crit} can be either analytical or experimental (Greitzer, 1981) for a particular system layout. Clearly, when an existing compression system is experiencing surge problems and strong oscillation in performance, its existing value of B can be used to determine an approach to fix the problem. Knowing B_{crit}, one can modify the geometry of the system or the speed of the machine to reduce B sufficiently to pull the system out of surge.

10.6 CFD in Turbomachinery

Possibly the most significant advance in turbomachinery engineering in the late twentieth and early twenty-first century has been the development and use of a wide array of CFD methods. Defined loosely, CFD is "the field of study devoted to the solution of the equations of fluid flow through the use of a computer" (Cèngel and Cimbala, 2010). Most would agree that this is too wide a definition as it would include, say, using a spreadsheet program to iteratively calculate the flow rate in a pipe from the Darcy–Weisbach equation and the Colebrook friction factor formula. For most engineers, CFD means using computer hardware and software to simulate a flow field by solving, at least approximately, the differential equations describing fluid flow. The solution may be for steady or unsteady flow and in either two, three, or "quasi-three" dimensions. The flow modeled may be compressible or incompressible, and inviscid, laminar, or turbulent. CFD was once exclusively a topic for graduate studies and advanced research and development; however, one can now choose from a variety of codes available in the marketplace, such as ANSYS CFX and FLUENT, and undergraduate textbooks contain introductory, overview chapters, or appendixes on the subject (Cèngel and Cimbala, 2010; Munson et al., 2009).

The ultimate aim of CFD for turbomachinery has been to develop comprehensive numerical systems to simulate the detailed flow field in a turbomachine. This is no simple task as the typical turbomachine flow is three dimensional and unsteady, with a high level of turbulence and, in many cases, compressible. The work and results of the many investigators involved cover the range from preliminary design and layout to fully developed schemes that rigorously solve the discretized fully 3D, viscous/turbulent, compressible, and often unsteady flow. These methods, and others like them, provide the starting point for a very detailed flow field analysis that can be used to verify or modify the selected geometry. The design layout is thus improved or optimized in an iterative process to increase performance, boost efficiency, or achieve better off-design performance and stability.

On the other hand, CFD is no panacea for solving all flow problems in turbomachinery. One strong conclusion of the collection of papers on turbomachinery design edited by Denton (Denton, 1999) is that CFD analysis is most useful to refine designs that began as one-dimensional and mean-line studies of the type developed in the earlier chapters of this book. The overall schemes needed to produce a full numerical simulation of the flow fields in turbomachines or their components—from pumps to aircraft engines—will generally employ available, well-established CFD codes as an engineering tool integrated within the overall design framework. The more rigorous treatment of these flow fields through CFD allows an experienced designer to develop better engineered and more clearly understood designs at low engineering cost. The overall approach to design includes selection of machine types, overall sizing, mean-line analysis, experimental correlations, and other preliminary design steps as a precursor to assessment by CFD analysis and subsequent iterative refinement of a design (Denton, 1999).

The essential "operational" topics involved in applying CFD include choosing among finite-difference, finite-element and finite-volume formulations, grid generation (structured, unstructured, and transformed), and the key topics of accuracy, convergence, and validation. The "essential ingredients" for good CFD analysis are (Lakshminarayana, 1991) proper use of the governing equations, the imposition of realistic boundary conditions, adequate modeling of the discretized domain, relevant modeling of the flow turbulence, and the use of an appropriate numerical technique. The turbulence modeling must be reasonably capable of including the influence of end-wall flows, tip leakages, blade/vane wakes, and other features typical of the flow fields in a turbomachine, including low Reynolds number behavior (see Sections 10.3 and 10.4).

The methods of CFD have been developed over a period of several decades and have required simultaneous advances in computer capability, numerical algorithms, understanding and modeling of flow physics (especially turbulence), and, certainly not least important, development of methods for entering turbomachine geometry and displaying computed results. Many of the "older" methods may find good application in advanced design and analysis. For example, the use of inviscid flow analysis, perhaps coupled with boundary layer calculations, can be particularly appropriate in the early design phases or in regions where the flow is substantially inviscid. Although these flows may be revisited in a later, more rigorous study using 3D full Navier–Stokes methods, great reductions in time and expense can be achieved through the use of the more approximate technique at the appropriate time.

It is certainly beyond the scope of this book to include a qualitative discussion of CFD methods. The following represents a brief survey of some of the more important turbomachinery-oriented work, both historic and current. The authors are indebted to Dr. R. V. Chima, of the NASA Glenn Research Center, for his assistance in compiling this information (Chima, 2009a).

General references on CFD include the following:

- Lakshminarayana (1996)—This extensive work is specifically oriented toward turbomachinery.
- Tannehill, et al. (1997)—This is a detailed text on CFD for essentially all applications. The level of mathematical rigor is high.
- Hirsch (1990)—Hirsch specializes in turbomachinery and CFD in turbomachinery.
- Wilcox (2006)—Wilcox develops the all-important turbulence models for CFD in great detail.

As noted in Chapter 9, the Quasi-Three-Dimensional (Q3D) model represented the first practical effort to address 3D flows in turbomachines. For most Q3D methods, except SRE in axial flow, it is necessary to use a computer code to generate useful solutions. The equations of motion are solved on S1 and S2 stream surfaces, typically using an inviscid stream function formulation or a streamline curvature approach. Semiempirical loss models allow these methods to approach reality. Typically CFD codes based on the Q3D model are very fast on modern computers.

Throughflow (S2) models may use either finite-difference stream function formulations (Katsanis and McNally, 1977) or streamline curvature formulations that can operate on very coarse grids (Katsanis, 1964). These models are often coupled to geometry generators as part of design systems.

Blade-to-blade (S1) models can be used to optimize blade profile shapes. Older models may use panel methods (McFarland, 1993) or a stream function formulation (Katsanis, 1969) and be limited to inviscid, subsonic flow. Newer methods solve the full Navier–Stokes equations and can model shocks, losses, and heat transfer (Chima, 1986).

The "state of the art" at the time of this writing, at least for design analysis, is the fully three-dimensional Reynolds-averaged Navier–Stokes (3D-RANS) method, typically using two-equation turbulence models. These are commonly used in industry to predict the performance of isolated blade rows. These methods may use finite-difference or finite-volume methods and are capable of predicting shocks, efficiency, tip clearance flows, and heat transfer. To save computer time and storage, they usually assume blade-to-blade periodicity so that only one blade per row is analyzed. A good overview of current capabilities is found in Dunham (1998). Most codes use explicit time-marching schemes (Chima, 1991; Dawes, 1988; Denton, 1982; Heidmann et al., 2000). This type of scheme can handle compressible flows but may require preconditioning for low subsonic or incompressible flows such as pumps and fans. Some codes use pressure-correction schemes (Hah, 1984). 3D-RANS codes usually require that the users have some knowledge of CFD and prior experience to produce good results. The results can be sensitive to the turbulence model employed.

Possibly the greatest challenge is to accurately predict the flow in multistage machines. Most codes (e.g., Chima, 1998) use mixing plane schemes. Since

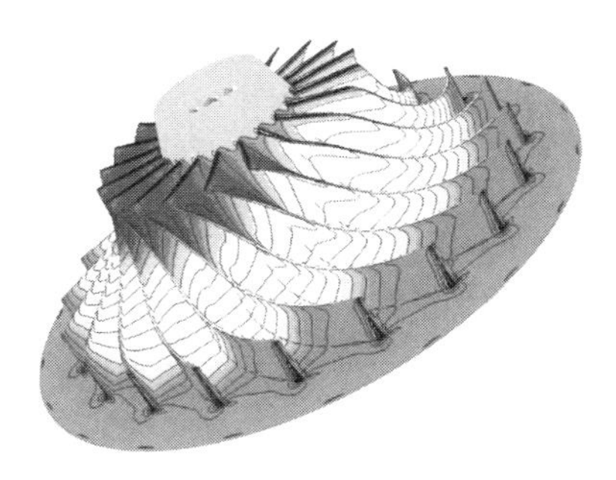

FIGURE 10.14 Pressure contours on a radial flow compressor impeller computed using the SWIFT 3D-RANS code. (From Chima, R. V., 2009b, website at http://www.grc.nasa.gov/WWW/5810/rvc/. With permission.)

blade rows are alternately rotating and stationary, these codes assume that each blade row sees the average flow from its immediate neighbors. Wakes, shocks, and other flow nonuniformities are mixed out at a grid interface between blade rows. This model usually gives a good approximation of the overall stage performance but misses details of rotor–stator interaction. The average passage method (Adamczyk, 1999) uses a formal time average of the Navier–Stokes equations in stationary and rotating frames of reference to show how blade rows interact. The averaging process introduces unknown terms similar to Reynolds stresses that must be modeled. Fully unsteady multistage calculations (Chen and Whitfield, 1990) model the complete unsteady interaction between blade rows. Since neighboring blade rows usually have unequal blade counts, periodicity cannot be used and all blade passages must be modeled. These codes require large computer clusters and sometimes months of computer time, but can give complete information about blade row interaction.

Computer codes for advanced CFD modeling of turbomachine flows are often proprietary property of design and manufacturing companies. On the other hand, general purpose CFD codes available on the open market can be expensive to license and might require considerable modification to be useful for turbomachine flow modeling. For the smaller user, CFD codes developed by Dr. R. V. Chima of the NASA Glenn Research Center are available free of charge to companies or universities in the United States (Chima, 2009b). Figure 10.14 shows a sample of computational results obtained with one of these, the SWIFT 3D-RANS code.

10.7 Summary

The models and analytical methods used in earlier chapters have all, to some degree, sidestepped or simplified many of the difficulties that arise in real turbomachinery flows. Although the simpler methods seek to model the dominant physical phenomena in the flow and often provide a starting point for design, a search for additional rigor or more in-depth understanding requires that some of these problem areas be addressed. Several such topics are briefly introduced in this chapter and discussed.

The inherently high levels of freestream turbulence intensity can dominate the development and behavior of the boundary layers and other strongly viscous regions of flow. Transition from laminar to turbulent flow depends very strongly on the turbulence intensity, and transition in a turbomachine is usually quite different from that in a typical external flow. Further, development and growth of separated flows, stalls, and flow mixing in the flow passages are also strongly dependent on the level and scale of turbulence imbedded in the flow.

A complete understanding of flow in the blade channels must include the influence of the channel secondary flows. These develop as a result of strong pressure gradients from blade to blade (circumferentially) and from hub to shroud (radially or normally). The flow is additionally complicated by the growth of vortices that may be triggered by the movement of flow through the end wall or tip gaps in machines without integral shrouds. These end-wall gap and leakage flows can lead to a strong rotational motion superposed on the basic core flow in the passage. The dichotomy of convective and diffusive mixing of the flow in a channel was discussed briefly, and the need to include both mechanisms in realistic flow modeling was noted.

One of the more difficult flow problems is the influence of very low Reynolds numbers on the flow turning and losses in blade rows. The experimental information on these flows in the open literature is limited, so that effective modeling is hampered by some uncertainty. The problem is discussed, and a rough predictive model is developed on the basis of some extensions to the original model of Roberts (1975a) and additional experimental data. The resulting model provides adequate estimation of the effect of low Reynolds number on the deviation angles, but it yields more uncertain results for the prediction of total pressure losses.

A brief introduction of the problems of static and dynamic stability was given in terms of the models developed over the years by Greitzer (1981, 1996). Operation of a machine at the "zero slope or positive slope" condition on the pressure-flow characteristic of the machine will lead to instability of the flow. Development of rotating stall cells can, for certain system conditions, trigger a potentially destructive onset of surge in the system. The models presented in the chapter can help the user or designer of turbomachines avoid these unacceptable regions of operation.

Finally, the basic concepts of the application of CFD methods in turbomachinery design and flow analysis were introduced. The literature in this field was briefly discussed, with a discussion of the basic models and their capability. Finally, codes currently available to interested users were discussed.

References

Abbott, I. and von Denhoff, A., 1959. *Theory of Wing Sections*, Dover Books, New York.
Adamczyk, J. J., 1999. "Aerodynamic analysis of multistage turbomachinery flows in support of aerodynamic design," ASME Paper No. 99-GT-80, June.
Addison, J. S. and Hanson, H. P., 1992. "Modeling of unsteady transitional layers," *ASME Journal of Turbomachinery*, 114 (July), pp. 550–589.
Adenubi, S. O., 1975. "Performance and flow regimes of annular diffusers with axial-flow turbomachinery discharge inlet conditions," ASME Paper No. 75-WA/FE-5.
Adkins, G. G. and Smith, L. H., Jr., 1982. "Spanwise mixing in axial-flow turbomachines," *ASME Journal of Engineering for Power*, 104, 97–100.
Adkins, R. C., 1975. "A short diffuser with low pressure loss," *ASME Journal of Fluids Engineering*, 93, 297–302.
Ainley, D. G. and Mathieson, G. C. R., 1951. "A method of performance estimation of axial-flow turbines," ACR R&M 2974.
Allis-Chalmers, 1986. "Catalog of general purpose pumps," Industrial Pump Division.
AMCA, 1999. *Laboratory Methods of Testing Fans for Aerodynamic Performance Rating*, ANSI/AMCA 210-99; ANSI/ASHRAE 51-99; Air Moving and Conditioning Association.
AMCA, 2002. "*Industrial process/power generation fans: Establishing performance using laboratory models*," Publication 802-02, Air Moving and Conditioning Association.
Anderson, J. D., 1984. *Modern Compressible Flow with Historical Perspectives*, McGraw-Hill, New York.
ANSYS, *ANSYS-CFX*, ANSYS Inc., Cannonsburg, PA.
ASME, 1990. "*Centrifugal pumps*," ASME–PTC-8.2, American Society of Mechanical Engineers.
ASME, 1997. "*Performance test code on compressors and exhausters*," ASME–PTC-10, American Society of Mechanical Engineers.
ASME, 2004. "*Flow measurement*," ASME–PTC-19.5, American Society of Mechanical Engineers.
ASME, 2008. "*Fans*," ASME–PTC-11, American Society of Mechanical Engineers.
Baade, P. K., 1982. Private communication.
Baade, P. K., 1986. Mathematical models for noise generation: An appraisal of the literature, *Proceedings of the Air Movement and Distribution Conference*, Purdue University.
Balje, O. E., 1968. "Axial cascade technology and application to flow path designs," *ASME Journal of Engineering for Power*, 81.
Balje, O. E., 1981. *Turbomachines*, Wiley, New York.
Bathie, W. W., 1996. *Fundamentals of Gas Turbines*, 2nd ed., Wiley, New York.
Baumeister, T., Avallone, E. A., and Baumeister, T., III, 1978. *Marks' Standard Handbook for Mechanical Engineers*, McGraw-Hill, New York.

Beckwith, T. G., Marangoni, R. D., and Lienhard, J. H., 1993. *Mechanical Measurements*, Addison-Wesley, New York.

Beranek, L. L. and Ver, I., 1992. *Noise and Vibration Control Engineering Principles and Applications*, Wiley, New York.

Bolz, R. E. and Tuve, G. L., 1973, "Section 1.2, Properties of liquid," in *Handbook of Tables for Applied Engineering Science*, CRC Press, Boca Raton, FL.

Broch, J., 1971. *Acoustic Noise Measurement, The Application of Bruel & Kjaer Measuring Systems*, 2nd ed., Bruel & Kjaer, Denmark.

Burdsall, E. A. and Urban, R. H., 1973. "Fan-compressor noise: Prediction, research and reduction studies," Final Report FAA-RD-71-73.

Busemann, A., 1928. "The head ratio of centrifugal pumps with logarithmic spiral blades" ("Das Forderhohenverhaltnis Radialer Krielselpumpen mit Logarithmisch-Spiraligen Schaufeln"), *Z. agnew Math. Mech.* 8. (in Balje, 1981). (See p. 161 in Balje, 1981).

Carter, A. and Hughes, H., 1946. "A theoretical investigation into the effect of profile shape on the performance of aerofoils in cascade," R&M 2384, British ARC, March.

Casey, M. V., 1985. "The effects of Reynolds number on the efficiency of centrifugal compressor stages," *ASME Journal for Gas Turbines and Power*, 107, 541–548.

Cebeci, T., 1983. "Essential ingredients of a method for low Reynolds number airfoils," *AIAA Journal*, 27(12), 1680–1688.

Cèngel, Y. and Cimbala, J., 2010. *Fluid Mechanics: Principles and Applications*, 2nd ed., McGraw-Hill, New York.

Chen, J.-P. and Whitfield, D. L., 1990. "Navier–Stokes calculations for the unsteady flowfield of turbomachinery," AIAA Paper 1990-0676.

Chicago Blower Company, 1998. *Fan Catalogs.*

Chima, R. V., 1986. "Development of an explicit multigrid algorithm for quasi-three-dimensional viscous flows in turbomachinery," NASA TM 87128.

Chima, R. V., 1991. "Viscous three-dimensional calculations of transonic fan performance," NASA TM 103800, May.

Chima, R. V., 1998. "Calculation of unsteady multistage turbomachinery using steady characteristic boundary conditions," AIAA Paper 98-0968 or NASA TM-1998-206613.

Chima, R. V., 2009a. Personal communication with the authors.

Chima, R. V., 2009b. Website at http://www.grc.nasa.gov/WWW/5810/rvc/

Citavy, J. and Jilek, J., 1990. "The effect of low Reynolds number on straight compressor cascades," ASME Paper No. 90-GT-221.

Compte, A., Ohayon, G., and Papailiou, K.D., 1982. "A method for the calculation of the wall layers inside the passage of a compressor cascade with and without tip clearance," *ASME Journal of Engineering for Power*, 104.

Constant, H., 1939. "Note on the performance of cascades of aerofoils," Note E-3996, British RAE.

Cordier, O., 1955. "Similarity considerations in turbomachines," VDI Reports, 3.

Crocker, M. J., 2007. *Handbook of Noise and Vibration Control Engineering*, Wiley, New York.

Csanady, G. T., 1964. *Theory of Turbomachines*, McGraw-Hill, New York.

Cumpsty, N. A., 1977. "A critical review of turbomachinery noise," *ASME Journal of Fluids Engineering*, 99, 278–293.

Davis, R. C. and Dussourd, J. L., 1970. "A unified procedure for calculation of off-design performance of radial turbomachinery," ASME Paper No. 70-GT-64.
Dawes, W. N., 1988. "Development of a 3D Navier–Stokes solver for application to all types of turbomachinery," ASME Paper No. GT-88-70.
de Haller, P., 1952. "Das Verhalten von Traflugol gittern im Axialverdictern und im Windkanal," *Brennstoff-Warme-Kraft*, 5, 333–336.
Denton, J. D., 1982. "An improved time marching method for turbomachinery flow calculation," ASME Paper No. 82-GT-239, 1982.
Denton, J. D. (Ed.), 1999. *Developments in Turbomachinery Design*, Professional Engineering Publishing, Bury St. Edmunds, England.
Denton, J. D. and Usui, S., 1981. "Use of a tracer gas technique to study mixing in a low speed turbine," ASME Paper No. 81-GT-86.
Dixon, S. L., 1998. *Fluid Mechanics and Thermodynamics of Turbomachinery*, 5th ed., Elsevier, Amsterdam.
Dullenkopf, K. and Mayle, R., 1994. "The effect of incident turbulence and moving wakes on laminar heat transfer in gas turbines," *ASME Journal of Turbomachinery*, 116, 23–28.
Dunham, J. (Ed.), 1998. "CFD validation for propulsion system components," AGARD-AR-355, May.
Emery, J. C., Herrig, L. J., Erwin, J. R., and Felix, A. R., 1958. *Systematic Two-Dimensional Cascade Tests of NACA 65-Series Compressor Blades at Low Speed*, NACA-TR-1368.
Emmons, H. W., Pearson, C. E., and Grant, H. P., 1955. "Compressor surge and stall propagation," *Transactions of the ASME*, 77, 455–469.
FLUENT, ANSYS Inc., Cannonsburg, PA, http://www.ansys.com
Fox, R. W., Prichard, P. J., and McDonald, A. T., 2009. *Introduction to Fluid Mechanics*, 7th ed., Wiley, New York.
Gallimore, S. J. and Cumpsty, N. A., 1986a. "Spanwise mixing in multi-stage axial flow compressors: Part I—experimental investigations," *ASME Journal of Turbomachinery*, 8, 1–8.
Gallimore, S. J. and Cumpsty, N. A., 1986b. "Spanwise mixing in multi-stage axial flow compressors: Part II—Throughflow calculations including mixing," *ASME Journal of Turbomachinery*, 8, 9–16.
Gerhart, P. M., Gross, R. J., and Hochstein, J. I., 1992. *Fundamentals of Fluid Mechanics*, 2nd ed., Addison-Wesley; Reading, MA.
Goldschmeid, F. R., Wormley, D. M., and Rowell, D., 1982. "Air/gas system dynamics of fossil fuel power plants—system excitation," EPRI Research Project 1651, Vol. 5, CS-2006.
Gostelow, J. P. and Blunden, A. R., 1989. "Investigations of boundary layer transition in an adverse pressure gradient," ASME Journal of Turbomachinery, 111.
Goulds, 1982. *Pump Manual: Technical Data Section*, 4th ed., Goulds Pump, Inc., Seneca Falls, NY.
Graham, J. B., 1972. "How to estimate fan noise," *Journal of Sound and Vibration*, 6, 24–27.
Graham, J. B., 1991. "Prediction of fan sound power," *ASHRAE Handbook: HVAC Applications*, American Society of Heating, Refrigerating and Air-Conditioning Engineers, Inc.
Granger, R. A. (Ed.), 1988. *Experiments in Fluid Mechanics*, Holt, Rinehart and Winston, New York.
Granville, P. S., 1953. "The calculations of viscous drag on bodies of revolution," David Taylor Model Basin Report No. 849.

Greitzer, E. M., 1981. "The stability of pumping systems—the 1980 Freeman scholar lecture," *ASME Journal of Fluids Engineering*, 103, 193–242.

Greitzer, E. M., 1996. "Stability in pumps and compressors," in *Handbook of Fluid Dynamics and Turbomachinery* (J. Schetz and A. Fuhs, Eds), Wiley, New York.

Hah, C., 1984. "A Navier–Stokes analysis of three-dimensional turbulent flows inside turbine blade rows at design and off-design conditions," *ASME Journal of Engineering for Power*, 106, 421–429.

Hansen, A. G. and Herzig, H. Z., 1955. "Secondary flows and three-dimensional boundary layer effects," (in NASA SP-36, Johnson and Bullock, 1965).

Hanson, D. B., 1974. "Spectrum of rotor noise caused by atmospheric turbulence," *Journal of the Acoustical Society of America*, 55, 53–54.

Heidmann, J., Rigby, D. L., and Ameri, A. A., 2000. "A three-dimensional coupled internal/external simulation of a film cooled turbine vane," *ASME Journal of Turbomachinery*, 122, 348–359.

Heywood, J. B., 1988. *Internal Combustion Engine Fundamentals*, McGraw-Hill, New York.

HI, 1994. *Test Standard for Centrifugal Pumps*; HI-1.6; Hydraulic Institute.

Hirsch, C., 1990. *Numerical Computations of Internal and External Flows*, Vols. 1 and 2, Wiley, New York.

Hirsch, Ch., 1974. "End-wall boundary layers in axial compressors," *ASME Journal of Engineering for Power*, 96.

Holman, J. P. and Gadja, W. J., Jr., 1989. *Experimental Methods for Engineers*, 7th ed., McGraw-Hill, New York.

Horlock, J. H., 1958. *Axial Flow Compressors, Fluid Mechanics and Thermodynamics*, Butterworth Scientific, London.

Howard, J. H. G., Henseller, H. J., and Thornton, A. B., 1967. "Performance and flow regimes for annular diffusers," ASME Paper No. 67-WA/FE-21.

Howden Industries, 1996. Axial Fan Performances Curves.

Howell, A. R., 1945. "Fluid mechanics of axial compressors," *Proceeding of the Institute of Mechanical Engineers*, 153.

Hunter, I. H. and Cumpsty, N. A., 1982. "Casing wall boundary layer development through an isolated compressor rotor," *ASME Journal of Engineering for Power*, 104, 805–818.

Idel'Chik, I. E., 1966. *Handbook of Hydraulic Resistance*, AEC-TR-6630 (also Hemisphere, New York, 1986).

Industrial Air, 1986. *General Fan Catalog*, Industrial Air, Inc.

Jackson, D. G., Jr. and Wright, T., 1991. "An intelligent learning axial fan design system," ASME Paper No. 91-GT-27.

Japikse, D. and Baines, N., 1994. *Introduction to Turbomachinery*, Concepts ETI, White River Junction, VT.

Johnsen, I. A. and Bullock, R. O., 1965. "Aerodynamic design of axial flow compressors," NASA SP-36.

Johnson, M. W. and Moore, J., 1980. "The development of wake flow in a centrifugal impeller," *ASME Journal of Turbomachinery*, 102, 383–390.

Jorgensen, R. (Ed.), 1983. *Fan Engineering—An Engineer's Handbook on Fans and Their Applications*, Buffalo Forge Company, New York.

Kahane, A., 1948. "Investigation of axial-flow fan and compressor rotor designed for three-dimensional flow," NACA TN-1652.

Karassik, I. J., Messina, J. P., Cooper, P., and Heald, C. C., 2008. *Pump Handbook*, 4th ed., McGraw-Hill, New York.

Katsanis, T., 1964. "Use of arbitrary quasi-orthogonals for calculating flow distribution in the meridional plane of a turbomachine," NASA TN D-2546.

Katsanis, T., 1969. "Fortran program for calculating transonic velocities on a blade-to-blade surface of a turbomachine," NASA TN D-5427.

Katsanis, T. and McNally, W. D., 1977. "Revised Fortran program for calculating velocities and streamlines on the hub-shroud midchannel stream surface of an axial-, radial-, or mixed flow turbomachine or annular duct, I—user's manual," NASA TN D-8430.

Kestin, J., Meader, P. F., and Wang, W. E., 1961. "On boundary layers associated with oscillating streams," *Applied Science Research*, A 10. (See Schlichting, 1979.)

Kittredge, C. P., 1967. "Estimating the efficiency of prototype pumps from model tests," ASME Paper No. 67-WA/FE-6.

Koch, C. C., 1981. "Stalling pressure capability of axial-flow compressors," *ASME Journal of Engineering for Power*, 103(4), 645–656.

Koch, C. C. and Smith, L. H., Jr., 1976. "Loss sources and magnitudes in axial-flow compressors," *ASME Journal of Engineering for Power*, 98(3), 411–424.

Ladson, C., Brooks, C., Jr., Hill, A., and Sproles, D., 1996. "Computer program to obtain ordinates for NACA airfoils," NASA-TM-4741, 1996.

Laguier, R., 1980. "Experimental analysis methods for unsteady flow in turbomachinery," *Measurement Methods in Rotating Components of Turbomachinery*, ASME, New York, 1980.

Lakshminarayana, B., 1979. "Methods of predicting the tip clearance effects in axial flow turbomachinery," *ASME Journal of Basic Engineering*, 92.

Lakshminarayana, B., 1991. "An assessment of computational fluid dynamic techniques in the analysis and design of turbomachinery—the 1990 Freeman scholar lecture," *ASME Journal of Fluids Engineering*, 113, 315–352.

Lakshminarayana, B., 1996. *Fluid Dynamics and Heat Transfer of Turbomachinery*, Wiley, New York.

Leylek, J. H. and Wisler, D. C., 1991. "Mixing in axial-flow compressors: Conclusions drawn from three-dimensional Navier–Stokes analyses and experiments," *ASME Journal of Turbomachinery*, 113, 139–160.

Li, Y. S. and Cumpsty, N. A., 1991a. "Mixing in axial flow compressors: Part 1—test facilities and measurements in a four-stage compressor," *ASME Journal of Turbomachinery*, 113, 161–165.

Li, Y. S. and Cumpsty, N. A., 1991b. "Mixing in axial flow compressors: Part 2—measurements in a single-stage compressor and duct," *ASME Journal of Turbomachinery*, 113, 166–174.

Lieblein, S., 1959. "Loss and stall analysis of compressor cascades," *ASME Journal of Basic Engineering*, 81, 387–400.

Lieblein, S., 1957. "Analysis of experimental low-speed loss and stall characteristics of two-dimensional blade cascades," NACA RM E57A28.

Lieblein, S. and Roudebush, R. G., 1956. "Low speed wake characteristics of two-dimensional cascade and isolated airfoil sections," NACA TN 3662.

Longley, J. P. and Greitzer, E. M., 1992. "Inlet distortion effects in aircraft propulsion systems," AGARD Lecture Series LS-133.

Mayle, R. E., 1991. "The role of laminar-turbulent transition in gas turbine engines," ASME Paper No. 91-GT-261.

Mayle, R. E. and Dullenkopf, K., 1990. "A theory for wake induced transition," *ASME Journal of Turbomachinery*, 112, 188–195.
Madhavan, S., DiRe, J., and Wright, T., 1984. "Inlet flow distortion in centrifugal fans," ASME Paper No. 84-JPGC/GT-4.
Madhavan, S. and Wright, T., 1985. "Rotating stall caused by pressure surface flow separation in centrifugal fans," *ASME Journal of Engineering for Gas Turbines and Power*, 107:33.
Mattingly, J., 1996. *Elements of Gas Turbine Propulsion*, McGraw-Hill, New York.
McDonald, A. T. and Fox, R. W., 1965. "Incompressible flow in conical diffusers," ASME Paper No. 65-FE-25.
McDougall, N. M., Cumpsty, N. A., and Hynes, T. P., 1989. "Stall inception in axial compressors," Presented at the *ASME Gas Turbine and Aeroengine Conference and Exposition*, June, Toronto, Ontario.
McFarland, E. R., 1993. "An integral equation solution for multi-stage turbomachinery design," NASA TM-105970, NASA Lewis Research Center.
Mellor, G. L., 1959. "An analysis of compressor cascade aerodynamics, Part I: Potential flow analysis with complete solutions for symmetrically cambered airfoil families," *ASME Journal of Basic Engineering*, 81, 362–378.
Mellor, G. L. and Wood, G. M., 1971. "An axial compressor end-wall boundary layer theory," *ASME Journal of Basic Engineering*, 93, 300–316.
Moody, L. F., 1925. "The propeller type turbine," *Proceedings of the American Society of Civil Engineers*, 51, 628–636.
Moore, J. and Smith, B. L., 1984. "Flow in a turbine cascade—Part 2: Measurement of flow trajectories by ethylene detection," *ASME Journal of Engineering for Gas Turbines and Power*, 106, 409–413.
Moran, M. J. and Shapiro, H. N., 2008. *Fundamentals of Engineering Thermodynamics*, 6th ed., Wiley, New York.
Moreland, J. B., 1989. "Outdoor propagation of fan noise," ASME Paper No. 89-WA/NCA-9.
Morfey, C. L. and Fisher, M. J., 1970. "Shock wave radiation from a supersonic ducted rotor," *Aeronautical Journal*, 74(715), 579–585.
Munson, B. R., Young, D. F., Okishii, T. H., and Huebsch, W. W., 2009. *Fundamentals of Fluid Mechanics*, 6th ed., Wiley, New York.
Myers, J. G., Jr. and Wright, T., 1993. "An inviscid low solidity cascade design routine," ASME Paper No. 93-GT-162.
Nasar, S. A., 1987. *Handbook of Electric Machines*, McGraw-Hill, New York.
NCFMF, 1963. *Flow Visualization* (S. J. Kline, principal), National Committee for Fluid Mechanics Films (Encyclopedia Britannica, Educational Corporation) (redirects to http://web.mit.edu/hml/ncfmf.html).
Neise, W., 1976. "Noise reduction in centrifugal fans: A literature survey," *Journal of Sound and Vibration*, 45(3), 375–403.
New York Blower, 1986. *Catalog on Fans*, New York Blower Co.
Novak, R. A., 1967. "Streamline curvature computing procedure for fluid flow problems," *ASME Journal of Engineering for Power*, 89, 478–490.
O'Brien, W., Jr., Cousins, W., and Sexton, M., 1980. "Unsteady pressure measurements and data analysis techniques in axial flow compressors," *Measurement Methods in Rotating Components of Turbomachinery*, Ed. B. Lakshminarayana and P. Runstadler ASME, New York.
Obert, E. F., 1973. *Internal Combustion Engines, and Air Pollution*, Intext, New York.

O'Meara, M. M. and Mueller, T. J., 1987. "Laminar separation bubble characteristics on an airfoil at low Reynolds numbers," *AIAA Journal*, 25(8), 1033–1041.

Papailiou, K. D., 1975. "Correlations concerning the process of flow decelerations," *ASME Journal of Engineering for Power*, 97, 295–300.

Pfenninger, W. and Vemura, C. S., 1990. "Design of low Reynolds number airfoils," *AIAA Journal of Aircraft*, 27(3), 204–210.

Ralston, S. A. and Wright, T., 1987. "Computer-aided design of axial fans using small computers," *ASHRAE Transactions*, 93(Part 2), Paper No. 3072.

Rao, S. S., 1990. *Mechanical Vibrations*, 2nd ed., Addison-Wesley, New York.

Reneau, L. R., Johnson, J. P., and Kline, S. J., 1967. "Performance and design of straight two-dimensional diffusers," *ASME Journal of Basic Engineering*, 89(1), 141–150.

Richards, E. J. and Mead, D. J., 1968. *Noise and Acoustic Fatigue in Aeronautics*, Wiley, New York.

Roberts, W. B., 1975a. "The effects of Reynolds number and laminar separation on cascade performance," *ASME Journal of Engineering for Power*, 97, 261–274.

Roberts, W. B., 1975b. "The experimental cascade performance of NACA compressor profiles at low Reynolds number," *ASME Journal of Engineering for Power*, 97, 275–277.

Schetz, J. A. and Fuhs, A. E. (Eds), 1996. *Handbook of Fluid Dynamics and Turbomachinery*, Wiley, New York.

Schlichting, H., 1979. *Boundary Layer Theory*, 7th ed., McGraw-Hill, New York.

Schmidt, G. S. and Mueller, T. J., 1989. "Analysis of low Reynolds number separation bubbles using semi-empirical methods," *AIAA Journal*, 27(8), 993–1001.

Schubauer, G. B. and Skramstad, H. K., 1947. "Laminar boundary layer oscillations and stability of laminar flow, National Bureau of Standards Research Paper 1772," *Journal of Aeronautical Sciences*, 14, 69–78.

Sharland, C. J., 1964. "Sources of noise in axial fans," *Journal of Sound and Vibration*, 1(3), 302–322.

Shigley, J. S. and Mischke, C. K., 1989. *Mechanical Engineering Design*, McGraw-Hill, New York.

Shepherd, D. G., 1956. *Principles of Turbomachinery*, Macmillan and Company, New York.

Smith, L. C., 1976. "A note on diffuser generated unsteadiness," *ASME Journal of Fluids Engineering*, 97, 327–379.

Smith, L. H., Jr., 1955. "Secondary flows in axial-flow turbomachinery," *Transactions of the ASME*, 177, 1065–1076.

Soderberg, C., 1949. *Gas Turbine Laboratory*, Massachusetts Institute of Technology, 1949 (cited in Dixon, 1999).

Sovran, G. and Klomp, E. D., 1967. "Experimentally determined optimum geometries for diffusers with rectilinear, conical, or annular cross-sections," *Fluid Mechanics of Internal Flow*, Elsevier Publications, Amsterdam, The Netherlands.

Stanitz, J. D. and Prian, V. D., 1951. "A rapid approximate method for determining velocity distributions on impeller blades of centrifugal compressors," NACA TN 2421.

Stepanoff, A. J., 1948. *Centrifugal and Axial Flow Pumps*, Wiley, New York.

Stodola, A., 1927. *Steam and Gas Turbines*, Springer, New York.

Strub, R. A., Bonciani, L., Borer, C. J., Casey, M. V., Cole, S. L., Cook, B. B., Kotzur, J., Simon, H., and Strite, M. A., 1984. "Influence of Reynolds number on the performance of centrifugal compressors," *ASME Journal of Turbomachinery*, 106(2), 541–544.

Tannehill, J. C., Anderson, D. A., and Pletcher, R. H., 1997. *Computational Fluid Mechanics and Heat Transfer*, Taylor & Francis, Washington.

Thumann, A., 1990. *Fundamentals of Noise Control Engineering*, 2nd ed., Fairmont Press, Atlanta.

Thoma, D. and Fischer, K., 1932. "Investigation of the flow conditions in a centrifugal pump," *Transactions of the ASME*, 58, 141–155.

Topp, D. A., Myers, R. A., and Delaney, R. A., "TADS: A CFD-based turbomachinery analysis and design system with GUI, Vol. 1: Method and results," NASA CR-198440, NASA Lewis Research Center.

Van Wylen, G. J. and Sonntag, R. E., 1986. *Fundamentals of Classical Thermodynamics*, 3rd ed., Wiley, New York.

Vavra, M. H., 1960. *Aerothermodynamics and Flow in Turbomachines*, Wiley, New York.

Ver, I. L. and Beranek, L. L., 2005. *Noise and Vibration Control Engineering Principles and Applications*, 2nd ed., Wiley, New York.

Verdon, J. M., 1993. "Review of unsteady aerodynamics methods for turbomachinery aeroelastic and aeroacoustic applications," *AIAA Journal*, 31(2), 235–250.

Wang, R. E., 1993. *Predicting the losses and deviation angles with low Reynolds separation on axial fan blades*, Masters degree thesis in Mechanical Engineering, The University of Alabama at Birmingham.

Weber, H. E., 1978. "Boundary layer calculations for analysis and design," *ASME Journal of Fluids Engineering*, 100, 232–236.

Wiesner, F. J., 1967. "A review of slip factors for centrifugal impellers," *ASME Journal of Engineering for Power*, 89, 558–576.

White, F. M., 2005. *Viscous Fluid Flow*, 3rd ed., McGraw-Hill, New York.

White, F. M., 2008. *Fluid Mechanics*, 6th ed., McGraw-Hill, New York.

Wislicenus, G. F., 1965. *Fluid Mechanics of Turbomachinery*, Dover Publications, New York.

Wilcox, D., 2006. *Turbulence Modeling for CFD*, DCW Industries, LaCanada, CA.

Wilson, C. E., 1989. *Noise Control*, Harper & Row, New York.

Wilson, D. G., 1984. *The Design of High-Efficiency Turbomachinery and Gas Turbine Engines*, MIT Press, Cambridge.

Wisler, D. C., Bauer, R. C., and Okiishi, T. H., 1987. "Secondary flow, turbulent diffusion, and mixing in axial-flow compressors," *ASME Journal of Turbomachinery*, 109, 455–471.

Wittig, S., Shultz, A., Dullenkopf, K., and Fairbanks, J., 1988. "Effects of turbulence and wake characteristics on the heat transfer along a cooled gas turbine blade," ASME Paper No. 88-GT-179.

Wormley, D. M., Rowell, D., and Goldschmied, F. R., 1982. "Air/gas system dynamics of fossil fuel power plants—pulsations," EPRI Research Project 1651, Vol. 5, CS-2006.

Wright, S. E., 1976. "The acoustic spectrum of axial flow machines," *Journal of Sound and Vibration*, 85, 165–223.

Wright, T., 1974. "Efficiency prediction for axial fans," *Proceedings of the Conference on Improving Efficiency in HVAC Equipment for Residential and Small Commercial Buildings*, Purdue University.

Wright, T., 1982. "A velocity parameter for the correlation of axial fan noise," *Noise Control Engineering*, 19, 17–25.

Wright, T., 1984c. "Centrifugal fan performance with inlet clearance," *ASME Journal for Gas Turbines and Power*, 106(4), 906–912.

Wright, T., 1984d. "Optimal fan selection based on fan–diffuser interactions," ASME Paper No. 84-JPGC/GT-9.

Wright, T., 1988. "A closed-form algebraic approximation for quasi-three-dimensional flow in axial fans," ASME Paper No. 88-GT-15.

Wright, T., 1989. "Comments on compressor efficiency scaling with Reynolds number and relative roughness," ASME Paper No. 89-GT-31.

Wright, T., 1996. "Low pressure axial fans," Section 27.6, *Handbook of Fluid Dynamics and Turbomachinery* (J. Shetz and A. Fuhs, Eds), Wiley, New York.

Wright, T., Baladi, J. Y., and Hackworth, D. T., 1985. "Quiet cooling system development for a traction motor," ASME Paper No. 85-DET-137.

Wright, T., Madhavan, A., and DiRe, J., 1984a. "Centrifugal fan performance with distorted inflows," *ASME Journal for Gas Turbines and Power*, 106(4), 895–900.

Wright, T. and Ralston, S., 1987. "Computer aided design of axial fans using small computers," *ASHRAE Transactions*, 93(Part 2), ASHRAE Paper No. 3072.

Wright, T., Tzou, K. T. S., and Madhavan, S., 1984b. "Flow in a centrifugal fan impeller at off-design conditions," *ASME Journal for Gas Turbines and Power*, 106(4), 913–919.

Wright, T., Tzou, K. T. S., Madhavan, S., and Greaves, K. W., 1982. "The internal flow field and overall performance of a centrifugal fan impeller—experiment and prediction," ASME Paper No. 82-JPGC-GT-16.

Wu, C. H., 1952. "A general theory of three-dimensional flow in subsonic and supersonic turbomachine in radial, axial and mixed flow types," NACA TR 2604.

Yang, T. and El-Nasher, A. M., 1975. "Slot suction requirements for two-dimensional Griffith diffusers," *ASME Journal of Fluids Engineering*, 97, 191–195.

Zurn, 1981, *General Fan Catalogs*.

Zweifel, O., 1945. "The spacing of turbomachine blading, especially with large angular deflection," *The Brown Boveri Review*, p. 436, December (p. 132 in Balje, 1981).

Appendix A

TABLE A.1

Approximate Gas Properties at 20°C, 101.3 kPa

Gas	M	γ	$\rho(\text{kg/m}^3)$
Air	29.0	1.40	1.206
Carbon dioxide	44.0	1.30	1.831
Helium	4.0	1.66	0.166
Hydrogen	2.0	1.41	0.083
Methane	16.0	1.32	0.666
Natural gas (typical)	19.5	1.27	0.811
Propane	44.1	1.14	1.835

Source: From Baumeister, T., Avallone, E. A., and Baumeister, T., III. 1978. *Marks' Standard Handbook for Mechanical Engineers*, McGraw-Hill, New York. With permission.

TABLE A.2

Approximate Values for Dynamic Viscosity of Gases at $T_0 = 273$ K

Gas	$\mu_0((\text{kg/m s}) \times 10^5)$	n
Air	1.71	0.7
CO_2	1.39	0.8
H_2	0.84	0.7
N_2	1.66	0.7
Methane	1.03	0.9

Source: Based on data from Baumeister, T., Avallone, E. A., and Baumeister, T., III. 1978. *Marks' Standard Handbook for Mechanical Engineers*, McGraw-Hill, New York. With permission.

Note: Values at other temperatures may be computed using $\mu = \mu_0(T/T_0)^n$.

TABLE A.3

Approximate Values of Specific Gravity and Dynamic Viscosity for Liquids at 25°C and 1 atm

Liquid	Specific Gravity (S.G.)	Viscosity, μ (Centipoise, cP = 10^{-3} N s/m^2)
Water	0.997	0.891
Seawater	1.02	0.9
Gasoline	0.68	0.51
Kerosene	0.78	1.64
Turpentine	0.86	1.38
Heating oil	0.80	1.38
Ethylene glycol	1.13	16.2
Glycerine	1.26	950

Source: From Bolz, R. E. and Tuve, G. L., 1973, *Handbook of Tables for Applied Engineering Science*, CRC Press, Boca Raton, FL. With permission.

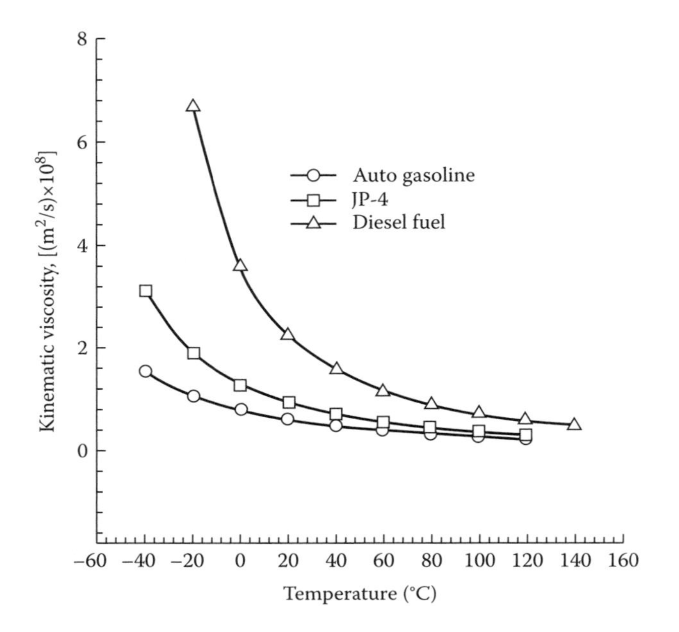

FIGURE A.1 Variation of kinematic viscosity of fuels with temperature. (Data from Schetz, J. A. and Fuhs, A. E. (Eds). 1996. *Handbook of Fluid Dynamics and Turbomachinery,* Wiley, New York.)

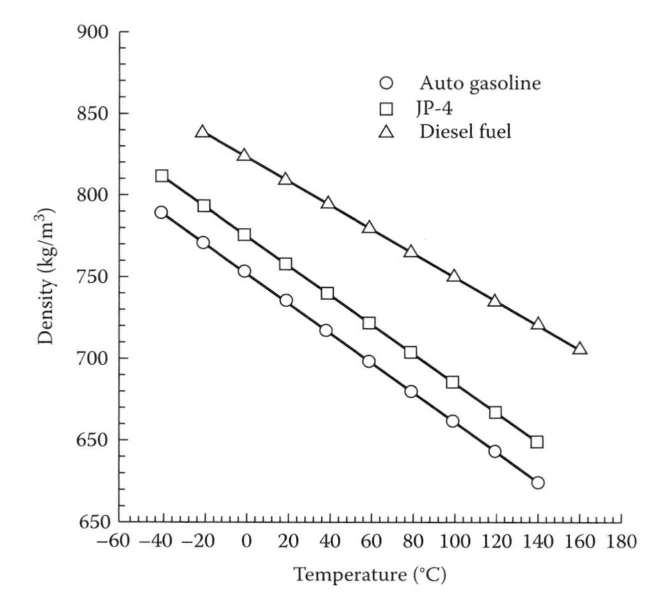

FIGURE A.2 Variation of density of fuels with temperature. (Data from Schetz, J. A. and Fuhs, A. E. (Eds). 1996. *Handbook of Fluid Dynamics and Turbomachinery,* Wiley, New York.)

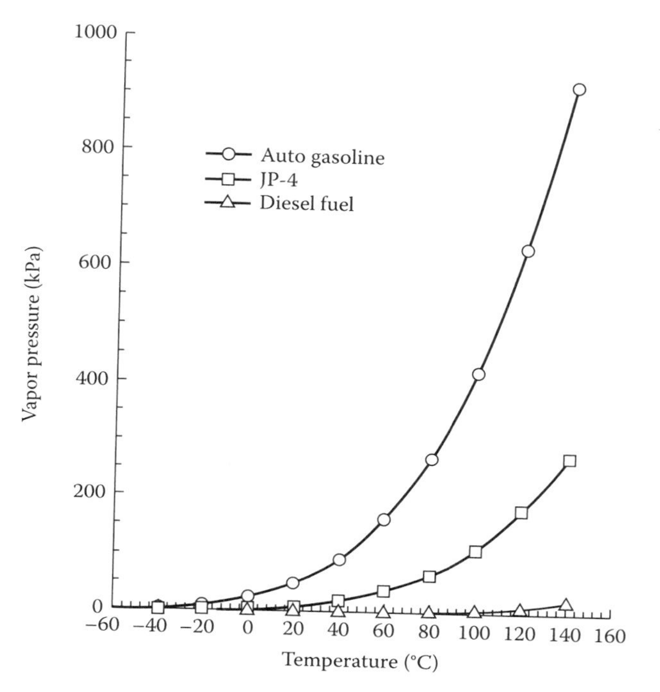

FIGURE A.3 Variation of fuel vapor pressure with temperature. (Data from Schetz, J. A. and Fuhs, A. E. (Eds). 1996. *Handbook of Fluid Dynamics and Turbomachinery*, Wiley, New York.)

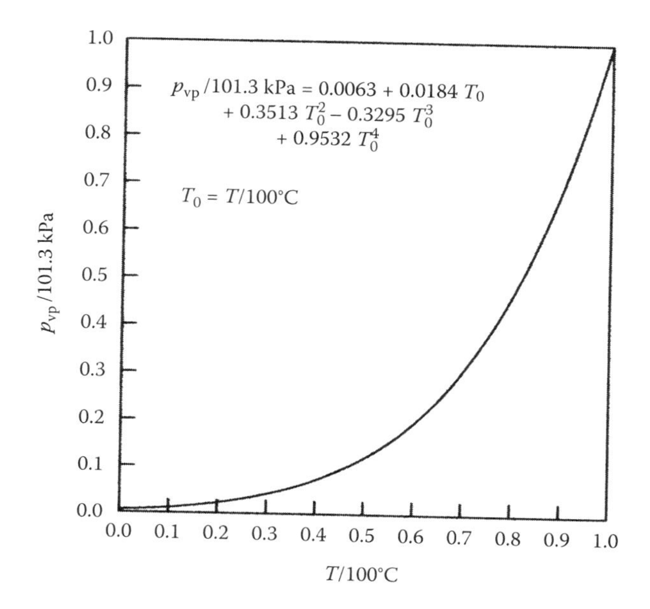

FIGURE A.4 Normalized vapor pressure for water with a forth order polynomial curve fit. (Data from Baumeister, T.,Avallone, E. A., and Baumeister, T., III. 1978. *Marks' Standard Handbook for Mechanical Engineers*, McGraw-Hill, New York.)

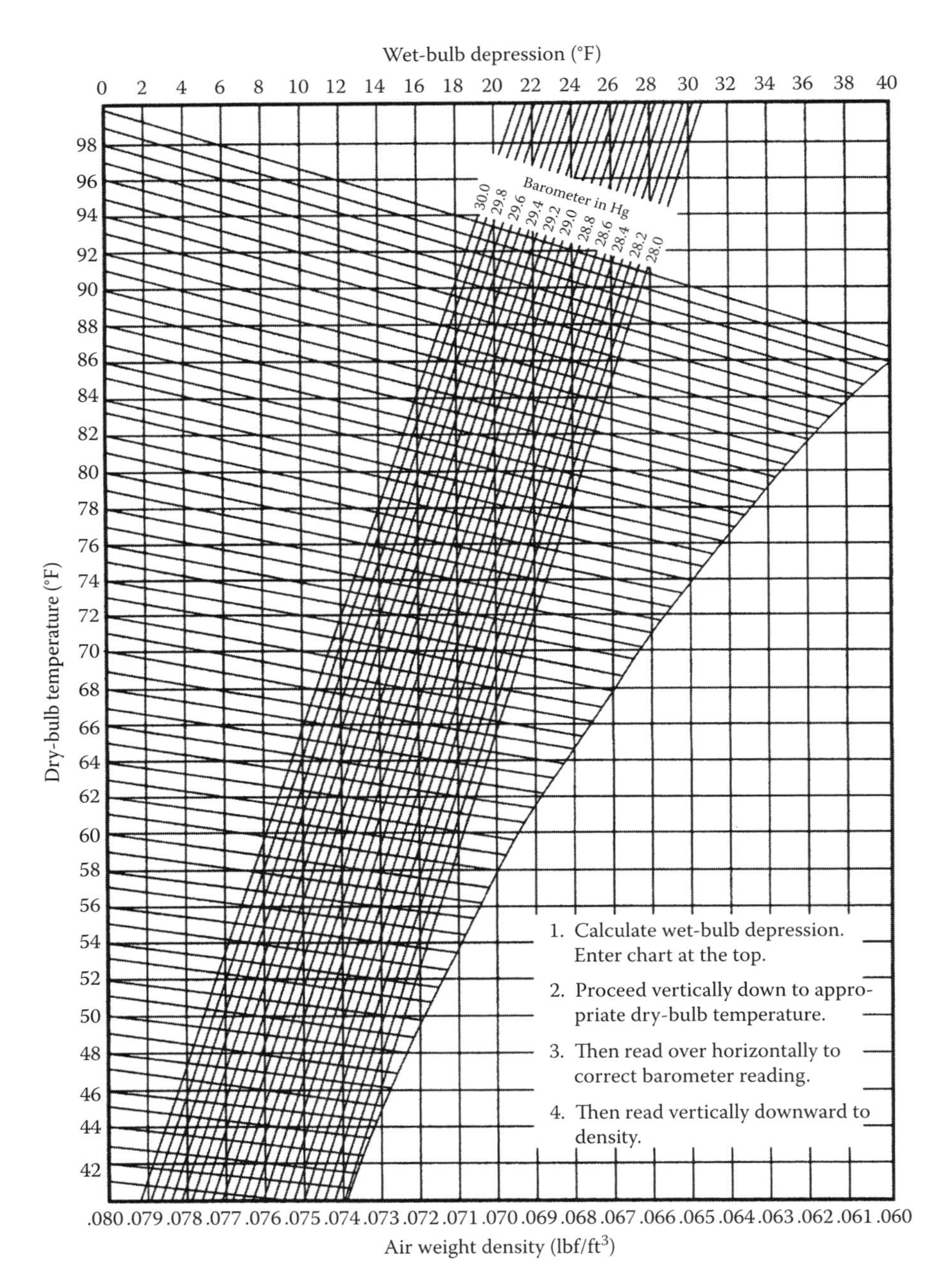

FIGURE A.5 A psychometric density chart. (From ANSI/AMCA Standard 210-85; ANSI/ASHRAE Standard 51-1985.)

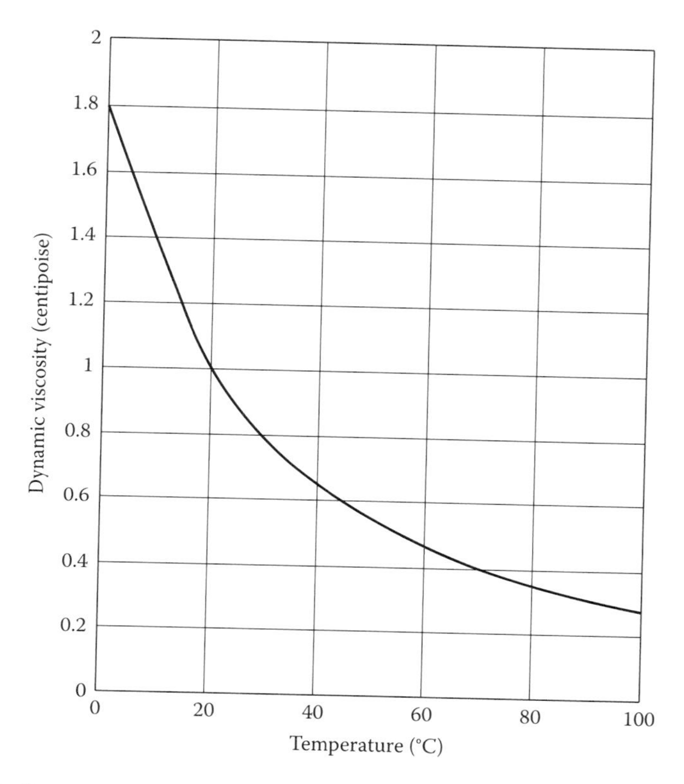

FIGURE A.6 Dynamic viscosity of liquid water.

Appendix B: Conversion Factors

Instructions: Locate the two units that you need to convert. The conversion factor is at the intersection of the row and column. Multiply or divide by the conversion factor as necessary

Slug	**lbm**	**MASS**
6.838E−2 slug/kg	2.2 lbm/kg	**kg**
	32.174 lbm/slug	**slug**

cm^2	0.01 cm^2/mm^2			
m^2	1E−6 m^2/mm^2	1 E-4 m^2/cm^2		
$in.^2$	1.55E−3 $in.^2/mm^2$	0.155 $in.^2/cm^2$	1.55E3 $in.^2/m^2$	
ft^2	1.076E−5 ft^2/mm^2	1.076E−3 ft^2/cm^2	10.76 ft^2/m^2	6.944E−3 $ft^2/in.^2$
AREA	**mm^2**	**cm^2**	**m^2**	**$in.^2$**

m^2/s	**ft^2/s**	**KINEMATIC VISCOSITY**
1E−6 $(m^2/s^2)/cSt$	1.076E−5 $(ft^2/s)/cSt$	**Centistoke (cSt)**
	9.29E−2 $(ft^2/s)/(m^2/s)$	**m^2/s**

cm	0.1 cm/mm			
m	1E−3 m/mm	0.01 m/cm		
in.	0.03937 in./mm	0.3937 in./cm	39.37 in./m	
ft	3.281E−3 ft/mm	3.2808E−2 ft/cm	3.2808 ft/m	8.333E−2 ft/in.
LENGTH	**mm**	**cm**	**m**	**in.**

DYNAMIC VISCOSITY	lb s/ft²	N s/m²
Centipoise (cP)	2.09E−5 (lb s/ft²)/cP	1E−3 (N s/m²)/cP
lb s/ft²		47.88 (N s/m²)/(lb s/ft²)

cm³	1E−3 cm³/mm³					
m³	1E−9 m³/mm³	1E−6 m³/cm³				
L	1E−6 L/mm³	1E−3 L/cm³	1E3 L/m³			
in.³	6.102E−5 in.³/mm³	6.102E−2 in.³/cm³	6.102E4 in.³/m³	61.02 in.³/L		
ft³	3.531E−8 ft³/mm³	3.53E−5 ft³/cm³	35.31 ft³/m³	3.534E−2 ft³/L	5.787E−4 ft³/in.³	
gal	2.641E−7 gal/mm³	2.642E−4 gal/cm³	264.2 gal/m³	0.2642 gal/L	4.329E−3 gal/in.³	7.48 gal/ft³
VOLUME	**mm²**	**cm²**	**m²**	**L**	**in.²**	**ft³**

in. Hg	in. wg	lb/in.^{2} (psi)	lb/ft^2	bar	PRESSURE
2.95E−4 in. Hg/Pa	4.01E−3 in. wg/Pa	1.45E−4 psi/Pa	2.09E−2 (lb/ft^2)/Pa	1.0E−5 bar/Pa	**Pa**
	13.6 in. wg/in. Hg	0.491 psi/in. Hg	70.7(lb/ft^2)/in. Hg	3.39E−2 bar/in. Hg	**in. Hg**
		0.0361 psi/in. wg	5.2 (lb/ft^2)/in. wg	2.49E−3 bar/in.wg	**in. wg**
			144 (lb/ft^2)/psi	6.89E-2 bar/psi	**lb/in^2(psi)**

L/s	1.0E3 (L/s)/(m^3/s)				
L/min	6.0E4 (L/min)/(m^3/s)	60 (L/min)/(ℓ/s)			
ft^3/s	35.31 (ft^3/s)/(m^3/s)	3.534E−3 (ft^3/s)/(L/s)	2.119 (ft^3/s)/(L/min)		
ft^3/min(cfm)	2119 cfm/(m^3/s)	0.2119 cfm/(L/s)	3.534E−2 cfm/(L/min)	1.667E−2 cfm/(ft^3/s)	
gal/min(gpm)	1.585E4 gpm/m^3/s	15.85 gpm/L/s	0.2642gpm/L/min	449 gpm/(ft^3/s)	7.481 gpm/cfm
VOLUME FLOW	**m^3/s**	**L/s**	**L/min**	**ft^3/s**	**ft^3/min**

J/kg	ft lb/slug	ft lb/lbm	Btu/slug	ft^2/s^2	SPECIFIC ENERGY
2.32 (J/kg)/(Btu/lbm)	2.50E4 (ft lb/slug)/(Btu/lbm)	778 (ft lb/lbm)/(Btu/lbm)	32.2(Btu/slug)/(Btu/lbm)	2.50E5 (ft^2/s^2)/(Btu/lbm)	**Btu/lbm**
	10.76 (ft lb/slug)/(J/kg)	0.334 (ft lb/lbm)/(J/kg)	1.38E−2 (Btu/slug)/(J/kg)	1.079E5 (ft^2/s^2)/(J/kg)	**J/kg**
		3.11E−2 (ft lb/lbm)/(ft lb/slug)	1.29E−3(Btu/slug)/(ft lb/slug)	1.0 (ft^2/s^2)/(ft lb/slug)	**ft lb/slug**
				32.2 (ft^2/s^2)/(ft lb/lbm)	**ft lb/lbm**
				778 (ft^2/s^2)/(Btu/slug)	**Btu/slug**

hp	0.746 hp/kW		
ft lb/s	738 (ft lb/s)/kW	550 (ft lb/s)/hp	
Btu/s	0.949 (Btu/s)/kW	0.707 (Btu/s)/hp	1.29 (Btu/s)/(ft lb/s)
POWER	**kW**	**hp**	**ft lb/s**

Index

V

W